ESSAYS IN LOGIC AND ONTOLOGY

POZNAŃ STUDIES
IN THE PHILOSOPHY OF THE SCIENCES AND THE HUMANITIES

VOLUME 91

Poznań Studies in the Philosophy of the Sciences and the Humanities
is partly sponsored by SWPS and Adam Mickiewicz University

Address: dr Katarzyna Paprzycka · Instytut Filozofii · SWPS · ul. Chodakowska 19/31
03-815 Warszawa · Poland · fax: ++48 22 517-9625
E-mail: PoznanStudies@swps.edu.pl · Website: http://PoznanStudies.swps.edu.pl

ESSAYS IN LOGIC AND ONTOLOGY

Edited by

Jacek Malinowski
and
Andrzej Pietruszczak

Rodopi

Amsterdam - New York, NY 2006

The paper on which this book is printed meets the requirements of "ISO 9706:1994, Information and documentation - Paper for documents - Requirements for permanence".

ISSN 0303-8157
ISBN-10: 90-420-2130-6
ISBN-13: 978-90-420-2130-3

Printed in The Netherlands

Dedicated to
Professor Jerzy Perzanowski

CONTENTS

Jacek Malinowski
Andrzej Pietruszczak

EDITORIAL INTRODUCTION

LOGIC IN TORUŃ: 1992–2003

The aim of this book is to present the papers centered upon the subjects of Formal Ontology and Logical Philosophy. The idea of investigating philosophical problems by means of logical methods was intensively promoted in Toruń by the Department of Logic of Nicolaus Copernicus University during last decade. The main driving force behind the initiative and subsequent realizer of this ambition was Jerzy Perzanowski, who headed the Torunian Department of Logic since October 1992 until June 2003. Another aim of this book is to present to the philosophical and logical audience the activities of the Torunian Department of Logic during this decade.

The papers in this volume contain the results concerning Logic and Logical Philosophy, obtained within the confines of the projects initiated by the Department of Logic and other international research projects in which the Torunian Department of Logic took part.

To sketch the general setting of the Torunian enterprises in the area of Logic and Philosophy we would like to present here a brief history of the Department.

The Department of Logic was founded by Tadeusz Czeżowski in the very beginning of the existence of Nicolaus Copernicus University in 1945. In October 1945 he took over the position of professor at the First Chair of Philosophy in the Philosophical Seminar at the Faculty of Humanities. Six years later this chair changed its name to "Chair of Logic".

Nicolaus Copernicus University is a direct successor of Jan Kazimierz University in Vilnius. For this reason, the research in the Department was from the beginning strongly influenced by the extraordinarily rich and fruitful tradition of logical investigations which had thrived in Poland between the World Wars and had been established by such prominent logicians and philosophers as Jan Łukasiewicz, Alfred Tarski, Stanisław Leśniewski, Leon Chwistek, Tadeusz Kotarbiński, Kazimierz Ajdukiewicz, Mordechaj Wajsberg, Andrzej Mostowski and Adolf Lindenbaum. All the logicians and philosophers mentioned above belonged to the famous Lvov-Warsaw School of Philosophy founded by Kazimierz Twardowski. Tadeusz Czeżowski belonged to the most prominent representatives of this School.

In: J. Malinowski and A. Pietruszczak (eds.), *Essays in Logic and Ontology* (*Poznań Studies in the Philosophy of the Sciences and the Humanities*, vol. 91), 9–16. Amsterdam/New York, NY: Rodopi, 2006.

After Tadeusz Czeżowski had retired in 1960 the position of the Head of Department was taken over in 1965 by Leon Gumański who directed it until his retirement in 1991. The next year Professor Jerzy Perzanowski was appointed the Head of the Department of Logic of Nicolaus Copernicus University in Toruń and he steered it for a bit more than a decade, until 2003.[1] As we mentioned it above our main goal is to present the activities of the Torunian Department of Logic and the results of the projects initiated during this decade.

The period of last thirteen years was the time of the intensive growth of the ventures and achievements of the Department. Many logicians and philosophers contributed to them in the form of considerable results of research as well as by initiating and promoting valuable projects. We believe however that the activity and cordiality of Profesor Perzanowski was crucial for all the accomplishments of the Department during this decade.

We would like to stress that each of the papers presented in this volume is inspired by some of numerous projects of the Department.

The Journal "Logic and Logical Philosophy"

In 1993 there appeared the first volume of *Logic and Logical Philosophy*, its publisher being the Nicolaus Copernicus University Press. The leading ambition of the journal was, and still is, to promote application of logical methods in Philosophy, Cognitive and Computer Sciences, Linguistics and related disciplines. Along with the first issue of the journal the gap has been filled between purely logical journals publishing formal results on logical systems and journals publishing informal considerations concerning problems that could be considered by means of logical tools.

In the period of thirteen years thirteen volumes of *Logic and Logical Philosophy* appeared. About seventy authors published more than one hundred original research papers. The editorial policy of the journal has been successful in attracting a broad audience of logicians and philosophers.

In this volume we present eight papers contributed by the authors cooperating with the journal. None of these papers is devoted to purely logical investigations. The following three of them represent the mainstream of investigations in Formal Ontology — the subject intended to belong to the core of the scope of *Logic and Logical Philosophy*. The paper "At the crossroads of logic and ontology: A modal combinatorial analysis of why there is something rather than nothing" by Dale Jacquette contains a comprehensive analysis of the idea of making logic the basis of ontology. Uwe Meixner's paper "An onto-nomological theory of modality" presents a logical theory of ontological modality. Roberto

[1]Between those periods prof. T. Szczurkiewicz (1960–1965) and prof. J. Pawlak (1991/1992) were the wardens of Department of Logic.

Poli contributed with the short essay with self-explicable title "The ontology of what is not there".

The paper "Contradiction and contrariety. Priest on negation" by Heinrich Wansing, as one of many papers in this volume, concerns the problem of adopting inconsistency. The author investigate the relations between contradiction and negation in paraconsistent logics.

The following four papers concern the logical analysis of the issues coming from four different areas: Theory of Action, Linguistics, Sociology and Set Theory. Krister Segerberg's paper "Moore problems in Full Dynamic Doxastic Logic" consider the seminal Moore's paradox in the setting of the Logic of Action. Jacek Malinowski in "On the formalization of Strawson's presupposition" develops the formal tools for systematic investigation of Strawson's notion of presupposition. Janusz Mucha in "The concept of 'social relations' in classic analytical interpretative sociology: Weber and Zaniecki" contribute to the clarification of the concept of "social relations". Andrzej Pietruszczak in the paper "On applications of truth-value connectives for testing arguments with natural connectives" proves that the methods from standard first courses in logic can be used for testing arguments with strict implications, strict equivalences and other connectives from natural language.

Stanisław Jaśkowski Memorial Symposium

Friday, March 19, 1948 is a very important date in the history of Logic and Philosophy. On that day at a meeting of the Mathematical and Natural Science Department of the Societatis Scientiarum Torunensis, Stanisław Jaśkowski presented his paper "A propositional calculus for inconsistent deductive systems" which familiarized the broad audience with a new, extraordinary idea. The paper turned out to be crucial to the origin and emergence of the whole new field of logical investigations known today as Parainconsistent Logics. To commemorate that event and to celebrate the 50-th anniversary of Parainconsistent Logics, the Department of Logic — inspired by Professor Perzanowski — organized in Toruń Stanisław Jaśkowski Memorial Symposium "Parainconsistent Logic, Logical Philosophy, Mathematics & Informatics". It held in 1998 from 15th to 18th July. The conference was very successful in attracting prominent scholars as well as a number of young researchers interested in the investigations of parainconsistent reasonings. The papers presented during the conference as well as a number of papers inspired by it were published in the volumes 7, 8 and 9 of *Logic and Logical Philosophy*. In this volume we publish two papers inspired by the discussions held during that Symposium.

The paper contributed by Itala M. Loffredo D'Ottaviano and Hércules Araujo Feitosa, "Translating from Łukasiewicz's logics into classical logic: Is it possible?" presents some results concerning translations between logics and a variant of Łukasiewicz's logics. The paper by Luis Fariñas del Cerro and Olivier Gasquet "Modal tableaux for reasoning about diagrams" concerns tableaux methods for diagrammatic reasoning, the area being of high interest in the knowledge representation.

Polish-Flemish Cooperation

Long term cooperation with Flemish logicians was one of most important among the enterprises of the Department. Since September 1998 until August 2001 there were realized an international research projects concentrated under common title "Ontological Foundations of Paraconsistency". In the research participated: the Center for Logic and Philosophy of Science, Ghent Univeristy; the Center for Logic and Philosophy of Science, Free University of Brussels and Torunian Department of Logic. The coordinators of the project were: Diderik Batens (Ghent), Jean Paul van Bendegem (Brussels) and Jerzy Perzanowski. The project was funded within the Bilateral Scientific and Technological Cooperation Project BIL98/37 by the Science, Innovation and Media Department of the Ministry of the Flemish Community (Belgium) and by the State Committee for Scientific Research of the Republic of Poland.

The first project found its natural continuation in the second one: "The Dynamics of Reasoning in the Sciences: Adaptive and Interrogative Perspectives". It started in September 2001 and finished in August 2004. In the second project participated: the Center for Logic and Philosophy of Science, Ghent University; the Center for Logic and Philosophy of Science, Free University of Brussels; the Chair of Logic and Philosophy of Science, University of Zielona Góra and a group of participants from the Torunian Department of Logic. The coordinators of the second project were: Erik Weber (Ghent), Jean Paul van Bendegem (Brussels) and Adam Grobler (Zielona Góra). The project was funded within the Bilateral Scientific and Technological cooperation Project BIL01/80 by the Science, Innovation and Media Department of the Ministry of the Flemish Community (Belgium) and by the State Committee for Scientific Research of the Republic of Poland.

Nine International Workshops were organized within those two projects:

VlaPoLo1 Pierwsze Flamandzko-Polskie Warsztaty Logiczno-Filozoficzne (First Flemish-Polish Logical-Philosophical Workshop) Nicolas Copernicus University, Toruń 28th – 29th of October, 1999.

VlaPoLo2 The Second Flemish-Polish Workshop on the Ontological Foundations of Paraconsitency, University of Ghent, Belgium, 27th – 29th of April 2000.

VlaPoLo3 The Third Flemish-Polish Workshop on the Ontological Foundations of Paraconsitency, Free University of Brussels (VUB), Belgium 12th – 14th of October, 2000.

VlaPoLo4 The Fourth Flemish-Polish Workshop on the Ontological Foundations of Paraconsitency, University of Ghent, Belgium, 9th – 11th of April 2001.

VlaPoLo5 I Flemish-Polish Workshop on Adaptive and Erotetic Logics and their Applications to the Philosophy of Science, University of Zielona Góra (Poland), 14th – 18th November, 2001.

VlaPoLo6 International Workshop "The Dynamics of Reasoning in the Sciences: Adaptive and Interrogative Perspectives", Ghent University, Belgium, 17th – 19th of October, 2002.

VlaPoLo7 International Workshop "Problem Solving in the Sciences: Adaptive and Interrogative Perspectives", Free University of Brussels (VUB), Belgium, 8th – 10th of May, 2003.

VlaPoLo8 International Workshop "Problem Solving in the Sciences: Adaptive and Interrogative Perspectives", 20th – 22nd of November, 2003, University of Zielona Góra.

VlaPoLo9 International Workshop "The Dynamics of Reasoning in the Sciences: Adaptive and Interrogative Perspectives", Ghent University, Belgium, 6th – 8th of May, 2004.

The result of research within the above projects were about 30 original research papers devoted to Parainconsistent Logics and containing considerable results. These were published in two special volumes of *Logique et Analyse* (nos. 165–166 and 173–175) and in *Logic and Logical Philosophy* (no. 11/12). Besides, the mutual discussions and the exchange of knowledge during the workshops and other cooperation within the confines of this project inspired a number of further research. In this volume we present six papers sparked by the ideas that this fruitful cooperation gave rise to.

The papers contributed by Diderik Batens, Joke Meheus, Sergei Odintsov and Guido Vanackere fall in the mainstream of investigations in the area of Paraconsitent Logics. Diderik Batens' paper "Narrowing down suspicion in inconsistent premise sets" concerns the adaptive approach to the problem of paraconsitency. The paper presents some approaches for handling new, possibly inconsistent premises in the context of the original premises. The aim of this approach is to narrow down suspicion in inconsistent premise sets and hence locates the real problems among the possible ones. Joke Meheus' paper "Discussive adaptive logics: Handling internal and external inconsistencies" investigates the discussive adaptive logic $\mathrm{DLI}^{\mathrm{r}}$ closely connected to the original ideas of Jaśkowski with the aim to adopt them and compare to the adoptive approach to Paraconsitency. The paper by Sergei P. Odintsov "Absurdity as unary operator" contains results presented in author's lecture at the VlaPoLo2 concerning non-trivial extensions of minimal Johansson's logic. Guido Vanackere in the "A world of experiences, an adequate language, and self-reference revisited" presents a formal system in which experiences are central entities.

Sonja Smets contributed the paper "From intuitionnistic to dynamic operational quantum logic" containing the general review of relations between Intuitionistic Logic and the operational approach to the logical foundations of physics.

Erik Weber in his short essay "Are there ontological explanations?" describes some physical experiments which could be interpreted as a kind of ontological explanations.

Autumn Workshops

The Department of Logic organized the series of Workshops for Polish logical and philosophical audience their aim being presentation of the original results concerning Logical Philosophy. The idea was very successful in promoting logical research methods in research projects of young philosophers from many polish scientific centers. Since 1994 until 2002 eight Logical-Philosophical Workshops were organized. The first one held in Szklarska Poręba in September 25–29, 1994. The second Workshop, held in Toruń September 10–14, 1996, The third one in Górzno in September 2–6, 1997. Five subsequent Workshops were organized each Year in September in Zawoja.

The best results presented during the workshops were published (in Polish) in the book series "Philosophy/Logic: Logical Philosophy" published by Nicolaus Copernicus University Press.[2] In this volume we publish seven papers contributed by polish authors closely cooperating with the Torunian Department of Logic. The following three of them belong to the mainstream of ontological investigations.

Marek Rosiak in his paper "Formal and existential analysis of subject and properties" presents a contribution to the object ontology. Tadeusz Szubka in "The metaphysical realism debata: What is at stake?" raises some doubts about pure metaphysical realism. Tomasz Jarmużek, Maciej Nowicki, and Andrzej Pietruszczak in "An outline of the Anselmian theory of God" present an exact

[2]Perzanowski *et al.* (1995); Perzanowski and Pietruszczak (1997); Perzanowski and Pietruszczak (2000).

translation of the Anselmian argument into the realm of logic and give a basis for further logical research within the Anselm's theory presented in *Proslogion*.

Three papers concern the ontology of mathematics and physics: Roman Murawski's "Philosophy of mathematics in the 20th century: Main trends and doctrines", Krzysztof Wójtowicz's "Independence and justification in mathematics" and Tomasz Bigaj's "Do quantum-mechanical systems always possess definite properties dictated by their states?".

Renata Ziemińska in her contribution "Two notions of the internal and Goldman's epistemic externalism" analyzes two concepts of the internal considered in the current epistemic internalism/externalism debate.

Other projects

There were many other activities of the Torunian Department of Logic, for example, the Sixth Polish Philosophical Congress held in Toruń in September 1995[3] and the conference devoted to the memory of Roman Ingarden[4]. However the most important activity of the Department which is not mentioned here concerns the idea intensively promoted by the Torunian Department of Logic — the idea of Cognitive Science. A big deal of work of the people from the Torunian Department of Logic has been done to create in Toruń M.A. programm in the area of Cognitive Science. The activity of Max Urchs who was a member of the Department since 1994 until 1999 is most important here. During his time in Toruń Max Urchs created, developed and promoted the Cognitive Seminar (Konwersatorium Kognitywne). Within the confines of the Seminar about forty scientific meetings were organized. It induced a big number of scientists and students to pay their attention to the area of Cognitive Science and revealed close relation of this area to Logic and Philosophy. Unfortunately the task of creating the Master Degree Programm in Cognitive Science appeared to be harder than we all had expected. Nevertheless, some considerable (although partial) success has been achieved here and the cognitive specialization within philosophical M.A. programm at Nicolaus Copernicus University is one of them. The journal *Kognitywistyka i media w edukacji* [Cognitive Science and Media in Education] was established as a result of this activity. Last but not least, The Polish Association for Cognitive Science was founded in 2002.

To all the activities presented above, those successful and those less successful, a number of people contributed their time, knowledge and energy. Besides members of the Department many others could be mentioned here. Nevertheless, the founding idea for all these activities originated with Jerzy Perzanowski.

---

[3]The logical aspects of the Congress are presented in details in Perzanowski and Pietruszczak (1997).

[4]See the book Perzanowski and Pietruszczak (2003). This book was distributed to some people with a rather confusing letter by professor Perzanowski.

By publishing this volume, we would like to emphasize it in the clear way. On behalf of all who cooperated with the Department from 1992 until 2003 we would like to thank Professor Perzanowski for his work and hospitality. We hope that we will manage to continue his work in the way he would accept.

Polish Academy of Sciences
Institute of Philosophy and Sociology
ul. Nowy Świat 72
00-330 Warszawa
Poland

Nicholaus Copernicus University
Department of Logic
ul. Asnyka 2
87-100 Toruń
Poland
e-mail: jacekm@uni.torun.pl

REFERENCES

Perzanowski, J., A. Pietruszczak and C. Gorzka, eds. (1995). *Filozofia/Logika: Filozofia Logiczna 1994* [Philosophy/Logic: Logical Philosophy 1994]. Toruń: The Nicolaus Copernicus University Press.

Perzanowski, J. and A. Pietruszczak, eds. (1997). *Byt, Logos, Matematyka* [Being, Logos, Mathematics]. Toruń: The Nicolaus Copernicus University Press.

Perzanowski, J. and A. Pietruszczak, eds. (2000). *Logika & Filozofia Logiczna. FLFL 1996–1998* [Logic & Logical Philosophy]. Toruń: The Nicolaus Copernicus University Press.

Perzanowski, J. and A. Pietruszczak, eds. (2003). *Od teorii literatury do ontologii świata* [From Theory of Literature to Ontology of the World]. Toruń: The Nicolaus Copernicus University Press.

Dale Jacquette

CROSSROADS OF LOGIC AND ONTOLOGY: A MODAL-COMBINATORIAL ANALYSIS OF WHY THERE IS SOMETHING RATHER THAN NOTHING*

ABSTRACT. Although it is frequently said that logic is a purely formal discipline lacking any content for special philosophical subdisciplines, I argue in this essay that the concepts of predication, and of the properties of objects presupposed by standard first-order logic are sufficient to address many of the traditional problems of ontology. The concept of an object's having a property is extended to provide an intensional definition of the existence of an object as the object's possessing a maximally consistent property combination, consisting, for any property or its complement. Nonexistent objects by the proposed definition are those that either lack both some property and its complement, or have both a property and its complement included in its corresponding property combination. The definition in turn makes it possible to offer solutions based on purely logical concepts of such longstanding metaphysical problems as why there is something rather than nothing, and why there exists exactly one logically contingent actual world, itself a maximally consistent combination of true predication instances or facts. Additionally, the ontology upholds an argument in support of modal actualism and against modal realism, treating all logically possible worlds other than the actual world as predicationally incomplete nonexistent objects or semantic fictions.

1. Logic and Ontology

What can logic teach us about ontology? Symbolic logic in and of itself is not that philosophically interesting. Logic nevertheless reveals something important when, appropriately interpreted, it helps to answer longstanding metaphysical problems, by answering such questions as what it means to say that something exists, why there is something rather than nothing, and why there exists exactly one logically contingent actual world.

Classical logic embodies a set of concepts and philosophical commitments about what it means for something to exist that can be recovered from the expressive capabilities of logic. The fundamental questions of ontology are explained implicitly by symbolic logic when we understand the implications of the availability to logic of all logically possible predications of all logically possible

*A more complete examination of these topics is presented in my book, *Ontology* (Acumen Books and McGill-Queen's University Press, 2003). I am grateful to the Alexander von Humboldt-Stiftung for supporting this and related research projects during my 2000–2001 sabbatical year on leave from The Pennsylvania State University at the Franz Brentano Forschung, Bayerische-Julius-Maximilians-Universität, Würzburg, Germany.

In: J. Malinowski and A. Pietruszczak (eds.), *Essays in Logic and Ontology* (*Poznań Studies in the Philosophy of the Sciences and the Humanities*, vol. 91), 17–46. Amsterdam/New York, NY: Rodopi, 2006.

properties to all logically possible objects. Symbolic logic, concerned primarily with the formal requirements of deductively valid inference, does not entail that any given specimen of actual reasoning conform to its truth conditions, let alone that any contingently existent entity in the actual world actually exists. Logic, devoted to the rules of correct inferential form, seems ontologically impotent to the extent that we focus exclusively on the deductive mechanisms of logical systems. For purposes of pure philosophical ontology, there is more to be learned from the expressive capabilities of classical logic, whether Aristotelian syllogistic, Boolean algebra, or modern symbolic first-order predicate logic. Any system of logic worthy of the name must have available for its operations the complete set of all logically possible propositions. The metaphysical presuppositions of this logical requirement in turn contain solutions to the most difficult problems of ontology.[1]

2. Combinatorial Pure Philosophical Ontology

The idea of making logic the basis of ontology is not new. A variation of one standard type of combinatorial pure philosophical ontology can be offered on the foundation of the predicational requirements of symbolic logic. The analysis of the concept of being which the theory makes possible depends on the totality of logically possible combinations of predications involving all logically possible combinations of logical objects with all logically possible properties.

We begin by considering the concept of a logically possible object, a logically possible property, and the ways in which they can be combined to produce different logically possible objects, states of affairs, and worlds. We do not say that all of these combinations are or correspond to actually existent entities, but we define the concept of existence as a logical condition on the full range of combinations of logically possible objects with logically possible properties, logically possible states of affairs, and logically possible worlds.

Combinatorial pure philosophical ontology is thus identical with a certain part of pure philosophical logic. The conceptual roots of ontology are the same as those of classical first-order formal symbolic or other formal logics in which logically possible properties are truly or falsely combined with or predicated of logically possible objects. We expect any and all logical possibilities to derive from no other source than logic. We uncover the same logical-ontological concepts when from the resources of pure logic we try to answer the question why there is something rather than nothing and why there exists exactly one logically contingent actual world.

---

[1] The fundamental questions of ontology are clearly formulated but inadequately addressed by Martin Heidegger in his treatises *Being and Time: The Question of Being*, and *An Introduction to Metaphysics*.

When we look at any system of formal symbolic logic, they all have one conspicuous feature in common. All involve the possibility of predicating properties of objects. In Aristotelian logic we have what by modern standards are relatively opaque ways of formulating such predications as “All men are mortal”, “Socrates is a man”, and “Socrates is mortal”. Symbolic logic permits formulations that exhibit the internal logical structure of these sentences, that makes it even more perspicuous that the basis of all logic is the possibility of true or false predications of logically possible properties to logically possible objects. Standard first-order or predicate-quantificational logic permits a detailed formal representation of the way in which object terms are combined with property terms, regardless of the ontic status of objects and properties. In defining a semantic domain of logically possible predications, which is to say of logically possible propositions, predicational combinatorial possibilities provide the propositional building blocks for a formal theory of deductively valid inference.

Logical formalisms express more rigorously and precisely, some better than others, a correspondingly basic concept, without which we cannot undertake the study of ontology or any other subject. Nothing is meaningful, true or false, except by virtue of a property being truly or falsely predicated of an object or objects. We cannot even say that a combinatorial ontology is correct or incorrect without attributing a property to an object. The logical possibility of combining objects with properties, represented in the most articulate logical notations as the combination of object terms with property terms, is the foundation of every system of logic. To the extent that logic represents the structure of reasoning, it equally underlies the possibility of thought, and, as we shall see, of reality insofar as it is thinkable. If we are to consider pure philosophical ontology exclusively as a discipline, then this is the explanatory level at which we should expect to discover its most fundamental concepts.

We can think of predication minimally as the logical possibility that there could be something that has a property. The presupposition of logic by which properties can be predicated of objects makes it intelligible to refer to logically possible predication subjects and the logically possible properties that can be predicated of them, without which there could be no philosophical problem of their ontic status. We shall accordingly assume that we can speak at least provisionally of objects as logically possible predication subjects and of the logically possible properties that can be predicated of them, as among the minimal combinatorial requirements of pure logic. We further assume that when an object, whatever an object is and whether or not objects exist, whether or not it is not something more than a purely logical predication subject, has a property, whatever a property is and whether or not properties exist, then a logically possible object’s having a logically possible property is a logically possible state of affairs. A logically possible world, in turn, including the actual world, is constituted by logically possible states of affairs, and ultimately by logically possible objects combining with logically possible properties. Logic, we can say, has as its subject matter all such logical possibilities in the form of all such logically possible states of affairs involving all logically possible predications of all logicallly possible properties to all logically possible objects.[2]

If we are grappling with the problems of pure philosophical ontology, then we are not yet ready to commit ourselves to any particular domain of existent objects, but are trying to answer more fundamental metaphysical questions about the nature of being. We should correspondingly adopt as ontologically neutral and noncommital a logical vocabulary as possible, in order to see what can be learned from logic alone. We shall speak intensionally of properties and property combinations, which is to say a choice of properties considered together, as the properties of a particular object. We leave open the possibility of interpreting the ontology of objects in alternative ways, none of which is excluded by the logical explanation we offer later of why there is something rather than nothing, and why there is only one logically contingent actual world. By presupposing the possibility of predication, pure logic presupposes only the logical possibility of something, a logically possible object, having a logically possible property.

The ontic status of properties and objects, predication subjects, the "something" capable of having a property, remains a topic for applied scientific rather than pure philosophical ontology, in the course of which objects or properties might in principle be eliminated in favor of a radically different concept of predicability. In pure philosophical logic and pure philosophical ontology, precisely where these disciplines dovetail, we engage in discourse about logical possibilities to which we are not yet in any way ontically committed. The purely logical possibilities for objects having properties are conceptually prior to the deliberations of applied scientific ontology. There is no specific preferred existence domain for logic any more than there is for pure philosophical ontology. It follows that pure philosophical ontology, like pure philosophical logic, can only be interpreted in the sense of a discipline, and not as a preferred theoretical existence domain, let alone as the extant domain of real existent entities.

3. Predicational Maximal Consistency

We consider objects as logically possible predication subjects at some level of discourse independently of their ontic status. We need to do so in order to be able to say that some objects exist and others do not exist, without which distinction we can mean nothing significant when we say positively that an object exists.

[2]Perhaps the most famous prototype combinatorial ontology is Wittgenstein (1922). See also Russell (1985).

Objects as mere predication subjects turn out to exist or not to exist, depending on whether they satisfy the definition of the concept of existence, yet to be specified. Some objects pass the test, and so are said to exist, and are admitted to the preferred existence domain in applied ontology; others do not pass the test and are not admitted. What is an object, besides constituting a logically possible predication subject? We do not assume that all objects exist; an existent object is an *entity*. Logically possible objects must satisfy a correct analysis of the concept of being in order to exist, in order to be entities. The concept of an object and of an existent object or entity must be more thoroughly explicated.

We shall say that an object as a predication subject is identifiable in terms of all its true predications; in terms, in other words, of everything that is true of it, or the properties it has. To be an object is to be a predication subject, and to be this as opposed to that particular object, whether existent or not, is to have or to be associated with a distinctive combination of properties. The red ball, [red,ball], which, as far as logic is concerned, might exist or not exist, is different from the blue ball, [blue,ball], because the red ball is red and the blue ball is not red, and because the blue ball is blue and the red ball is not blue. The same identity conditions hold with appropriate modifications for states of affairs and logically possible worlds, independently and regardless of their ontic status as existent entities or nonexistent mere predication subjects. The state of affairs which is such that this ball is red is different from the state of affairs which is such that this ball is blue, because this state of affairs involves a red and not a blue ball and the predication of the property red rather than blue, while the second designated state of affairs involves a blue and not a red ball and the predication of the property blue rather than red. A logically possible world in which a red ball or the redness of a ball only obtains is different from a logically possible world in which a blue ball or the blueness of a ball only obtains, because the first world includes as its only state of affairs a red ball or the redness of a ball rather than a blue ball or the blueness of a ball. The more painfully obvious are these preliminary observations, the more secure the purely logical foundations of pure philosophical ontology.

If we consider objects and their properties independently of their ontic status, even allowing that in principle objects might turn out in some sense to be reducible to property combinations, we can appreciate the virtually unlimited possibilities for distributions of properties to be combined with or predicated of objects. If we have three properties, *F1*, *F2*, and *F3*, then logically there must be property combinations for objects that include these possibilities: [*F1*], [*F2*], [*F3*], [*F1*, *F2*], [*F2*, *F3*], [*F1*, *F3*], [*F1*, *F2*, *F3*]. We do not include the null combination consisting of no properties, which we do not consider a limiting case of predication but as a nonpredication, and we do not suppose that there need be any purely logical ordering of properties in an object's associated property combination. In general, if we are considering a total of *N* properties, then the number of combinations of properties will be precisely identical to the number of objects as logical predication subjects, which, excluding the null combination as nonpredicational, is the cardinality of the ontology, considered here in the abstract without benefit of the findings of applied scientific ontology, and expressed by the formula: $C = 2^N - 1$. This is the number of existent and nonexistent objects as logically possible predication subjects, of which only some are entities. Again, only some, presumably the same ones, belong to the actually existent states of affairs that collectively constitute the actually existent world. The equation by itself is not very helpful in establishing more meaningfully the cardinality of pure philosophical ontology, unless we know more precisely how many properties there are. In some applications, such as Aristotle's metaphysics of substance, there are only as many properties as actually inhere in existent spatiotemporal particulars. A problem for Aristotle's account of inherent properties as "secondary substances" is that we cannot know in advance of completing the project of applied scientific ontology how many properties there are, and hence of how many objects there could be. We do not want to prejudge the truth of Aristotle's theory of substance and essential and accidental properties, which is a topic for applied rather than pure philosophical ontology. We shall therefore not try to assign definite values to N or decide how many existent or nonexistent objects could possibly exist.

What do we mean when we say some objects considered as predication subjects can or do or must, or do or need not or cannot exist or have being? Objects considered as logically possible predication subjects are individuated one from another by virtue of their distinctive true predications, corresponding to their distinctive property combinations by which they are defined. We say that an object as a mere predication subject, existent or nonexistent, has or even is a certain property combination, which is to say a combination of true predications of properties, and that these identifying and individuating property combinations belong to several distinctive types, in terms of which we commonsensically characterize the existence or nonexistence of objects, states of affairs, and logically possible worlds. A property combination can be either predicationally complete or incomplete, internally consistent or inconsistent. These are purely logical characteristics of objects considered as mere predication subjects that might in principle exist or not exist.

A property combination is complete when, for any property and property complement, the property combination includes the property or its complement. The red ball, considered only as such, has an incomplete property combination that includes the property red and the property of being a ball, but does not include or exclude the property of being ten centimeters in diameter or the complementary property of not being ten centimeters in diameter. The object itself, we can say, the red ball, considered only as such, is predicationally undetermined. If by contrast we consider an actual ball in the actual world, it will have and be correctly defined only as a complete property combination by which the actually existent object, an entity, is fully predicationally determined. In addition to being red or blue or some other color, it will either be ten centimeters in diameter or not ten centimeters in diameter; it will either be made of rubber or it will not be made of rubber; and so on. There are no gaps in the total combination of properties or property complements that hold true of an existent as opposed to merely logically possible entity.

A unicorn does not exist, although it might; untold numbers must frolic about alternative logically possible worlds. No unicorn exists because there is no actual entity that has among its properties the characteristics associated with being a unicorn roughly, being equine with a single horn sprouting naturally from the forehead. The property combinations of unicorns relative to the actual world are inevitably missing at least one property or property complement. Completeness and consistency are already taken to be hallmarks for the "existence" of logically possible worlds in modal semantics. Whether they are indeed so depends on whether as putative abstract objects they satisfy the existence requirements for entities generally and abstract entities in particular in applied scientific ontology. The existence of logically possible worlds on the proposed analysis boils down to the question of whether or not they can be correctly characterized as predicationally maximally or only submaximally consistent. It can be argued that the existence of logically possible worlds even as an abstract set of propositions with a specifiably consistent and complete distribution of truth values is logically dependent on the prior existence of objects as maximally consistent property combinations. If even one such object does not exist, then any state of affairs in which it is supposed to be involved and any logically possible world to which it is attributed does not exist.

Similarly with respect to the existence of abstract entities, if the concept of being admits any, and if there are strong reasons for including them in the existence domain. An abstract entity like the number π, supposing it for the moment to exist, or the property red, if properties should turn out to exist, will be predicationally complete in its property combination. It will not be red, it will not be blue, but it will be an irrational real number, and it will be the ratio of the circumference to the diameter of a circle, and so on. In principle, a nonexistent object might turn out to be incomplete with respect to only one property and property complement, although in practice the nonexistent objects that are considered in ontology are generally the imaginary concoctions of fiction and myth, which have only an easily surveyable number of predications that excludes all other property and property complement combinations. The objects of fiction are more like the red ball considered only as such than the dynamic predicationally inexhaustible objects that we encounter in sensation or study in logic, mathematics, and philosophy. The Eiffel Tower is predicationally complete; the fabled golden city of Eldorado is not.

Property combinations, whether complete or incomplete, can also be consistent or inconsistent. An example of a complete and consistent property combination is that of an actually existent physical or abstract entity. An incomplete and consistent predication combination belongs to a fictional object. There need be no contradictions in its description, but the most that can be said about it leaves innumerable questions unanswered. Did Eldorado have champagne fountains? Was it paved with 18 or 24 karat gold? Was a monkey ever made its king? We cannot say yes or no. What we know of it may be logically consistent, but the most we can possibly know of it will be logically incomplete with respect to at least one property and property complement. Thus, unlike the Eiffel Tower, Eldorado does not exist; it has no being, even as an abstract entity. The Eiffel Tower makes it into the preferred existence domain, but Eldorado as an incomplete and hence nonexistent object does not. This is as it should be. Difficulties arise when we try to generalize an analysis of the concept of being or existence from these data. Shall we say that an object has being if and only if it has a consistent and complete property combination?

We argue that being or existence is nothing more or less than predicational maximal consistency. The definition, properly understood and applied, provides a unified analysis of the concept of being or actual existence for all physical and abstract entities, actual states of affairs, and the uniquely existent logically contingent actual world, as distinct from logically consistent but nonactual merely logically possible worlds.

MAXIMAL CONSISTENCY

A property combination C for any logically possible object O is *maximally consistent* if and only if, for any logically possible extraontological property F, either F is in C or non-F (the complement of F), and not both property F and complementary property non-F, is in C.

We need only consider logically possible objects in ontology. For an extraontology demanded by a fully general philosophical semantics, we would also need to discuss predicationally internally inconsistent property combinations, such as [red, nonred], [round, square], and the like. These objects, unlike their predicationally consistent counterparts, identify intensionally and correspond to metaphysically impossible but still thinkable and linguistically designatable objects. They are the bearers of properties in predications that are true by the very definition of *impossibilia*. The same semantic license that is required of logically impossible objects generally must then apply equally to logically impossible states of affairs and logically impossible worlds. In ontology, as the theory specifically of entities or existent objects, objects of all kinds with being, we limit theory to the intensional characterization of actually existent entities, states of affairs, and the actual world, and merely logically possible objects, states of affairs, and worlds.

COMBINATORIAL CONCEPT OF BEING (EXISTENCE)
ANALYSIS OF THE CONCEPT OF BEING

For any object *O*, *O exists*, has *being*, or is an *entity* if and only if *O* has a maximally consistent property combination.

The actual world can accordingly be defined equivalently as a maximally consistent combination of actual states of affairs, or maximally consistent states of affairs combination. An actual state of affairs is defined in turn as the true predication to or possession of a property by an actual object, and where an actual object is defined as an intrinsically and extrinsically predicationally maximally consistent property combination. A particular maximally cosistent property combination already determines a maximally consistent states of affairs combination, and thereby also determines an actual world complete in all its actual states of affairs involving all and only actually existent entities.

4. Why There is Something Rather Than Nothing

We can now say that there exists something rather than nothing because there is certain to be a logically possible combination of all logically possible properties with all logically possible objects in which every object is intrinsically and extrinsically predicationally maximally consistent. Such a combination is logically combinatorially necessitated as only one of indefinitely many logical possibilities. There is thus sure to be at least one combination of properties and objects that satisfies the proposed definition of being as maximal consistency.

An analysis of the concept of being can be given combinatorially in terms of maximal consistency. What it is to be, to exist or be real or actual or to have being, whether the entity in question is a physical or abstract object, state of affairs, or world, is to be predicationally maximally consistent. This condition, although purely logical and purely a matter of logical possibility, when properly interpreted, is sufficiently strong to guarantee that there is only one actual world consisting of and constituted by all and only the actually existent states of affairs, involving all and only the actually existent objects. If applied ontology declares that there are abstract objects, then these can be added to the existence domain only if they are predicationally maximally consistent. Since they are not going to include spatiotemporal properties in their property combinations, they are not going to be predicationally incomplete by virtue of existing in any logically possible world that also contains any predicationally incomplete objects or states of affairs. If there are abstract entities, they are spatiotemporally and in every other way relationally untouched by the predicational completeness or incompleteness of any physical objects belonging to any logically possible world, and so generally by their ontic status.

We summarize the three combinatorial categories of existence in the following sets of diagrams, beginning with a general division between logically possible objects, states of affairs, and worlds, and then defining the concepts of entities, existent states of affairs, and the actual world.

COMBINATORIAL EXISTENCE CANDIDATES

Object

Property combination

Most general category; subsumes all objects, including existent and merely logically possible nonactual objects, states of affairs, worlds, and the actual world

State of Affairs

Inclusion of property or predication in a property combination Propertyhood, Quality or Relation, of an Object

(Logically Possible) World

Consistent states of affairs combination, maximal or submaximal; ultimately logically possible combination of logically possible objects with logically possible properties

The ontically neutral divisions between objects, logically possible states of affairs and logically possible worlds are now more precisely specified to represent a parallel three-part division between actually existent entities, including actually existent states of affairs and the actual world. The definitions add the requirement of maximal consistency to objects in each of the three original categories. Then we have:

COMBINATORIAL CATEGORIES OF EXISTENCE

Entity

Maximally consistent property combination

Most general category, subsumes all existent objects, including existent objects, states of affairs and the actual world

Existent State of Affairs

Inclusion of property or predication in a maximally consistent property combination where entity $a = [F, \ldots]$, Fa is an existent state of affairs

Propertyhood, Quality or Relation, of an Entity

(The) Actual World

Maximally consistent combination of all and only existent states of affairs; ultimately maximally consistent combination of all logically possible objects with all logically possible properties

Total Reality

We are prepared within this combinatorial framework as a result to broach the problem of why there exists something rather than nothing. The argument has the following structure:

WHY THERE IS SOMETHING RATHER THAN NOTHING

1. It is logically combinatorially necessary that the totality of combinations of all logically possible combinationss of all logically possible states of affairs include at least one maximally consistent combination. (Combinatorial Necessity)
2. An actual world is and is nothing more than a maximally consistent states of affairs combination. (Definition)

---

3. An actual world is logically combinatorially necessary; it is logically combinatorially necessary that there be or exist an actual world, that an actual world exist. (1,2)
4. Logically necessarily, there must exist an actual world. The reason why there exists something rather than nothing is that logic combinatorially necessitates a maximally consistent states of affairs combination, which satisfies the definition of the concept of an actual world. (3)

The deductive necessity and *a priori* truth of the argument's assumptions are essential in maintaining an *a priori* answer to the question of being. We can know, without relying on any experience of the world, that out of all the logically possible states of affairs combinations, and of all the logically possible combinations of all logically possible objects with all logically possible properties, there must be at least one that is maximally consistent.

How do we know this? It is an intuitive consequence of the idea of the totality of combinations of states of affairs or of objects and properties, together with the fact that there seems to be no reason to exclude a maximally consistent combination from the unimaginably large totality of all logically possible combinations. By definition, a maximally consistent states of affairs combination is logically consistent, and so cannot be ruled out on the only relevant purely logical grounds. As we have defined the concept of an actual world, it follows that a maximally consistent states of affairs combination just is an actual world.

There is no special need to prove that logically there must be a maximally consistent combination of states of affairs involving maximally consistent logical object-property combinations. We need only suppose that a maximally consistent states of affairs combination is logically possible. The only obstacle to the possibility of a maximally consistent states of affairs combination would be if it were somehow logically inconsistent, which is evidently not a problem for anything that by hypothesis is supposed to be predicationally maximally consistent. Among all the imaginable combinations of objects with properties, whether existent or nonexistent, constituting all imaginable states of affairs, whether existent or nonexistent, and among all the imaginable combinations of states affairs, there logically must be at least one that is maximally consistent. We know this *a priori* from the logical requirements for combination and the fact that there can be no logical inconsistency in a combination designated not only as internally predicationally logically consistent, but predicationally maximally consistent. We can therefore conclude as a matter of logical combinatorial necessity that there must be such a possible maximally consistent states of affairs combination, implied by pure classical logic among all the possibilities of combining logically possible objects with logically possible properties.

Although it may seem paradoxical or even modally invalid to put the point this way, we can therefore say that, since it is logically necessary that it is logically possible for the totality of all logically possible combinations of all logically possible states of affairs to include a maximally consistent combination, then it is logically necessary for an actual world to exist. The logical necessity of the logical possibility of a maximally consistent states of affairs combination logically guarantees that at least one actual world exists, and therefore that something rather than nothing exists. Later, we consider the modality of the inference, and conclude that it is logically correct, given the definitions of concepts of combinatorial pure philosophical ontology, to deduce the actuality of a logically possible world or existence of an actual world from the logical necessity of the logical possibility that the full range of all logically possible combinations of objects and properties include a maximally consistent states of affairs combination, which we have also defined as an actual world.

As a dividend for our labors in the vineyards of logic and predication, we clarify the concept of being as we consider the analysis of existence in purely logical terms. We define an actual world as a predicationally maximally consistent states of affairs or object-property combination. There is certain to be included at least one such combination among all logically possible combinations of logically possible states of affairs. A maximally consistent states of affairs combination logically must exist; but since that is what we mean by the existence of an actual world, logically there must exist an actual world. Something rather than nothing exists because it is logically combinatorially necessary for at least one maximally consistent states of affairs combination to be included among all logically possible object-property combinations, and hence for something, at least one actual world, to exist.

When we spin out all the combinations of distributions of properties over objects considered as logical predication subjects, we can be sure that there must be at least one combination in which the property combinations of all logically possible objects are maximally consistent. That combination, or combinations if there are more than one, satisfy on the present definition the requirements for actual existence, or, more simply but with less of the conventional emphasis, are actual, real, exist, or have being.

Something rather than nothing exists because "something" is a logically general way of designating *an* actual world that is not necessarily *the* actual world in all its particularity. There is something rather than nothing because: (1) To be means to be maximally consistent, a purely logical qualification, whether we are speaking of the being or existence of an individual object, a state of affairs, or an entire actual world, including *the* actual world in all its particularity. (2) We know combinatorially that there is certain to be at least one maximally consistent states of affairs or object-property combination that constitutes and is linguistically representable as an actual world, a world all of whose states of affairs actually exist, involving all and only actually existent objects. There must logically be such a combination because logically there is no contradiction in the idea of there being such a combination.

It is from the logical possibilities for combinations of properties, the distribution of all logically possible properties over all logically possible objects understood as logical predication subjects, that the logical necessity of the existence of an actual world is deduced. There must be a logically possible predicationally maximally consistent combination of just the properties that constitute a world of logically possible predicationally maximally consistent actual objects in logically possible predicationally maximally consistent actual states of affairs. What it is to be is therefore in a sense to be one of these so defined actual entities, one of these so defined actual states of affairs, or the so defined actual world. The analysis of being as maximal consistency thus applies down the line. An actual object is a predicationally maximally consistent property combination; an actual state of affairs is the possession by any actual entity or entities of any of the properties in their respective maximally predicationally consistent property combinations. An actual world is a maximally consistent combination of actual states of affairs. The actual world is a particular maximally consistent combination of actual states of affairs. From the fact that all actual states of affairs and all actually existent entities can belong at most only to one truly maximally consistent states of affairs combination, it follows as a further implication that there uniquely exists only and at most one actual world, so that, as we shall see, *an* actual world can only be *the* actual world. To be is to belong to one of these three kinds of entities, actual objects, actual states of affairs, or an actual world in its entirety, and hence ultimately of logicallly possible objects as logically possible predication subjects within the purview of logic having predicationally maximally consistent property combinations.

Logically, there must be just one combination of properties to objects that exactly instantiates the entities and states of affairs that constitute the actual world. In the distribution of all logically possible properties over all logically possible objects or logically possible predication subjects, there is certain to be at least one maximally consistent combination. Should we call such a combination, or combinations, if there are more than one, the actual world or actual worlds? The question is cantilevered. If we are satisfied with the explanation of why there exists something rather than nothing which the concept affords, then we may also be inclined to accept the definition. The trouble is that we are not yet in a good position to judge whether the explanation is satisfactory. We can probably agree no matter what view we take with respect to the relation between logic and ontology that the explanation is unacceptable if it does not support a further combinatorial argument to show that there can at most exist one actual world, that it is the logically possible world we happen to inhabit, and that the actual world is logically contingent.

We may need to remind ourselves at this turn that pure philosophical ontology is a discipline only, and not a domain. We are not trying to use logic to deduce the particular predications that are true of the actual world, but only to understand that there must exist an actual world, if what we mean by that term is a maximally consistent states of affairs combination. As such, we make a valuable contribution to pure philosophical logic by explaining why there exists something rather than nothing. There must exist something, because the maximally consistent combination of predications that constitutes an actual world is logically possible; there is logically certain to be at least one such combination of predications in the full range of all logical possibilities. What it means for something other than an actual world in its entirety to exist is then for it to be in some sense a part of or entity within an actual world, as or as a constitutive part of an actually existent state of affairs.

5. Logical Analysis of Combinatorial Ontology

We have defined the concept of being in purely logical terms, arguing on the basis of a combinatorial analysis of existence that there must exist exactly one logically contingent actual world. The inferences we have proposed have been informally presented, relying on an intuitive understanding of the implications of the idea that to exist is to be maximally consistent.

The modality of this intuitive argument to answer the question of being, to explain why there exists something rather than nothing, and why there exists exactly one logically contingent actual world, should now be formalized. Doing so provides an opportunity to review all three of the explanations that have now been offered in combinatorial pure philosophical ontology, to see precisely how they are related, and to reconsider their conclusions.[3]

A critic might imagine that the inference used to explain why there exists something rather than nothing contains a modal fallacy. The claim that the combinatorial possibilities for a logical object to combine with a logical property in pure classical logic logically necessitate the existence of an actual world that can then be shown to be the uniquely existent logically contingent actual world may seem modally problematic. The uniquely existent actual world is supposed to be logically contingent, despite the fact that its existence is proved by the combinatorially logically necessary existence of an actual world. Is there not something wrong here?

The fallacy seems to be that, for any proposition p, if it is logically necessary that it is logically possible that p, then it is logically necessary that p. It appears that by claiming that from the fact that it is logically necessary that there be a maximally consistent combination of states of affairs in the totality of all logically possible states of affairs combinations we can logically infer that combinatorially an actual world necessarily exists. It is a modal fallacy to permit the deductively invalid inference, $\Box\Diamond p \vdash \Box p$, or the false conditional, $\Box\Diamond p \rightarrow \Box p$, even in the strongest sound systems of iterated modal logic. A counterexample to remind us of how the statement goes astray can be given in modal system S_5. There we are permitted to infer $\Box\Diamond p$ from $\Diamond p$. Let p be the logically possible proposition that pigs fly. Then it follows in S_5 that $\Box\Diamond p$, but not that $\Box p$, that it is logically necessary that pigs fly. This, fortunately, is *not* the modal structure of the argument we have advanced, which is modally and in other ways logically more complicated.[4]

---

[3]Useful standard expositions of modal logic are found in Hughes & Cresswell (1968) and Chellas (1980). See also the section on "Modal Logic and Semantics" in Jacquette (2002, pp. 391–509).

[4]Even strong modal systems like S_5 do not generally support the inference of logical necessity, $\Box p$, from the iterated logical possibility of logical necessity, $\Box\Diamond p$. By contrast, S_5 characteristically permits the valid inference $\Diamond\Box p \vdash \Box p$.

To explain how we avoid the fallacy, we formalize the underlying modality of the arguments by which we propose to demonstrate combinatorially that logically there must exist something rather than nothing, that an actual world logically must exist, but that the condition of the actual world is not logically necessary. We refer to the actual world as an individual entity, $w_@$, and to apply a special modal operator to represent the actuality of any logically possible world. For this purpose, we adopt a modal symbolism in which actuality is intepreted as a special modal operator, Δ, within a standard weakest alethic modal logic. We assume as we have throughout that for a logically possible world to be actual means for it to be a maximally consistent states of affairs combination, which we symbolize as the metalogical predicate, MaxCon. We allow modal operators to attach directly to individual propositions or to any conjunction of propositions, even to a conjunction of propositions that completely describe a logically possible world. The distinction makes it possible to formalize the proposition that the actual world is actual, which we write as: $\Delta\, w_@$. We can then more formally consider the logic of the previous inferences used to explain why there exists something rather than nothing, and why there exists exactly one logically contingent actual world. Through much of the logical symbolizations to follow we make intuitive use of the Δ operator, and only near the end propose inference rules supported by the same intuitions for its introduction in standard first-order predicate logic.

We assume that being or actuality is maximal consistency. We begin by appealing only to the conditional, $\forall w[\texttt{MaxCon}\, w \rightarrow \Delta\, w]$ (or, $\forall w[\texttt{MaxCon}\, w \rightarrow \exists x(\Delta\, x \wedge w = x)]$). The converse conditional appears in other arguments, involving the biconditional, whereby we more generally formulate the combinatorial answer to the question of being, $\forall x\, [\texttt{MaxCon}\, x \leftrightarrow \Delta\, x]$. We refer to the conditional or complete biconditional analysis of being respectively as Df. $\Delta \rightarrow$ and $\Delta \leftrightarrow$. Here, in answering the question why there exists something rather than nothing, it is important to see that we can deduce actuality from maximal consistency alone. We note that in deducing the instantiated consequent, $\Delta\, \alpha$, from the instantiated antecedent, $\texttt{MaxCon}\, \alpha$, we *deduce* the *actuality* of α. The universal implies that we can deduce the actuality of *any* α, including any real existent entity or state of affairs or combination of states of affairs or the entire actual world, provided only that, according to the principle, α is maximally consistent. We can in principle soundly substitute for "α" anything we can *name* that is maximally consistent, including the actual world, $w_@$. We prove the actuality of *the* actual world as a substitution instance, $\Delta\, w_@$, *if* we know that the actual world is maximally consistent, $\texttt{MaxCon}\, w_@$. We cannot avoid circularity if we invoke the converse conditional, $\Delta\, w_@ \rightarrow \texttt{MaxCon}\, w_@$, at this stage of the argument, in order to conclude that the actual world is maximally consistent, unless we are in a position to assert what is now supposed to be proved, that the actual world is actual. What, then, can we do?

It is tempting to try to establish the maximal consistency of the actual world by means of an argument that we should consider only to reject. We should not assume that only a maximally consistent world can be perceived, and then cite the empirical fact that we do in fact experience a world, whether in sensation or in the experience of the flow of time in reflecting self-consciously on the fact of thinking, in order to conclude that the actual world which we perceive must be maximally consistent. Experiential data cannot count as evidence in noncircular reasoning to show that the actual world is maximally consistent. Thankfully, inspection of the states of affairs we experience is not the only methodology to which we are limited. We are no more required to survey each state of affairs and consider them all in all their complex relations in order to know that the actual world is maximally consistent than we are required to go door-to-door with a questionnaire in order to find out if all bachelors are unmarried. A single glimpse of the outside world by any of the senses is sufficient to uphold this argument, if we also hold that only a maximally consistent states of affairs combination could possibly or even logically possibly be perceived. It will not do to dig in our heels and declare that existent thought cannot be related to what is nonexistent. That assumption, in the first place, might not even be true; it disguises, in any case, rather than exposes, the interesting conceptual connections that need to be articulated in order to explain why only a maximally consistent world is experienceable. If there is just an eye (and brain), and nothing else outside it, not even a single lightray, the eye might still be able to perceive by virtue of internal changes of state.

We only make things worse if we proceed to argue that there must be an actual world in order for anything to be perceived, on the assumption that the fact of perception implies a change of state, and that there can be no changes of state except in physical space and time, the lonely eye and brain changing internal states in the impenetrable darkness. The trouble is that with each step we are wandering farther and farther away from pure philosophical ontology into applied scientific ontology. If we simply stipulate that any experienceable world is actual, and then argue as above that the experienced world is the only uniquely existent actual world, then we seem merely to be insisting on the fact of a connection between experience and actuality without analyzing its meaning or explaining its conceptual structure. An argument to the effect that we do in fact experience the world and that we could only experience a maximally consistent world is not absurd, but it is also not the best we can do.

Ultimately, we want to be able to maintain that any logically possible world is actual if and only if it is identical with the actual world, and that the actual world is nevertheless logically contingent. We quantify alike over the actual world and nonactual merely logically possible worlds, even though, as a consequence of our reform of modal semantics, only the actual world exists. In doing so, we in effect quantify over one maximally consistent and indefinitely many submaximally consistent states of affairs combinations, which is equivalently to quantifying over one actualized and many unactualized logical possibilities.

As might be expected, combinatorial pure philosophical ontology has a non-empirical nonexperiential way of proving that the actual world is maximally consistent. We know that there must exist *an* actual world, and we know that there can exist at most one actual world. We do not try to prove that the actual world, in particular the world we experience, is maximally consistent, regardless of whatever conceptual linkages might be adduced. Accordingly, we shall not try to justify the choice of "$w_@$" as a substitution instance for "α" in the definition of actuality, except by arguing that $w_@$ is the one and only actual world, and hence the only logically possible truth-preserving substitution for "α", together with designations for any particular actual states of affairs or existent entities in $w_@$. We require a principle to mediate the application of alethic modalities of necessity, possibility, and actuality, from propositions to worlds, particularly because we do not interpret worlds as proposition sets. Accordingly, we offer the following equivalence to justify the modality of a logically possible world in terms of the modality of the propositions true in or of that world, $\forall w[\Box w \leftrightarrow \forall p(p^w \rightarrow \Box p)]$. Similarly for the modal qualification of a logically possible world as logically possible or actual, substituting "$\Diamond$" or "Δ" respectively for "$\Box$". It is sufficient for present purposes that it is logically necessary for there to be included in the totality of all states of affairs combinations at least one combination that is maximally consistent. This shows, assuming that only the actual world is maximally consistent, that the actual world actually exists, $\mathtt{MaxCon}\, w_@ \rightarrow \Delta\, w_@$. We argue that $\Box\, \mathtt{MaxCon}\, w_@$, from which and the assumption that $\Box\forall x\, [\mathtt{MaxCon}\, x \rightarrow \Delta\, x]$, we deduce the necessary actuality of the actual world, $\Box\, \Delta\, w_@$ and $\Box\, \Delta\, \exists w\, \Delta\, w$. We can then prove $\forall w[\mathtt{MaxCon}\, w \leftrightarrow w = w_@]$ and $\forall w[w \neq w_@ \leftrightarrow \neg\, \mathtt{MaxCon}\, w]$.[5]

6. Logical Solution to the Question of Being in Combinatorial Pure Philosophical Ontology

We argue that logically necessarily if a maximally consistent states of affairs combination possibly exists, then it exists. The combination exists as a logical possibility in the first instant, as a logically possible combination of logically

[5]The proposition in assumption (6), $\Box\Diamond\exists w\, \mathtt{MaxCon}\, w$, expresses the combinatorial implication whereby logically it must be logically possible for there to exist a maximally consistent logically possible world, which is to say at least one maximally consistent states of affairs or object-property combination among all logical possibilities. The inference is presented in terms of the weakest modal systems; in stronger modal logics, like modal S_5 and the Brouwersche system, the assumption is derivable, relying on combinatorial possibilities for the simpler modally unqualified assumption, $\exists w\, \mathtt{MaxCon}\, w$. (I am grateful to John Divers for pointing out this alternative route to the conclusion in (7), without directly appealing to the iterated modality in the original assumption (6).) The essential

possible objects and logically possible properties. It is then shown to be actual by virtue of satisfying the definition of actuality as maximal consistency.

We are logically guaranteed the logical possibility of every logically consistent states of affairs combination. We prove that there must be a maximally consistent logically possible world, and hence an actual world. Then we prove that there can be only one uniquely actual world, from the intermediate conclusion that logically there can be only one maximally consistent states of affairs combination. The maximally consistent states of affairs combination by virtue of being maximal eats up all the maximal consistency available, depriving other nonactual merely logically possible worlds of the (especially relational) object-property combinations they would need in order to be independently predicationally complete.

For convenience in what follows, we adopt a standard notation involving restricted quantification over worlds. We consider the modal status of individual propositions and logically possible worlds alike, on the grounds that a logically possible world can be construed as a long conjunction of propositions that collectively express all the states of affairs included in a world's states of affairs combination. We permit modal qualifications of some of these restrictedly quantified sentences, where, in an obvious sense, to speak of logical possibility is already to speak of truth in at least one logically possible world, and to speak of logical necessity is already to speak of truth in every logically possible world. Such retranslations are naturally possible, but not as illuminating as allowing the iterations of both forms of reference over all or some worlds, to remind us that some of our assumptions and conclusions are supposed to be logically necessary, and that these are interpreted by reference to worlds as maximally consistent states of affairs combinations.

We can now further formalize some of these inferences. We add abbreviations for a function of logically possible combinations (`Combo`) of logically possible predication objects (`Object`) and logically possible properties (`Property`), indicating membership in a combination by the inclusion sign ($\in$). We begin by formalizing the argument that logically combinatorially there must exist an actual world, something rather than nothing must exist.

---

modal principle for these stronger logical systems appears in assumption (5c):

5.	$\Box[\Diamond\exists w\,\mathtt{MaxCon}\,w \rightarrow \exists w\,\mathtt{MaxCon}\,w]$	Df `MaxCon`
5a.	$\Box\Diamond\exists w\,\mathtt{MaxCon}\,w \rightarrow \Box\exists w\,\mathtt{MaxCon}\,w$	(5, Modal logic)
5b.	$\Diamond\neg\exists w\,\mathtt{MaxCon}\,w \rightarrow \Diamond\Box\neg\exists w\,\mathtt{MaxCon}\,w$	(6, Modal duality)
5c.	$\Diamond\Box\neg\exists w\,\mathtt{MaxCon}\,w \rightarrow \neg\exists w\,\mathtt{MaxCon}\,w$	(S_5 or Brouwersche)
5d.	$\Diamond\neg\exists w\,\mathtt{MaxCon}\,w \rightarrow \neg\exists w\,\mathtt{MaxCon}\,w$	(5b, 5c)
5e.	$\exists w\,\mathtt{MaxCon}\,w \rightarrow \Box\exists w\,\mathtt{MaxCon}\,w$	(5d, Modal duality)
6′.	$\exists w\,\mathtt{MaxCon}\,w$	(Combinatorics)
7.	$\Box\exists w\,\mathtt{MaxCon}\,w$	(5e, 6)

Logically Necessarily, There Exists a Maximally Consistent World:

1. $\Box\exists w[[\mathtt{MaxCon}\, w \wedge w \in \forall x \mid \mathtt{Combo}(\forall y \mid \mathtt{Object}\ y, \forall z \mid \mathtt{Property}\, z)x] \leftrightarrow$
$\Diamond\exists w[\mathtt{MaxCon}\, w \wedge w \in \forall x \mid \mathtt{Combo}(\forall y \mid \mathtt{Object}\ y, \forall z \mid \mathtt{Property}\, z)x]]$ Df. MaxCon
2. $\Box\Diamond\exists w[\mathtt{MaxCon}\, w \wedge w \in \forall x \mid \mathtt{Combo}(\forall y \mid \mathtt{Object}\ y, \forall z \mid \mathtt{Property}\, z)x]$ Combinatorics

---

3. $\Box\exists w[\mathtt{MaxCon}\, w \wedge w \in \forall x \mid \mathtt{Combo}(\forall y \mid \mathtt{Object}\ y, \forall z \mid \mathtt{Property}\, z)x]$ (1, 2)
4. $\Box\exists w\ \mathtt{MaxCon}\, w$ (3)

The core of the combinatorial analysis of the concept of being is that it authorizes deducing the necessary existence of a world on the grounds that necessarily it is logically possible for there to be a maximally consistent states of affairs or object-property combination. This, the theory maintains, is what it means for a world to exist. We expand the inference from (1)–(4) above by citing the logical necessity that the mere logical possibility of a maximally consistent states of affairs combination implies its existence, together with the proposition that it is logically necessary that it is logically possible that there exist a maximally consistent states of affairs combination.[6] The argument states:

5. $\Box[\Diamond\exists w\ \mathtt{MaxCon}\, w \rightarrow \exists w\ \mathtt{MaxCon}\, w]$ Df. MaxCon
6. $\Box\Diamond\exists w\ \mathtt{MaxCon}\, w$ Combinatorics

---

7. $\Box\exists w\ \mathtt{MaxCon}\, w$ (5, 6 Modal logic)

We note that assumption (5) is not universal, but applies specifically to worlds or maximally or submaximally consistent states of affairs combinations. Now we apply either conclusion (4) or (7) above to prove:

[6]It might appear, in light of the combinatorial analysis of nonactual merely logically possible worlds as submaximally consistent, that it is obligatory to define logically necessary truth, $\Box p$, nonstandardly as a proposition p not being false in any logically possible world, rather than being true in every logically possible world. The grounds for such a claim are that in a modal actualist semantics there are countless propositions that will not describe states of affairs in any given nonactual merely logically possible world, for which the worlds are incomplete, submaximal, or for which they have constitutive predicational gaps. Redefining the concept of logical necessity in this way is not required, however, as we see when we reflect hypothetically on a dilemma concerning the applied ontology of abstract entities, whose ontic status is yet to be determined. If there exist abstract entities, then by definition they exist in every logically possible world, including nonactual merely logically possible worlds; in which case, necessary truths concerning only the properties of abstract entities remain true of every logically possible world. If, on the contrary, no abstract entities exist, then there are no logically necessary truths, but only logically contingent truths involving the variably existent objects and states of affairs differentially distributed across some but not all logically possible worlds. This situation again supports the standard definition of logically necessary truth as truth in every logically possible world, where by hypothesis there are no logically necessary truths and equally no propositions that are true in every logically possible world.

Why There is Something Rather Than Nothing
— Logically, There Must Exist an Actual World:

8. $\Box\forall x\,[\texttt{MaxCon}\,x \rightarrow \Delta\,x]$ Df. $\Delta \rightarrow$

---

9. $\Box\exists w\,\Delta\,w$ (4 or 7, 8 Modal logic)

10. $\exists w\,\Delta\,w$ (9 $\Box p \rightarrow p$, *a fortiori*)

11. $\Delta\,\exists w\,\Delta\,w$ (10 $p \rightarrow \Delta\,p$, *a fortiori*)

12. $\Box\,\Delta\,\exists w\,\Delta\,w$ (8, 9, 11 Modal logic)

In conclusion (11) we argue somewhat redundantly that an actual world actually exists. In principle, there need be no limit to how many times we iterate the actuality operator. Similarly, there need be no limit to how many times we iterate a possibility operator; in practice, it is not especially helpful to continue beyond statement (11), by deriving, as we could, $\Delta\,\Delta\,\exists w\,\Delta\,w$, $\Delta\,\Delta\,\Delta\,\exists w\,\Delta\,w$, etc. Instead, we prove (12), relying only on the necessary truth of propositions (8) and (9), to derive conclusion (11), which, if the inference is deductively valid, must be as logically necessary as the two assumptions from which it is inferred. Thus, we prove that it is logically necessary that there exists an actual world. Now we formalize the previous argument to establish the unique actuality of the actual world.[7]

7. Unique Existence of the Actual World

The proof depends on the assumption that a maximally consistent combination of all logically possible objects and all logically possible properties into all logically possible states of affairs is unique. If there were other states of affairs that could be consistently added to such a combination, it would not after all be maximally consistent, but at best submaximal. We do not propose to prove formally the uniqueness of maximal consistency, but to formalize it as an assumption from which to derive the conclusion that necessarily an actual world is uniquely actual. This is just to say that logically necessarily there is only one actual world.

Above, we demonstrated the necessary existence of an actual world using only the conditional definition, Df. $\Delta \rightarrow$, $\Box\forall x\,[\texttt{MaxCon}\,x \rightarrow \Delta\,x]$, and avoided the circularity with which we were concerned. We can therefore avail ourselves now of the full biconditional combinatorial analysis of the concept of being, Df. $\Delta \leftrightarrow$, $\Box\forall x\,[\texttt{MaxCon}\,x \leftrightarrow \Delta\,x]$, as provided by definition, to prove the uniqueness of the actual world. As a lemma, we thereby establish the principal conclusion of

---

[7] Inference rules for the delta Δ actuality operator are specified as needed for proofs further below.

modal actualism in opposition to the modal realism of conventional model set theoretical modal semantics with its definition of all worlds, including the actual world and nonactual merely logically possible world, as maximally consistent proposition sets. The proof is this:

Why There is Only One Actual World
— There Can Be at Most One Maximally Consistent World:

13. $\Box\forall x[\texttt{MaxCon}\, x \leftrightarrow \Delta\, x]$ (Df. $\Delta \leftrightarrow$)
14. $\Box\forall w[\, \texttt{MaxCon}\, w \rightarrow \neg\exists x[\texttt{MaxCon}\, x \wedge w \neq x]]$ ($\texttt{MaxCon}$ Unique)

---

15. $\Box\forall w[\, \Delta\, w \rightarrow \forall x[\Delta\, x \rightarrow w = x]]$ (13, 14)

8. Logical Contingency of the Actual World

It remains to relate the necessary actual existence of an actual world, $\Box\,\Delta\,\exists w\,\Delta\, w$, to the modality of the actual world, $w_{@}$. Can we validly substitute "$w_{@}$" for "w" in conclusion (12) to obtain $\Box\,\Delta\,\Delta\, w_{@}$ or $\Box\,\Delta\, w_{@}$?

To prove, on the contrary, that the actuality of the actual world is logically contingent, $\Delta\, w_{@} \wedge \neg\Box\,\Delta\, w_{@}$, we appeal to the multiplicity of logically possible worlds distinct from $w_{@}$. We can be sure of their multiplicity because combinatorially we cannot collapse all logically possible worlds into one, there are logically possible combinations of objects and properties into nonexistent merely logically possible states of affairs that do not exist in the actual world. If the combinatorial analysis of being is correct, and it is combinatorially logically necessary that a logically contingent actual world uniquely exists, then intuitively we expect to be able to prove the logically necessary propositions:

$\Box\,\Delta\,\exists w\,\Delta\, w$

$\Box\exists w\,\Delta\, w$

$\Box\,\Delta\, w_{@}$

As, indeed, we can; but *not*:

$\exists w\,\Delta\,\Box w$

$\exists w\Box w$

$\Delta\,\exists w\Box w$

$\Delta\,\exists w\,\Delta\,\Box w$

$\Delta\,\Box w_{@}$

$\Box w_{@}$

As another way of arriving at the same conclusion, consider the following argument. We say that a world is maximally consistent when we write $\exists w\ \mathtt{MaxCon}\, w$. This is a shorthand way of saying:

$$\exists x\,[\mathtt{MaxCon}\, x \wedge x = \mathtt{Combo}(\forall y \mid \mathtt{Object}\ y, \forall z \mid \mathtt{Property}\, z)]$$

Or, equivalently, where $\mathtt{SA}$ denotes a state of affairs:

$$\exists x\,[\mathtt{MaxCon}\, x \wedge x = \mathtt{Combo}\,\forall y \mid \mathtt{SA}\, y]$$

It is a conclusion of combinatorial pure philosophical ontology that:

$$\mathtt{MaxCon}\,\mathtt{Combo}\,\forall y \mid \mathtt{SA}\, y \leftrightarrow \Delta\, w_{@}$$

We are accordingly entitled to make the following inference, deriving the *de dicto* logically necessary existence of something (rather than nothing, an actual world); the *de dictio* logically necessary existence of the actual world; and the *de dicto* logically necessary actuality of the actual world:[8]

[1]	$\Box\forall x[[\mathtt{MaxCon}\, x \wedge x = \mathtt{Combo}\,\forall y \mid \mathtt{SA}\, y] \rightarrow [x = w_{@} \wedge \Delta\, x]]$	Actuality
[2]	$\Box\exists x[\mathtt{MaxCon}\, x \wedge x = \mathtt{Combo}\,\forall y \mid \mathtt{SA}\, y]$	Combinatorics
[3]	$\Box\exists x\,\Delta\, x$	[1, 2]
[4]	$\Box\exists x[x = w_{@} \wedge \Delta\, x]$	[1, 2]
[5]	$\Box\,\Delta\, w_{@}$	[4]

Again, what cannot be proved from any of the assumptions we have considered is the *de re* logical necessity of the actual world. It is true in combinatorial pure philosophical ontology, in other words, that: $\neg\exists x\Box[x = w_{@} \wedge \Delta\, x]$ or $\neg\,\Delta\,\Box w_{@}$ or $\neg\Box w_{@}$. We shall now see how the required restrictions permit the above inferences of the *de dicto* necessary existence of the actual world, and prohibit inferences of the *de re* necessary existence of the actual world, which would collapse logical contingency and freedom of action and will into logical determinacy.

The excluded inferences in the second list contradict the logical contingency of the actual world considered as one among innumerably many logically possible worlds that might have existed instead. We introduce inference rules for the modal actuality opertor, Δ, formally defining the modality. Proof of the

[8] A difference in modal scope is relevant to the argument. *De re* or *wide modal scope* directly qualifies the modality of an *object*, a thing or *res*, as in the sentence, "Object *a* is necessarily *F*", whereas *de dicto* or *narrow modal scope* qualifies the modality of a *proposition*, literally a saying or *dictum*, in which the object is mentioned, such as, "Necessarily, object *a* is *F*". *De dicto* contexts, furthermore, including modal contexts, are generally intensional, whereas *de re* contexts are extensional.

antitheorem $\neg[\Box \Delta p \rightarrow \Delta \Box p]$ that safeguards the logical contingency of the logically necessarily actual world, is then routine but still instructive. Up until this point, we have relied on an intuitive understanding of the laws of logical inference governing the delta modal operator Δ, and have not needed to support any implications involving its logical properties. We now present formal inference rules in support of the subsequent demonstration:

INFERENCE RULES FOR MODAL ACTUALITY OPERATOR Δ

$$\frac{p}{\Delta p} \qquad \frac{\Delta p}{p} \qquad \frac{\Delta \neg p}{\neg \Delta p} \qquad \frac{\neg \Delta p}{\Delta \neg p} \qquad \frac{\Delta \exists x p}{\exists x \Delta p} \qquad \frac{\exists x \Delta p}{\Delta \exists x p}$$

Proof of Antitheorem: $\neg[\Box \Delta p \rightarrow \Delta \Box p]$

[1]	$\neg\Box p$	Assumption
[2]	$\Box \Delta p$	Hypothesis for *reductio*
[3]	$\Box \Delta p \rightarrow \Delta \Box p$	Hypothesis for *reductio*
[4]	$\Delta \neg\Box p$	[1 Δ Rule]
[5]	$\neg \Delta \Box p$	[4 Δ Rule]
[6]	$\neg\Box \Delta p$	[3, 5 *Modus tollendo tollens*]
[7]	$\neg[\Box \Delta p \rightarrow \Delta \Box p]$	[2, 3, 6 *Reductio ad absurdum*]
[8]	$\Delta p \rightarrow p$	[Δ Rule]
[9]	$\Box[\Delta p \rightarrow p]$	[8, Necessitation]
[10]	$\Box \Delta p \rightarrow \Box p$	[9, Modal Logic]
[11]	$\Box p \rightarrow \Delta \Box p$	[Δ Rule]
[12]	$\Box \Delta p \rightarrow \Delta \Box p$	[10, 11]
[13]	$\neg\Box \Delta p$	[2, 7, 12 *Reductio ad absurdum*]
[14]	$\neg\Box p \rightarrow \neg\Box \Delta p$	[1, 13]

Let p be a contingently true proposition, that pigs wallow. Then we have $\neg\Box p$ in the second assumption, without any argument. It would be circular, even in this elementary proof context, simply to assume that $\neg\Box w_{@}$, even though our immediate purpose is not to prove that $\neg\Box w_{@}$, but only negatively that $\Box w_{@}$ cannot validly be deduced from $\Box \Delta w_{@}$. It is therefore enough to suppose that there is at least one logically non-necessary proposition that is necessarily actual, in order to show that there is no valid deduction from necessary actuality to actual necessity. This is enough to establish the logical contingency of the condition of the actual world, despite the logical necessity of its existence.

What about $\Box \Delta p$? Is it obvious that it is logically necessary that actually pigs wallow? Not in the absence of a combinatorial modal ontological theory perhaps. We recall that we are dealing not only with any random proposition p in the case we are interested in, but with the proposition Δp that in context has been validly derived from two logically necessary assumptions. As long as the proposition that or state of affairs in which pigs wallow is true of or exists in the actual world, then we are justified in the problem at issue to infer $\Box \Delta p$ from Δp, which we stipulate accordingly above as the first hypothesis. We prove, even on the strength of the first hypothesis, proclaiming the logical necessity of the actuality of p, that we cannot validly derive the logical necessity of p itself, in the form $\Delta \Box p$ or $\Box p$, for any proposition p. Thus, an inference that would threaten the logical contingency of the condition as opposed to the existence of the actual world, in the sense in the sense of the particular states of affairs that constitute the actual world, is blocked.

What can we say more positively about the modal status of the actual world's constituent states of affairs? Is the logically necessarily existent uniquely actual world also logically necessary in all of its constituent states of affairs, or is it logically contingent? There are at least some different states of affairs that distinguish the states of affairs combinations to which they belong as world-variable states. These are effectively spatiotemporal physical states, which must vary from logically possible world to logically possible world, discounting nonexistent or universally distributed abstract states of affairs as before. The only way for distinct logically possible worlds to collapse into a single logically possible world would be if the one and only logically possible world consisted of nothing but abstract logically necessary states of affairs, such as the truths of logic and mathematics.

It follows in combinatorial ontology that there must be distinct logically possible worlds. Only one among all logically possible worlds, the uniquely actual world, actually exists. All others are only potentially existent, in the sense that they could and would have existed if they had been maximally instead of submaximally consistent. The possibility of logically possible worlds different in the condition of their physical states from the actual world is sufficient to uphold the actual world's logical contingency. This is an important implication, because if there were no worlds distinct from the actual world, then the actual world would be logically necessary in all its constituent states of affairs, including all its physical states of affairs. This would clearly not be a conclusion in which to rejoice. The implication is unsavory not only because it does not agree with our intuitive interpretation of the contingency of events in the actual world, but more importantly because it involves an intrusion of pure ontology this time on applied ontology's turf. Pure philosophical ontology, aimed at answering the question of being, should not dictate terms to applied scientific ontology concerning the existence of any particular states of affairs. If pure philosophical ontology determines that the physical state of the actual world is logically necessary, then it crosses the line. It implies that the particular states of affairs that exist in the actual world are the only ones that could logically possibly be recognized as existent by any competent applied scientific ontology. In effect, it tries to usurp applied ontology's work in advance of completing its own project of explaining what it means for something to exist.

The physical and mental events with which we are familiar, however we may later sort out their applied ontology, may or may not be causally necessary. It would be straining to interpret such occurrences as anything but logically contingent. I may have blue eyes, but it is logically possible that, like sixty percent of all human beings, I have brown eyes instead. In another world, we can easily stipulate a states of affairs combination in which I have differently pigmented irises. A little logically possible chromosomal cut-and-paste at the right stage of my embryonic development is all it should take. The logically possible brown-eyed me does not actually exist, because that logically possible me is not maximally consistent in all his (or her?) predications. What is my height? What is my gender? What is my native language? Am I friends with Napoléon in that world? Do I learn to fly a helicopter? What are the further genetic and historical ramifications of altering my eye color in another logically possible world? It is one thing to wave a hand and say as conventional modal logicians do that there exists a logically possible world in which I have brown rather than blue eyes. Do we ever thereby manage to refer to a predicationally complete world? Or, do we not rather inevitably leave almost all properties of the objects in such a world unspecified? Why in modal logic should we ever feel the need to pretend otherwise?

If the actual world should turn out to be the only logically possible world, if all logical possibility were to collapse into actuality, we would still have a logical answer to the question of being. We could continue even in that extreme case to say that there is something rather than nothing because there must exist an actual world, where the actual world in all its physical states is logically necessary rather than logically contingent. This, we must re-emphasize, is the very opposite conclusion indicated by a combinatorial pure philosophical ontology. In a combinatorial framework, the logical possibility of transworld-variant world-distinguishing physical properties must be taken into account as among the totality of all logically possible combinations of all logically possible objects with all logically possible properties in all logically possible states of affairs.

The physical states of affairs that we might know about from our experience of the actual world are real, they happen to exist, but are not logically necessary. If we trace effect to cause back far enough we find more and more contributing events explainable only as an outcome of pure chance. Consider only that your mother happened to meet your father. Almost every couple has a story of sheer coincidence to tell about how they came together or how they produced their offspring. Additional chance occurrences determine whether or not a woman with child survives the kinds of disease or accidents that threaten their destruction. The same is true of all spatiotemporal occurrences involving physical entities and states of affairs in the actual world. We can, in other words, to answer a previous question, and we should, authorize the substitution of "$w_@$" for "w" in conclusion (12), on the grounds that "$w_@$" is our name for the actual world, whose reference we are entitled to stipulate once we have proved that there is only one actual world. That the actual world is logically necessarily actual does not entail that the actual world is logically necessary. We can now formalize these inferences concerning the actuality and logical contingency of the actual world.

Why the Actual World is Logically Contingent
— There are Other Actually Nonexistent Logically Possible Worlds:

16. $\exists w[\Delta w \wedge w = w_@]$ (10 @-Nominalization)
17. $\forall w[\Diamond w \rightarrow \Diamond \Delta w]$ Possible actuality
18. $\exists w[\Diamond w \wedge w \neq w_@]$ Multiplicity of $\Diamond$ Worlds Distinct from $w_@$

---

19. $\exists w[\Diamond \Delta w \wedge w \neq w_@]$ (17, 18)
20. $\Delta w_@ \wedge \Box \Delta w_@ \wedge \neg \Box w_@$ (12, 15, 16, 19)

It is worth knowing that, even in the strongest standard systems of modal logic, we can validly derive $\Box \exists w[\Delta w \wedge w = w_@]$ and $\Box \Delta w_@$, but not $\Box w_@$. The latter would blatantly contradict the conclusion in (20). Logically necessarily, there exists an actual world; logically necessarily there exists at most one actual world, which is to say *the* actual world, $w_@$. The actual world is nevertheless not logically necessary but logically contingent in all its constituent states of affairs, and in particular in all its transworld-variant physical states of affairs.[9]

---

[9]There is no difficulty in obtaining conclusion (20) by Δ inference rules. It might appear that we could advance the following counter-derivation:

[1] $\Delta w_@ \rightarrow w_@$ [Δ rule]
[2] $\Box[\Delta w_@ \rightarrow w_@]$ [[1] Necessitation]
[3] $\Box \Delta w_@ \rightarrow \Box w_@$ [[2] Modal logic]
[4] $\Box w_@ \wedge \neg \Box w_@$ [[3], 20]

However, the inference is invalid. The term "$w_@$" does not designate a proposition, but a world, which is to say a (maximally consistent) states of affairs combination. The actual world $w_@$ has no truth value, as the consequent of the conditional in assumption [1] requires. The Δ inference rules explicitly apply only to propositions as truth value vehicles, and not to worlds, thus blocking the first step in the inference at step [1]. A counterpart version of the corresponding Δ rule holds for worlds, of the form: $\forall w[\Delta w \rightarrow \exists x(x = w)]$. With this provision, we preserve the ordinary linguistic equivocation between the actuality of the actual world, which is not itself a proposition, even if it is conventionally identified with or described by a set of propositions, and the actual truth of corresponding propositions. The alternative rule for the actual world does not support a version of the respective inference that would otherwise contradict the conclusion in (20). The modified

9. Logic, Ontology, and the Combinatorial Analysis of Being

The combinatorial theory of pure philosophical ontology which we have now developed is not intended as a contribution to applied scientific ontology. It makes no ontological commitment to any particular preferred theoretical domain of existent entities, states of affairs, or worlds. All such important discoveries are left to applied scientific ontology.

To criticize the reduction of ontology to logic misses the mark if it merely states the obvious but irrelevant fact that logical form lacks sufficient substantive ontological content to derive a domain of existent entities. The questions of pure philosophical ontology for which logic offers insight are conceptual, asking what it means for something to exist, rather than what actually exists. We should not conclude that logic cannot offer the required explanation just because we cannot conjure an actually existent bunny, or even a theoretical commitment to its existence, wriggling out of a logical magician's silk hat. Logic, of course, can do no such thing. The objection misfires by assuming that a reduction of ontology to logical form is supposed to provide answers to problems of applied scientific ontology by deriving a preferred existence domain of actually existent entities. The possibilities of logical form on the contrary are intended instead to answer another very different specific question about the nature of being or meaning of existence in pure philosophical ontology considered exclusively as a discipline.

To avoid equivocation, we proceed carefully, distinguishing the questions and answers appropriate to divisions among the four meanings of ontology, avoiding conclusions that stand in cross-categorial purposes to its distinctions. If and when we answer the problems of pure philosophical ontology, we are entitled to move on to the task of identifying a preferred existence domain as the goal of applied scientific ontology. If we do our work well, then the preferred

---

inference states:

[1′] $\Delta w_@ \rightarrow \exists w[w = w_@$ [Δ Worlds rule]
[2′] $\Box[\Delta w_@ \rightarrow \exists w(w = w_@)]$ [[1′] Necessitation]
[3′] $\Box \Delta w_@ \rightarrow \Box\exists w(w = w_@]$ [[2′] Modal logic]
[4′] $\Box\exists w[w = w_@]$ [[3′], 20]

Conclusion [4′] is harmless, particularly when we recall that $\forall w[\Box w \leftrightarrow \forall p(p^w \rightarrow \Box p)]$. The proposition that $\Box\exists w(w = w_@)$ states only that the actual world is *de dicto* logically necessarily identical to some logically possible world. This is not only not paradoxical, but intuitively true, since it does not entail that the actual world is logically necessary *de re*. To say that necessarily the actual world is identical to some world is not to say that the content or condition or any constitutive proposition of the world is logically necessary or necessarily true — as always, leaving aside propositions about transworld universal abstract entities, if there are any. Thus: $\neg[\Box\exists w(w = w_@) \rightarrow \Box w_@]$. Equally, there is no contradiction in the conjunction of the implications of conclusion (20) with [4′]:

[5′] $\neg\Box p^w \wedge \Box\exists w(w = w_@)$ [[4’],20]

existence domain we recommend on paper as a result of our theoretical deliberations will also be the extant domain consisting of the actual world and all actually existent entities. We produce a system of categories of existent entities adequate to explain the meaning of thought and discourse. We then commit applied scientific ontology as a discipline to a particular applied scientific domain, the theoretical content of which should ideally correspond perfectly to the real extant order of existent entities. Logic assumes a secondary role in applied ontology as a discipline. It is useful in developing a correct criterion of ontological commitment, and as a watchdog supervising the inferences we make in trying to establish a preferred theoretical existence domain, particularly with regard to the cardinality of infinitary and transinfinitary logical domains, but it does not decide any more interesting substantive questions of applied scientific ontology.

Why is there something rather than nothing? Why is there only one actual world? If the reason is supposed to be logical rather than experiential, is the actual world itself logically necessary or logically contingent? If abstract objects and relations are indistinguishably the same for every logically possible world, and if we can distinguish logically possible worlds only as distinct totalities of physical states of affairs, are the actual physical states of affairs that characterize the actual world logically necessary or logically contingent? These questions are not purely academic. If events in the actual world are logically necessary, as some philosophers have believed, then it becomes a problem to uphold commonsense belief in free will, freedom of action, and moral responsibility.

The presupposition of logic that we can predicate properties to objects makes it intelligible to refer to predication subjects and the properties predicated of them, without which there would be no philosophical problem of their ontic status. We can at least provisionally speak of objects as possible predication subjects and the properties predicated of them, by virtue of which the problem of the ontology of objects and properties makes sense. When an object, whatever an object is and whether or not it exists, whether it is or is not something more than a logical predication subject, has a property, whatever a property is and whether or not properties exist, then the object's having a property is a state of affairs, while a logically possible world, including the actual world, and even a logically impossible model, is constituted by states of affairs, ultimately by objects having properties. We have used this minimalist logical vocabulary to gain footing among the difficult questions of pure philosophical ontology. We reserve until afterward, when it is appropriate at last to address the problems of applied scientific ontology, to consider the ontic status of objects and specific types of objects and properties, states of affairs, and nonactual merely logically possible worlds.

We have argued that to exist is to be predicationally maximally consistent. The combinatorial analysis of the concept of being implies that there is something rather than nothing because in the ontically neutral predication sense there must be a maximally consistent combination of logical objects with properties. We have said that combinatorially there must be variations among the physical states in distinct logically possible worlds, and that the actual world is logically contingent because all physical states of the actual world are logically contingent. Pure philosophical ontology has now done its job, and can hand over its analysis of the concept of being for use in applied scientific ontology. The existence of physical states of affairs and the potential convenience of referring to abstract entities including logical objects and properties, logical states of affairs, and the like, as candidate items to include in a preferred existence domain, provide an obvious bridge to the ontic sciences. We are ready to begin applying the answer to the question of being, to the analysis of the concept of being as maximal consistency which we have proposed, to some of the most interesting specialized fields of applied scientific ontology.

The Pennsylvania State University
Department of Philosophy
246 Sparks Building
University Park, PA 16802
e-mail: dlj4@psu.edu

REFERENCES

Chellas, B. F. (1980). *Modal Logic: An Introduction.* Cambridge: Cambridge University Press.

Hughes, G. E. and M. J. Cresswell (1968). *An Introduction to Modal Logic.* London: Methuen & Co. Ltd.

Jacquette, D., ed. (2002). *A Companion to Philosophical Logic.* London: Blackwell Publishers.

Russell, B. (1985). *The Philosophy of Logical Atomism* (Lectures 1917–1918). Edited by David Pears. LaSalle: Open Court.

Wittgenstein, L. (1922). *Tractatus Logico-Philosophicus.* Edited by C. K. Ogden. London: Routledge & Kegan Paul Ltd.

Uwe Meixner

AN ONTO-NOMOLOGICAL THEORY OF MODALITY

ABSTRACT. This paper is dedicated to the formulation of a *restricted* theory of ontic modality (for example, I do not address questions that arise when modal operators interact with quantifiers, although some of the theoretical developments presented here certainly suggest such questions). As will be seen, notwithstanding its restrictions, the theory has a pleasing richness to it, as well as formal rigor and intuitive satisfactoriness. It also offers an unusual perspective on modality.

1. What Statements of Possibility and Necessity Mean

Besides a *semantical problem* there is also an *epistemological* and an *ontological problem* connected with modality: How can we justifiably determine which statements of necessity and possibility are true, and which false? What are the ontological requirements of modal theory? Answers to these questions I will leave almost totally implicit. (Draw your own conclusions.)

The semantical problem for ontic modalities can be formulated as the following question:

> What are the (necessary and sufficient) truth conditions of sentences with objective meaning having the forms "it is possible that *A*", "it is necessary that *A*," and "if *A*, then *B*"?

Here is an answer to this question in the form of an explicit definition:

$$\begin{aligned}
\Diamond^n(x) &:= \mathsf{S}(x) \wedge \neg\, \mathsf{P}(\mathtt{neg}(x), b^n),\\
\Box^n(x) &:= \mathsf{P}(x, b^n),\\
{}^n\!\!\rightarrow(x, y) &:= \mathsf{P}(y, \mathtt{conj}(b^n, x)),\\
\Diamond^n A &:= \Diamond^n(\mathtt{that}\ A),\\
\Box^n A &:= \Box^n(\mathtt{that}\ A),\\
B\ {}^n\!\!\rightarrow A &:= {}^n\!\!\rightarrow(\mathtt{that}\ B, \mathtt{that}\ A).
\end{aligned}$$

Of course, this, at the moment, is quite incomprehensible. So let me explain.

The predicate $\mathsf{P}(x, y)$ expresses that *the state of affairs x is an intensional part of the state of affairs y*. Hence $\forall x \forall y(\mathsf{P}(x, y) \supset \mathsf{S}(x) \wedge \mathsf{S}(y))$ is analytically true. What is meant by a state of affairs x being an intensional part of a state of affairs y can be effectively illustrated by an example: the state of affairs that this object has a surface is an intensional part of the state of affairs that this object is colored.

In: J. Malinowski and A. Pietruszczak (eds.), *Essays in Logic and Ontology* (*Poznań Studies in the Philosophy of the Sciences and the Humanities*, vol. 91), 47–72. Amsterdam/New York, NY: Rodopi, 2006.

The notion of parthood — or its inverse: containment — in the intensional sense is readily comprehensible with respect to states of affairs (and also with respect to properties). No antecedent comprehension of the term "possible world" is necessary for grasping that notion, and there is no checking on possible worlds in applying that notion either positively or negatively. The concept of intensional parthood is more basic than the concept of possible world. For the sake of brevity, I will usually omit the modifier "intensional", and simply speak of a state of affairs being a part of another state of affairs (or of itself).

"b^n" designates *the Nth basis of necessity* (hence "b^1" designates the *first* basis of necessity, "b^2" the *second* basis of necessity, etc.). A basis of necessity is always some state of affairs or other. Hence $\mathsf{S}(b^n)$ is analytically true.

The functor "$\mathtt{conj}(x, y)$" designates *the conjunction of x and y*; for *states of affairs x* and y, their conjunction, $\mathtt{conj}(x, y)$, is identical with $\iota z\,[\mathsf{S}(z) \wedge \mathsf{P}(x, z) \wedge \mathsf{P}(y, z) \wedge \forall u(\mathsf{P}(x, u) \wedge \mathsf{P}(y, u) \supset \mathsf{P}(z, u))]$ ("the unique state of affairs of which x and y are parts, and which is a part of every state of affairs of which x and y are parts"). This definition does of course have a theoretical background: there are principles that guarantee that there is exactly one state of affairs that fulfills the defining predicate, given that x and y are states of affairs; this theoretical background will be stated shortly.

The functor "$\mathtt{neg}(x)$" designates *the negation of x*, and is defined as follows: $\mathtt{neg}(x) := \mathtt{CONJ}\ y(\mathsf{QA}(y) \wedge \neg\,\mathsf{P}(y, x))$ ("the conjunction of all quasi-atomic states of affairs that are not parts of x"). This definition will be elucidated shortly, when its theoretical background has been stated.

We are now in a position to grasp the content of the above series of definitions. The first three definitions define modal predicates, the three other definitions define the corresponding modal sentence connectives.

According to the first and fourth definition, it is possible *in the Nth sense* that A if, and only if, the negation of the state of affairs that A is not a part of *the Nth basis of necessity*. In other words: $\Diamond^n A \equiv \neg\,\mathsf{P}(\mathtt{neg}(\mathtt{that}\ A), b^n)$ is a logical truth (due to the first and fourth definition).

According to the second and fifth definition, it is necessary *in the Nth sense* that A if, and only if, the state of affairs that A is part of *the Nth basis of necessity*. In other words: $\Box^n A \equiv \mathsf{P}(\mathtt{that}\ A, b^n)$ is a logical truth (due to the second and fifth definition).

According to the third and sixth definition, "if B, then A" *in the Nth sense* is true if, and only if, the state of affairs that A is part of the conjunction of *the Nth basis of necessity* with the state of affairs that B. In other words: $B \overset{n}{\rightarrow} A \equiv \mathsf{P}(\mathtt{that}\ A, \mathtt{conj}(b^n, \mathtt{that}\ B))$ is a logical truth (due to the third and sixth definition).

Clearly, the above definitions do not fully specify modal concepts: *modalities*. For this, one needs to specify *a basis of necessity*. If we intend to speak about an *ontic* modality, then the corresponding basis of necessity needs to be specified *ontically*. But there certainly seems to be not only one ontically specifiable basis of necessity. For the time being, let me indicate two salient ontic bases of necessity:

b^1: the basis of logical (or conceptual) necessity — this is simply the minimal (or "tautological") state of affairs (for further clarification, see the next section).

b^2: the basis of nomological (or natural) necessity — this is the state of affairs which is the conjunction of all states of affairs that are laws of nature.

Given the above definitions, it is now clear that $\Diamond^1 A$ means as much as "it is logically possible that A", $\Diamond^2 A$ as much as "it is nomologically possible that A" (and correspondingly the intended sense of $\Box^1 A$ and $\Box^2 A$ is likewise clear). Even so, as long as the general theory of states of affairs that provides the background for the above definitions is not specified, the content of an assertion of $\Diamond^1 A$ remains vague, and the content of an assertion of $\Diamond^2 A$ remains vague even after that specification (which is provided in the next section). For giving precise content to $\Diamond^2 A$, one has to specify, in addition to the general theory of states of affairs, the concept of law of nature, *and* one has to specify which states of affairs are laws of nature in the sense of the specified concept of law of nature. These tasks generate considerable philosophical difficulties, which are just conspicuous aspects of a more general problem: *Can just any state of affairs be a basis of necessity, and if not, what distinguishes a state of affairs that can be a basis of necessity from a state of affairs that cannot?* This question, and to some extent the problem of nomological necessity, will be addressed in due course.

Remark. Note that the bases of necessity implicitly invoked in making assertions of the form "if A, then B" are singularly unstable, even if those assertions are meant ontically, and vary from context to context much more so than the bases of monadic modalities do; in fact, they may even vary in the same context (of utterance). This generates the illusion that inferences that are prima facie logically valid for "if A, then B" — for example, the transitivity-inference: *If A, then B. If B, then C. Therefore: If A, then C* — are, *enfin*, not logically valid for it. (The transitivity-inference will be logically valid for "if A, then B" if the same basis of necessity is employed for all three conditionals involved; the transitivity-inference can easily fail to be logically valid if the basis of necessity is allowed to vary.) □

2. The Mereology of States of Affairs

The semantics of modality outlined above (without the use of model theory, simply by presenting explicit definitions of modal terms) is incomplete without stating the ontological background theory that is connected with it. Here is this background theory:

(P0) $\forall x \forall y(\mathrm{P}(x,y) \supset \mathrm{S}(x) \wedge \mathrm{S}(y))$,

(P1) $\forall x \forall y \forall z(\mathrm{P}(x,y) \wedge \mathrm{P}(y,z) \supset \mathrm{P}(x,z))$,

(P2) $\forall x(\mathrm{S}(x) \supset \mathrm{P}(x,x))$,

(P3) $\forall x \forall y(\mathrm{P}(x,y) \wedge \mathrm{P}(y,x) \supset x = y)$,

(P4) $\exists z [\mathrm{S}(z) \wedge \forall x(\mathrm{S}(x) \wedge A[x] \supset \mathrm{P}(x,z)) \wedge \forall y(\mathrm{S}(y) \wedge \forall x(\mathrm{S}(x) \wedge A[x] \supset \mathrm{P}(x,y)) \supset \mathrm{P}(z,y))]$,

(P5) $\forall z \forall z'(\mathrm{S}(z) \wedge \mathrm{S}(z') \wedge \forall x(\mathrm{QA}(x) \wedge \mathrm{P}(x,z) \supset \mathrm{P}(x,z')) \supset \mathrm{P}(z,z'))$,

(P6) $\forall x [\mathrm{P}(x, \mathtt{CONJ}\, y A[y]) \wedge \neg \mathtt{M}(x) \supset \exists k'(\mathrm{P}(k',x) \wedge \neg \mathtt{M}(k') \wedge \exists z(\mathrm{P}(k',z) \wedge A[z]))]$,

(P7) $w^* \neq k^*$,

(P8) $\mathrm{QC}(w^*)$,

(P9) $w^* \neq t^*$,

(P10) $A \equiv \mathrm{O}(\mathtt{that}\ A)$.

These eleven principles state the most basic principles of a mereology of states of affairs; they contain some defined terms which I will explain as I go through them.

The first four principles describe the basic properties of P as a relation of (proper or improper) intensional parthood between states of affairs. (Note that P3 formulates an identity criterion for states of affairs.)

P4 states that the states of affairs falling under an arbitrary description $A[x]$ have a smallest state of affairs that comprises them all. In other words (in view of P3), for any description $A[x]$, there is the state of affairs which is the conjunction of all states of affairs that satisfy $A[x]$. And accordingly we can formulate the following definition:

(D1) $\mathtt{CONJ}\, x A[x] := \iota z [\mathrm{S}(z) \wedge \forall x(\mathrm{S}(x) \wedge A[x] \supset \mathrm{P}(x,z)) \wedge \forall y(\mathrm{S}(y) \wedge \forall x(\mathrm{S}(x) \wedge A[x] \supset \mathrm{P}(x,y)) \supset \mathrm{P}(z,y))]$.

The fulfillment of the so-called *existence-condition*[1] for this definition (using the operator of definite description, ι) is guaranteed for all predicates $A[x]$ by P4. The fulfillment of the uniqueness-condition[2] for this definition is guaranteed for all predicates $A[x]$ by P3. Here are two prominent states of affairs that can be defined as conjunctions:

(D2) $t^* := \mathtt{CONJ}\, x \neg \mathrm{S}(x)$,

(D3) $k^* := \mathtt{CONJ}\, x\ \mathrm{S}(x)$.

[1] One should rather call it "the at-least-one-condition".

[2] One could also call it "the at-most-one-condition".

One can easily prove that $\mathtt{CONJ}\,x \neg \mathtt{S}(x)$ is identical with $\iota z[\mathtt{S}(z) \wedge \forall y(\mathtt{S}(y) \supset \mathtt{P}(z,y))]$ ("the state of affairs that is a part of every state of affairs"), and that $\mathtt{CONJ}\,x\ \mathtt{S}(x)$ is identical with $\iota z[\mathtt{S}(z) \wedge \forall x(\mathtt{S}(x) \supset \mathtt{P}(x,z))]$ ("the state of affairs of which every state of affairs is a part"). Thus, for obvious reasons, k^* can be called "the maximal [or total] state of affairs," and because intensional parthood is the ontological counterpart of logical implication, k^* can also be called, metaphorically, "the *self-contradictory* state of affairs". In turn, for obvious reasons, t^* can be called "the minimal state of affairs", and again because intensional parthood is the ontological counterpart of logical implication, a metaphorical designation of t^* is "the *tautological* state of affairs". It is useful to have the following defined predicates (although they provably apply only to one state of affairs: t^*, respectively k^*):

(D4) $\mathtt{M}(z) := \mathtt{S}(z) \wedge \forall y(\mathtt{S}(y) \supset \mathtt{P}(z,y))$,

(D5) $\mathtt{T}(z) := \mathtt{S}(z) \wedge \forall x(\mathtt{S}(x) \supset \mathtt{P}(x,z))$.

The fact (provable on the basis of P3) that there is only one "tautological" state of affairs, and only one "self-contradictory" one, reveals that the conception of states affairs *here* employed is *coarse-grained.* A coarse-grained conception of states of affairs is not adequate if the aim is to employ states of affairs as *meanings* of sentences that are either true or false; for if states of affairs are to be the meanings of such sentences, then, under the coarse-grained conception of states of affairs, all logically true sentences turn out to have the same meaning (i.e., t^*), and all logically false sentences turn out to have the same meaning too (i.e., k^*), which does not seem to be at all desirable. Nevertheless, although coarse-grained states of affairs cannot serve adequately as the meanings of true or false sentences, they *can* serve in a respectable semantical function: namely, as the *intensions* of true or false sentences, where intensions are taken to be rough approximations to meanings.[3] For the logic of *ontic* modalities, and quite generally for the purposes of a philosophical theory of *ontic* modalities, coarse-grained states of affairs are entirely sufficient; fine-grained states of affairs need not be considered.

The following predicate is useful to the point of being indispensable:

(D6) $\mathtt{O}(x) := \mathtt{S}(x) \wedge \mathtt{A}(x)$.

In other words, *to obtain* (*to be the case, to be a fact*) is nothing else than *to be an actual state of affairs*. Hence it is clear what is meant by P10, and hence it is also clear which other prominent state of affairs is defined by the following definition:

(D7) $w^* := \mathtt{CONJ}\,x\ \mathtt{O}(x)$.

[3] Identity of meaning implies identity of intension, but not vice versa. Identity of intension implies identity of truth-value, but not vice versa.

It is *the world* in Wittgenstein's sense: the totality of all obtaining states of affairs. The import of principles P7 and P9 is now clear; according to them, the world is neither the minimal (or "tautological") nor the maximal (or "self-contradictory") state of affairs.

Note, furthermore, that the "small" conjunction of states of affairs can be defined on the basis of the "big":

(D8) $\mathtt{conj}(x, y) := \mathtt{CONJ}\, v(v = x \vee v = y)$.

As can easily be seen, $\mathtt{CONJ}\, v(v = x \vee v = y)$ is for all states of affairs x and y identical with $\iota z\,[\mathrm{S}(z) \wedge \mathrm{P}(x, z) \wedge \mathrm{P}(y, z) \wedge \forall u(\mathrm{P}(x, u) \wedge \mathrm{P}(y, u) \supset \mathrm{P}(z, u))]$, in other words: the conjunction of states of affairs x and y is the smallest state of affairs of which both x and y are parts.

Moving on to the next principle after P4, it becomes clear that P5 is an *atomistic* principle, once the predicate $\mathtt{QA}(x)$ is defined:

(D9) $\mathtt{QA}(x) := \mathrm{S}(x) \wedge \forall y(\mathrm{P}(y, x) \supset y = x \vee \mathtt{M}(y))$.

According to D9, a *quasi-atomic state of affairs* is a state of affairs whose only proper part (if it has a proper part) is the minimal state of affairs (considering that $\forall y(\mathtt{M}(y) \equiv y = t^*)$). One does well to distinguish *quasi-atomic* states of affairs, *atomic* states of affairs — states of affairs that have no proper part — and *elemental* states of affairs — states of affairs that have exactly one proper part. It can easily be shown that all atomic or elemental states of affairs are quasi-atomic, and vice versa: that all quasi-atomic states of affairs are either atomic or elemental. In the mereology of states of affairs, there is exactly one atom: t^*, and thus, in the mereology of state of affairs, atomicity coincides with minimality. For this reason, I do not here introduce a formally defined predicate for expressing atomicity of states of affairs. But I do formally introduce a predicate for expressing elementalness of states of affairs:

(D10) $\mathtt{EL}(x) := \mathtt{QA}(x) \wedge \neg\, \mathtt{M}(x)$.

While there is exactly one atomic state of affairs, which is a consequence of P4 and P3, the above eleven principles do not determine the exact number of elemental states of affairs. Some mereologists have argued that entities that have the same proper parts must be identical — which, if correct, would imply that all elemental states of affairs are identical with each other, since they all have exactly one proper part, and the very same proper part, namely t^*. But the invoked identity-principle is false for *nonmaterial mereologies*, like the mereology of states of affairs, or the mereology of sets.[4] Thus there can easily be more than

[4] Its validity is also dubious for material mereologies: Since material atoms have no proper parts (in the relevant material sense), they all have the same proper parts, and hence, according to the

one elemental state of affairs, and in fact the eleven principles already require that there be at least two elemental states of affairs.

P5 states that states of affairs are, in a manner, *exhausted* by the quasi-atomic states of affairs that are parts of them: if all the quasi-atomic states of affairs that are parts of one state of affairs are also parts of the other, then this is already sufficient for concluding that the former state of affairs is itself a part of the latter. Equivalently, one can say that states of affairs are exhausted by the *elemental* states of affairs that are parts of them (for the quasi-atomic states of affairs nearly coincide with the elemental states of affairs, and the one state of affairs that blocks their coincidence — that is, t^* — is a part of every state of affair, and therefore does not make a difference regarding the exhaustion of states of affairs by parts of them that belong to a certain kind).

Moving on to the next principle after P5, P6 is a very important principle that regulates how the parts of a conjunction of states of affairs stand to the description (or the property) used in specifying that conjunction. P6 states that every nonminimal part of a conjunction of states of affairs has a nonminimal overlap with some state of affairs that satisfies the description used for specifying that conjunction. Thus stated, P6 is a perfectly evident principle for a mereology of states of affairs.

Given the principles P0–P6, one can prove: $\forall x(\mathtt{S}(x) \supset x = \mathtt{CONJ}\, y(\mathtt{QA}(y) \wedge \mathtt{P}(y, x))$, and $\forall x(\mathtt{S}(x) \supset x = \mathtt{CONJ}\, y(\mathtt{EL}(y) \wedge \mathtt{P}(y, x))$. In fact, as is suggested by the second theorem, the number of elemental states of affairs determines the total number of states of affairs according to the simple equation: $\text{card}(\mathtt{S}) = 2^{\text{card}(\mathtt{EL})}$ ("The cardinal number of states of affairs is 2 put to the power of the cardinal number of elemental states of affairs"). But for proving this the system has to be set-theoretically embedded.

Furthermore, given the principles P0–P6, it can be shown that if the functor of negation is defined as follows:

(D11) $\mathtt{neg}(x) := \mathtt{CONJ}\, y(\mathtt{QA}(y) \wedge \neg\, \mathtt{P}(y, x))$,

then the general truths that one would expect from a negation of states of affairs become provable (for example, $\forall x(\mathtt{S}(x) \supset \mathtt{neg}(\mathtt{neg}(x)) = x)$, $\mathtt{neg}(t^*) = k^*$, $\forall x \forall y(\mathtt{S}(x) \wedge \mathtt{S}(y) \wedge \mathtt{neg}(x) = \mathtt{neg}(y) \supset x = y)$). It should not go unmentioned — as a further demonstration of the great definitional power of the proposed mereology of states of affairs — that the "big" disjunction can be defined on the basis of the "big" conjunction as follows:

(D12) $\mathtt{DISJ}\, x A[x] := \mathtt{CONJ}\, y(\forall x(A[x] \supset \mathtt{P}(y, x))$.

identity-principle at issue, all material atoms are identical with each other — a consequence which is simply absurd, *unless there are no material atoms*. If, on the other hand, material atoms are excepted from the identity-principle at issue, the obvious question is: why not make more exceptions? If material atoms can differ from each other although they all have no proper parts, why may not elemental states of affairs differ from each other although they all have the same proper part?

Then the "big" disjunction can be used for defining the "small" disjunction:

(D13) $\mathtt{disj}(x, y) := \mathtt{DISJ}\, v(v = x \vee v = y)$.

As can easily be seen, $\mathtt{DISJ}\, v(v = x \vee v = y)$ is for all states of affairs x and y identical with $\iota z\,[\mathrm{S}(z) \wedge \mathrm{P}(z, x) \wedge \mathrm{P}(z, y) \wedge \forall u(\mathrm{P}(u, x) \wedge \mathrm{P}(u, y) \supset \mathrm{P}(u, z))]$, in other words: the disjunction of states of affairs x and y is the largest state of affair that is a part both of x and of y. Given the definitions of the three functors $\mathtt{neg}(x)$, $\mathtt{conj}(x, y)$, and $\mathtt{disj}(x, y)$, all the well-known and less well-known Boolean principles that form the stock of truth-functional propositional logic can be proved on the basis of P0–P6.[5]

The one principle that has not yet been touched on is P8, which contains one more as yet undefined predicate: $\mathrm{QC}(x)$. Here is its definition:

(D14) $\mathrm{QC}(x) := \mathrm{S}(x) \wedge \forall y(\,\mathrm{P}(x, y) \supset x = y \vee \mathrm{T}(y))$.

According to D14, a *quasi-complete state of affairs* is a state of affairs which is a proper part (if it is a proper part of anything) only of the total (or maximal) state of affairs (considering that $\forall y(\mathrm{T}(y) \equiv y = k^*)$). The predicate $\mathrm{QC}(x)$ is the counterpart of the predicate $\mathrm{QA}(x)$. As the quasi-atomic states of affairs are divided into the one minimal state of affairs, t^*, and all the other quasi-atomic states of affairs: the elemental state of affairs, so the quasi-complete state of affairs are divided into the one maximal state of affairs, k^*, and all the other quasi-complete states of affairs: the *maximally consistent states of affairs*:

(D15) $\mathrm{MC}(x) := \mathrm{QC}(x) \wedge \neg\, \mathrm{T}(x)$.

It can be shown that the maximally consistent states of affairs are precisely the negations of the elemental states of affairs (and therefore the elemental states of affairs precisely the negations of the maximally consistent states of affairs). It is easily verified on the basis of P8 and P7 that w^* — *the world* — is a maximally consistent state of affairs. Because of this fact, it is justified to call the states of affairs that are maximally consistent "possible worlds". The following remarkable theorems are provable on the basis of P0–P6: $\forall x \forall y[\mathrm{MC}(x) \wedge \mathrm{S}(y) \supset (\mathrm{P}(y, x) \equiv \neg\, \mathrm{P}(\mathtt{neg}(y), x))]$; $\forall x \forall y \forall z[\mathrm{MC}(x) \wedge \mathrm{S}(y) \wedge \mathrm{S}(z) \supset (\mathrm{P}(\mathtt{disj}(y, z), x) \equiv \mathrm{P}(y, x) \vee \mathrm{P}(z, x))]$.

This concludes my brief survey of the *basic* mereology of states of affairs. (It is explored in much greater detail in my book *Axiomatic Formal Ontology*.) Some philosophers would prefer that the present theory of states of affairs be

---

[5] To be precise: Every logically true formula $\phi[p_1, \ldots, p_n]$ of truth-functional propositional logic can be translated (in the obvious way) into a functional term $\phi^*[x_1, \ldots, x_n]$ of the mereology of states of affairs such that $\forall x_1 \ldots \forall x_n(\mathrm{S}(x_1) \wedge \cdots \wedge \mathrm{S}(x_n) \supset \phi^*[x_1, \ldots, x_n] = t^*)$ is provable on the basis of P0–P6.

not called "a mereology". There may be historical reasons for restricting the use of the term "mereology" to theoretical systems that deal with *individuals* and their part-relations, but there are no compelling systematical reasons for this restriction. The present theory of states of affairs is indeed a powerful Boolean algebra, but at the same time it is a *mereology*: a *nonmaterial* mereology, and a mereology that has *a minimal element*, indeed *exactly one minimal element*, which one might call "its center".

This *centered* or *Boolean* mereology of coarse-grained states of affairs requires *universes of states of affairs* that have a spindle-shaped structure, with k^* at one end and t^* at the other. If, for example, we put the number of elemental states of affairs at *three*,[6] and call the elemental states of affairs that we are considering "*a*", "*b*" and "*c*", then we obtain the universe U_3:

$$[k^* :]abc$$

$$ab \quad ac \quad [w^* :]bc$$

$$a \quad b \quad c$$

$$t^*$$

If $S(x)$ is interpreted to be true of exactly the eight elements in U_3 and if $P(x, y)$ is interpreted to hold true of exactly those ordered pairs $\langle x, y\rangle$ of elements in U_3 such that x is part of y (in the obvious sense suggested by the diagram), then this universe of states of affairs[7] fulfills the principles P0–P9.

There are smaller universes than U_3 that also fulfill P0–P9. In fact, three such universes are contained in U_3. One of them is U_2:

$$[k^* :]ac$$

$$a \quad [w^* :]c$$

$$t^*$$

There is no smaller universe of states of affairs than U_2 that fulfills the principles P0–P9.

3. The Actuality of States of Affairs in the Mereology of States of Affairs

The stated mereology of states of affairs is basic, but it is far from complete. First of all, two principles for actuality have to be added. They are obviously

[6]This number is wildly unrealistic if we are talking about *reality*; but it may be entirely appropriate if we are talking about some very restricted "virtual reality", for example the reality of a very simple game.

[7]I omit, as obvious, any arrow that indicates that the state of affairs from which the arrow originates is a part of the state of affairs at which the arrow ends (with its head). Inserting the arrows would, by the way, also be misleading, because it would suggest that there are more states of affairs in the universe U_3 than are designated by the eight designators used. Note that the universe U_3 can be a small part of a much larger *universe of discourse*.

true, but nevertheless not provable on the basis of the principles already stated:

(P11) $\forall x \forall y(\mathtt{S}(x) \wedge \mathtt{A}(x) \wedge \mathtt{P}(y, x) \supset \mathtt{A}(y))$,

(P12) $\mathtt{A}(w^*)$.

P11 states that every intensional part of an actual state of affairs is itself actual (an actual *state of affairs*, according to P0); P12 states that *the world* is actual.

Given the mereology of states of affairs as it now stands, there is a sober truth about the nonactuality of states of affairs. It is this: If the number of elemental states of affairs is N (and the number of elemental state of affairs is $\geqslant 2$ according to the stated principles), then the number of nonactual (or nonobtaining) states of affairs is this: $2^N - 2^{N-1}$. And this implies that *actualism* about states of affairs — the doctrine that there are no nonactual (or nonexistent) states of affairs, that all states of affairs are facts — is not merely false, it is also *incoherent*.

But how is the equation $\text{card}(\mathtt{S} \wedge \neg \mathtt{A}) = 2^{\text{card}(\mathtt{EL})} - 2^{\text{card}(\mathtt{EL})-1}$ obtained? Central for obtaining it is the following theorem:

$\forall x[\mathtt{S}(x) \supset (\mathtt{A}(x) \equiv \mathtt{P}(x, w^*))]$ —
"A state of affairs is actual if, and only if, it is a part of the world".

Proof. It is easily seen — given P3, P4, the definition of w^* and the definition of $\mathtt{O}(x)$ — that $\forall x[S(x) \supset (\mathtt{A}(x) \supset \mathtt{P}(x, w^*))]$ is true. It remains to be seen that $\forall x[\mathtt{S}(x) \supset (\mathtt{P}(x, w^*) \supset \mathtt{A}(x))]$ is also true. Suppose, therefore, $\mathtt{P}(x, w^*)$ (the additional supposition $\mathtt{S}(x)$ is not needed, since it follows from the supposition already made by P0). From this, it follows on the basis of P11 and P12: $\mathtt{A}(x)$, because we (provably) have $\mathtt{S}(w^*)$.

The number of nonactual states of affairs is obtained by subtracting from the number of all states of affairs — that is, from $2^{\text{card}(\mathtt{EL})}$ — the number of all actual states of affairs. According to the theorem just proven, this number is the number of all states of affairs that are parts of w^*. How many states of affairs are there that are parts of w^*? For determining their number, one needs to determine the number of the elemental states of affairs that are parts of w^*. This number, as can be proven, is equal to $\text{card}(\mathtt{EL}) - 1$ (since it can be proven that there is exactly one elemental state of affairs that is not a part of w^*, namely, $\mathtt{neg}(w^*)$). Since every subset of the set of elemental states of affairs which are parts of w^* determines a different part of w^*, and vice versa, the number of parts of w^* is: $2^{\text{card}(\mathtt{EL})-1}$. Hence one obtains that the number of nonactual states of affairs is: $2^{\text{card}(\mathtt{EL})} - 2^{\text{card}(\mathtt{EL})-1}$. □

Looking at the five principles that (disregarding P10, which has a special status) govern the actuality of states of affairs: P7, P8, P9, P11, and P12, one may well wonder which of them are true for conceptual reasons only, and which are indeed true, but not for conceptual reasons only. Incidentally, there can be no reasonable doubt about the purely conceptual nature of the truth of the principles P0–P6. Who finds this doubtful should consider that the truth of P0–P6 is compatible with there being only one state of affairs: the prodigious definitional power of the subsystem P0–P6 is combined with an utter ontological weakness. Not even $t^* \neq k^*$ can be deduced from these principles. This ontological weakness should go a long way towards removing doubts about the truth of P0–P6 being of a purely conceptual nature. *True*, one can deduce from these principles: $\exists x\, \mathrm{S}(x)$. But that seems to be a safe conceptual truth.

In my view, indeed, numbering-statements *of any kind* are not excluded from being conceptually true. And in fact, in my view, "card(S) $\geqslant N$" and "card(S) $= N'$" are conceptually true for whatever numbers N and N' they are true (therefore, since "card(S) $\geqslant$ 1" is true — being deducible from true principles — it is also conceptually true; and since "card(S) $\geqslant$ 2" is true as well — being deducible from true principles — it is conceptually true as well). The idea that true numbering-statements cannot be conceptually true is rooted in the positivistic prejudice that conceptual truths cannot say anything informative about the universe of discourse (also sometimes called "the world", which designation, however, is here reserved for something else), cannot say, for example, how many entities of a certain kind are (*at least* or *exactly*) in it.

But back to P7, P8, P9, P11, and P12. The only principle out of these five whose purely conceptual truth is, I think, *evident* is P11. The purely conceptual truth of the other four is not evident, and perhaps some of them are, though true, not conceptually true after all. But it seems clear (1) that k^* is purely for conceptual reasons a nonactual state of affairs, and (2) that t^* is purely for conceptual reasons an actual state of affairs (accepting (1) and (2) is, by the way, another way to establish that it is a conceptual truth that there are at least two states of affairs).[8] Moreover, one can plausibly hold (3) that the conjunction of all actual states of affairs is purely for conceptual reasons itself actual. (1) and (3) have the consequence that P12 and P7 are conceptual truths. (Unlike P12, P7 is

---

[8] The propositions (1) and (2) can be shown to be true; they are, however, not immediately forced upon us.

The statement $\mathtt{A}(t^*)$ is on the basis of P0–P6, P11 equivalent with $\exists x(\mathrm{S}(x) \wedge \mathrm{A}(x))$. Is this *indubitably* a *conceptual* truth? Is it *indubitably* a *conceptual* absurdity to suppose that no state of affairs is actual? Yes, indeed, it is: Suppose $\neg\exists x(\mathrm{S}(x) \wedge \mathrm{A}(x))$; hence on the basis of the conceptually true principle P10: $\mathrm{O}(\mathtt{that}\ \neg\exists x(\mathrm{S}(x) \wedge \mathrm{A}(x)))$; hence by applying D6: $\mathrm{S}(\mathtt{that}\ \neg\exists x(\mathrm{S}(x) \wedge \mathrm{A}(x))) \wedge \mathrm{A}(\mathtt{that}\ \neg\exists x(\mathrm{S}(x) \wedge \mathrm{A}(x)))$; hence: $\exists x(\mathrm{S}(x) \wedge \mathrm{A}(x))$.

In turn, the statement $\neg\,\mathrm{A}(k^*)$ is on the basis of P0–P6, P11, P12 equivalent with $\exists x(\mathrm{S}(x) \wedge \neg\,\mathrm{A}(x))$, with $\neg\,\mathrm{P}(k^*, w^*)$, and with $w^* \neq k^*$. Is $\exists x(\mathrm{S}(x) \wedge \neg\,\mathrm{A}(x))$ *indubitably* a *conceptual* truth? Is it *indubitably* a *conceptual* absurdity to suppose that all states of affairs are actual? Yes, indeed, it is: Suppose $\forall x(\mathrm{S}(x) \supset \mathrm{A}(x))$; $\mathrm{S}(\mathtt{that}\ \exists x(\mathrm{S}(x) \wedge \neg\,\mathrm{A}(x)))$ is a conceptual truth (it is an instance of a conceptually true general principle stated in the next section: P13); hence, according to the supposition made: $\mathrm{A}(\mathtt{that}\ \exists x(\mathrm{S}(x) \wedge \neg\,\mathrm{A}(x)))$; hence by applying D6: $\mathrm{O}(\exists x(\mathrm{S}(x) \wedge \neg\,\mathrm{A}(x)))$; hence on the basis of P10: $\exists x(\mathrm{S}(x) \wedge \neg\,\mathrm{A}(x))$.

not an independent principle, since not only $\mathtt{A}(t^*)$ but also $\neg\,\mathtt{A}(k^*)$, and therefore $\neg\,\mathtt{P}(k^*, w^*)$, and consequently $w^* \neq k^*$, are theorems logically deducible from the rest of the given conceptually true principles; see footnote 8.) But we still have no hint why P8 and P9 should also be conceptual truths.

We can, however, make a case for P8 ⊃ P9 — i.e., $\mathtt{QC}(w^*) \supset w^* \neq t^*$ — being a conceptual truth. I have said above that "card(S) ⩾ N" is a conceptual truth *if* it is a truth. But "card(S) ⩾ 4" is certainly true; and therefore it is conceptually true. Now, from this conceptual truth ("card(S) ⩾ 4", or in other words: "$\exists^{\geqslant 4} x\,\mathtt{S}(x)$") it follows on the basis of the conceptually true principles P0–P6: $\mathtt{QC}(w^*) \supset w^* \neq t^*$. And therefore this latter statement is also a conceptual truth.

But this leaves it open whether (a) $w^* \neq t^*$ is a conceptual truth, and $\mathtt{QC}(w^*)$ is not, or whether (b) both sentences are conceptual truths, or whether (c) neither sentence is a conceptual truth. How one decides in this matter depends on one's theory of actuality, in particular, on one's theory of the actuality of states of affairs. One might, for example, hold that a certain quasi-complete state of affairs is actual for purely conceptual reasons. But personally I do not believe that this position is very plausible. In my view, not even $\exists y(\mathtt{S}(y) \wedge y \neq t^* \wedge \mathtt{A}(y))$ is a conceptual truth. But then $w^* \neq t^*$ (in view of $\mathtt{S}(w^*)$ and $\mathtt{A}(w^*)$ being conceptual truths) and $\mathtt{QC}(w^*)$ cannot be conceptual truths either.

4. Further Principles for "That"

In the previous section, I have added to the basic principles of the mereology of states of affairs, i.e., P0–P10, two further actuality-principles: P11 and P12. Thus there are now in addition to the seven *principles of intensional parthood*, P0–P6, five *principles of actuality*: P7, P8, P9, P11 and P12, of which P7 proved to be redundant — last but not least on the basis of the conceptual truth P10 (see footnote 8), which is, so far, the only principle for "that".

More principles than P10 are needed for "that":

(P13) $\mathtt{S}(\mathtt{that}\ A)$,

(P14) $\mathtt{that}\ \neg A = \mathtt{neg}(\mathtt{that}\ A)$,

(P15) $\mathtt{that}\,(A \wedge B) = \mathtt{conj}(\mathtt{that}\ A, \mathtt{that}\ B)$,

(P16) $\mathtt{that}\ \forall x A[x] = \mathtt{CONJ}\ y \exists x(y = \mathtt{that}\ A[x])$.

Remark. Using the definition $A \vee B := \neg(\neg A \wedge \neg B)$ one can prove on the basis of P14 and P15: $\mathtt{that}\,(A \vee B) = \mathtt{neg}(\mathtt{conj}(\mathtt{neg}(\mathtt{that}\ A), \mathtt{neg}(\mathtt{that}\ B))) = \mathtt{disj}(\mathtt{that}\ A, \mathtt{that}\ B)$. Using the definition $A \supset B := \neg A \vee B$ one can prove in addition: $\mathtt{that}\,(A \supset B) = \mathtt{disj}(\mathtt{neg}(\mathtt{that}\ A), \mathtt{that}\ B)$. Using the definition $\exists x A[x] := \neg \forall x \neg A[x]$, one can prove on the basis of P14 and P16: $\mathtt{that}\ \exists x A[x] = \mathtt{neg}(\mathtt{CONJ}\ y \exists x(y = \mathtt{neg}(\mathtt{that}\ A[x]))) = \mathtt{DISJ}\ y \exists x(y = \mathtt{that}\ A[x])$. □

These four principles and P10—each of which is rather obvious in itself—are, in conjunction, consistent with the twelve principles *not concerning "that"* that have already been stated. Those twelve principles—i.e., P0–P9, P11, and P12—are consistent in themselves, since they have a verifying model (see above, at the end of Section 2, the universe of states of affairs U_2; in order to accommodate also the principles P11 and P12, imagine circles—symbolizing the property of actuality—drawn around c and t^*, but not around ac and a). And the consistency of all stated principles follows, because it is easily seen that the principles P10 and P13–P16 can be proven on the basis of the principles P0–P9, P11 and P12 if the following definition of "`that` A" is adopted:

(D*) $\quad \texttt{that}\ A := \texttt{CONJ}\ y(y = y \wedge \neg A).$

Thus, the universe U_2 is not only a verifying model for P0–P9, P11 and P12, it is also a verifying model for P0–P16. This series of principles, or in other words: this theory of states of affairs, has herewith been proven consistent. (Of that theory, P13–P16 are conceptual truths, just like all the other principles, excepting P8 and P9.)

The above definition of "`that` A" allows to deduce this pleasing result. But of course it is not a definition that captures the intended meaning of "`that` A," because, according to it, "`that` A" always designates either t^* or k^*, whatever true or false sentence A we are looking at. Interesting enough, though, what is sometimes thought to initiate all by itself an inflation of intensional entities: the introduction of "that" as a name-forming operator applicable to sentences, does not necessarily do any such thing.

But the equation "$\texttt{that}\ A = \texttt{CONJ}\ y(y = y \wedge \neg A)$" and the utter ontological restriction to which this equation gives rise (of the states of affairs that serve as intensions of sentences: according to it, all true sentences have the same intension, namely t^*, and all false sentences also the same intension, namely k^*)[9] *is not provable if* "`that` A" *is a basic (and not a defined) operator*—although one might think it to be provable according to an argument that is reminiscent of an argument of some notoriety, called "the slingshot":

Suppose A is a *true* sentence, hence (provably) ID1: $\iota x(x = b) = \iota x(A \wedge x = b)$, and (provably) ID2: $\texttt{CONJ}\ y(y = y \wedge \neg A) = t^*$. Furthermore:

1. $A \equiv \iota x(x = b) = \iota x(A \wedge x = b)$ — a (provable) logical truth[10]

[9]Note that the intension of a statement that is designated by "A" (for A, substitute the statement) is designated by "`that` A," which, without further determinations, is an ambiguous designator that can also be taken to designate at least two other things: *naturally, the meaning* of the statement, and *artificially, the truth-value* of the statement.

[10]1 is a provable logical truth if the singular term b is appropriately chosen: b must be different from c^*—the singular term that serves as designating the artificial referent of all definite descriptions $\iota x A[x]$ for which $\exists^{=1} x A[x]$ is not true.

2. $\mathtt{that}\ A = \mathtt{that}\ (\iota x(x = b) = \iota x(A \wedge x = b))$ from 1 by EQU (see below)

3. $\mathtt{that}\ A = \mathtt{that}\ (\iota x(x = b) = \iota x(x = b))$ from 2 and ID1

4. $\mathtt{that}\ (\iota x(x = b) = \iota x(x = b)) = t^*$ a plausible assumption

5. $\mathtt{that}\ A = \mathtt{CONJ}\ y(y = y \wedge \neg A)$ from 3, 4, and ID2

In this argument, the principle designated by "EQU", which is used in line 2, is the principle *that logically equivalent sentences have the same intensions*.

Suppose now that A is a *false* sentence, hence $\neg A$ is a true sentence, and therefore according to the above argument: $\mathtt{that}\ \neg A = \mathtt{CONJ}\ y(y = y \wedge A)$. But, according to P14, $\mathtt{neg}(\mathtt{that}\ A) = \mathtt{that}\ \neg A$; and we also have: $\mathtt{CONJ}\ y(y = y \wedge A) = t^* = \mathtt{neg}(k^*) = \mathtt{neg}(\mathtt{CONJ}\ y(y = y \wedge \neg A))$ (because $\mathtt{CONJ}\ y(y = y \wedge \neg A) = k^*$, since A, according to supposition, is false). Hence: $\mathtt{neg}(\mathtt{that}\ A) = \mathtt{neg}(\mathtt{CONJ}\ y(y = y \wedge \neg A))$. And therefore we obtain the same result as above: $\mathtt{that}\ A = \mathtt{CONJ}\ y(y = y \wedge \neg A)$.

This entire deduction of the equation corresponding to definition (D^*) is *almost* impeccable. EQU, indeed, is not deducible in the system P0–P16. But one can add the provability-rule EQU* to the proof-rules of the system: If $A \equiv B$ is logically provable, then $\mathtt{that}\ A = \mathtt{that}\ B$ is also logically provable.[11] This proof-rule serves the same purpose as EQU, and there is nothing at all wrong with it. There is also nothing at all wrong with the *plausible assumption* in line 4; it, too, is not deducible in the system P0–P16, but it can be added to that system without scruples.

Well, what, then, is wrong with the deduction? — The false step occurs in moving from 2 to 3 on the basis of ID1. But there is nothing whatever wrong with ID1, the problem is the implicitly applied inference-rule: *substitution of identicals*. This rule is not universally valid; there are occasions when its application leads from a true sentence to a false one, and above we have such an occasion: 2 and ID1 are true (2 purely logically, and ID1 on the basis of assumption), but 3 is *not* true — at least not for any sentence A that is not logically true. The rule of substitution of identicals is safe as long as identicals are not substituted for each other in contexts that are ruled by an occurrence of "that"; for the name-forming operator "that" creates *intensional contexts* — contexts in which the *topical* (or *contextual*) referential function of some singular terms can shear way from their *factual* referential function, creating an ambiguity that had better not be ignored. (Such singular terms are called "(referentially) unstable"; for more on this matter, see Section 9 below.)

P13–P16 and P10 are still not all the principles that are needed for "that". In Section 7 and Section 9 below, more "that"-principles will follow.

---

[11]On the concept of (broadly) logical provability, see footnote 15.

5. The Number of States of Affairs

The ontological weakness of the mereology of states of affairs has still been preserved up to this point. No principle demanding a dramatically high number of states of affairs has so far been added, and the principles already assumed do merely require that there be at least four states of affairs: according to them, there need not be more states of affairs than this. From one point of view, this is as it should be, since the mereology of states of affairs should be applicable to finite artificial universes of states of affairs, with more or less restricted numbers of states of affairs (with 4, or 8, or 16, or 32, or 64 states of affairs, for example). If the mereology of states of affairs is to remain thus applicable (as a *multipurpose* abstract machine), then it must remain open to having any *principle of finite number* added to it that is appropriate to the occasion of application, i.e., some principle or other that looks like this: $\exists^{=N}x\ \mathrm{EL}(x)$ ("There are exactly N elemental states of affairs", where N is some natural number $\geqslant 2$).

From the *realistic* point of view, however — i.e., if the mereology of states of affairs is to decribe the *real* universe of states of affairs — there are only two envisageable *principles of the number of states of affairs* (via specifying the number of elemental states of affairs) that are not entirely arbitrary: (1) There is *exactly one* elemental state of affairs (in other words, there are *exactly two* states of affairs, namely, t^* and k^*). (2) There are *infinitely many* elemental states of affairs (in other words, there are *infinitely many* states of affairs). On the basis of the principles already stated (P0–P16), (1) has already been ruled out. This leaves us with (2) — every other envisageable principle of the number of states of affairs would be entirely arbitrary as an assertion about the *real* universe of states of affairs ((1) having been ruled out). And (2) can be formulated by employing the means of a first-order language, as follows:

(P17) $\exists^{\geqslant N}x\ \mathrm{EL}(x) \supset \exists^{\geqslant N+1}x\ \mathrm{EL}(x)$, for every natural number $N \geqslant 2$.[12]

6. Modal Bases and Modal Principles in the Mereology of States of Affairs

In Section 1 above, two bases of necessity were indicated: b^1 — the basis of logical necessity, and b^2 — the basis of nomological necessity. For characterizing the basis of logical necessity completely we merely need to add the principle:

[12]The initial assertion $\exists^{\geqslant 2}x\ \mathrm{EL}(x)$ is already provable. Note that the reference to natural numbers is not necessary but can be eliminated, since $\exists^{\geqslant N}x\ \mathrm{EL}(x)$ is definable without reference to natural numbers: $\exists^{\geqslant 1}x\ \mathrm{EL}(x) := \exists x\, \mathrm{EL}(x)$; $\exists^{\geqslant 2}x\ \mathrm{EL}(x) := \exists x\, \exists x'(\mathrm{EL}(x) \wedge \mathrm{EL}(x') \wedge x \neq x')$, $\exists^{\geqslant 3}x\ \mathrm{EL}(x) := \exists x \exists x' \exists x''(\mathrm{EL}(x) \wedge \mathrm{EL}(x') \wedge \mathrm{EL}(x'') \wedge x \neq x' \wedge x' \neq x'' \wedge x \neq x'')$; etc.

(P18) $\mathrm{S}(b^n) \wedge b^1 = t^*$.[13]

For characterizing the basis of nomological necessity completely, one would have to indicate exactly which states of affairs are laws of nature, which I will not do here (and who *can* do it?). But the following principles clearly hold for the basis of nomological necessity, independently of any specification of the laws of nature:

(P19) $\mathrm{P}(b^2, w^*)$,

(P20) $b^2 \neq t^*$,

(P21) $b^2 \neq w^*$.

In general we have: If $\mathrm{P}(b^n, w^*)$ is true of the basis of necessity b^n, then the (corresponding) necessity $\Box^n$ is *faithful to truth*, and vice versa; if $\mathrm{P}(b^n, w^*)$ is false of the basis b^n, then the necessity $\Box^n$ is *unfaithful to truth*, and vice versa. For necessities $\Box^n$ that are faithful to truth one can prove in the mereology of states of affairs: $\Box^n B \supset B$.

PROOF. Assume $\Box^n B$, where $\Box^n$ is a necessity faithful to truth. Hence according to the definitions at the beginning of Section 1: $\mathrm{P}(\mathtt{that}\ B, b^n)$. Hence because of $\mathrm{P}(b^n, w^*)$ and P1: $\mathrm{P}(\mathtt{that}\ B, w^*)$. Hence because of $\forall x[\mathrm{S}(x) \supset (\mathrm{A}(x) \equiv \mathrm{P}(x, w^*))]$ and P13: $\mathrm{A}(\mathtt{that}\ B)$. Hence because of D6 and P13: $\mathrm{O}(\mathtt{that}\ B)$. Hence because of P10: B. □

One can also prove in the mereology of states of affairs: $\mathrm{P}(b^n, w^*) \equiv \forall x(\Box^n(x) \supset \mathrm{O}(x))$ (where $\forall x(\Box^n(x) \supset \mathrm{O}(x))$ is the obvious predicate-logical correlate of $\Box^n B \supset B$ if the latter is taken as a general schema). A question I leave open for the time being is whether one can deduce, in the mereology of states of affairs, $\mathrm{P}(b^n, w^*)$ from assuming $\Box^n B \supset B$ as a general schema (but see Section 9).

Note that the fact that a necessity is faithful to truth does not by itself mean that the necessity is ontic or alethic, for the basis which is an intensional part of w^*, making the corresponding necessity faithful to truth, may have been picked out according to criteria which are wholly or partly epistemic (for example). Nor does it seem necessary that every ontic necessity is faithful to truth.

A property of necessities which is weaker than *faithfulness to truth* (in view of the provable statement $\mathrm{P}(w^*, k^*)$, P3, P7) is *consistency*: If $b^n \neq k^*$ is true of the basis of necessity b^n, then the necessity $\Box^n$ is *a consistent necessity*, and vice

[13] Strictly speaking, $\mathrm{S}(b^n)$ does nothing to characterize b^1 beyond what is already asserted by $b^1 = t^*$ (since one can already prove $\mathrm{S}(t^*)$). But $\mathrm{S}(b^n)$ will be needed frequently (for bases of necessity b^n that are not described in any further way, so that their being states of affairs does not emerge as a consequence of their description), and it is better to assert it here explicitly (in as good as place as any other) than to make unacknowledged use of it.

versa; if $b^n \neq k^*$ is false of the basis b^n, then $\Box^n$ is *an inconsistent necessity*, and vice versa. For necessities $\Box^n$ that are consistent one can prove in the mereology of states of affairs: $\Box^n B \supset \neg\Box^n\neg B$.

Proof. Assume $\Box^n B$, where $\Box^n$ is a consistent necessity. Hence: $\mathrm{P}(\mathtt{that}\ B, b^n)$. Assume also $\Box^n\neg B$. Hence: $\mathrm{P}(\mathtt{that}\ \neg B, b^n)$. Hence: $\mathrm{P}(\mathtt{neg}(\mathtt{that}\ B), b^n)$, according to P14. It is provable in the mereology of states of affairs: $\forall x\forall y(\mathrm{P}(x, y) \wedge \mathrm{P}(\mathtt{neg}(x), y) \equiv y = k^*)$. Hence: $b^n = k^*$ — contradicting the consistency of $\Box^n$. Therefore: $\neg\Box^n\neg B$. □

One can also prove in the mereology of states of affairs: $b^n \neq k^* \equiv \forall x[\Box^n(x) \supset \neg\Box^n(\mathtt{neg}(x))]$ (where $\forall x[\Box^n(x) \supset \neg\Box^n(\mathtt{neg}(x))]$ is the obvious predicate-logical correlate of $\Box^n B \supset \neg\Box^n\neg B$ if the latter is taken as a general schema). *Another* question I leave open for the time being is whether one can deduce, in the mereology of state of affairs, $b^n \neq k^*$ from assuming $\Box^n B \supset \neg\Box^n\neg B$ as a general schema (but see again Section 9).

We also have in general: If $b^n \neq t^*$ is true of the basis of necessity b^n, then the necessity $\Box^n$ is *a contingent necessity*, and vice versa; if $b^n \neq t^*$ is false of the basis b^n, then $\Box^n$ is *a noncontingent necessity*, and vice versa. The designation "contingent necessity" is confusing, and still remains so when the gloss is added that what is meant by "contingent necessity" is *logically contingent* necessity. In what sense contingent necessities $\Box^n$ are called "(logically) contingent" becomes clear when one considers that the following is deducible in the mereology of states of affairs: $b^n \neq t^* \equiv \exists x(\Box^n(x) \wedge \neg\Box^1(x))$.

Finally we have in general: If $b^n \neq w^*$ is true of a basis of necessity b^n, then the necessity $\Box^n$ is a *proper necessity*, and vice versa; if $b^n \neq w^*$ is false of the basis b^n, then $\Box^n$ is an *improper necessity*, and vice versa. It is deducible in the mereology of states of affairs for every *consistent* necessity $\Box^n$: $b^n \neq w^* \equiv \exists x(\mathrm{O}(x) \wedge \neg\Box^n(x))$.

Proof. Let $\Box^n$ be a consistent necessity. Assume $b^n = w^*$, and $\mathrm{O}(x)$. Hence according to D7, D1, P3 and P4: $\mathrm{P}(x, w^*)$. Hence $\mathrm{P}(x, b^n)$. Hence $\Box^n(x)$. Assume conversely: $\forall x(\mathrm{O}(x) \supset \Box^n(x))$. Hence because of $\mathrm{S}(w^*)$ (a consequence of D7, D1, P3 and P4), P12, D6: $\mathrm{O}(w^*)$. Hence: $\Box^n(w^*)$. Hence: $\mathrm{P}(w^*, b^n)$. According to P8: $\mathrm{QC}(w^*)$, and hence (according to D14): $w^* = b^n \vee \mathrm{T}(b^n)$. Since $\Box^n$ is a consistent necessity, we have: $b^n \neq k^*$, and therefore: $\neg\,\mathrm{T}(b^n)$. Hence: $b^n = w^*$. □

Since there are, according to P17, infinitely many states of affairs, there are infinitely many necessities faithful to truth, just as many necessities faithful to truth as there are bases for them. The extremes are marked, on the one side, by *logical necessity*, with the basis t^* ($= b^1$), which is a proper and noncontingent necessity faithful to truth, and, on the other side, by *factuality*, with the basis w^*, which is an improper and contingent necessity faithful to truth. There are infinitely many proper and contingent necessities faithful to truth in between, *nomological necessity*, with the basis b^2, being one of them. Most of these necessities do not have a designation. The sequence $\Box^1, \Box^2, \Box^3, \Box^4, \Box^5, \Box^6, \ldots$ – even if prolonged to infinity – does not contain enough terms to give a designation to every proper and contingent necessity faithful to truth. (But there are certainly enough designations in the sequence for every necessity we specifically refer to here.) It is a matter of fact that we do not give much thought to most of the infinitely many proper and contingent necessities faithful to truth. But this should not mislead us into thinking that somehow they are not "real" necessities.

The *strength (or force) of a necessity* is inverse to the (intensional) strength of its basis. *Logical necessity* – having the absolutely weakest basis: the state of affairs t^* which has no intensional content – is the absolutely strongest necessity. *Factuality* – having the strongest basis which is faithful to truth (and which is *almost* the *absolutely* strongest basis): the state of affairs w^* – is the weakest necessity faithful to truth, so weak that it is quite rightfully called an improper necessity. We have the following correlation: $\Box^n$ is a stronger necessity than $\Box^m$ if, and only if, b^n is a proper intensional part of b^m. There are of course many pairs of necessities which are such that the basis of each pair-member does not contain the basis of the other pair-member: neither one of the paired necessities is stronger than the other.

An important fact should be noted. *General determinism* is sometimes asserted as the thesis that *every obtaining state of affairs is necessary*, where "is necessary" is thought to express a necessity $\Box^n$ which is faithful to truth (and therefore consistent) *and* has more force than "obtains". But as, a matter of fact, it cannot have more force. Who accepts $\forall x(O(x) \supset \Box^n(x))$ for a consistent necessity $\Box^n$ must, according to the theorem stated and proven above, also accept $b^n = w^*$, and therefore the necessity $\Box^n$ *is not* stronger than *factuality*, which is the weakest necessity faithful to truth. This shows that general determinism is an incoherent position: it intends to spread a *proper* necessity over more states of affairs than it can apply to, namely, over all obaining states of affairs.

Leibniz famously asserted that *general logical determinism* is true, i.e., that $\forall x(O(x) \supset \Box^1(x))$ is true. But if $\forall x(O(x) \supset \Box^1(x))$ is true, then, according to the theorem $b^1 \neq w^* \equiv \exists x(O(x) \wedge \neg\Box^1(x))$, $b^1 = w^*$ is also true ($\Box^1$ being a consistent necessity), or in other words: $w^* = t^*$ is true (since $b^1 = t^*$, according to P18) – contradicting P9. General logical determinism is, therefore, false – provably false in the mereology of states of affairs. Leibniz is, therefore, provably wrong. But it should be noted that P9 is one of the two principles of the mereology of states of affairs (the other being P8) which are not conceptually true. Leibniz, therefore, did not commit a logical mistake in asserting the truth of general logical determinism. It seems, however, that Leibniz would have endorsed $\mathtt{MC}(w^*)$ — "the world is maximally consistent" (which is equivalent to the conjunction of P7: $w^* \neq k^*$, and P8: $\mathtt{QC}(w^*)$). And the only way to square $w^* = t^*$ (to which Leibniz is committed by being a logical determinist) with $\mathtt{MC}(w^*)$ is by assuming that there are less than four states of affairs (see the end of Section 3; $\mathtt{MC}(w^*)$ definitionally implies $\mathtt{QC}(w^*)$), and more than one state of affairs. Therefore, the only way to square $w^* = t^*$ with $\mathtt{MC}(w^*)$ is by assuming that *there are exactly two states of affairs* (because the mereology of states of affairs forbids that there are exactly three states of affairs): t^* (= w^*) and k^*. Leibniz, therefore, as a logical determinist and "maximal-consistentist," is committed to the view that there are only two states of affairs. It would be striking if there were any evidence in his works that he did believe this.

7. The Classical Modal Principles

We have already seen that for nomological necessity, $\Box^2$, two classical modal principles are true: $\Box B \supset B$, and $\Box B \supset \neg\Box\neg B$. (I leave out the index attached to "$\Box$" if reference is made to some determinable necessity or other; I also leave out the index attached to "b" if reference is made to some determinable basis or other.) The same two principles are also true for logical necessity, $\Box^1$.

What about the following other classical principles: (1) $\Box(A \supset B) \supset (\Box A \supset \Box B)$, (2) $\Box B \supset \Box\Box B$, (3) $\neg\Box B \supset \Box\neg\Box B$? Let us see whether we need further principles for proving them, and which principles.

$\Box(A \supset B) \supset (\Box A \supset \Box B)$ is provable without further assumptions.

Proof. Assume $\Box(A \supset B)$, and assume $\Box A$. Hence $\mathtt{P}(\mathtt{that}\,(A \supset B), b)$, and $\mathtt{P}(\mathtt{that}\ A, b)$. Hence: $\mathtt{P}(\mathtt{disj}(\mathtt{neg}(\mathtt{that}\ A), \mathtt{that}\ B), b)$, according to P14, P15, etc., and $\mathtt{P}(\mathtt{that}\ A, b)$. Hence $\mathtt{P}(\mathtt{that}\ B, b)$ — because $\mathtt{S}(\mathtt{that}\ A)$ and $\mathtt{S}(\mathtt{that}\ B)$, according to P13, and because of the theorem $\forall x \forall y \forall z[\mathtt{S}(x) \wedge \mathtt{S}(y) \wedge \mathtt{P}(\mathtt{disj}(\mathtt{neg}(x), y), z) \wedge \mathtt{P}(x, z) \supset \mathtt{P}(y, z)]$. Hence $\Box B$. □

But the attempt to prove $\Box B \supset \Box\Box B$ reveals that there is need of further principles: Assume $\Box B$. Hence $\mathtt{P}(\mathtt{that}\ B, b)$ (b being the state of affairs which is the appropriate basis of necessity). From this one must derive: $\mathtt{P}(\mathtt{that}\ \Box B, b)$, which, purely definitionally, implies $\Box\Box B$. The step from $\mathtt{P}(\mathtt{that}\ B, b)$ to $\mathtt{P}(\mathtt{that}\ \Box B, b)$ is, by definition alone, equivalent to the step from $\mathtt{P}(\mathtt{that}\ B, b)$ to $\mathtt{P}(\mathtt{that}\ \mathtt{P}(\mathtt{that}\ B, b), b)$, which, in turn, is a consequence of the following general principle:

Principle A $\quad \forall x \forall y(\mathtt{P}(x, y) \supset \mathtt{P}(\mathtt{that}\ \mathtt{P}(x, y), y))$.

Let us now see which further principle is necessary for proving $\neg\Box B \supset \Box\neg\Box B$. It is this (which is easily found, given Principle A):

Principle B $\quad \forall x\forall y(\mathrm{S}(x) \wedge \mathrm{S}(y) \wedge \neg\,\mathrm{P}(x,y) \supset \mathrm{P}(\mathtt{neg}(\mathtt{that}\ \mathrm{P}(x,y)),y))$.

Both Principle A and Principle B are easily provable if one assumes the following stronger principle:

(P22) $\quad \forall x\forall y(\mathrm{P}(x,y) \supset \mathtt{that}\ \mathrm{P}(x,y) = t^*) \wedge \forall x\forall y(\neg\,\mathrm{P}(x,y) \supset \mathtt{that}\ \mathrm{P}(x,y) = k^*)$.

I adopt this latter principle, since a relationship of intensional parthood, or a negation of such a relationship, is clearly a matter of logical necessity.[14]

Another classical modal principle is not really a principle but a *provability-rule*: If B is logically provable,[15] then $\Box B$ is logically provable. On the basis of the definition of $\Box$, this inference-rule amounts to: If B is logically provable, then $\mathrm{P}(\mathtt{that}\ B, b)$ is logically provable (b being the state of affairs that is the appropriate basis of necessity). I put this, more compactly, into the following form:

(P23) $\quad \vdash B \Rightarrow\ \vdash \mathrm{P}(\mathtt{that}\ B, b)$.

The provability-rule $\vdash B \Rightarrow\ \vdash \Box B$ is an easy consequence of P23.

One should note that the modal propositional system S5, that is, classical truth-functional propositional logic (axiomatized in some way) *plus*

$\vdash \Box B \supset B$,

$\vdash \Box(A \supset B) \supset (\Box A \supset \Box B)$,

$\vdash \Box B \supset \Box\Box B$,

$\vdash \neg\Box B \supset \Box\neg\Box B$,

$\vdash B \Rightarrow\ \vdash \Box B$,

(where "$\vdash$" stands for "logically provable") has now been justified as *the* correct system of modal propositional logic for all necessities that (a) can be represented as founded on a basis of necessity, and that (b) are faithful to truth. The foundations of justification are the mereology of states of affairs and a conception

[14]An equivalent formulation of P22 is this: $\forall x\forall y(\mathrm{P}(x,y) \supset \Box^1\,\mathrm{P}(x,y)) \wedge \forall x\forall y(\neg\,\mathrm{P}(x,y) \supset \Box^1\neg\,\mathrm{P}(x,y))$.

[15]"Logically provable" means: *provable purely on the basis of principles and inference-rules that are conceptually (broadly logically, analytically) valid.* Thus, if P8 or P9 are necessary for deducing a certain theorem in the mereology of states of affairs, then that theorem is not logically provable — because P8 and P9 are not conceptually true (see Section 3).

of necessity that is expressed by the following two definitions: $\Box(x) := \mathrm{P}(x, b)$, $\Box A := \Box(\texttt{that}\ A)$. These definitions are expressive of a *basis-theory of necessity*, since "b" refers to a state of affairs which is *the basis* of the necessity concerned.

Remark. The basis-theory of necessity is not a theory of necessity that is applicable to all kinds of necessity. There are some necessities for which the analysis in terms of a basis of necessity is inadequate, for example, *epistemic necessity*: knowledge. What somebody knows is not what follows from his basis of knowledge (there is no such thing); it is what follows from his *basis of belief* and satisfies in addition several further conditions. Take the simplest concept of knowledge (which is implied by all other concepts of knowledge): *true conviction*. In this sense, what somebody knows is what follows from his basis of belief and is true. The principle $\neg\Box B \supset \Box\neg\Box B$ is not valid for knowledge in this sense, for the following situation frequently obtains:

B is false, and therefore $\neg\Box B$ is true; but (in this same situation) $\texttt{that}\ B$ is an intensional part of the basis of belief of the person concerned, and therefore that person is convinced that B (i.e., $\Box_c B$), hence convinced that she is convinced that B (i.e., $\Box_c\Box_c B$, according to P22), hence convinced that she knows that B (because of $\Box_c\Box_c B \supset \Box_c(\Box_c B \wedge B)$). Hence $\Box\neg\Box B$ must be false. (For if it were true, then the person concerned would know, and therefore be convinced, that she does not know that B. Hence the person concerned would be both convinced that she knows that B (as we have already seen) and convinced that she does not know that B — which cannot be.)

Note that conviction — or *doxastic necessity* — *can* very well be treated according to the basis-theory of necessity. Since one will not require that conviction be faithful to truth but only that it be consistent, $\Box B \supset B$ needs to be replaced by $\Box B \supset \neg\Box\neg B$ in the above system; this replacement turns it into a system which defines a defensible modal propositional logic for — *dispositional, rational* — conviction. □

The theory of necessity advocated here is, moreover, an *onto-nomological theory of necessity*, since it is grounded in ontological laws for the realm of states of affairs. By being an onto-nomological theory of necessity, it is automatically also an adequate onto-nomological theory of possibility, since possibility can be adequately defined on the basis of necessity ($\Diamond A := \neg\Box\neg A$) and is, therefore, completely determined by the latter concept. (But the onto-nomological theory of possibility can also be developed *independently* of the theory of necessity: on the basis of the mereology of states of affairs, and the definitions $\Diamond^n(x) := \mathrm{S}(x) \wedge \neg\, \mathrm{P}(\mathtt{neg}(x), b^n)$ and $\Diamond^n A := \Diamond^n(\texttt{that}\ A)$.)

It needs a separate paper to handle the question how far *conditionals* (or *relational necessities*) can be treated within the mereology of states of affairs on the basis of the definitions ${}^n\!\!\rightarrow(x, y) := \mathrm{P}(y, \mathtt{conj}(b^n, x))$ and $B\, {}^n\!\!\rightarrow A := {}^n\!\!\rightarrow(\texttt{that}\ B, \texttt{that}\ A)$.

8. Necessity and Possible Worlds

The theory of modality developed here is a theory of modality in which the concept of possible world plays no essential role. Nevertheless, in view of the popularity of the possible-worlds-theory of modality, it will be good to see how that theory is derivable from, and therefore reducible to, the present theory.

We have seen above (in Section 2) that maximally consistent states of affairs can be regarded as possible worlds (and that there are exactly as many maximally consistent states of affairs as there are elemental states of affairs: *negation* is a function that maps elemental states of affairs onto maximally consistent states of affairs). This can only mean that *the possible worlds* (the entire set of them) *can be identified* with *the maximally consistent states of affairs* (the entire set of them). (It would hardly make sense to allow that the maximally consistent states of affairs are possible worlds, but to demand that there are in addition some other possible worlds that are not maximally consistent states of affairs.)

We need to define another important concept of the mereology of states of affairs:

(D16) $\mathtt{O}(x, y) := \mathtt{P}(x, y)$.

$\mathtt{O}(x, y)$ is read as "(the state of affairs) x obtains in (the state of affairs) y". One can then prove the following theorems:

$$\forall x[\,\mathtt{S}(x) \supset (\,\mathtt{O}(x) \equiv \mathtt{O}(x, w^*))],$$
$$\forall x[\,\mathtt{S}(x) \supset (\Box^n(x) \equiv \forall y(\mathtt{MC}(y) \wedge \mathtt{O}(b^n, y) \supset \mathtt{O}(x, y)))],$$
$$\Box^n A \equiv \forall y(\,\mathtt{MC}(y) \wedge \mathtt{O}(b^n, y) \supset \mathtt{O}(\mathtt{that}\ A, y)),$$
$$\forall x[\,\mathtt{S}(x) \supset (\Diamond^n(x) \equiv \exists y(\mathtt{MC}(y) \wedge \mathtt{O}(b^n, y) \wedge \mathtt{O}(x, y)))],$$
$$\Diamond^n A \equiv \exists y(\,\mathtt{MC}(y) \wedge \mathtt{O}(b^n, y) \wedge \mathtt{O}(\mathtt{that}\ A, y)).$$

The first theorem states that a state of affairs obtains (*simpliciter*) if, and only if, it obtains in *the world*. The second theorem states that a state of affairs is n-necessary if, and only if, it obtains in every possible world in which the basis of the n-necessity obtains. The third theorem is a corollary of the second theorem: for the sentence connective of necessity (not the predicate). The fourth theorem states that a state of affairs is n-possible if, and only if, it obtains in some possible world in which the basis of the n-necessity obtains. The fifth theorem is a corollary of the fourth theorem: for the sentence connective of possibility (not the predicate).

Proof. As an example, here is the (not altogether easy) proof of the second theorem:

1. Suppose: $\mathrm{S}(x)$, $\Box^n(x)$, $\mathrm{MC}(y)$, $\mathrm{O}(b^n, y)$. Hence $\mathrm{P}(x, b^n)$ and $\mathrm{P}(b^n, y)$ (according to the definition of $\Box^n(x)$, and D16). Hence by P1: $\mathrm{P}(x, y)$, hence by D16: $\mathrm{O}(x, y)$.

2. Suppose: $\mathrm{S}(x), \forall y(\mathrm{MC}(y) \wedge \mathrm{O}(b^n, y) \supset \mathrm{O}(x, y))$. Hence (because of P3, P4, D1): $\mathrm{P}(x, \mathrm{CONJ}\, z\forall y(\mathrm{MC}(y) \wedge \mathrm{O}(b^n, y) \supset \mathrm{O}(z, y)))$. But $\mathrm{CONJ}\, z\forall y(\mathrm{MC}(y) \wedge \mathrm{O}(b^n, y) \supset \mathrm{O}(z, y)) = b^n$, because (applying D16) $\mathrm{CONJ}\, z\forall y(\mathrm{MC}(y) \wedge \mathrm{P}(b^n, y) \supset \mathrm{P}(z, y)) = b^n$. Therefore: $\mathrm{P}(x, b^n)$, and consequently: $\Box^n(x)$.

$\mathrm{CONJ}\, z\forall y(\mathrm{MC}(y) \wedge \mathrm{P}(b^n, y) \supset \mathrm{P}(z, y)) = b^n$ remains to be proved. It is a consequence of:

(†) $\mathrm{P}(b^n, \mathrm{CONJ}\, z\forall y(\mathrm{MC}(y) \wedge \mathrm{P}(b^n, y) \supset \mathrm{P}(z, y)))$,

(‡) $\mathrm{P}(\mathrm{CONJ}\, z\forall y(\mathrm{MC}(y) \wedge \mathrm{P}(b^n, y) \supset \mathrm{P}(z, y)), b^n)$,

according to P3.

(†) holds, because $\forall y(\mathrm{MC}(y) \wedge \mathrm{P}(b^n, y) \supset \mathrm{P}(b^n, y))$ and P4, P3, D1, P18.

For (‡) assume $\mathrm{QA}(u)$, $\mathrm{P}(u, \mathrm{CONJ}\, z\forall y(\mathrm{MC}(y) \wedge \mathrm{P}(b^n, y) \supset \mathrm{P}(z, y)))$; what is in question is established according to P5 if $\mathrm{P}(u, b^n)$ can be deduced from this assumption.

If $\mathrm{M}(u)$, then $\mathrm{P}(u, b^n)$. ($\mathrm{S}(b^n)$ according to P18.)

If $\neg\mathrm{M}(u)$, then (according to P6): $\exists k'[\mathrm{P}(k', u) \wedge \neg\mathrm{M}(k') \wedge \exists z'(\mathrm{P}(k', z') \wedge \forall y(\mathrm{MC}(y) \wedge \mathrm{P}(b^n, y) \supset \mathrm{P}(z', y)))]$. Because of $\mathrm{QA}(u)$, $\mathrm{P}(k', u)$, $\neg\mathrm{M}(k')$: $k' = u$ (according to D9), and also $\mathrm{EL}(u)$ (according to D10). Hence: $\exists z'(\mathrm{P}(u, z') \wedge \forall y(\mathrm{MC}(y) \wedge \mathrm{P}(b^n, y) \supset \mathrm{P}(z', y)))$. Hence: $\exists z'(\mathrm{P}(u, z') \wedge \forall y(\mathrm{S}(y) \wedge \mathrm{EL}(\mathrm{neg}(y)) \wedge \neg\mathrm{P}(\mathrm{neg}(y), b^n) \supset \neg\mathrm{P}(\mathrm{neg}(y), z')))$.[16] Hence: $\exists z'(\mathrm{P}(u, z') \wedge \forall y(\mathrm{S}(y) \wedge \mathrm{EL}(\mathrm{neg}(y)) \wedge \mathrm{P}(\mathrm{neg}(y), z') \supset \mathrm{P}(\mathrm{neg}(y), b^n)))$. Hence: $\exists z'(\mathrm{P}(u, z') \wedge \forall y(\mathrm{S}(y) \wedge \mathrm{EL}(\mathrm{neg}(\mathrm{neg}(y))) \wedge \mathrm{P}(\mathrm{neg}(\mathrm{neg}(y)), z') \supset \mathrm{P}(\mathrm{neg}(\mathrm{neg}(y)), b^n)))$. Hence: $\exists z'(\mathrm{P}(u, z') \wedge \forall y(\mathrm{EL}(y) \wedge \mathrm{P}(y, z') \supset \mathrm{P}(y, b^n)))$.[17] Hence because of $\mathrm{EL}(u)$: $\mathrm{P}(u, b^n)$. □

Note that the conjunct $\mathrm{O}(b^n, y)$ drops out of the five theorems for $n = 1$, since $b^1 = t^*$ (P18) and $\forall y(\mathrm{S}(y) \supset \mathrm{O}(t^*, y))$. Thus one obtains, for example:

$$\forall x[\,\mathrm{S}(x) \supset (\Box^1(x) \equiv \forall y(\mathrm{MC}(y) \supset \mathrm{O}(x, y)))],$$
$$\forall x[\,\mathrm{S}(x) \supset (\Diamond^1(x) \equiv \exists y(\mathrm{MC}(y) \wedge \mathrm{O}(x, y)))].$$

The Leibnizian conception of modality emerges as a pair of theorems of the mereology of states of affairs (which theorems also show that Leibnizian necessity is logical necessity, Leibnizian possibility logical possibility). I add the

[16]The theorems employed are $\forall y[\mathrm{MC}(y) \supset \forall x(\mathrm{S}(x) \supset (\mathrm{P}(x, y) \equiv \neg\mathrm{P}(\mathrm{neg}(y), x)))]$ and $\forall y[\mathrm{MC}(y) \equiv \mathrm{EL}(\mathrm{neg}(y))]$. b^n is a state of affairs according to P18. The additional conjunct "$\mathrm{S}(y)$" is simply a definitional consequence of "$\mathrm{MC}(y)$".

[17]The theorem employed is $\forall y(\mathrm{S}(y) \supset \mathrm{neg}(\mathrm{neg}(y)) = y)$.

following two theorems that rather strikingly illuminate the relation between *modality* and *basis of modality*[18]:

$$b^n = \texttt{CONJ}\, y\, \Box^n(y),$$
$$b^n = \texttt{DISJ}\, y\, \Diamond^n(y).$$

9. Final Observations: Another "That" — Principle and the Trickiness of "w^*"

In Section 6, I left it open whether one can deduce, in the mereology of states of affairs, $\mathtt{P}(b^n, w^*)$ from assuming $\Box^n B \supset B$ as a general schema, and likewise whether one can deduce $b^n \neq k^*$ from assuming $\Box^n B \supset \neg\Box^n\neg B$ as a general schema. Let us try to make these deductions.

(1) Assume $\Box^n B \supset B$ as a general schema. Hence: $\Box^n\, \mathtt{O}(b^n) \supset \mathtt{O}(b^n)$. According to the definition of $\Box^n(x)$: $\Box^n(b^n)$, because of $\mathtt{P}(b^n, b^n)$, which in turn is true because of P18 and P2. Now, there is another very plausible "that"-principle — a principle complementing P22 and P23:

(P24) $\forall x(\mathtt{S}(x) \supset x = \mathtt{that}\ \mathtt{O}(x))$.

Applying P24, one obtains $\Box^n(\mathtt{that}\ \mathtt{O}(b^n))$ from $\Box^n(b^n)$, and therefore: $\Box^n\, \mathtt{O}(b^n)$. Hence, since we already have $\Box^n\, \mathtt{O}(b^n) \supset \mathtt{O}(b^n)$: $\mathtt{O}(b^n)$. Therefore: $\mathtt{P}(b^n, w^*)$ (according to $\forall x[\mathtt{S}(x) \supset (\mathtt{A}(x) \equiv \mathtt{P}(x, w^*))]$ and D6).

(2) Assume $\Box^n B \supset \neg\Box^n\neg B$ as a general schema. Hence $\Box^n\mathtt{O}(b^n) \supset \neg\Box^n\neg\mathtt{O}(b^n)$. Just as in (1), one obtains $\Box^n\, \mathtt{O}(b^n)$. Therefore: $\neg\Box^n\neg\, \mathtt{O}(b^n)$, and hence (by definition) $\neg\, \mathtt{P}(\mathtt{that}\ \neg\, \mathtt{O}(b^n), b^n)$, hence (applying P14) $\neg\, \mathtt{P}(\mathtt{neg}(\mathtt{that}\ \mathtt{O}(b^n)), b^n)$, hence (applying P24) $\neg\, \mathtt{P}(\mathtt{neg}(b^n), b^n)$, hence $b^n \neq k^*$ (because $\mathtt{P}(\mathtt{neg}(k^*), k^*)$).

The designator "w^*" has been defined above as "$\mathtt{CONJ}\ x\ \mathtt{O}(x)$" (D7). The designator "$t^*$", on the other hand, has been defined as "$\mathtt{CONJ}\ x \neg\, \mathtt{S}(x)$" (D2). Though the two definitions have a very similar structure, there is a striking difference between the two defined designators. Of "t^*," one would say that it could not, not even in principle, have designated a different state of affairs than it designates in fact. Of "w^*", however, one would indeed say that it could, in principle, have designated a different state of affairs than it designates in fact.[19] This is a consequence of the predicate "$\mathtt{O}(x)$" occurring in its definition: one would say that this predicate could have applied differently than it does in fact apply, because other states of affairs could have obtained (i.e., could have been actual) than in fact

[18] A basis of necessity b^n is likewise a *basis of possibility*, in short: it is a *basis of modality*.

[19] Note that these are *meta*-modal statements. They can, however, be represented in the object-language: $\forall y'(t^* \neq y' \supset \Box^1(t^* \neq y'))$, $\exists y'(w^* \neq y' \wedge \Diamond^1(w^* = y'))$.

obtain, and if they had, "w^*" would have designated a different state of affairs than it designates in fact.

This *instability* of "w^*" leads to certain problems:

(1′) Let y be an obtaining (therefore, according to D6, actual) state of affairs that is neither w^* nor t^*. Hence according to a previously proven theorem (in Section 3): $\mathtt{P}(y, w^*)$. Hence according to P22: $\mathtt{that}\ \mathtt{P}(y, w^*) = t^*$, and therefore: $\Box^1\, \mathtt{P}(y, w^*)$.

(2′) Since $y \neq t^*$, it follows according to P24: $\mathtt{that}\ \mathtt{O}(y) \neq t^*$. But it seems that we have the following identity: $\mathtt{that}\ \mathtt{O}(y) = \mathtt{that}\ \mathtt{P}(y, w^*)$. Therefore: $\mathtt{that}\ \mathtt{P}(y, w^*) \neq t^*$, and consequently: $\neg\Box^1\, \mathtt{P}(y, w^*)$.

We are confronted with a contradiction. It seems the only way of escaping from it — short of a modification of principles — is to reject the identity statement used in the second deduction, namely, the statement "$\mathtt{that}\ \mathtt{O}(y) = \mathtt{that}\ \mathtt{P}(y, w^*)$". But, unfortunately, this is not a statement that can easily be rejected: it has intuition on its side, and what is more important: *purely on the basis of principles that are conceptually true* one can prove $\forall y(\mathtt{O}(y) \equiv \mathtt{P}(y, w^*))$ in the mereology of states of affairs, and therefore also the particular case $\mathtt{O}(y) \equiv \mathtt{P}(y, w^*)$.

Remark. See the proof of $\forall y(\mathtt{S}(y) \supset (\mathtt{A}(y) \equiv \mathtt{P}(y, w^*)))$ in Section 3. One merely needs to additionally consider P0 and that "$\mathtt{O}(y)$" is defined as "$\mathtt{S}(y) \wedge \mathtt{A}(y)$" (according to D6) in order to obtain a proof of $\forall y(\mathtt{O}(y) \equiv \mathtt{P}(y, w^*))$ from the already proven theorem. No principle or inference-rule that is not conceptually (or broadly logically) valid is used in the proofs. □

Using the provability-rule EQU*: "If $A \equiv B$ is logically provable, then $\mathtt{that}\ A = \mathtt{that}\ B$ is also logically provable" (see Section 4), one obtains: $\mathtt{that}\ \mathtt{O}(y) = \mathtt{that}\ \mathtt{P}(y, w^*)$. There is no escaping this conclusion.

The real source of the contradiction derived above is not the statement "$\mathtt{that}\ \mathtt{O}(y) = \mathtt{that}\ \mathtt{P}(y, w^*)$" (which cannot be rejected, as has just been shown). The real source of the contradiction is an illicit step of inference, which is so unobtrusive as to be easily overlooked. In Section 4 an absurd conclusion was obtained by substituting, according to the rule of substitution of identicals, the term "$\iota x(A \wedge x = b)$", which is *unstable* for every sentence A that is neither conceptually true nor conceptually false,[20] into a "that"-context. Above, *another* absurd conclusion is obtained by substituting *another* unstable term — "w^*" — into *another* "that"-context. But not by applying the rule of substitution of identicals. Rather, the rule used in the second case is *universal instantiation*: from the first conjunct of P22 [$\forall x \forall y(\mathtt{P}(x, y) \supset \mathtt{that}\ \mathtt{P}(x, y) = t^*)$],

[20] If A is a sentence that is neither conceptually true nor conceptually false, then "$\iota x(A \wedge x = b)$" could have designated a different entity then it designates in fact: If A is true, then the factual referent of "$\iota x(A \wedge x = b)$" is b, but it could have been c^*; if A is false, then the factual referent of "$\iota x(A \wedge x = b)$" is c^*, but it could have been b. (Concerning c^*, see footnote 10.)

"$\mathrm{P}(y, w^*) \supset \mathtt{that}\ \mathrm{P}(y, w^*) = t^*$" is obtained (in (1′)) by universal instantiation. (And further: from the already established statement "$\mathrm{P}(y, w^*)$" and the newly won "$\mathrm{P}(y, w^*) \supset \mathtt{that}\ \mathrm{P}(y, w^*) = t^*$," one obtains by applying *modus ponens*: "$\mathtt{that}\ \mathrm{P}(y, w^*) = t^*$".) But universal instantiation, just like substitution of identicals, is not a universally valid inference-rule: it can lead from a true sentence to a false one when being applied in such a manner that substitution into a "that"-context occurs.[21]

University of Regensburg
Institute of Philosophy
93040 Regensburg
Germany
e-mail: uwe.meixner@psk.uni-regensburg.de

---

[21]Note that all modal operators $\mathtt{Q}$ — ontic, epistemic, or otherwise — have "that"-contexts connected to them, since they all can be defined on the basis of the corresponding predicate for states of affairs as follows: $\mathtt{Q}(B) := \mathtt{Q}(\mathtt{that}\ B)$. This is the reason why failures of the rules of substitution of identicals and of universal instantiation were first noticed in connection with substitutions into the scopes of modal operators.

Roberto Poli

THE ONTOLOGY OF WHAT IS NOT THERE

ABSTRACT. After discussion of the mainstream definition of ontology, and a short analysis of the Aristotelian's principles, the paper addresses the problem of the categorial nature of the future by distinguishing (a) possibility from potentiality and (b) forward from upward emergence.

Introduction

Contemporary "common-sense" readings of ontology often rely on the idea that ontology is the theory of what exists. The following are just a few of the many quotes in defence of that thesis:

- "The word 'ontology' is used to refer to philosophical investigation of existence or being" (*Routledge Encyclopaedia of Philosophy*, 1998)
- "[Ontology is] the branch of philosophy concerned with the nature of existence and the relations between things" (*Longman Dictionary of Contemporary English*, 1981)
- "[Ontology is] the philosophical investigation of the nature, constitution, and structure of reality" (*The Cambridge Dictionary of Philosophy*, 1999; under the entry "metaphysics", to which one is directed from "ontology").

Its wide acceptance notwithstanding, the thesis that equates ontology and reality is nevertheless wrong. Better, it is ruinous for any scientifically and philosophically well-structured understanding of ontology, and its wide acceptance only testifies the miserable level attained by contemporary awareness of philosophical matters.

A few exemplary cases immediately show that "existence" and "being" are not synonymous, and that "material existence" is but one of the many forms of "temporal extendedness".

CASE 1. The simplest case concerns mental (psychological) and social entities. The former do not exist in the same way as material beings are said to exist. Moreover, they present both forms of dependence on some material bearer and forms of independence from them.[1] Detailed analysis of the laws of dependence

[1] The idea of form as bearer of temporal information is due to M. Leyton. See his (1992, 2001).

In: J. Malinowski and A. Pietruszczak (eds.), *Essays in Logic and Ontology* (*Poznań Studies in the Philosophy of the Sciences and the Humanities*, vol. 91), 73–80. Amsterdam/New York, NY: Rodopi, 2006.

and independence between intentional items and their bearer is a demanding undertaking that requires intensive study and cannot be decided by postulation.

Social entities raise similar problems: they both present forms of dependence on material bearers and independence from them. Moreover, mental and social entities display forms of mutual dependence. Boldly stated, societies require minds as their environment, and minds require society as theirs. Proper treatment of these topics requires a well-developed theory of levels of reality and their dependences.[2] Considering material entities as the only ones that exist seems at best an ideologically oriented decision. If mental and social entities are legitimate entities as well, their analyses from an ontological viewpoint (call them "mental ontology" and "social ontology") are perfectly acceptable areas of research.

CASE 2. Properly speaking, ideal entities do not exist, because they are not temporally bounded like truly "real" entities are. On the other hand, both (ideal) values and mathematical entities are structured entities. This means that they offer "resistance" to improper treatment. Moreover, both values and formal constructs influence and shape "normal" entities. In short: non-existing entities exert influence on existing ones. Something has obviously gone wrong. Accepting the oversimplified understanding of reality put forward by the mainstream definition of ontology is the main source of the above difficulties. If ideal entities are legitimate entities, their ontological analysis is a perfectly acceptable area of research.

CASE 3. Past and future entities obviously do not exist. At most they existed and will possibly exist. Logicians have invented a number of clever tricks to deal with these cases. This is not to gainsay their relevance and usefulness from a formal point of view, but there is no reason to believe that formal tricks are free, neutral or transparent from an ontological viewpoint. In the rest of this paper I shall begin analysis from a categorical viewpoint of one class of the ontological entities that are not there, namely future entities.

An Aristotelian Intermezzo

Strange though it may seem, Aristotle still has interesting surprises in store. Everyone knows that, for Aristotle, individual substances have a basically twofold structure: they are composed of matter and form. Form in its turn has a number of different meanings, the most relevant of which is the interpretation of form as essence. Now, what is essence?

Aristotle's espression for what is nowadays called essence was *to ti en einai*, rendered in Latin by *quod quid erat esse*: "what was (considered) to be". That is

[2] For the theory of levels of reality, see Poli (2001).

"the essence of x is what x was considered to be" This definition raises a number of interpretive problems, the most relevant being the meaning of the past sense. Why did Aristotle formulate his definition of (what we call) essence in the past?

My understanding is as follows: The form of the actually existing x depends on what x was before. In contemporary jargon: the form of the actual x is *the result* of iteratively applying a number of transformation rules to the original x.

The definition advanced by Aristotle can therefore be explicated as follows: (1) x is a temporally extended entity; (2) the actual nature of x depends on x's previous nature. Aristotle's definition can be further improved by adding a third clause not explicitly contained in his original wording: (3) there *are rules of transformation* connecting the actual nature of x to its previous nature(s). In one sentence, the concept of essence carries temporal information.

Aristotle's definition of matter seems much more difficult to decipher. Matter for Aristotle is the indeterminate, the unknowable, what offers resistance to the understanding. This does not explain much until a further aspect is taken into account, namely that the above features of matter (indeterminability, unknowability, etc.) depend on its role as substratum of the becoming, of changes. I have found that the best way to understand Aristotle's definition of matter is to apply the same criterion used for deciphering form, but shifting from the past to the future. That is to say: The matter of the actual x is what x will become. Otherwise stated: matter is the bearer of the future developments of x. Matter is therefore a *principle of openness*. This explains why matter is indeterminate and unknowable. If my reading is accepted, all the classical features ascribed to matter become immediately transparent.

The actuality of a given individual substance comprises form and matter. In my rewording, this amounts to saying that any entity has a past (form) and a future (matter). The past is what is maximally informative: the actual (internal or esternal) configuration of the entity presents traces of its past history (better: the entity's past history is retrievable as far as it has left traces in the entity's actual configuration). The future is minimally informative: its actual configuration excludes some of the possible choices, but for the rest almost everything is open.

Enlarging the Aristotelian Framework

Two moves are relevant. Firstly, and properly speaking, the actually given entity is the only existing entity. Past and future entities do not exist. This can be accepted without further ado. What really matters is the subsequent idea that actually existing entities contain *traces* of their past and *tendencies* toward the future. Traces and tendencies are legitimate existing components or aspects of the actually existing entity. This first move of a Leibnizian flavour is nevertheless not enough.

The point is that, upon a deeper reflection, matter shows two different components. The first component of matter is the reflection of form around the present. The same rules used to obtain the actual from the past can be used to project what is actually given into the future. In its simplest form, this is the idea embedded in the geometry of state (or phase) spaces so widely developed in science. The problem with the theory of state spaces is that all the possibilities should be calculated in advance. Which is unlikely, to say the less, and simply unfeasible as far as complexity issues are considered.

The second and truly demanding component of matter is embedded in the creativity of causality, i.e. permanent openness towards novelty. Entities change: they are born, grow, develop, die. A different conceptual framework is required for these cases. To mark the difference between the two components of matter, I shall distinguish between possibility and potentiality.

The Categorical Structure of the Future

We can retrieve (fragments of) the past from its traces. Can we do the same for the future? Otherwise stated: can we categorically analyze the tendencies actively present in actually existing entities in order to foresee (fragments of) the future? The answer is definitely "yes".

Unfortunately, the task of providing a categorical analysis of the future is absent from the agenda of mainstream philosophy. This means that we cannot rely on the work of other colleagues: everything must be done from scratch. On the other hand, there have been and are thinkers trying to describe the future. Utopian thinkers, fantasy writers, cyber literates do so. Leaving aside their literary merits, which are not our concern here, we may try to consider them as sources for categorical analyses of what is not. Unfortunately, from a categorical viewpoint, such literature is usually poor, fragmented and astonishingly boring. Generally speaking, utopian and cyber works are make up of the already seen. The simplest way to figure out the situation is to consider Freud's theory of dreams and related topics (repression, etc.): they always concern what happened before. Utopian and cyber writer more often than not embellish on already seen pictures. By resorting to our previous distinction between the two dimensions of matter (the continuation of the past in the future dealt with by the category of possibility and the originally new taken into account by the category of potentiality), utopian and cyber literatures are as a norm patently confined within the boundaries of the category of possibility. The past is projected on the future, possibly with some make up. The motto "magistra vitæ" (history teaches us) well represent the underlying idea. This said, it is clear that possibility is not of much interest to us.

As soon as we turn to the category of potentiality, we immediately realise that the idea of potentiality as described above depends on a strong ontological thesis, namely that *reality is open*. This means that the ontological nature of entities is not thoroughly established. Something new can always happen. Entities are never totally given in advance. All of them present some kind of tendency toward the future.

Understanding the past often helps understanding of actually given situations. Likewise, prefiguring the future may also help in understanding actually given situations.

Not many thinkers have tried to clarify these issues. A number of sociologists and psychologists, generally gathered under the heading of field and dynamic thinkers, have done some preliminary work.[3] I have in mind scholars like Sorokin, Coutu, and Dewey. The concepts of invitation-character (Lewin), demand-character (Koffka), affordance (Gibson) and implicit order and its unfoldings (Bohm) are all relevant. From a philosophical viewpoint, Uschenko did some good work. Important hints can occasionally be found in Leibniz (the concept of tendency is his), Husserl, Bergson, Hartmann, Peirce and Whitehead. But the thinker that developed the most extensive categorical analyses of the future has undoubtedly been Ernst Bloch. His *The Principle of Hope* is an encyclopaedia of the categories of the future.[4]

In what follows I shall present a minimal selection of some aspects of Bloch's work.

Potentiality

Ontology, as opposed to epistemology, is what interests us. This implies that the problem of missing information (the entity is imperfectly or only partially known) must be distinguished from the problem of the entity's categorical openness. The former is an epistemological problem, one which does not concern us. The latter is a truly ontological problem.

"Categorical openness" means that the entity is only partially determined, some of its aspects are still hidden. Better: some of its determination may be latent. The difference between being hidden and being latent can be clarified as follows: hidden components are there, waiting for proper triggers to activate them. On the other hand, latent components do not exist at all in the entity's actual state.

Hidden and open components interact with each other. They form the entity's space of possibilities.

---

[3]To my knowledge, the widest discussion of field theorists is provided by the first volume of Rummel (1975–1981) entitled *The Dynamic Psychological Field*.

[4]E. Bloch's opus magnum has been translated into English as Bloch (1995).

Latent components relate to incompletely present conditions and aspects. Their incompleteness may be ascribed either to still maturing conditions or to new conditions that may subsequently arise. The constantly open space of potentialities is constructed around them.

The whole of the entity therefore comprises both tendencies and latencies, possibilities and potentialities.

On submitting the category of potentiality to closer scrutiny, a number of new categories emerge, the most important being those of horizon and front.

Everything that is, being temporal, tends towards something else. From the point of view of the space of possibilities the development is towards the entity's stable points, if any. On the other hand, from the point of view of the space of potentialities, the existence of something like stable points cannot be accepted. What is needed is the idea of something that is neither reachable nor crossable. The concept of *horizon* — as opposed to the concept of boundary, used to distinguish the system of the entity from its environment — suits our purpose precisely.

The entity's horizon sketches that fragment of the space of potentialities whose conditions are maturing. As its conditions mature, the entity's horizon moves ahead. The category of horizon is a structurally teleological category. It encompasses the whole of the entity, its *entelechy*. Without horizons, entities are dead. They may still continue along their trajectories, but no real novelty can ever appear: everything is forever fixed.

Horizons have fronts, "areas" where novelties appear. The front is the growing, maturing, changing section of the horizon.

Emergence

In many respects, Bloch's mystical theory of what I have called potentiality (real possibility, in his terms) fades into obscurity. In my understanding, it may be given a chance to develop into a more tenable theory only if it is conjugated with a theory of the levels of reality. I have presented elsewhere a broad outline of a theory of levels (see references below). In a nutshell, the main tenet of the theory foresees a categorical segmentation of entities into four main strata (material, psychological, social, ideal). The various strata are subsequently segmented into a number of sublevels, called layers. The organization of the layers of each stratum differs as we pass from one stratum to the next. A complex network of laws of dependence and independence structures the connections between strata and within strata.

In this framework, emergence intervenes in at least two different fashions:

- *Forward emergence*: the interplay of phenomena at one level of reality gives birth to categorically new (emergent) phenomena at the same level of reality.

- *Upward emergence*: the interplay of phenomena at one level of reality gives birth to categorically new (emergent) phenomena at a higher level of reality.

Reference interplay among phenomena enables me to link emergence with causation. The two above-distinguished forms of emergence can therefore be translated into two different forms of causation. This means that causation can be creative and open to novelty. The last step is to add the idea that higher levels of reality can possibly constrain lower levels of reality, what is nowadays known as downward causation.[5] This third form of causation closes the circle, reorienting forward causation into a form of feedback. All this amounts to saying that the idea of causality should be properly updated: the traditional understanding of causation as only forward causation should be enriched with (at least) upward causation and downward forms of causation.

This feedback circle of causes may therefore throw some light on at least some of the features of potentiality.

All things considered, we are falling back on the Aristotelian idea that there are different types of causes, variously interrelated and reciprocally dependent on each another. The source of the difference between the space of possibility and the space of potentiality may therefore depend on the difference between linear and non-linear forms of causality. The deplorable lack of an adequate conceptual framework, and in particular of a theory of the levels of reality, has made it almost impossible to work on themes of this kind, and forced interested people to lapse into mystical and prescientific jargons.

University of Trento
Department of Sociology and Social Research
26, Verdi street
38100 Trento
Italy
e-mail: roberto.poli@soc.unitn.it

REFERENCES

Andersen, P. B., C. Emmeche, N. O. Finnemann, P. V. Christiansen, eds. (2000). *Downward Causation: Minds, Bodies and Matter*. Aarhus: University Press.

Bloch, E. (1995). *The Principle of Hope*. 3 vols. Cambridge, MA: The MIT Press.

Leyton, M. (1992). *Symmetry, Causality, Mind*. Cambridge, MA: The MIT Press.

Leyton, M. (2001). *A Generative Theory of Shape*. Berlin: Springer.

[5] On downward causation, see Andersen *et al.* (2000).

Poli, R. (2001). The Basic Problem of the Theory of Levels of Reality. *Axiomathes* **12** (3/4), 261–283.

Poli, R. (2006). Levels of Reality and the Psychological Stratum. *Revue Internazionale de Philosophie*, 165–182.

Poli, R. (2006). First Steps in Experimental Phenomenology. In: A. Loula, R. Gudwin, J. Queiroz (eds.), *Artificial Cognition Systems*, pp. 359–387. Hersey: Idea Group Publishing.

Rummel, R. J. (1975–1981). *Understanding Conflict and War*. NY: J. Wiley and Sons.

Heinrich Wansing

CONTRADICTION AND CONTRARIETY. PRIEST ON NEGATION

ABSTRACT. Although it is not younger than other areas of non-classical logic, paraconsistent logic has received full recognition only in recent years, largely due to the work of, among others, Newton da Costa, Graham Priest, Diderik Batens, and Jerzy Perzanowski. A logical system Λ is paraconsistent if there is a set of Λ-formulas $\Delta \cup \{A\}$ such that (i) in Λ one may derive from Δ both A and its negation, and (ii) the deductive closure of Δ with respect to Λ is different from the set of all formulas. If from Δ one may derive a formula and its negation, Δ is said to be syntactically inconsistent. But is every syntactically inconsistent set of formulas contradictory? In classical logic and many non-classical logics, every syntactically inconsistent set is unsatisfiable, that is, semantically inconsistent. If contradictoriness means semantical inconsistency, there is, up to logical equivalence, only one contradiction. In paraconsistent logics, there are usually many non-equivalent formulas representing the semantically unique contradiction in a non-paraconsistent logic. So when is a formula the contradiction of another formula, and, moreover, how does the notion of contradiction relate to the notions of contrariety and negation?

1. When is a Statement the Contradictory of its Negation?

Usually, the notion "contradiction" is defined using the notion "negation". Moshé Machover (1996, p. 124), e.g., explains that "a set of two formulas $\{\alpha, \neg\alpha\}$, one of which is the negation of the other, is called a *contradictory pair*". But what if it is the notion of negation that is to be elucidated? Recently, Graham Priest (1999) suggested that theories of negation are, primarily, theories about the relationship of contradiction. This claim is contentious, but the discussion of it becomes even more complicated, if there is no consensus on the notion of contradictoriness. Whereas Priest (1999) claims that intuitionistic negation fails to form contradictory pairs, but forms contrary pairs, the present writer (Wansing 2001) holds that intuitionistic negation fails to form contrary pairs, but forms contradictory pairs. It seems that either Priest or the present writer is wrong, or that different notions of contradictoriness and contrariety are used. The disagreement concerns also logics other than intuitionistic logic. Whereas the present author holds that strong negation in Nelson's three-valued constructive logic **N3** and Nelson's four-valued constructive logic **N4** gives rise to contrary pairs, Priest's notion of contrariety does not apply to strong negation in **N3** or **N4**. The following essay is devoted first of all to clarifying these issues concerning contradictoriness and contrariety.

In: J. Malinowski and A. Pietruszczak (eds.), *Essays in Logic and Ontology* (*Poznań Studies in the Philosophy of the Sciences and the Humanities*, vol. 91), 81–93. Amsterdam/New York, NY: Rodopi, 2006.

It should be emphasized that the ensuing considerations are rather limited in the sense that we shall not discuss the syntactic categories in which negation(s) or negation-like expressions may or may not occur or ought to be represented formally. In particular, we shall not discuss the advantages or disadvantages of developing an Aristotelian or Neo-Aristotelian or Neo-Leibnizian term logic (see Englebretsen 1981; Horn 1989; Sommers 1982), in which a distinction is drawn between predicate denial and predicate term negation. Our concern is contradictoriness, propositional contrariety, and *sentential* negation in various (propositional) logics. For other negation related issues not dealt with in the following note, the reader may, for example, consult Fine (1989), Gabbay & Wansing (1999), Horn (1989), Wansing (1996), Wansing (2001) and references therein.

2. Notions of Contradictoriness

In *Categories* (13b1-3) Aristotle explains that two statements are contradictories of each other if necessarily one is true and the other is false. Against the background of *classical* logic, this characterization has many equivalents, and, notably, it also has an equivalent such that intuitionistic negation emerges as a contradictory-forming operation.

According to Priest (1999, p. 103), theories of negation are, in the first place, theories about the relationship of contradiction. Priest (1999, p. 104) then reasons as follows:

> [I]f α is any statement, let $\neg\alpha$ represent its contradictory. [...] What relationship holds between these? Traditional logic and common sense are both very clear about the most important one: we must have at least one of the pair, but not both.

Now, what does it mean to have at least one of the pair, but not both? Priest (1999, p. 104) expounds that the quoted "fact about contradictories obviously gives immediately two of the traditional laws of negation, the law of excluded middle (LEM), $\alpha \vee \neg\alpha$, and the law of non-contradiction (LNC), $\neg(\alpha \wedge \neg\alpha)$". The set $\{\alpha, \neg\alpha\}$ thus forms a contradictory pair if the LNC and the LEM are provable:

(1) $\vdash \neg(\alpha \wedge \neg\alpha)$ and $\vdash \alpha \vee \neg\alpha$.

A reformulated statement of the contradictoriness of $\{\alpha, \neg\alpha\}$ corresponding to the reading Priest attributes to traditional logic and common sense is

1. α and $\neg\alpha$ cannot both be true.
2. One of α and $\neg\alpha$ must be true.

On another reading, $\{\alpha, \neg\alpha\}$ is a contradictory pair if

1. Necessarily, if α is true, then $\neg\alpha$ is false.
2. α and $\neg\alpha$ cannot both be false.

If the Deduction Theorem is assumed (to avoid using implication), this reading gives

(2) $\alpha \vdash \neg\neg\alpha$ and $\vdash \neg(\neg\alpha \wedge \neg\neg\alpha)$.

If the contradictoriness of $\{\alpha, \neg\alpha\}$ is expressed by (1), intuitionistic negation fails to be contradictory-forming, and if the contradictoriness of $\{\alpha, \neg\alpha\}$ is expressed by (2), intuitionistic negation *does* give rise to contradictions.

As Priest (1999, p. 107) observes, in order to consider the truth conditions of negated statements, we

> need a definition of falsity. Let us define "α is false" to mean that $\neg\alpha$ is true. This is not the only plausible definition: one might also define it to mean that α is not true. However, to assume so here would be to beg too many important questions. And the present definition is one that all parties can agree upon, classical, intuitionist and paraconsistent.

Since Priest regards intuitionistic negation to be not a contradictory-forming connective but a contrary-forming one, his explanation cannot mean that α is false iff its contradictory is true. Otherwise the intuitionist as Priest conceives of him or her should not agree upon the definition of falsity. Thus, the notion of *falsity in a logic* Λ is expressed by the negation operation used in Λ (assuming that Λ comprises only one "official" negation and taking into account that this may not be a contradictory-forming connective and hence not a negation in Priest's sense). The question thus is: When is $\{\alpha, \neg\alpha\}$ a contradictory pair?

Among the various classically equivalent Priest-style formulations of Aristotle's characterization of contradictoriness, there are three pairs of conditions that may be set aside:

(a) $\alpha \vdash \neg\neg\alpha, \quad \neg\alpha \vdash \neg\alpha$
(Necessarily, if α is true, then $\neg\alpha$ is false. Necessarily, if α is false, then $\neg\alpha$ is true.)

(b) $\neg\alpha \vdash \neg\alpha, \quad \neg\neg\alpha \vdash \alpha$
(Necessarily, if $\neg\alpha$ is true, then α is false. Necessarily, if $\neg\alpha$ is false, then α is true.)

(c) $\vdash \neg(\alpha \wedge \neg\alpha), \quad \vdash \neg(\neg\alpha \wedge \neg\neg\alpha)$
(α and $\neg\alpha$ cannot both be true. α and $\neg\alpha$ cannot both be false.)

In (a) and (b) the condition $\neg\alpha \vdash \neg\alpha$ holds because of a general property of derivability. In (c), one condition is an instantiation of the other. We may ask whether or not the remaining classically valid and classically equivalent pairs of conditions, exhibiting in addition to $\neg$ at most $\wedge$ and $\vee$, hold in various non-classical propositional logics. The logics listed in Table 1 are intuitionistic propositional logic **IPL**, the three-valued constructive logic **N3**, the four-valued constructive paraconsistent logic **N4** (see Almukdad & Nelson 1984;

Dunn 2000; Wansing 2001) and Johansson's paraconsistent minimal logic **MIN**.[1]

Table 1. Notions of contradictoriness.

	$\{\alpha, \neg\alpha\}$ *is contradictory iff*	**CPL**	**IPL**	**N3**	**N4**	**MIN**
(1)	$\vdash \neg(\alpha \wedge \neg\alpha), \vdash \alpha \vee \neg\alpha$	√	–	–	–	–
	$\vdash \neg(\alpha \wedge \neg\alpha), \vdash \neg\alpha \vee \neg\neg\alpha$	√	–	–	–	–
	$\vdash \neg(\alpha \wedge \neg\alpha), \neg\neg\alpha \vdash \alpha$	√	–	–	–	–
	$\alpha \vdash \neg\neg\alpha, \vdash \alpha \vee \neg\alpha$	√	–	–	–	–
	$\alpha \vdash \neg\neg\alpha, \vdash \neg\alpha \vee \neg\neg\alpha$	√	–	–	–	–
(2)	$\alpha \vdash \neg\neg\alpha, \vdash \neg(\neg\alpha \wedge \neg\neg\alpha)$	√	√	–	–	–
	$\alpha \vdash \neg\neg\alpha, \neg\neg\alpha \vdash \alpha$	√	–	√	√	–
	$\vdash \neg(\neg\alpha \wedge \neg\neg\alpha), \alpha \vee \neg\alpha$	√	–	–	–	–
	$\vdash \neg(\neg\alpha \wedge \neg\neg\alpha), \vdash \neg\alpha \vee \neg\neg\alpha$	√	–	–	–	–
	$\vdash \neg(\neg\alpha \wedge \neg\neg\alpha), \neg\neg\alpha \vdash \alpha$	√	–	–	–	–
	$\neg\neg\alpha \vdash \alpha, \alpha \vee \neg\alpha$	√	–	–	–	–
	$\neg\neg\alpha \vdash \alpha, \vdash \neg\alpha \vee \neg\neg\alpha$	√	–	–	–	–
	$\neg\neg\alpha \vdash \alpha, \vdash \neg(\neg\alpha \wedge \neg\neg\alpha)$	√	–	–	–	–

In view of the comparison made in Table 1, we may conclude that Priest and the present author assume different notions of contradictoriness. Whereas Priest commits himself to conditions (1), the present author is, on the present analysis, committed to conditions (2). If attention is restricted to the non-classical logics listed in Table 1, the present author's conception of contradictoriness is wider than Priest's. Is one of the two pairs of conditions correct (for systematic reasons not necessarily in accordance with tradition or common sense)? Is one of the two pairs of conditions to be preferred over the other? One could, for example, note that on all except one of the classically equivalent accounts listed in Table 1, intuitionistic negation fails to be contradictory-forming. However, if it were claimed that conditions (1) are more appropriate than conditions (2), because according to (2), intuitionistic negation is a contradictory-forming operation, this would be a position biased towards the "traditional" viewpoint expressed by (1). One may, of course, ask whether or not intuitionistic logicians themselves think of intuitionistic negation as contradictory-forming. For Heyting, the founder of intuitionistic logic, "contradiction must be taken as a

[1] I assume the reader's familiarity with minimal and intuitionistic logic. Since Nelson's systems are, perhaps, less well-known, a semantical presentation is offered in the Appendix. There are other important and interesting systems of non-classical logic, but for the present purposes considering the logics listed in Table 1 is enough. Note that sometimes Johansson's logic is said to be only partially paraconsistent (or to be partially explosive), because in this logic, from a set $\{\alpha \wedge \neg\alpha\}$ arbitrary negations $\neg\beta$ may be deduced and not completely arbitrary formulas β.

primitive notion" (Heyting 1971, p. 102). Moreover, there are passages showing that Heyting conceives of intuitionistic negation as contradictory-forming, for example, when he explains that "[i]f $a = b$ is contradictory (that means: if the supposition that $a = b$ leads to a contradiction)" (Heyting 1971, p. 17). In other words, for Heyting, a statement is contradictory, only if it is the intuitionistic negation of another statement, since "$\neg p$ can be asserted if and only if we possess a construction which from the supposition that a construction p were carried out, leads to a contradiction". Also Brouwer, the founder of intuitionistic mathematics, thought of intuitionistic negation as contradictory-forming. In Brouwer (1954), Brouwer considers possible values of certain functions $\psi(x)$ originating from some real number $\rho = k_1, k_2, k_3 \ldots$ and an infinite sequence of functions satisfying a certain condition. The assertion that ρ is rational is abbreviated by "α". The following quotation clearly reveals that, for Brouwer, intuitionistic negation gives rise to contradictions:

> Hence in the second case it is neither possible that α should some time turn out to be true. This means that in the second case α is contradictory (Brouwer 1954, p. 205).

3. Notions of Contrariety

It ought to be emphasized that the notion of contrariety did not emerge in the context of propositional or predicate logics containing a unary negation operation, but in Aristotelian term logic. In term logic, various notions of contrariety have been distinguished, see Horn (1989, p. 36–40). "Logically contrary" terms, for example, give rise to contradictory pairs of statements. Englebretsen (1974, p. 614) explains that the contrariety between two statements derives from the contrariety between their predicates. But, also, if we take seriously that propositional contrariety is a derived notion, on no reasonable conception of contrariety, intuitionistic negation is both a contrary-forming and not a contrary-forming operation. Under which conditions is $\{\alpha, \neg\alpha\}$ a contrary pair? Concerning the intuitionistic account of negation Priest explains that

> $\neg\alpha$ is true (= assertable) just if there is a proof that there is no proof that α. This is obviously a *contrary* of α.[2] (Priest 1999, p. 105).

This explanation presuppose a notion of contrariety, and indeed Priest writes the following:

> If we have two contraries, e.g., "Socrates was black" and "Socrates was white", it is necessarily false that Socrates was black $\wedge$ Socrates was white; but it is not necessarily false that Socrates was black $\vee$ Socrates was white (Priest 1999, p. 104).

[2] Moreover, Priest refers to the faithful embedding of **IPL** into the modal logic **S4** and emphasizes that in terms of this embedding, "$\neg\alpha$ is intuitionistically true iff the [classical] negation of α is *provable*" (Priest 1999, Footnote 8).

In *De Interpretatione* (24b7-10), for example, Aristotle states that a contrary pair is a pair of statements that cannot both be true. The elements of a contrary pair may, however, both fail to be true. If we follow the characterizations of contradictoriness in Table 1, we obtain the characterizations of propositional contrariety in Table 2.

Table 2. Notions of contrariety.

	$\{\alpha, \neg\alpha\}$ *is contrary iff*	**CPL**	**IPL**	**N3**	**N4**	**MIN**
(1*)	$\vdash \neg(\alpha \wedge \neg\alpha), \nvdash \alpha \vee \neg\alpha$	–	√	–	–	–
	$\vdash \neg(\alpha \wedge \neg\alpha), \nvdash \neg\alpha \vee \neg\neg\alpha$	–	√	–	–	–
	$\vdash \neg(\alpha \wedge \neg\alpha), \neg\neg\alpha \nvdash \alpha$	–	√	–	–	–
	$\alpha \vdash \neg\neg\alpha, \nvdash \alpha \vee \neg\alpha$	–	√	√	√	–
	$\alpha \vdash \neg\neg\alpha, \nvdash \neg\alpha \vee \neg\neg\alpha$	–	√	√	√	–
(2*)	$\alpha \vdash \neg\neg\alpha, \nvdash \neg(\neg\alpha \wedge \neg\neg\alpha)$	–	–	√	√	–
	$\alpha \vdash \neg\neg\alpha, \neg\neg\alpha \nvdash \alpha$	–	√	–	–	–

Table 2 clarifies the disagreement between Priest and the present author concerning the notion of contrariety. Whereas Priest commits himself to conditions (1*), the present author is, on the present analysis, committed to conditions (2*). Can the commitment to conditions (2*) be justified, in particular the commitment to strong negation in **N3** and **N4** as contrary-forming? To do this, we shall take up Englebretsen's observation that propositional contrariety is a derived notion. In Aristotelian term logic, it is assumed that every statement consists of exactly one subject and exactly one predicate. The subject and the predicate are possibly complex terms. In the schematic statement "*S* is *P*", "*S*" is the subject, "is *P*" is the predicate, and "*P*" is the predicate term. If $\{P, Q\}$ is a pair of contrary terms, in both "*S* is *P*" and "*S* is *Q*" what is denoted by the predicate term is affirmed of what is denoted by the subject. The sentence "John is happy", for example, affirms of John that he is happy and the sentence "John is unhappy" affirms of John that he is unhappy. The transition from "*S* is *P*" to "*S* is *Q*", or vice versa, is *not* a transition from a positive statement to its sentential negation, even if there is some systematic device of obtaining from a given term "*P*" a contrary term "*Q*", e.g., if "*Q*" has the shape "*P**". In the Aristotelian tradition, there simply is no notion of propositional contrariety.[3]

Another issue, however, is whether the transition from "*S* is *P*" to "*S* is *P**" can appropriately be represented by translating one statement as the sentential

[3]We shall not here go into Horn's (1989, p. 39) distinction between contrariety simpliciter, immediate (strong or logical) contrariety, and mediate (weak or non-logical) contrariety. Neither shall we deal with complications arising form considering categorially mistaken sentences. Note, however, that logical contraries are taken to be unique, whereas a term may have more than one non-logical contrary.

negation of the other, using a suitable unary negation connective. In the natural language example considered, a simple non-compound term was subjected to predicate term negation resulting in the contrary pair {"John is happy", "John is unhappy"}. A strategy for deriving a notion of propositional contrariety that not merely postulates that α and $\neg\alpha$ cannot both be true, but can both fail to be true if $\{\alpha, \neg\alpha\}$ is a contrary pair, consists in considering complex, Boolean predicate terms. In natural language the counterparts of the Boolean operations not only occur as unary or binary sentential connectives, but exhibit what in Categorial Grammar is called variable polymorphism. In the surface syntax, the expression "or", for example, combines not only pairs of sentences, but also pairs of names, intransitive verbs, transitive verbs, adverbs etc. If the expression "or" is variably polymorphic, what would be the predicate term negation of "John is happy or rich"? It seems that this is '1John is unhappy and poor". Similarly, the predicate term negation of "John is happy and rich" appears to be "John is unhappy or poor". If the predicate term negation of compound Boolean terms is to be represented by a sentential negation $\neg$, the verification conditions of $\neg\alpha$ should be the falsification conditions of α.[4] If the negated predicate term "unhappy" in "John is unhappy" is subjected to predicate term negation, the resulting sentence should have the same truth and falsity conditions as "John is happy". If we now look at the semantics of strong negation in **N3** and **N4**, it turns out that the verification conditions of the (strong) negation of a compound formula α are exactly the falsification conditions of α. With the extra semantical requirement that if $\neg$ is contrary-forming, the verification conditions of $\neg\alpha$ are the falsification conditions of α, it is clear that intuitionistic negation fails to be a contrary-forming negation, because $\neg\neg\alpha \vdash \alpha$ fails.

4. Another View of Contrariety and Contradictoriness

So far we have been looking at provable formulas, but logic is the study of valid *inferences*. If we assume that "happy" and "unhappy" are contrary terms, and that "even" and "odd" (applied to natural numbers) are terms such that {"n is odd", "n is even"} is a contradictory pair, then it appears that contraposition as

[4] A bit more problematic is the question whether there are conditional predicate terms. Is "John is happy if rich" well-formed? In the present author's mother tongue, the corresponding expression "Johannes ist glücklich falls reich" sounds quite ungrammatical. Anyway, note that I do not intend to read the ill-formed "John is happy if rich" as "John is happy, if he is rich". The term logician will read the later sentence as something like

(3) Every case of John being rich is a case of John being happy.

The predicate term negation of (3) is "Every case of John being rich is a case of John being unhappy". If "John is happy if rich" were grammatical, its predicate term negation would seem to be expressible as "John is rich but unhappy".

a rule ($\alpha \vdash \beta$ implies $\neg\beta \vdash \neg\alpha$) marks a difference between contrariety and contradictoriness. If

If n is even, then $f(n)$ is odd

is true in a model, then so is

If $f(n)$ is even, then n is odd.

However, if

If John is happy, then John's mother is happy

is true, the contrapositive

If John's mother is unhappy, then John is unhappy

need not be true. Intuitionistic logic satisfies contraposition as a rule, but **N3** and **N4** do not. Therefore, according to this criterion, intuitionistic negation is contradictory-forming. If negation in Nelson's logics is taken to be a genuine negation, it is contrary-forming. The criterion allows for contradictoriness in paraconsistent logics: contraposition as a rule holds, for example, in the paraconsistent logic **MIN**.

5. Negation and the LNC

Priest (1999, p. 108 ff.) defends a dialetheic conception of negation. According to him, negation not only satisfies the LNC and other laws, but also for some "αs we have both α and $\neg\alpha$" (Priest 1999, p. 110). In other words, taking up our earlier interpretation, there are statements α such that $\vdash \alpha \wedge \neg\alpha$. Priest perceives no conflict with the LNC and offers the following argument:

> [I]f some contradictions are true, we may well have both $\beta \wedge \neg\beta$ and $\neg(\beta \wedge \neg\beta)$. Hence, the fact that some contradictions are true does not, of itself, refute the LNC (at least in the form in question here) (Priest 1999, p. 104).

On one possible reading, the argument is meant to show that the statement that there are formulas α such that both α and $\neg\alpha$ are provable does not imply that the LNC fails. This reading is supported by Priest's explanation that the possible truth of α and $\neg\alpha$ for some α "does not imply that the LNC fails, as we have already seen" (Priest 1999, p. 110). However, a weaker reading seems to be intended, namely that if it is assumed that some contradictions are provable, then it is still *possible* that for all formulas β, $\vdash \neg(\beta \wedge \neg\beta)$. In this sense, the LNC is compatible with the existence of provable contradictions. Semantically, it has to be required that formulas may both be true and false in (a state from the non-empty set of states of) a model. However, the quoted argument does not show that the assumption that some contradictions are provable does not imply that the LNC fails. Showing that this implication is not true, requires showing that (i) there are provable contradictions and that (ii) the LNC holds. Even if one grants Priest that the semantical paradoxes show that there are provable contradictions, the argument in the quoted passage does not prove that $\vdash \neg(\beta \wedge \neg\beta)$. The LNC is *assumed*, since $\neg$ is taken to be a negation in Priest's sense.

Note that negation in Priest's sense satisfies not only the LEM and the LNC, but also the two double negation laws

(4) $\alpha \vdash \neg\neg\alpha, \quad \neg\neg\alpha \vdash \alpha$

and the familiar De Morgan laws. The LNC and the LEM thus become provably interderivable. The LNC or rather its provable equivalent, the LEM, ensures that every statement is either true or false, and there is no conflict with having a pair of valid and hence, assuming completeness, provable contradictories. In **IPL**, the LNC and the LEM fall apart (the LNC holds, the LEM does not), and the assumption that for some β, $\vdash \beta \wedge \neg\beta$ is incompatible with the LNC. In **N4**, the LNC and the LEM fail to hold, but the two laws are provably interderivable.

6. Negation and Contradictoriness

Let us briefly come back to the idea of characterizing negation in terms of contradictoriness. It has often been observed and it is confirmed by the present paper that in studies on the notion of negation, disagreement abounds. Of course, one may take different attitudes toward characterizing the notion of negation. If attention is limited to a study of sentential forms or representations of negation, one may wonder whether there are some properties shared by all or almost all recognized unary negation connectives. This approach aims at a broad notion of negation and may be justified by a broad range of phenomena of semantic opposition to be found in natural languages. Priest (1999, p. 102) doubts that this is a promising approach. However, Priest's own approach leads to a rather narrow conception. The unary connectives, which constructive logicians call intuitionistic negation and strong negation, fail to be contradictory-forming in Priest's sense and hence fail to be negations at all in Priest's sense, at least in the first place. I agree with Priest that this is not a terminological controversy. It is established wisdom that there exists more than one form of semantic opposition. If phenomena of semantic opposition are negation phenomena, then certainly not only contradictoriness, but also contrariety is a negation phenomenon.

7. Appendix

The relational possible worlds semantics for Nelson's constructive logics was developed by Routley (1974) and Thomason (1969). In this semantics, verification and falsification are taken as equally important basic semantical notions. The possible worlds may be understood as information states or information pieces that are partially ordered by a relation of "possible expansion of information states". Information states differ from the possible worlds in the relational semantics of modal logics based on classical non-modal logic in that an information state s may verify (support the truth of) an atomic formula p ($s \models^+ p$) or falsify (support the falsity of) p ($s \models^- p$) or, in the case of the semantics for **N4** even both support the truth and the falsity of p.

A *Nelson model* is a structure $\langle \mathbf{I}, \sqsubseteq, v^+, v^- \rangle$, where $\langle \mathbf{I}, \sqsubseteq \rangle$ is a partially ordered set and both v^+ and v^- are valuation functions assigning to every propositional variable p a subset of $\mathbf{I}$. Intuitively, v^+ maps atomic formulas to the information states at which they are verified, whereas v^- maps atomic formulas to the information states at which they are falsified. Moreover, it is required that for every sentence letter p and every $t, u \in \mathbf{I}$:

$$\begin{array}{ll} (\text{persistence}^+) & \text{if } t \sqsubseteq u, \text{ then } t \in v^+(p) \text{ implies } u \in v^+(p); \\ (\text{persistence}^-) & \text{if } t \sqsubseteq u, \text{ then } t \in v^-(p) \text{ implies } u \in v^-(p). \end{array}$$

Let $\mathfrak{M} = \langle \mathbf{I}, \sqsubseteq, v^+, v^- \rangle$ be a Nelson model, $t \in \mathbf{I}$ and α a formula in the language with strong negation $\neg$, intuitionistic implication $\supset$, conjunction $\wedge$, and disjunction $\vee$ over the denumerable set *Atom* of sentence letters. The notions $\mathfrak{M}, t \models^+ \alpha$ (α is verified at t in $\mathfrak{M}$) and $\mathfrak{M}, t \models^- \alpha$ (α is falsified at t in $\mathfrak{M}$) are inductively defined as follows:

$$\begin{array}{lll} \mathfrak{M}, t \models^+ p & \text{iff} & t \in v^+(p), \quad p \in Atom \\ \mathfrak{M}, t \models^- p & \text{iff} & t \in v^-(p), \quad p \in Atom \\[1ex] \mathfrak{M}, t \models^+ \beta \wedge \gamma & \text{iff} & \mathfrak{M}, t \models^+ \beta \text{ and } \mathfrak{M}, t \models^+ \gamma \\ \mathfrak{M}, t \models^- \beta \wedge \gamma & \text{iff} & \mathfrak{M}, t \models^- \beta \text{ or } \mathfrak{M}, t \models^- \gamma \\[1ex] \mathfrak{M}, t \models^+ \beta \vee \gamma & \text{iff} & \mathfrak{M}, t \models^+ \beta \text{ or } \mathfrak{M}, t \models^+ \gamma \\ \mathfrak{M}, t \models^- \beta \vee \gamma & \text{iff} & \mathfrak{M}, t \models^- \beta \text{ and } \mathfrak{M}, t \models^- \gamma \\[1ex] \mathfrak{M}, t \models^+ \beta \supset \gamma & \text{iff} & (\forall u \in \mathbf{I}) \text{ if } t \sqsubseteq u, \text{ then } \mathfrak{M}, u \models^+ \beta \text{ implies } \mathfrak{M}, u \models^+ \gamma \\ \mathfrak{M}, t \models^- \beta \supset \gamma & \text{iff} & \mathfrak{M}, t \models^+ \beta \text{ and } \mathfrak{M}, t \models^- \gamma \\[1ex] \mathfrak{M}, t \models^+ \neg\beta & \text{iff} & \mathfrak{M}, t \models^- \beta \\ \mathfrak{M}, t \models^- \neg\beta & \text{iff} & \mathfrak{M}, t \models^+ \beta \end{array}$$

With this definition, verification and falsification of arbitrary formulas are persistent with respect to $\sqsubseteq$. Semantic consequence is defined as follows: $\Gamma \models_{\mathbf{N4}} \alpha$ iff for every Nelson model $\mathfrak{M}$, $= \langle \mathbf{I}, \sqsubseteq, v^+, v^- \rangle$ and every $t \in \mathbf{I}$, if $\mathfrak{M}, t \models^+ \beta$ for every $\beta \in \Gamma$, then $\mathcal{M}, t \models^+ \alpha$. Nelson's propositional logic **N4** is the theory of the class of all Nelson models in the given language. **N4** conservatively extends positive propositional logic, the positive part of intuitionistic logic. An axiomatic presentation of **N4** can be obtained by adding to an axiomatization of positive logic the following axiom schemata:

$$A1 \;\; \neg\neg\alpha \equiv \alpha$$
$$A2 \;\; \neg(\alpha \wedge \beta) \equiv (\neg\alpha \vee \neg\beta)$$
$$A3 \;\; \neg(\alpha \vee \beta) \equiv (\neg\alpha \wedge \neg\beta)$$
$$A4 \;\; \neg(\alpha \supset \beta) \equiv (\alpha \wedge \neg\beta),$$

where $\alpha \equiv \beta$ is defined as $(\alpha \supset \beta) \wedge (\beta \supset \alpha)$.

As mentioned already, contraposition as a rule does not hold in **N4**.[5] Also, provable equivalence fails to be a congruence relation on the set of formulas. If formulas α and β are defined as being *strongly equivalent* iff both α, β and their strong negations $\neg\alpha$, $\neg\beta$ are provably interderivable, then it can be shown that strong equivalence is a congruence relation in **N4**, that is, there is a replacement theorem for strongly equivalent formulas. In view of treating positive and negative information on a par, this is welcome observation. The logic **N4** is a system of paraconsistent logic, because not for every β, $\{\alpha, \neg\alpha\} \vdash_{\mathbf{N4}} \beta$. It can easily be seen that **N4** is a system of four-valued logic.[6] Every pair (v^+, v^-) of valuations induces a valuation v: $\mathbf{I} \times Atom \longrightarrow \{1, 0, \emptyset, \{1, 0\}\}$ by defining:

$$\begin{array}{ll} v(t,p) = 1 & \text{iff } (t \in v^+(p) \text{ and } t \notin v^-(p)) \\ v(t,p) = 0 & \text{iff } (t \in v^-(p) \text{ and } t \notin v^+(p)) \\ v(t,p) = \{1,0\} & \text{iff } (t \in v^+(p) \text{ and } t \in v^-(p)) \\ v(t,p) = \emptyset & \text{iff } (t \notin v^+(p) \text{ and } t \notin v^-(p)) \end{array}$$

The model $\mathfrak{M} = \langle \mathbf{I}, \sqsubseteq, v^+, v^- \rangle$ and the induced model $\mathfrak{M}' = \langle \mathbf{I}, \sqsubseteq, v \rangle$ validate the same formulas of the language under consideration, if for every $p \in Atom$, we define

$$\mathfrak{M}', t \models^+ p \text{ iff } v(t,p) = 1;$$
$$\mathfrak{M}', t \models^- p \text{ iff } v(t,p) = 0.$$

The non-paraconsistent, three-valued logic **N3** is the theory of the class of all Nelson models $\langle \mathbf{I}, \sqsubseteq, v^+, v^- \rangle$, where for every atom p, $v^+(p) \cap v^-(p)$ is empty. **N3** can be axiomatized by adding to an axiomatization of **N4** the *ex contradictione* schema $\neg\alpha \supset (\alpha \supset \beta)$. Strong negation in Nelson's logics **N3** and **N4** is also referred to as *constructive* negation, since in both systems negation satisfies

$$\text{(constructible falsity)} \quad \vdash \neg(\alpha \wedge \beta) \text{ iff } (\vdash \neg\alpha \text{ or } \vdash \neg\beta).$$

---

[5]Contraposible strong negation is studied in (Nelson 1959).

[6]A comprehensive study of four-valued logic including Nelson's **N4** can be found in (Dunn 1997); see also (Belnap 1977).

In **N3**, the contradictory forming intuitionistic negation $\neg_h$ can be defined using the primitive, contrary forming strong negation $\neg$:

(5) $\neg_h\alpha$ iff $\alpha \supset \neg\alpha$.

The logic **N3** provides a natural example of a logical system with two kinds of negation suitable for representing both predicate denial ($\neg_h$) and predicate term negation ($\neg$) as unary connectives.[7]

Acknowledgement. I would like to thank Graham Priest for critical remarks on an earlier version of this note.

Dresden University of Technology
Institute of Philosophy
01062 Dresden
Germany
e-mail: Heinrich.Wansing@mailbox.tu-dresden.de

REFERENCES

Almukdad, A. and D. Nelson (1984). Constructible Falsity and Inexact Predicates. *Journal of Symbolic Logic* **49**, 231–233.

Belnap, N. D. (1977). A Useful Four-Valued Logic. In: J. M. Dunn and G. Epstein (eds.), *Modern Uses of Multiple-Valued Logic*, pp. 8–37. Dordrecht: Reidel.

Brouwer, L. E. J. (1954). An Example of Contradictority in Classical Theory of Functions. *Indagationes Mathematicæ* **16**, 204–205.

Dunn, J. M. (2000). Partiality and its Dual. *Studia Logica* **66**, 5–40.

Englebretsen, G. (1974). A Note on Contrariety. *Notre Dame Journal of Formal Logic* **15**, 613–614.

Englebretsen, G. (1981). *Logical Negation*. Assen: Van Gorcum.

Fine, K. (1989). The Justification of Negation as Failure. In: J. E. Fenstad *et al.* (eds.), *Logic, Methodology and Philosophy of Science VIII*, pp. 263–301. Amsterdam: Elsevier.

Gabbay, D. and H. Wansing, eds. (1999). *What is Negation?* Dordrecht: Kluwer Academic Publishers.

Heyting, A. (1971). *Intuitionism: An Introduction*. 3rd revised edition. Amsterdam: North-Holland.

Horn, L. (1989). *A Natural History of Negation*. Chicago: Chicago UP.

[7] Other application areas in which the need for logics with more than one kind of negation arises include database theory, logic programming, and nonmonotonic reasoning (see, for instance, Wagner 1991, 1994; and Wansing 1995).

Johansson, I. (1936). Der Minimalkalkül, ein reduzierter intuitionistischer Formalismus. *Composito Mathematicæ* **4**, 119–136.

Machover, M. (1996). *Set Theory, Logic and Their Limitations*. Cambridge: Cambridge UP.

Nelson, D. (1949). Constructible Falsity. *Journal of Symbolic Logic* **14**, 16–26.

Nelson, D. (1959). Negation and Separation of Concepts in Constructive Systems. In: A. Heyting (ed.), *Constructivity in Mathematics*, pp. 208–225. Amsterdam: North-Holland.

Priest, G. (1999). What Not? A Defence of Dialethic Theory of Negation. In: D. Gabbay and H. Wansing (eds.), *What is Negation?*, pp. 101–120. Dordrecht: Kluwer Academic Publishers.

Routley, R. (1974). Semantical Analyses of Propositional Systems of Fitch and Nelson. *Studia Logica* **33**, 283–298.

Sommers, F. (1982). *The Logic of Natural Language*. Oxford: Clarendon Press.

Thomason, R. (1969). A Semantical Study of Constructive Falsity. *Zeitschrift für Mathematische Logik und Grundlagen der Mathematik* **15**, 247–257.

Wagner, G. (1991). Logic Programming with Strong Negation and Inexact Predicates. *Journal of Logic and Computation* **1**, 835–859.

Wagner, G. (1994). *Vivid Logic: Knowledge-Based Reasoning with Two Kinds of Negation*. Lecture Notes in AI 764. Berlin: Springer-Verlag.

Wansing, H. (1995). Semantics-based Nonmonotonic Inference. *Notre Dame Journal of Formal Logic* **36**, 44–54.

Wansing, H., ed. (1996). *Negation: A Notion in Focus*. Berlin: de Gruyter.

Wansing, H. (2001). Negation. In: L. Goble (ed.), *The Blackwell Guide to Philosophical Logic*, pp. 415-436. Cambridge/MA: Basil Blackwell Publishers.

Krister Segerberg

MOORE PROBLEMS IN FULL DYNAMIC DOXASTIC LOGIC

ABSTRACT. Dynamic doxastic logic (DDL) is the modal logic of belief change. In basic DDL a modal operator $[*\varphi]$ carries the informal meaning "after the agent has revised his beliefs by φ" or "after the agent has accepted the information that φ"; it is assumed that the arguments of the star operator $*$ are pure Boolean formulæ. That assumption is discarded in full DDL where any pure doxastic formula may be an argument. As noted by other authors, a straight-forward extension of the theory from basic DDL to full DDL invites problems of the kind first discussed by G. E. Moore. In this paper it is argued that a way to escape those problems is to redefine revision in a way that seems appropriate for this semantically richer context. The paper deals only with the one-agent case, but the approach can be extended to the case of multiple agents.

1. Moore Problems

The phenomenon called Moore's Paradox has attracted considerable attention. For example, Wittgenstein is reported to have considered this Moore's best achievement: "the only work of Moore's that greatly impressed him was his discovery of the peculiar kind of nonsense involved in such a sentence as, e.g. 'It is raining but I don't believe it' " (Malcolm 1958). After Hintikka, in doxastic logic such sentences are usually formalised on the format

(1) $\varphi \wedge \neg \mathbf{B}\varphi$,

although the formalisation

(2) $\varphi \wedge \mathbf{B}\neg\varphi$

also comes to mind. What is "peculiar" is that, while (1) and (2) may describe perfectly plausible situations, they cannot be believed by the agent associated with the operator **B** if the latter is "rational" — for example, is a human being under normal conditions. As Hintikka puts it, they are "necessarily *unbelievable* by the speaker" (Hintikka 1962, p. 67).

Let us say that an agent has a *Moore problem* (*of rank* 0) if $\mathbf{B}(\varphi \wedge \neg \mathbf{B}\varphi)$ or $\mathbf{B}(\varphi \wedge \mathbf{B}\neg\varphi)$ is true (in a certain situation and with respect to his beliefs). In the former case, we say (with a terminology usually applied to accents) that the problem is *acute*, in the latter that it is *grave*. We shall also be concerned with agents who, whether or not they have a Moore problem, believe that they have one, or believe that they believe that they have one, and so on. For this reason, let

In: J. Malinowski and A. Pietruszczak (eds.), *Essays in Logic and Ontology* (*Poznań Studies in the Philosophy of the Sciences and the Humanities*, vol. 91), 95–110. Amsterdam/New York, NY: Rodopi, 2006.

us say, more generally, that the agent has a *Moore problem of rank n*, where n is a nonnegative integer, if, for some formula φ, either $\mathbf{B}^n(\varphi \wedge \neg \mathbf{B}\, \varphi)$ or $\mathbf{B}^n(\varphi \wedge \mathbf{B}\, \neg\varphi)$ is true. Again, the problem is *acute* in the former case, *grave* in the latter. It is probably impossible to compile a complete list of all the ways in which a doxastic agent may be incoherent or exhibit some degree of inconsistency, but certainly an agent with a Moore problem of any rank is not perfect. Doxastically ambitious agents will stay clear of Moore problems as far as possible!

2. Full Dynamic Doxastic Logic

The principal operator of simple doxastic logic is **B** ("the agent believes that"); there may also be an operator **K** ("the agent knows that" or, perhaps better, "the agent «knows» that"). In dynamic doxastic logic (DDL) there are, in addition, change operators, whose task it is to formalise operations such as revision and contraction. In basic DDL, the object language is quite restricted; for example, there is no iteration of either doxastic operators or change operators. By contrast, in full DDL there is no such restriction. Basic DDL is a suitable tool for work in what is popularly known as the theory of belief revision. However, if one is interested in higher-order belief within a DDL framework, then it becomes necessary to proceed to full DDL.

The present paper may be seen as a sequel to Segerberg (1999), in which a certain system of full DDL was defined. The object language may be described as a language of modal propositional logic, where the box operators are **B**, **K**, and, for all formulæ φ, $[*\,\varphi]$. Here $*$, let us call it the *star operator*, is a change operator. The term $*\,\varphi$ is understood as standing for a certain doxastic action or event. The operator $[\ldots]$ is well-known from dynamic logic; read "after $*\,\varphi$, it is the case that" for $[*\,\varphi]$. The duals of **B** and **K** are written **b** and **k**, respectively; the dual of $[*\,\varphi]$ is $\langle *\,\varphi\rangle$. The formal semantics laid down in Segerberg (1999) will not be needed here, but in order to make the paper reasonably self-contained, we repeat the axiom system. The rules of derivation included *modus ponens* and necessitation for all the box operators. In addition there was a rule of congruence:

> if $\varphi \equiv \psi$ is derivable, then, for all formulæ θ, you may derive $[*\,\varphi]\theta \equiv [*\,\psi]\theta$.

There were the following axioms:

$(*\,2)$	$[*\,\varphi]\,\mathbf{B}\,\varphi$,
$(*\,3)$	$[*\,\top]\,\mathbf{B}\chi \supset \mathbf{B}\chi$,
$(*\,4)$	$\mathbf{b}\top \supset (\mathbf{B}\chi \supset [*\,\top]\,\mathbf{B}\chi)$,
$(*\,5)$	$[*\,\varphi]\,\mathbf{B}\perp \supset \mathbf{K}\neg\varphi$,

$(*\,6)$	$\mathbf{K}(\varphi \equiv \psi) \supset ([*\,\varphi]\,\mathbf{B}\,\chi \equiv [*\,\psi]\,\mathbf{B}\,\chi)$,
$(*\,7)$	$[*(\varphi \wedge \psi)]\,\mathbf{B}\,\chi \supset [*\,\varphi]\,\mathbf{B}(\psi \supset \chi)$,
$(*\,8)$	$\langle *\,\varphi\rangle\,\mathbf{b}\,\psi \supset ([*\,\varphi]\,\mathbf{B}(\psi \supset \chi) \supset [*(\varphi \wedge \psi)]\,\mathbf{B}\,\chi)$,
$(*\,\mathrm{D})$	$[*\,\varphi]\chi \supset \langle *\,\varphi\rangle\chi$,
$(*\,\mathrm{F})$	$\langle *\,\varphi\rangle\chi \supset [*\,\varphi]\chi$,
$(\mathbf{K\,B})$	$\mathbf{K}\,\varphi \supset \mathbf{B}\,\varphi$,
$(\mathbf{K}\,*_1)$	$\mathbf{k}\,\varphi \supset (\mathbf{K}\,\chi \equiv [*\,\varphi]\,\mathbf{K}\,\chi)$,
$(\mathbf{K}\,*_2)$	$\mathbf{K}\,\neg\varphi \supset [*\,\varphi]\,\mathbf{K}\,\bot$.

van Linder, van der Hoek and Meyer were the first to observe that Moore sentences cause problems for theories of belief revision:

> [C]hanges of belief concerning doxastic or epistemic formulæ are not well understood. It is for instance not at all clear what it means to revise the beliefs of some agent i with the formula $\varphi \wedge \neg\,\mathbf{B}_i\,\varphi$: does or doesn't the agent believe $\varphi \wedge \neg\,\mathbf{B}_i\,\varphi$ after its beliefs are revised with this formula? (van Linder *et al.* 1995, p. 111; notation slightly modified).

This observation made them concentrate on logics at the level of what we call basic DDL. The challenge to extend the theory to full DDL was instead taken up by Lindström and Rabinowicz. They recognised in Lindström & Rabinowicz (1997) that there are two conditions, both provided by our calculus, whose combined power brings us into contact with Moore's Paradox:

(S)	$[*\,\varphi]\,\mathbf{B}\,\varphi$,
(P)	$(\mathbf{b}\,\varphi \wedge \mathbf{B}\,\chi) \supset [*\,\varphi]\,\mathbf{B}\,\chi$.

In the literature, (S) is usually referred to as the Success Postulate and (P) as the Preservation Postulate. (The former is simply our axiom schema $(*\,2)$, while the latter is derivable with the help of $(*\,*\,3, 4, 8)$, congruence, and modal logic). Now consider an agent who revises his beliefs in accordance with a normal theory satisfying Preservation and Success and for whom $*\,\varphi$ means "revision by φ". Such an agent risks running into a Moore problem if he ever holds second-order beliefs about something that is compatible with what he believes. Suppose, for example, that $\mathbf{b}\,\varphi$ and $\mathbf{B}\,\neg\,\mathbf{B}\,\varphi$ are true in a certain situation:

1.	$\mathbf{b}\,\varphi$	premise
2.	$\mathbf{B}\,\neg\,\mathbf{B}\,\varphi$	premise
3.	$[*\,\varphi]\,\mathbf{B}\,\neg\,\mathbf{B}\,\varphi$	1, 2, (P)
4.	$[*\,\varphi]\,\mathbf{B}\,\varphi$	(S)
5.	$[*\,\varphi]\,\mathbf{B}(\varphi \wedge \neg\,\mathbf{B}\,\varphi)$	ML: 3, 4

("ML" means "by modal logic"; remaining notation should be self-explanatory.) For some reason both the Dutch and the Swedish authors were concerned only with acute Moore problems. But grave problems are equally close at hand. Suppose, for example, that $\mathbf{b}\,\varphi$ and $\mathbf{B}\,\mathbf{B}\,\neg\varphi$ are true in a certain situation:

1. $\mathbf{b}\,\varphi$	premise
2. $\mathbf{B}\,\mathbf{B}\,\neg\varphi$	premise
3. $[*\varphi]\,\mathbf{B}\,\mathbf{B}\,\neg\varphi$	1, 2, (P)
4. $[*\varphi]\,\mathbf{B}\,\varphi$	(S)
5. $[*\varphi]\,\mathbf{B}(\varphi \wedge \mathbf{B}\,\neg\varphi)$	ML: 3, 4

The trouble here is that, for many choices of φ, the situation described by the premises may well be plausible (that is, the set of premises is consistent). Thus an agent, operating in accordance with the theory described in this section and believing that he does not believe that it is raining (or believing that he believes that it is not raining) will, when informed that is raining, accept this but continue to believe that he does not believe that it is raining (or that he believes that it is not raining). This does not sound right; there is a need for a better theory.

3. Digression on Modal Logic

This section is set aside to explain the modal logical terminology used in this paper. In particular, we wish to name certain systems of modal logic used in the sequel. Since we are mainly interested in doxastic logic, we write **B** for the box operator and **b** for the diamond operator. You may read "the agent believes that φ" for $\mathbf{B}\,\varphi$ and "it is consistent with all that the agent believes that φ" for $\mathbf{b}\,\varphi$.

(4) $\mathbf{B}\,\varphi \supset \mathbf{B}\,\mathbf{B}\,\varphi$,

(4^-) $\mathbf{B}\,\mathbf{B}\,\varphi \supset \mathbf{B}\,\varphi$,

(4!) $\mathbf{B}\,\varphi \equiv \mathbf{B}\,\mathbf{B}\,\varphi$,

(4^n) $\mathbf{B}^n\,\varphi \supset \mathbf{B}^{n+1}\,\varphi$,

(5) $\neg\,\mathbf{B}\,\varphi \supset \mathbf{B}\,\neg\,\mathbf{B}\,\varphi$,

(D) $\mathbf{B}\,\varphi \supset \mathbf{b}\,\varphi$,

(T) $\mathbf{B}\,\varphi \supset \varphi$,

($\mathrm{T}^\sharp$) $\mathbf{B}(\mathbf{B}\,\varphi \supset \varphi)$,

(V) $\varphi \equiv \mathbf{B}\,\varphi$.

We write K for the smallest normal modal logic. When $X_0, \ldots, X_k$ are schemata, we use E. J. Lemmon's convention of writing $\mathrm{K}X_0 \ldots X_k$ for the normal extension of K by $X_0, \ldots, X_k$; another name for the same system is $\mathrm{K} + X_0, \ldots, X_k$. The systems KT4 and KT45 are known in the literature as weak S4 and weak S5. Note that KD has the same theorems as $\mathrm{K} + \mathbf{b}\top$.

Lemma 3.1. For every n, $\mathbf{B}\,\mathbf{b}^n\,\top$ is derivable as a formal theorem in KD.

Proof. By induction on n. If $n = 0$, the claim holds. Suppose that $\mathbf{B}\,\mathbf{b}^n \top$ is a formal theorem, for some fixed number n. Then, by the (D)-schema, so is $\mathbf{b}\,\mathbf{b}^n \top$; in other words, $\mathbf{b}^{n+1} \top$ is derivable. By **B**-necessitation, therefore, $\mathbf{B}\,\mathbf{b}^{n+1} \top$ is derivable in KD. □

The schemata listed are well-known in the doxastic logical literature. Schemata (4) and (5) are nowadays often referred to as Positive Introspection and Negative Introspection, respectively; consequently, (4^n) is a weaker form of Positive Introspection when $n > 1$. Schema (4^-) is the converse of (4) and might be called the VIP-schema ("Veridicality of Introspection, Positive": cf. Lindström & Rabinowicz 1997). The schema (4!) has only a limited interest in the special context of DDL as it conflates higher-order beliefs and first-order beliefs. The schema (D), sometimes referred to as Doxastic Consistency, is regarded by many as a *sine qua non* for doxastic logic since it blocks contradictory beliefs. Schema (T) is of course inappropriate in doxastic logic — what agent holds only true beliefs? Schema ($T^\sharp$) may be overly optimistic but holds a discreet charm: it is characteristic of those confident agents who believe that their beliefs are true. For us, who will give much attention below to schema (4^-), it is interesting to note that it is derivable in the system $KT^\sharp$. Schema (V) is of course hopeless as a doxastic logic, obliterating as it does the distinction between belief and truth. (For a comprehensive and still readable philosophical discussion of doxastic logic, see Wolfgang Lenzen's survey (1978).)

Why are there so many doxastic logics? Perhaps because there are so many different ways of generating and processing beliefs. Humans, groups, and non-human agents are different, not least from a doxastic point of view. Instead of asking what doxastic logic is the correct one, one might try to identify the doxastic logic of a certain agent. Having said this, it has to be admitted that the existence of Moore problems in a certain logic would usually be regarded as a defect. An agent with a Moore problem is not as badly off as an agent who believes, at one and the same time, that it is raining and that it is not raining. But other things being equal he is worse off than an agent without Moore problems. If we are interested in modelling "rational" agents within normal doxastic logic, we should look for logics that are not susceptible to Moore problems.

4. Moore Problems at Level 0

Let us re-state with some care the problem noted by van Linder and his colleagues. Suppose that $\mathbf{b}\,\varphi$ is true (that is, φ is consistent with the agent's beliefs) and that the agent will revise them by φ. There are two cases. In the first case, $\mathbf{B} \neg \mathbf{B}\,\varphi$ is true *before the change* (the agent believes that he does not believe φ), and $\mathbf{B}(\varphi \wedge \neg \mathbf{B}\,\varphi)$ is true *after the change*. In the second case, $\mathbf{B}\,\mathbf{B} \neg\varphi$ is true *before the change* (the agent believes that he believes φ), and $\mathbf{B}(\varphi \wedge \mathbf{B}\,\neg\varphi)$ is true *after the change*. The two cases may be described in this way:

	before the change	after the change
first case	$\mathbf{b}\,\varphi$	$\mathbf{B}\,\varphi$
	$\mathbf{B}\neg\mathbf{B}\,\varphi$	$\mathbf{B}\neg\mathbf{B}\,\varphi$
second case	$\mathbf{b}\,\varphi$	$\mathbf{B}\,\varphi$
	$\mathbf{B}\,\mathbf{B}\,\neg\varphi$	$\mathbf{B}\,\mathbf{B}\,\neg\varphi$

In either case, after the change the agent faces a Moore problem.

There seem to be two strategies for avoiding these Moore problems. One is to accept that revision is properly formalized by the star operator $*$ (as we have tacitly done so far) and require the logic of **B** to be strong enough to make each of the sets $\{\mathbf{b}\,\varphi, \mathbf{B}\neg\mathbf{B}\,\varphi\}$ and $\{\mathbf{b}\,\varphi, \mathbf{B}\,\mathbf{B}\,\neg\varphi\}$ inconsistent; for inconsistent sets describe (what according to the logic are) impossible situations, and it is of no concern that Moore problems arise in impossible situations. The weakest normal logic of this kind is the normal extension of the normal logic K by the schemata

$$\mathbf{B}\,\mathbf{B}\,\varphi \supset \mathbf{B}\,\varphi,$$
$$\mathbf{B}\,\mathbf{b}\,\varphi \supset \mathbf{B}\,\varphi.$$

The former schema is (4^-), the well-known VIP-schema, while the latter is unusual. Of the few systems that provide both schemata — Verum is one of the few — no-one can claim plausibility as a doxastic logic.

Therefore it is desirable to look for another strategy for avoiding the Moore problems described above. The obvious alternative is to reject the assumption that the star operator formalises revision, and this is what we will do. As we try to re-define revision, we shall find ourselves embarking on a tour taking us through many levels. We are now leaving level 0.

5. Moore Problems at Level 1

If revision by φ is not to be formalized as $*\,\varphi$, then as an experiment suppose that it is formalized as $*(\varphi \wedge \mathbf{B}\,\varphi)$. This means that while the star operator continues to be part of our formal language, it is no longer appropriate to read $*\,\varphi$ as "revision by φ". To make this quite clear, imagine that we were to introduce a new operator R, the *revision operator*. At the level discussed in the preceding section, level 0, the revision operator is identified with the star operator: it is as if the definition

(df 0) $\quad \mathrm{R}\,\varphi =_{\mathrm{df}} *\,\varphi$

had been adopted. By contrast, we now propose to adopt as a new definition

(df 1) $\text{R}\varphi =_{\text{df}} *(\varphi \wedge \mathbf{B}\,\varphi)$.

Informally, read "revision by φ" for $\text{R}\varphi$ and "star change by φ" for $*\,\varphi$. The Preservation and Success Conditions are not affected by this definition. Thus the two derivations given in Section 2 continue to be valid. However, the conclusions are no longer an embarrassment, for the fact that, in a certain possible situation, a star change leads to a Moore problem is not embarrassing, however plausible the situation — why would one want to perform a star change anyway? What would be embarrassing is if

$$(\mathbf{b}\,\varphi \wedge \mathbf{B} \neg \mathbf{B}\,\varphi) \supset [\text{R}\varphi]\,\mathbf{B}(\varphi \wedge \neg \mathbf{B}\,\varphi),$$
$$(\mathbf{b}\,\varphi \wedge \mathbf{B}\,\mathbf{B} \neg\varphi) \supset [\text{R}\varphi]\,\mathbf{B}(\varphi \wedge \mathbf{B} \neg\varphi)$$

were derivable; but they are not (as can be seen with the help of results in Segerberg (1999)).

But while the new definition of revision avoids the old problems, unfortunately it leads to new problems. Suppose that φ is such that, *before revision* by φ, $\mathbf{b}(\varphi \wedge \mathbf{B}\,\varphi)$ is true, and that, *before revision*, either $\mathbf{B} \neg \mathbf{B}\,\mathbf{B}\,\varphi$ or $\mathbf{B}\,\mathbf{B} \neg \mathbf{B}\,\varphi$ or $\mathbf{B}\,\mathbf{B}\,\mathbf{B} \neg\varphi$ is true. By noting two instances of Preservation and Success, respectively,

$$(\mathbf{b}(\varphi \wedge \mathbf{B}\,\varphi) \wedge \mathbf{B}\chi) \supset [*(\varphi \wedge \mathbf{B}\,\varphi)]\,\mathbf{B}\chi,$$
$$[*(\varphi \wedge \mathbf{B}\,\varphi)]\,\mathbf{B}(\varphi \wedge \mathbf{B}\,\varphi),$$

it is easy to see that, in our example, *after revision* by φ — as we now understand revision — either $\mathbf{B}(\mathbf{B}\,\varphi \wedge \neg \mathbf{B}\,\mathbf{B}\,\varphi)$ or $\mathbf{B}\,\mathbf{B}(\varphi \wedge \neg \mathbf{B}\,\varphi)$ or $\mathbf{B}\,\mathbf{B}(\varphi \wedge \mathbf{B} \neg\varphi)$, respectively, is true.

So there are Moore problems also at level 1, even though they are different Moore problems. As at level 0, the Moore problems that we have identified can be avoided by assuming that the agent's doxastic logic is sufficiently strong to make each of the sets $\{\mathbf{b}(\varphi \wedge \mathbf{B}\,\varphi), \mathbf{B} \neg \mathbf{B}\,\mathbf{B}\,\varphi\}$ and $\{\mathbf{b}(\varphi \wedge \mathbf{B}\,\varphi), \mathbf{B}\,\mathbf{B} \neg \mathbf{B}\,\varphi\}$ and $\{\mathbf{b}(\varphi \wedge \mathbf{B}\,\varphi), \mathbf{B}\,\mathbf{B}\,\mathbf{B} \neg\varphi\}$ inconsistent. The weakest normal logic of the requisite strength is the normal extension of K by the following schemata:

(1A) $\mathbf{B} \neg \mathbf{B}\,\mathbf{B}\,\varphi \supset \mathbf{B} \neg(\varphi \wedge \mathbf{B}\,\varphi)$,

(1B) $\mathbf{B}\,\mathbf{B} \neg \mathbf{B}\,\varphi \supset \mathbf{B} \neg(\varphi \wedge \mathbf{B}\,\varphi)$,

(1C) $\mathbf{B}\,\mathbf{B}\,\mathbf{B} \neg\varphi \supset \mathbf{B} \neg(\varphi \wedge \mathbf{B}\,\varphi)$.

We note without proof — for proofs, see the more general argument in the following section — that (1A) is derivable in K4, and that (1B) and (1C) are derivable in K4^- and as well as in KD4. This means that all three of (1A)–(1C) are derivable in $K44^- = K4!$ as well as in KD4. The former has a limited interest in the present context as it does not distinguish higher-order from first-order beliefs. The latter, on the other hand, is a favorite with many doxastic logicians.

6. Moore Problems at Level *n*

The examination in the two previous sections can be subsumed under a general analysis by consideration of the definition

(df *n*) $\quad \mathrm{R}\varphi =_{\mathrm{df}} *(\varphi \wedge \mathbf{B}\varphi \wedge \cdots \wedge \mathbf{B}^n \varphi)$,

where *n* is any number; both (df 0) and (df 1) are instances of this more general definition. Suppose that the agent revises his beliefs, in accordance with the new definition, by a formula φ such that, *before the revision*,

$$\mathbf{b}(\varphi \wedge \mathbf{B}\varphi \wedge \cdots \wedge \mathbf{B}^n \varphi) \text{ is true,}$$
$$\mathbf{B}^p \neg \mathbf{B}^q \varphi \text{ is true,}$$

for some numbers such that

(i) $p > 0$,
(ii) $q \geqslant 0$,
(iii) if $p = 1$ then $q > n$.

(If $p = 1$ and $q \leqslant n$, the situation described by (i)–(iii) would not be plausible.) Then, *after the revision*, thanks to Preservation (and the fact that $p > 0$),

$$\mathbf{B}^p \mathbf{B}^q \varphi \text{ is true,}$$

and, thanks to Success, $\mathbf{B}(\varphi \wedge \mathbf{B}\varphi \wedge \cdots \wedge \mathbf{B}^n \varphi)$ is true, whence

$$\mathbf{B}^m \varphi \text{ is true, for all } m \text{ such that } 0 < m \leqslant n + 1.$$

This means that the agent encounters a Moore problem whenever

(iv) $p + q \leqslant n + 2$.

To see this, consider the following cases. If $p = 1$, then $q = n + 1$; hence *after the revision* $\mathbf{B}(\mathbf{B}^n \varphi \wedge \neg \mathbf{B}(\mathbf{B}^n \varphi))$ is true (an acute problem of rank 0). If $p > 1$ and $q = 0$, then *after the revision* $\mathbf{B}^{p-1}(\varphi \wedge \neg \mathbf{B}\varphi)$ is true (a grave problem of rank $p - 1$). And if $p > 1$ and $0 < q$, then $q < n + 1$; hence *after the revision* both $\mathbf{B}^p(\mathbf{B}^{q-1} \varphi \wedge \neg \mathbf{B}(\mathbf{B}^{q-1} \varphi))$ and $\mathbf{B}^{p-1}(\mathbf{B}^q \varphi \wedge \mathbf{B} \neg(\mathbf{B}^q \varphi))$ are true (the former an acute problem of rank *p*, the latter a grave problem of rank $p - 1$). These Moore problems are grouped as follows, where A, G and AG indicate the existence of acute problems, grave problems, and both acute and grave problems, respectively:

	$q = 0$	$q = 1$	$1 < q < n + 1$	$q = n + 1$
$p = 1$	—	—	—	A
$p > 1$	G	A	AG	—

As before, there is a choice of strategy if we want to avoid the Moore problems just described. One is to require the agent's doxastic logic to rule as inconsistent every situation described by a set

$$\{\mathbf{b}(\varphi \wedge \mathbf{B}\,\varphi \wedge \cdots \wedge \mathbf{B}^n\,\varphi), \mathbf{B}^p \neg \mathbf{B}^q\,\varphi\} \qquad (p \text{ and } q \text{ as in (i)–(iv)}).$$

In other words, the logic would have to provide the schemata

(§) $\qquad \mathbf{B}^p \neg \mathbf{B}^q\,\varphi \supset \mathbf{B}\,\neg(\varphi \wedge \mathbf{B}\,\varphi \wedge \cdots \wedge \mathbf{B}^n\,\varphi) \qquad (p \text{ and } q \text{ as in (i)–(iv)}).$

Among logics that satisfy this requirement, we mention two: the logic $K4^-4^n$, involving the schemata

(4^-) $\qquad \mathbf{B}\,\mathbf{B}\,\varphi \supset \mathbf{B}\,\varphi,$

(4^n) $\qquad \mathbf{B}^n\,\varphi \supset \mathbf{B}^{n+1}\,\varphi,$

and the logic $KD4^n$, involving the schemata

(D) $\qquad \mathbf{B}\,\varphi \supset \mathbf{b}\,\varphi,$

(4^n) $\qquad \mathbf{B}^n\,\varphi \supset \mathbf{B}^{n+1}\,\varphi.$

In order to establish that $K4^-4^n$ provides all instances of (§), it is convenient two divide the latter into two groups, depending on whether $q \leqslant n$ or $q = n + 1$ (in the latter case, $p = 1$). The following two deductions suffice as proof:

1.	$\mathbf{B}^p \neg \mathbf{B}^q\,\varphi$	premise, where $q \leqslant n$
2.	$\mathbf{b}(\varphi \wedge \mathbf{B}\,\varphi \wedge \cdots \wedge \mathbf{B}^n\,\varphi)$	premise
3.	$\mathbf{B}\,\neg \mathbf{B}^q\,\varphi$	ML: 1, (4^-) (if $p > 1$)
4.	$\mathbf{b}\,\mathbf{B}^q\,\varphi$	ML: 2 (NB $q \leqslant n$)
5.	$\bot$	ML: 3, 4

1.	$\mathbf{B}\,\neg \mathbf{B}^{n+1}\,\varphi$	premise
2.	$\mathbf{b}(\varphi \wedge \mathbf{B}\,\varphi \wedge \cdots \wedge \mathbf{B}^n\,\varphi)$	premise
3.	$\mathbf{b}\,\mathbf{B}^n\,\varphi$	ML: 2
4.	$\mathbf{b}\,\mathbf{B}^{n+1}\,\varphi$	ML: 3, (4^n)
5.	$\bot$	ML: 1, 4

Next, in order to show that also $KD4^n$ provides all instances of (§), we divide them into two groups according to whether $p+q < n+2$ or $p+q = n+2$. Again, two deductions suffice to prove the contention:

1.	$\mathbf{B}^p \neg \mathbf{B}^q \varphi$	premise, where $p + q \leqslant n + 1$
2.	$\mathbf{b}(\varphi \wedge \mathbf{B}\varphi \wedge \cdots \wedge \mathbf{B}^n \varphi)$	premise
3.	$\mathbf{b}\,\mathbf{B}^{p+q-1} \varphi$	ML: 2 (NB $0 \leqslant p + q - 1 \leqslant n$)
4.	$\mathbf{b}(\mathbf{B}^{p-1} \neg \mathbf{B}^q \varphi \wedge \mathbf{B}^{p-1} \mathbf{B}^q \varphi)$	ML: 1, 3 (NB $p > 0$)
5.	$\mathbf{b}\,\mathbf{B}^{p-1} \bot$	ML: 4
6.	$\mathbf{B}\,\mathbf{b}^{p-1} \top$	(D), Lemma 3.1
7.	$\bot$	ML: 5, 6

1.	$\mathbf{B}^p \neg \mathbf{B}^q \varphi$	premise, where $p + q = n + 2$
2.	$\mathbf{b}(\varphi \wedge \mathbf{B}\varphi \wedge \cdots \wedge \mathbf{B}^n \varphi)$	premise
3.	$\mathbf{b}\,\mathbf{B}^n \varphi$	ML: 2
4.	$\mathbf{b}\,\mathbf{B}^{n+1} \varphi$	ML: 3, (4^n)
5.	$\mathbf{b}(\mathbf{B}^{p-1} \neg \mathbf{B}^q \varphi \wedge \mathbf{B}^{p-1} \mathbf{B}^q \varphi)$	ML: 1, 4 (NB $p > 0, p + q - 1 = n + 1$)
6.	$\mathbf{b}\,\mathbf{B}^{p-1} \bot$	ML: 5
7.	$\mathbf{B}\,\mathbf{b}^{p-1} \top$	(D), Lemma 3.1
8.	$\bot$	ML: 6, 7

The other strategy for avoiding these Moore problems is, once again, to look for a new definition of revision.

7. Moore Problems at Level ∞

A definition at this level, in the same spirit as at the finite levels, would look something like

(pdf ∞) $\textsc{r}\varphi =_{\mathrm{df}} *(\varphi \wedge \mathbf{B}\varphi \wedge \cdots \wedge \mathbf{B}^n \varphi \wedge \cdots)$.

The trouble with this definition is of course that it is a pseudo-definition: the definiens is not well-formed in our language. To amend this defect we now expand the latter by introducing a new primitive operator **C**. Notice that **C** is an element of the object language, unlike the symbol R which was introduced as an abbreviatory device. (It would of course have been possible to make also R an element of the object language; but it has seemed more didactic not to do so.) With the new operator comes the new rule of **C**-necessitation:

if φ is derivable, then so is $\mathbf{C}\varphi$,

as well as the following axiom schemata:

$\mathbf{C}\varphi \supset \mathbf{B}\varphi$,

$\mathbf{C}\varphi \supset \mathbf{C}\mathbf{C}\varphi$,

$(\mathbf{B}\varphi \wedge \mathbf{C}(\varphi \supset \mathbf{B}\varphi)) \supset \mathbf{C}\varphi$.

The pseudo-definition (pdf ∞) can now be replaced by a correct definition:

(df ∞) $\quad \mathrm{R}\varphi =_{\mathrm{df}} *(\varphi \wedge \mathbf{C}\,\varphi)$.[1]

Once again, the old Moore problems fall away. But even this time, new problems arise, in fact, more serious than those associated with Moore. Suppose that $p > 0$ and $q \geqslant 0$:

1. $\mathbf{b}(\varphi \wedge \mathbf{C}\,\varphi)$ — premise
2. $\mathbf{B}^p \neg \mathbf{B}^q \varphi$ — premise
3. $[*(\varphi \wedge \mathbf{C}\,\varphi)]\,\mathbf{B}^p \neg \mathbf{B}^q \varphi$ — 1, 2, (P) (NB $p > 0$)
4. $[*(\varphi \wedge \mathbf{C}\,\varphi)]\,\mathbf{B}^{p+q} \varphi$ — (S) (NB $p + q > 0$)
5. $[*(\varphi \wedge \mathbf{C}\,\varphi)]\,\mathbf{B}^p(\mathbf{B}^q \varphi \wedge \neg \mathbf{B}^q \varphi)$ — ML: 3, 4
6. $[*(\varphi \wedge \mathbf{C}\,\varphi)]\,\mathbf{B}^p \bot$ — ML: 5

To accept this result would be to accept that there are situations, intuitively plausible, such that revising one's beliefs by a certain (consistent) proposition would lead to a kind of doxastic inconsistency. In order to rule out those situtations, one needs a belief operator strong enough to make all situations of type

$$\{\mathbf{b}(\varphi \wedge \mathbf{C}\,\varphi), \mathbf{B}^p \neg \mathbf{B}^q \varphi : p > 0 \ \&\ q \geqslant 0\}$$

inconsistent. Weakest amongst such logics is the normal extension of K by the schemata

(§§) $\quad \mathbf{B}^p \neg \mathbf{B}^q \varphi \supset \mathbf{B} \neg(\varphi \wedge \mathbf{C}\,\varphi) \qquad (p > 0, q \geqslant 0)$.

Among the systems that provide those schemata are — as readers have probably anticipated — K4$^-$ and KD. Suppose that $p > 0$ and $q \geqslant 0$:

1. $\mathbf{B}^p \neg \mathbf{B}^q \varphi$ — premise
2. $\mathbf{b}(\varphi \wedge \mathbf{C}\,\varphi)$ — premise
3. $\mathbf{B} \neg \mathbf{B}^q \varphi$ — ML: 1, (4$^-$) (if $p > 1$)
4. $\mathbf{b}\,\mathbf{B}^q \varphi$ — ML: 2
5. $\bot$ — ML: 3, 4

1. $\mathbf{B}^p \neg \mathbf{B}^q \varphi$ — premise
2. $\mathbf{b}(\varphi \wedge \mathbf{C}\,\varphi)$ — premise

[1] For readers familiar with Segerberg (1999) we mention that the intended truth-condition of the new operator is

$$H \vDash \mathbf{C}\,\varphi \text{ iff, for all } n > 0 \text{ and all } H' \text{ such that } (H, H') \in B^n,\ H' \vDash \varphi.$$

The truth-condition makes it clear that (df ∞) is what we want in lieu of (pdf ∞) since it implies that $H \vDash \mathbf{C}\,\varphi$ if and only if, for all $n > 0$, $H \vDash \mathbf{B}^n\,\varphi$.

3. $\mathbf{b}\,\mathbf{B}^{p+q-1}\,\varphi$	ML: 2 (NB $p+q-1 \geqslant 0$)
4. $\mathbf{b}(\mathbf{B}^{p-1}\,\neg\,\mathbf{B}^{q}\,\varphi \wedge \mathbf{B}^{p-1}\,\mathbf{B}^{q}\,\varphi)$	ML: 1, 3 (NB $p > 0$)
5. $\mathbf{b}\,\mathbf{B}^{p-1}\,\bot$	ML: 4
6. $\mathbf{B}\,\mathbf{b}^{p-1}\,\top$	(D), Lemma 3.1
7. $\bot$	ML: 5, 6

It is somehow gratifying that, in the limit, the schema (4^n) that figured in the previous section is no longer needed. As n goes to infinity, (4^n) vanishes!

8. Two Questions

There are two questions that it is natural pose in connexion with our analysis. One we can answer, the other not.

The first question is this: Granted that we have shown how certain Moore problems can be avoided, how do we know that other Moore problems are not lurking? In other words, when is a doxastic logic such that the agent will never run the risk of having to face a Moore problem?

Let us define a logic as *Moore-problem free* if and only if, for every $n > 0$ and every φ, neither the set $\{\mathbf{B}^n(\varphi \wedge \neg\,\mathbf{B}\,\varphi)\}$ nor the set $\{\mathbf{B}^n(\varphi \wedge \mathbf{B}\,\neg\varphi)\}$ is consistent (in that logic). In other words, a logic is Moore-problem free if and only if it provides the schemata

($\lhd$)	$\mathbf{b}^n(\varphi \supset \mathbf{B}\,\varphi)$,	for all $n > 0$,
($\rhd$)	$\mathbf{b}^n(\mathbf{B}\,\varphi \supset \varphi)$,	for all $n > 0$.

In still other words, the Moore-problem free logics are exactly those that are normal extensions of K + ($\lhd$) + ($\rhd$). For example, KD4 is one of them, while neither KD nor K4 nor K44⁻ is.

This notion gives us the answer to the given question. If a logic is not Moore-problem free, then there is a logical possiblity of having a Moore problem. But, more interestingly, if a logic in Moore-problem free, then no sequence of revisions will lead to a Moore problem. To see this, suppose that all instances of ($\lhd$) and ($\rhd$) are added as new axioms to our formulation of DDL. Since all our operators are normal, it follows, for all formulæ φ and χ, that if χ is a theorem of that system, then (for any of the above definitions of R) $[\text{R}\,\varphi]\chi$ is also a theorem; in particular, $[\text{R}\,\varphi](\lhd)$ and $[\text{R}\,\varphi](\rhd)$ are derivable. Thus, if there can be no Moore problems before a revision — that is, the logic is Moore-problem free — then there will be none after the revision.

The other question is this: Moore problems are only one kind of doxastic incoherence. How do we know that even Moore-problem free logics do not exhibit other, equally unacceptable, kinds of incoherence? The short answer is: we don't. The question is vague: it seems unlikely that we shall ever be able to rule out the possibility that quaint, previously unnoticed phenomena are discovered that we might be reluctant to associate with our standards of doxastic rationality.

9. Blindspots

Before concluding the paper, we will make a comment on a topic discussed by Lindström and Rabinowicz. Is it always possible to revise one's beliefs by a true proposition? Yes, say Lindström and Rabinowicz:

> That the revision of a consistent doxastic state with a consistent proposition always leads to a consistent doxastic [state] is, in fact, one of the fundamental principles of any reasonable theory of belief revision (Lindström & Rabinowicz 1999a, p. 371).
>
> A consistent agent cannot become inconsistent simply by learning a true fact about the world (Lindström & Rabinowicz 1999b, p. 14).

This position they express as two slightly different, though related, formal "consistency principles" (Lindström & Rabinowicz 1999a, p. 373, 1999b, p. 14; notation modified):

(c1) $\varphi \supset [* \varphi]\, \mathbf{b} \top$,

(c2) $\mathbf{b} \top \supset (\varphi \supset [* \varphi]\, \mathbf{b} \top)$.

They then go on to note note that, together with other (acceptable) principles, untoward consequences follow. The conclusion Lindström and Rabinowicz draw is that we need a radically different approach to full DDL.

Both (c1) and (c2) would at first sight appear to be quite natural conditions. Already in the classic formulation of AGM is it assumed that whenever a proposition is entertainable (that is, consistent with everything the agent knows), revision by it leads to a consistent new belief set (Alchourrón *et al.* 1985). In our present theory, that assumption is reflected in postulate ($*$ 5), from which (with some help of ($*$ F) and modal logic) we can derive

(c0) $\mathbf{k} \varphi \supset [* \varphi]\, \mathbf{b} \top$

(Neither (c1) nor (c2) is derivable.[2] However, if $\mathbf{K} \varphi \supset \varphi$ is added as an extra axiom schema, then both (c1) and (c2) become derivable as theorems.)

Nevertheless, Lindström and Rabinowicz are surely right in arguing that there should be some valid consistency principle along the lines they draw. The difficulty is that the three conditions considered are violated by Moore sentences.

[2] This is readily proved with the help of Segerberg (1999).

Before trying to resolve it, let us remember that the position of van Linder *et al.* with regard to revision by a Moore sentence was one of hesitation: they seemed prepared to leave open the possibility that there are exceptional propositions with which revision cannot be carried out, or where the result of revision cannot be predicted. If so, they are denying the full validity of the consistency principles we have been considering so far.

The term "blindspot" has become popular philosophical terminology after the appearance of Sorensen (1988). It is not always clear exactly what a blindspot is, let alone how to give a rigorous definition of the term. But in this connexion we shall be able to kill two birds with one stone: give a modified formulation of Lindström and Rabinowicz's consistency condition that should be acceptable to all, and at the same time give a definition of 'blindspot' that is meaningful within the context of full DDL. Our proposal is this:

> Revision by an entertainable proposition *that is not a blindspot* always leads to a consistent doxastic state.

We mean to make this claim trivially true by defining "blindspot" as exactly what makes the claim true. Formally, let R be the revision operator of one of the recommended logics in previous sections, and introduce a temporary operator `blindspot` which applies to formulæ. Our principle then amounts to requiring general validity of the schema

$$(c') \qquad \neg\,\mathtt{blindspot}(\varphi) \supset (\mathbf{k}\,\varphi \supset [\mathrm{R}\,\varphi]\,\mathbf{b}\,\top).$$

Adopting the following definition guarantees the validity of (c′) by turning it into a tautology:

$$\mathtt{blindspot}(\varphi) \;=_{\mathrm{df}}\; \neg(\mathbf{k}\,\varphi \supset [\mathrm{R}\,\varphi]\,\mathbf{b}\,\top).$$

Thus the schema

$$\mathbf{k}\,\varphi \supset ([\mathrm{R}\,\varphi]\,\mathbf{b}\,\top \equiv \neg\,\mathtt{blindspot}(\varphi))$$

is valid in our logic. That is to say, revising one's beliefs with an entertainable proposition results in doxastic consistency if and only if it is not a blindspot. Note that, in all the recommended logics above, $[\mathrm{R}(\varphi \wedge \neg\,\mathbf{B}\,\varphi)]\,\mathbf{B}\,\bot$ and $[\mathrm{R}(\varphi \wedge \mathbf{B}\,\neg\varphi)]\,\mathbf{B}\,\bot$ are theorems, so in those logics $\varphi\wedge\neg\,\mathbf{B}\,\varphi$ and $\varphi\wedge\mathbf{B}\,\neg\varphi$ are blindspots.

10. Conclusion

The emphasis in this paper on Moore problems has overshadowed what is really a more interesting question, namely, how to model revision in full DDL. Our solution, different from that of Lindström and Rabinowicz, is to define revision in a way that differs from — or generalises — that in basic DDL. As in doxastic logic generally, it may be wrong to expect a unique definition of revision. Nevertheless, our survey has suggested two main candidates, (df 1) and (df ∞). There is no need here to choose between them; let it just be noted that the demands on the underlying doxastic logic are lower in the latter case, while the computational complication is higher.

On the technical side it should be observed that the completeness proof in Segerberg (1999), which assumes nothing about the underlying doxastic logic, is trivially extendable to cover all the doxastic logics definable in terms of the schemata listed in Section 3. Where the new operator **C** is involved, however, the extension is less trivial; it will have to be the topic of another paper.*

Uppsala University
Department of Pilosophy
Drottninggatan 4
753 10 Uppsala
Sweden
e-mail: krister.segerberg@filosofi.uu.se

REFERENCES

Alchourrón, C., P. Gärdenfors and D. Makinson (1985). On the Logic of Theory Change. *The Journal of Symbolic Logic* **50**, 510–530.

Hintikka, J. (1962). *Knowledge and Belief: An Introduction to the Logic of the two Notions*. Ithaca, N.Y.: Cornell University Press.

van Linder, B., W. van der Hoek and J.-J. Meyer (1995). Actions that Make you Change your Mind. In: A. Laux and H. Wansing (eds.), *Knowledge and Belief in Philosophy and Artificial Intelligence*, pp. 103–146. Berlin: Akademie Verlag GmbH.

Lenzen, W. (1978). *Recent Work in Epistemic Logic*. Acta Philosophica Fennica, vol. 30, no. 1. Amsterdam: North-Holland.

Lindström, S. and W. Rabinowicz (1997). Extending Dynamic Doxastic Logic: Accommodating Iterated Beliefs and Ramsey Conditionals within DDL. In: L, Lindahl *et al.* (eds.), *For Good Measure: Philosophical Essays Dedicated to Jan Odelstad*, pp. 126–153. Uppsala Philosophical Studies, vol. 46. Uppsala.

---

*Part of the work on this paper was done while I was a Fellow 2001–2002 at SCASSS (the Swedish Collegium for Advanced Study in the Social Sciences). Most of the material has been presented in two talks, one at the conference "Filosofidagarna 2001" (Stockholm, June 2001), the other at the First Australian Knowledge Representation Conventicle (Sydney, January 2002). I am indebted to Per Martin-Löf for having intimated, in the nicest possible way, that perhaps not all claims made in the Stockholm talk were correct. Thanks to his remarks, I have been able to identify and correct at least some erroneous claims.

Lindström, S. and W. Rabinowicz (1999a). DDL Unlimited: Dynamic Doxastic Logic for Introspective Agents. *Erkenntnis* **50**, 353–385.

Lindström, S. and W. Rabinowicz (1999b). Belief Change for Introspective Agents. In: *Spinning Ideas: Electronic Essays Dedicated to Peter Gärdenfors*. http://www.lucs.lu.se/spinning/. Lund.

Malcolm, N. (1958). *Ludwig Wittgenstein: A Memoir*. London: Oxford University Press.

Segerberg, K. (1999). A completeness proof in full DDL. In: R. Sliwinski (ed.), *Philosophical Crumbs: Essays Dedicated to Ann-Mari Henschen-Dahlquist*, pp. 195–207. Uppsala Philosophical Studies, vol. 49. Uppsala.

Sorensen, R. A. (1988). *Blindspots*. Oxford: Clarendon Press.

Jacek Malinowski

ON THE FORMALIZATION OF STRAWSON'S PRESUPPOSITION*

ABSTRACT. In this paper we analyze the Strawson's notion of presupposition proposed in his book Introduction to Logical Theory. Strawsonian notion of presupposition is dependent on the notion of logical entailment. We make use of the theory of logical consequence operation as a general framework to show that it is impossible to find a logical consequence operation which mirrors the philosophical intuitions of the Strawson's notions of presupposition. The aim of this paper is to present in details the philosophical backgrounds of the formal analysis presented in the author's paper "Strawsonian presuppositions and logical entailment".

The notion of presupposition is one of the most extensively investigated notions of the logical pragmatics. Since Frege a number of approaches to dealing with the phenomenon of presupposition have been proposed. It is not our aim to present them. For review of these we refer the reader to Levinson (1983) and Beaver (1997). In this paper we present a logical analysis one approach to the concept of presupposition — a concept known in the literature as the Strawson's notion of presupposition.

According to the original proposal presented by Strawson in (1949, p. 175–176): "a statement S presupposes a statement S' if and only if the truth of S' is a precondition of the truth-or-falsity of S". The *Strawsonian presupposition* is usually considered in the literature in the following form:

(1) P presupposes Q if and only if Q is true provided P is true or P is false.

According to (1) Q has to be true in order for P to be true or false. This means that Q is a logical consequence of P and at the same time Q is a logical consequence of the falsity of P. Assuming also the bivalence principle (5) we obtain that Q has to be true independently of the logical value of P. Assuming the bivalence principle one can then show that Q is a tautology. That fact is well known. Levinson (1983, pp. 175–176) presents it in detail and then states that: "It has been shown that perfectly well-behaved logic with three values can be constructed and it could be claimed that such logical systems are (by virtue of their ability to handle presupposition) a notable advance in models of natural

*This paper has been prepared during author's stay at Leipzig University as Marie Curie Fellow MCFI-2002-00079

In: J. Malinowski and A. Pietruszczak (eds.), *Essays in Logic and Ontology* (*Poznań Studies in the Philosophy of the Sciences and the Humanities*, vol. 91), 111–118. Amsterdam/New York, NY: Rodopi, 2006.

language semantics." We are going to show that Levinson is mistaken. Introducing a third logical value does not save the Strawson definition. Surprisingly, it appears that also without assuming (5) the only presuppositions admitted by (1) are classical tautologies.

Strawson's definition formulated in (1) makes the notion of presupposition depend on the notion of logical entailment. There exists a plethora of formal systems of logical entailment. Even if we limit ourselves to the language of classical sentential logic we still can choose from the large set of possible logical entailments. This set includes the classical logic. However it contains much more: the systems of non-classical logics, the intuitionistic and intermediate logics, the systems of relevant entailment and much more. If we consider the sentential languages extending the classical sentential language, for example the language with modal operators, we obtain a much larger set of possible logical entailments. Our main aim is to to look for such a formal system of logical entailment which, when applied to the Strawson's definition, mirrors the intuitions of presupposition. For this reason we need a general framework describing the notion of logical entailment in the general terms. Our investigations will then lead us to the main result of this paper: None of possible logical entailments meet the properties desired for presuppositions. A logical framework perfectly appropriate to our need was proposed by Tarski. It is commonly known as a notion of logical consequence operation. It has been then extensively investigated since the 30-ties. The monograph R. Wójcicki (1987) contains the full presentation of this subject.

By an *operation* we will mean a function C which to any set of sentences X of a given language S corresponds the set $C(X)$. The operation satisfying the condition if $X \subseteq Y$, then $C(X) \subseteq C(Y)$ will be called *monotonic*. It will be called *inclusive* (respectively *idempotent*) if and only if it satisfies the conditions: $X \subseteq C(X)$ (respectively $C(C(X)) = C(X)$). An operation will be called *structural* if and only if $e(C(X)) \subseteq C(e(X))$ for any substitution e in S and any set $X \subseteq S$. The condition of structurality can be equivalently defined in the following easier to understand way: Any substitutions of the consequences of X are also consequences of the substitutions of X. Any inclusive, monotonic, idempotent and structural operation will be called a *logic* (or *logical entailment*).

Suppose that C is a logic (logical entailment). Then the part of (1) consisting of "if P is true then Q is true" can be obviously formulated by means of $Q \in C(P)$. The remaining part "if P is false then Q is true" is more complicated. Any inference with a false premise is valid. For this reason, in order to use here the notion of entailment, we have to use the notion of negation. If we suppose, that the language of C contains a negation connective $\neg$ then (1) might be expressed as the following notion of the *presupposition via negation*:

(2) P C-presupposes Q if and only if $Q \in C(P)$ and $Q \in C(\neg P)$.

The minimal assumption on the negation operator $\neg$ is that it is an unary operator in a given language satisfying the condition (3) below. (3) is weaker than the condition (4) satisfied by the classical negation, as well as, among others, by the negation connectives in many-valued logics.

(3) If P is false then $\neg P$ is true.

(4) $\neg P$ is true if and only if P is false.

(5) Any sentence P is either true or false.

The definition (1) can be approached also in other, more general and perhaps more direct way by means of of bi-matrices — the notion closely related to the notion of a logical matrix.

By a *bi-matrix* we mean a triple $M = (A, D, E)$ where A is an abstract algebra, D and E are subsets of A — two sets of designated elements of M. We will say that two bi-matrices are similar if and only if their underlaying algebras are of the same type (i.e., have the same operations). By a *valuation* of S in a bi-matrix M we mean any homomorphism of the language S into the algebra A of the matrix M.

Every class of bi-matrices $\boldsymbol{K}$ determines a function $Fn_{\boldsymbol{K}}$ corresponding to any set of sentences X a set of sentences $Fn_{\boldsymbol{K}}(X)$, defined in the following way: $P \in Fn_{\mathbf{K}}(X)$ if and only if for every bi-matrix $M = (A, D, E) \in \boldsymbol{K}$ and every valuation v, if $v(X) \subseteq D$, then $v(P) \in E$. It is easy to check that $Fn_{\boldsymbol{K}}$ is a structural, monotonic operation.

The set D of the designated elements of M is a generalization of the logical value of the truth while the set E generalizes the notion of logical value of the falsity.

We say that an operation F is *strongly complete* relative to a class of bi-matrices $\boldsymbol{K}$ for S, if $F = Fn_{\boldsymbol{K}}$. A class $\boldsymbol{K}$ of bi-matrices is *strongly adequate* for an operation F if F is strongly complete relative to $\boldsymbol{K}$.

It appears that any structural monotonic operation F is strongly complete relative to some class of bi-matrices. For this very reason we can consider the notion of bi-matrices as a proper tool for investigating the possible candidates for the presupposition defined by means of (1).

Let us look how the notion of bi-matrices works in the case of three-valued logic. David Beaver (1997, p. 948) formulates *three-valued Strawsonian presupposition* by means of three-valued possible worlds semantics: P presupposes Q if and only if for all possible worlds w if P in the world w is true or false then Q is true in the world w. Any sentence P may take in a world w any of the three semantic values: true, false and undefined.

Let us adopt this definition for the case of strong Kleene three-valued logic defined by means of the following algebra:

sk3

∨	0	½	1
0	0	½	1
½	½	½	1
1	1	1	1

∧	0	½	1
0	0	0	0
½	0	½	½
1	0	½	1

→	0	½	1	¬
0	1	1	1	1
½	½	½	1	½
1	0	½	1	0

Let *Fnsk3* denote an operation determined by the bi-matrix $(\mathbf{sk3}, \{1\}, \{0\})$. Thus $P \in Fnsk3(X)$ if and only if for any bi-valuation v such, that $v(X) \subseteq \{0, 1\}$ we have $v(P) = 1$. This formal definition can be understand in the following way: P is a presupposition of the set of sentences X if and only if P can be false only if at least one of the sentences from X is neither true nor false.

The bi-matrix $\mathbf{M_2} = (\mathbf{2}, \{0, 1\}, \{1\})$, where **2** is two-element Boolean algebra corresponds to the classical logical matrix. It mirrors the classical understanding of the logical entailment in (1). Given a set of sentences X, suppose that a given sentence P belongs to the set $Fn_M(X)$ which means that for any valuation v such that $v(X) \subseteq \{0, 1\}$ we have that $v(P) = 1$. This simply means that $v(P) = 1$ for any valuation v, hence P is a classical tautology.

In this way we proved that in classical logic there are no contingent presuppositions. Obviously it contradicts elementary intuitions concerning presuppositions and proves that Strawson's notion of presupposition makes no sense in classical logic.[1] It is then necessary to reject the bivalence principle (5) and admit that besides of being true or false a given sentence can have some other logical value. We will not discuss here what the status of other logical values is and in what extent it is justified to use the name "logical value" for the third element of the algebra **sk3** or to other elements of a given bi-matrix. By the *classical logical value* we mean just 1 and 0 — the truth and the false, while by a *logical value* we mean any element of a given bi-matrix. Then classical logical values are logical values but there exist many non-classical logical values.

The original Strawsonian definition of presupposition (Strawson 1949, pp. 174–176) explicitly introduces a kind of third logical value into the trichotomy: true, false, meaningless. The idea of interpretation of Strawson's definition in many-valued logics has been then extensively elaborated. We refer the reader to (Beaver 1997) for a review of three and four-valued logical systems interpreting Strawson's presupposition.

It appears that[2] by introducing the third logical value we do not improve the Strawson's definition. The result presented above for the classical logic remains true. For the bi-matrix determined by means of strong Kleene three-valued logic

[1] A similar statement formulated in a slightly different formal setting has been proved in (Kracht 1993).

[2] See (Malinowski 2004) for technical details.

(and also for the bi-matrix determined by week Kleene three-valued logic and Łukasiewicz three-valued logic) there are not contingent presuppositions. This shows that any three-valued logics do not work better for the formalization of presupposition than classical logic does. On the contrary, in a sense, they work even worse than classical logic. Classical logic does not give us a good tool for formalization of presupposition not because any tautology is a presupposition of any sentence — this is in principle acceptable — but because no contingent sentence can be a presupposition. It appears that other three-valued logics lead us to a narrower class of presuppositions of a given sentence than classical logic does.

The result presented above can be generalized for the much larger class of logics than presented examples of three-valued logics. Let $\boldsymbol{K}$ be a class of bi-matrices such the classical bimatrix $Bm(\mathbf{2})$ is a sub-bi-matrix of some bi-matrix from the class $\boldsymbol{K}$. Then for any set of sentences X any element of $Fn_{\boldsymbol{K}}(X)$ is a classical tautology.

This result is negative. It allows us to exclude a number of large classes of elaborated logical consequences from the set of acceptable candidates for formalization of presupposition by means of Strawson's idea . All the many-valued as well as fuzzy logics have to be excluded. Also all the logical systems based on lattice semantics, for example intuitionistic and intermediate logic, as well as orthologics, are excluded for the same reason. Many kinds of relevant logics also possess a matrix semantics based on lattices. In fact it is hard to imagine a logical system formulated in a language with the connectives $\rightarrow$, $\wedge$, $\vee$ and $\neg$ which does not satisfy the assumption of the result presented above. The only such logic is trivial inconsistent logic, which, of course, for other reasons cannot serve as a tool for the formalization of the notion of presupposition.

Let us come back to the notion of presupposition via negation presented in (2). The notion of logical falsity is not involved here. For this reason the notion of logical consequence and the notion of logical matrices form sufficient formal tool for its investigation.

Logical consequences are usually semantically defined not by logical matrices but by means of others structures like, for example, possible worlds. It is true that, for such logics, matrix semantics might be artificial if the elements of a given matrix does not mirror any intuitively clear objects like classical logical values of truth and false. Nevertheless all the results and methods presented in this paper remains valid also for logics which are usually defined, for example, by means of some kind of possible world semantics. We can consider logical matrices just as a formal tool to investigate logical consequences.

A natural question arises. Perhaps the language of classical sentential logic is to poor to formalize the phenomenon of presupposition. What, if we extend it by adding new operators? We are going to show that also this way does not lead us to a proper solution. It appears that the results similar to those presented above can be proved for any sentential language which is an extension of the classical one, i.e., for any language containing the classical connectives and eventually also the other ones.

Let S_0 denote the language of the classical logic, i.e., the language with two binary and one unary connective corresponding respectively to conjunction, disjunction and negation. A sentential language S will be called an extension of S_0 if and only if among connectives of S there exists two binary and one unary operators.

According to this definition the modal language with the following connectives necessitation, conjunction, disjunction and negation is an extension of the classical language S_0. More general, any language obtained by adding new connectives to S_0 is an extension of S_0.

Given a sentential language S and a logical consequence operation C in the language S, suppose that S is an extension of S_0. It is easy to observe that any sentence of S_0 is also a sentence of S. By a *reduct* of C to S_0 we mean the consequence operation C_0 defined on S_0 in the following way: $P \in C_0(X)$ if and only if $P \in S_0$ and $P \in C(X)$.

The following notion is crucial for the remaining part of this section. A consequence operation C in the language S will be called *sub-classical* if and only if: a) S is an extension of S_0 and b) for any $P \in S_0$ and $X \subseteq S_0$ if $P \in C(X)$, then $P \in Cn_{\mathbf{M_2}}(X)$, where $\mathbf{M_2}$ denotes the matrix consisting of two-element Boolean algebra **2** with its unit element as the only designated element.

The main result concerning presupposition via negation is also negative. J. Malinowski (2004) presents the proof that for any sub-classical logic C and the sentence $P, Q \in S_0$, if P C-presupposes Q then Q is a classical tautology.

The notion of sub-classical logic is very broad. It is hard to find a logic which is not sub-classical. We will show this in the following example and theorems. By the classical sentential logic we mean the consequence operation Cl determined in the usual way by two-element Boolean algebra. Cl is obviously sub-classical. If C is weaker than Cl (i.e. $C(X) \subseteq Cl(X)$ for any set of sentences X) then C is also sub-classical.

By the trivial (inconsistent) logic Tr we mean the consequence operation in the langauge S defined in the following way: $Tr(X) = S$ for any set of sentence X. Obviously Tr is not sub-classical.

We will present a very broad class of modal logical consequences satisfying the condition of sub-classicality. However, we have to start by introducing some preliminary notions illustrating the definitions above

By *modal sentential language* we mean the language $S_\Box = (S, \vee, \wedge, \neg, \Box)$ with two binary connectives $\wedge, \vee$ of conjunction and disjunction, and with two unary connectives: negation $\neg$ and necessitation $\Box$. Thus $S_\Box$ is an extension of the classical language S_0 in the sense of the above definition. A *classical modal system* (see Segerberg 1971) is any set of sentences of the language $S_\Box$ containing all classical tautologies and closed under substitution and modus ponens and the following rule of extensionality:

$$\text{RE} \quad \frac{P \leftrightarrow Q}{\Box P \leftrightarrow \Box Q}$$

It is easy to observe that any normal modal system is classical, but also many of non-normal modal system are classical. For details we refer to the monograph G. E. Hughes and M. J. Cresswell (1996).

The notion of a modal system, in principle, does not equip us with the notion of entailment. We will define two types of modal consequence operation formalizing modal logical entailment. Suppose that a modal system L is given. By $L^{\rightarrow}$ we mean the consequence operation in the language $S_{\Box}$ defined in the following way: for any set $X \subseteq S_{\Box}$, $L^{\rightarrow}(X) \subseteq S_{\Box}$ is the least set of sentences containing L and X and closed with respect to modus ponens. Thus $L^{\rightarrow}(\emptyset) = L$ is closed under substitution while, in general $L^{\rightarrow}(X)$ does not need to be close under substitution.

In the literature one can also find another notion of modal entailment. Thus, by $L^{\rightarrow}_{\Box}$ we mean the consequence operation in the language $S_{\Box}$ defined in the following way: for any set $X \subseteq S_{\Box}$, $L^{\rightarrow}_{\Box}(X) \subseteq S_{\Box}$ is the least set of sentences containing L and X and closed with respect to modus ponens and the rule of necessitation:

$$\frac{P}{\Box P}$$

It is easy to check that for any classical modal system L the consequence operations $L^{\rightarrow}_{\Box}$ and $L^{\rightarrow}$ are sub-classical. It is also clear that all the classical tense logic, classical deontic logics and their combinations are sub-classical. More general, any extension of the classical logic is sub-classical. It would be a very interesting problem to find a reasonable logic which is not sub-classical.

There are some modal system in which the rule *RE* fail. Lewis' systems S1, S2, and many other can serve as examples of such systems. As a consequence no logical consequence based on that system is of the form $L^{\rightarrow}_{\Box}$ or of the form $L^{\rightarrow}_{\Box}$. Nevertheless, it is easy to check that modal consequences based on such systems are also sub-classical.

Conclusion. The results presented in this paper show that among inclusive, idempotent, monotonic and structural operations it seems to be impossible to find a good candidate for Strawsonian definitions of presupposition. It does not mean that Strawsonian definitions make no sens. To the contrary, it seems that this definitions mirror very important intuitions concerning presupposition. The results presented in this paper show that the apparent similarity of the operation involved in (1) and (2) to the notion of logical consequence is misleading.

The operation properly mirroring the intuitions of the notion of presupposition cannot be at the same time inclusive, idempotent, monotonic and structural.

Polish Academy of Sciences
Institute of Philosophy and Sociology
ul. Nowy Świat 72
00-330 Warszawa
Poland

Nicholaus Copernicus University
Department of Logic
ul. Asnyka 2
87-100 Toruń
Poland
e-mail: jacekm@uni.torun.pl

REFERENCES

Beaver, D. (1997). Presupposition. In: J. van Benthem, A. ter Meulen (eds.), *Handbook of Logic and Language*. MIT Press.

Hughes, G. E. and M. J. Cresswell (1996). *A New Introduction to Modal Logic*. Publisher: Routledge.

Kracht, M. (1993). Logic and Control: How They Determine the Behaviour of Presuppositions. In: J. van Eijck and A. Visser (eds.), *Logic and Information Flow*, pp. 89–111. MIT Press.

Levinson, S. C. (1983). *Pragmatics*. Cambridge: Cambridge University Press.

Malinowski, J. (1989). *Equivalence in Intensional Logics*. Warszawa: IFiS PAN.

Malinowski, J. (2004). Strawsonian Presuppositions and Logical Entailment. *Logique et Analyse* **185–188**, 123–138.

Segerberg, K. (1971). *An Essay in Classical Modal Logic*. Uppsala Universitet.

Strawson, P. F. (1949). *Introduction to Logical Theory*. London: Methuen.

Wójcicki, R. (1987). *Theory of Sentential Calculi: An Introduction*. Dordrecht: Reidel.

Janusz Mucha

THE CONCEPT OF "SOCIAL RELATIONS" IN CLASSIC ANALYTICAL INTERPRETATIVE SOCIOLOGY: WEBER AND ZNANIECKI*

ABSTRACT. Sociology has been often defined as a science of "social relations". The aim of this article is to contribute to the clarification of this concept. I take into account only two classic analytical sociologies — those developed by Max Weber and by Florian Znaniecki.

These sociologies seem to me only partly useful for the analysis of macroscale (ethnic, racial, industrial, and international) problems. They refer to human individual interactions within social collectivities, and not between them. If we follow *expressis verbis* the individualistic aspects of Weber's model and the analytical aspects of Znaniecki's theory, it would be difficult to base empirical investigations of social relations. The situation changes when we depart from the classics' ways of construction of the concepts of social relation and concentrate on the implication of the concept presented by the same authors. Fortunately, the research practices of both scholars encourage us to abandon a literal interpretation of their models. They present a variety of interesting typologies of social relations and social objects between which these relations can exist.

Introduction

Sociology has been often defined as a science of "social relations". A the same time, in the social science encyclopedia edited by Edwin R. A. Seligman and Alvin Johnson published during the inter-war period (1930), there is neither an entry for "relations" nor "social relations" (nor "relationships"). Nearly forty years later, in the international social science encyclopedia edited by David L. Sills (1968), there is no entry "social relations" or "relations" ("relationships") either.

Such important fields of social research as sociology of ethnicity, economic sociology, and political science are interested (respectively) in "ethnic relations", "industrial relations" and "international relations". There is a very large number of books on these topics, which proves that the concept of social relations is very vital in social science production. However, what is too often missing in the fields mentioned above is the analysis of the very concept of social relations. Therefore, we actually are not sure what the authors mean by

*Aleksander Manterys, Elżbieta Hałas and Włodzimierz Wincławski read the first draft of this article and gave me their comments. Needless to say, they are not responsible for my interpretation.
I am using the terms "relation" and "relationship" as equivalents.

In: J. Malinowski and A. Pietruszczak (eds.), *Essays in Logic and Ontology* (*Poznań Studies in the Philosophy of the Sciences and the Humanities*, vol. 91), 119–142. Amsterdam/New York, NY: Rodopi, 2006.

ethnic, industrial, international relations. The aim of this article is to contribute to the clarification of the concept of "social relations".

When dealing with the problematic of social ("human") relations, we should face the problem of the ways of conceptualization of similar social phenomena. The phenomenon which is the subject of this article is often conceptualized in terms of "social interaction". Even if we decide to ignore social psychology, we should take into account at least two interactionist traditions within contemporary sociology: exchange theories broadly understood, and symbolic interactionism. In this text, however, I am interested in stable and relatively durable phenomena and these cannot be reduced to "social exchange". Moreover, the sociology of ethnicity, economic sociology, political sciences and other fields of macrosociology conceptualize their problematic rather in terms of "social relations" than in terms of "interactions".

In this short text, I take into account only two classic ideas (out of necessity ignoring many important traditions, like for instance that coming from Karl Marx): those developed by Max Weber and Florian Znaniecki. I present similarities and differences between them. For me, these two ideas are examples of analytical sociology because of the way the authors construct the discipline: they looked for concepts referring to elementary units of the social realm and later built out of them, in a systematic way, concepts referring to larger social systems. Weber strongly influenced the tradition of what then became symbolic interactionism, and the ideas of Znaniecki can be treated as a variety of this interactionism. Therefore, both ideas belong to a kind of interpretative sociology which has been looking not only for a subjective sense of social phenomena, but also for their causal explanations. Both were developed more or less at the same time. Both were influenced by Georg Simmel, particularly by his *Soziologie* of 1908.

My intention in this article is first and foremost the reconstruction of Weber's and Znaniecki's conceptualization of social relations, and secondly the stressing of ideas which could help analyze relations between social groups on the macro scale.

Weber's Sociology of Social Relations

It is impossible to refer to the whole body of Weberian scholarship. I will limit myself to commenting on three well known examples. In the classic intellectual biography authored by Reinhard Bendix (1962), only Weber's macrosociology is taken into account. In the classic (and reprinted many times) collection of Weber's works edited by Hans Gerth and C. Wright Mills (1991), only such topics are covered as relations between science and politics; power; religion; and social structure. A well known German Weberian scholar, Dirk Kaesler, devotes, in his book on Weber's biography and work, eleven pages (1998, pp. 191–202) to his "general sociology", including a few pages to social actions (*Soziale Handeln*) and a few sentences to social relations (*Soziale Beziehungen*).

Weber's sociological ideas were presented first and foremost in his main posthumous sociological treatise, *Economy and Society*. His ideas of social relationships can be found in its Part One: "Conceptual Exposition" in Chapter I, "Basic Sociological Terms". The concept of social relationships is introduced *via* the concept of "social action", and it precedes the analysis of such concepts as usage, custom, self-interest, legitimate order, conflict, association, power and domination. In Chapter II, "Sociological Categories of Economic Action", we read of economic relationships. The scholar is interested, however, in social relations WITHIN communities like the household, neighborhood, commune, etc., rather than relations BETWEEN social collectivities or institutions, which would be more interesting in the present article. Bearing in mind Weber's chapters on economy and on ethnicity, I will concentrate here on the *Basic Sociological Terms*.

The book's chapter devoted to these terms is a presentation of the Weberian conceptual scheme (in the sense of a system of basic conceptual categories as well as the analytical relationships between them), or definitions of the concrete sociological concepts which are needed by the author. It also contains a broad elaboration of these definitions.[1] It should be stressed that Weber's is very far from a "monopolistic" approach and he states many times that he is interested only in a particular kind of sociology. Beside this kind there are other conceptualizations of this discipline.

What people do, or what they are involved in, is their "behavior — be it overt or covert, omission or acquiescence". Action is a behavior to which the actor "attaches a subjective meaning". "Action is 'social' insofar as its subjective meaning takes account of the behavior of others and is thereby oriented in its course" (1978, p. 4). Social action can be oriented to the past, present or expected future of one's behavior (1978, p. 22).

The "social" character of human action is not opposed by Weber to its potential "individual" character. On the contrary, Weber is interested first and foremost in the social character of individual action. Social character is ascribed to an action insofar as it takes into account the behavior of other people. In the above citation, Weber is using the plural form but it seems that this form is not important for him. An action of a human individual is of a social character regardless of whether or not it takes into account the behavior of another individual or of an aggregate of individuals.

Who is the subject (actor) of a social action? In the Weberian definition, we read of an individual. When he writes about "subjective meaning", however, we learn that the term may refer

> to a particular actor or to the average or approximate meaning attributable to a given plu-

---

[1] I am aware of the possibilities of misusing quotations but in order to be as close as possible to Weber conceptualization, I will very often use quotations of his work.

> rality of actors; or [...] to the theoretically conceived *pure type* of subjective meaning attributed to the hypothetical actor or actors in a given type of action (1978, p. 4).

"Understanding", so important in Weberian sociology,

> involves the interpretive grasp of the meaning present in one of the following contexts: (a) as in the historical approach, the actually intended meaning for concrete individual action; or (b) as in cases of sociological mass phenomena, the average of, or an approximation to, the actually intended meaning; or (c) the meaning appropriate to a scientifically formulated pure type (an ideal type) of a common phenomenon (1978, p. 9).

Therefore, we can conclude that when presenting a sociological approach and when presenting other cases than an action of an individual person, he is interested only in an aggregate of similar (from a given point of view), acting individuals, or in an ideal type of a certain action in a certain situation. For the action, "in the sense of subjectively understandable orientation of behavior exists only as the behavior of one or more *individual* human beings" (1978, p. 13). In another section of the book (in Volume II) we read:

> always [...] "social action" (*Gemeinschaftshandeln*) is for us an individual's behavior, either historically observable or theoretically possible or likely, in relationship to the actual or anticipated potential behavior of other individuals (1978, p. 1376).

Max Weber was obviously aware of the fact that there have existed social entities other than individuals. Examples are states, associations, business corporations, and foundations. Sometimes, they are treated by scholars, says Weber, as individual persons, as "subjects of rights or duties or as performers of legally significant actions" (1978, p. 13). This is understandable within the frameworks of juristic studies or for practical ends. In sociological analysis, for the subjective interpretation of action, these collectivities must be analyzed *solely* as

> resultants and modes of organization of the particular acts of individual persons, since these alone can be treated as agents in a course of subjectively understandable action (1978, p. 13).

However, sociology, even "understanding" sociology, cannot ignore collective concepts, even if they are interesting first and foremost to other social disciplines. There are three reasons for this necessity to deal with collective entities has three reasons. Firstly, it must if interpretive sociology wants to obtain an intelligible terminology. Secondly, the concepts of "collective entities" have a meaning in the minds of individual persons. They orient their actions to them, and they have a causal influence on their course of action. Thirdly, we must not ignore the "organic" or "functional" school in sociology which interprets individuals and their actions as organs of a social body (1978, p. 14). Therefore, the sociologist, even using the interpretive methodology, should accept collective entities. They are for him a point of departure for the understanding interpretation of individual actions (1978, p. 15).

After having analyzed four well known types of social actions (instrumentally rational, value-rational, affectual, and traditional), Weber presents his idea about what the topic of this article is: "social relationship". This term is used

> to denote the behavior of a plurality of actors insofar as, in its meaningful content, the action of each takes account of that of the others and is oriented in these terms. The social relationship thus consists entirely and exclusively in the existence of a probability that there will be a meaningful course of social action — irrespective [...] of the basis for this probability (1978, pp. 26–27).

Although Weber writes in other contexts about two partners of a social relationship, the citation presented above (as well as in other places in his works) shows clearly that the number of partners is not relevant. It is important in my opinion that social relationship does not have to be considered as a dyad. The dyad is only a simplification of analysis, useful for practical purposes. On the other hand, if there can be more than two partners in a relationship, we could ask the question if, for instance, in a triad we have one relationship or perhaps two or three of them. We could ask if a given person in a triad, taking into account the behavior of one partner, has to take into account the behavior of another at the same time, in order to create a social relationship. All these and many other similar questions were analyzed for instance by Georg Simmel, whose work was well known to Weber.

A "social relationship" is characterized by

> at least minimum of mutual orientation of the action of each to that of the others. Its content may be of the most varied nature: conflict, hostility, sexual attraction, friendship, loyalty, or economic exchange. [...] Hence, the definition does not specify whether the relation of the actors is co-operative or the opposite (1978, p. 27).

It is very important in my opinion to underline that Weber considers conflict or hostility as social relationships (and not, for instance, as the breaking of relationships) because, among other things, this approach must be based on the assumption that social relations appear not only within a unified social realm but can appear between two or more different parties. However, as we will see, Weberian stress on the unified entity is strong.

In the previous paragraph, the parties of social relationships were first and foremost human individuals. The scholar writes, however, also about other instances.

> Even in cases of such forms of social organization as a state, church, association, or marriage, the social relationship consists exclusively in the fact that there has existed, exists, or will exist a probability of action in some definite way appropriate to this meaning (1978, p. 27).

The context does not make it clear whether the author means relations between the "forms of social organization", within them, or perhaps both.

In the Weberian sense, social relationship does not mean that the partners understand each other very well, that they accept a common definition of situation.

> The subjective meaning need not necessarily be the same for all the parties who are mutually oriented in a given social relationship; there need not in this sense be "reciprocity" (1978, p. 27).

A social relation can, therefore, be unilateral or "asymmetrical". It still is a social relation, a case of mutual orientation, "insofar as, even though partly or wholly erroneously, one party presumes a particular attitude toward him on the part of the other and orients his action to this expectation" (1978, p. 27). According to Weber, a social relationship of completely and fully corresponding attitudes "is in reality a limiting case" (1978, p. 27).

Social relationships in Weberian sociology can be, in my opinion, typologized in various ways. The first typology would depend on the length of their duration. A social relationship can be of a fleeting character or be relatively permanent. In the second case, "there is a probability of the repeated recurrence of the behavior which corresponds to its subjective meaning and hence is expected" (1978, p. 28). It should be added that it is the outside observer and not the participants who, in Weber's view, judges that the above mentioned probability exists (1978, p. 28). Therefore, the social relation is a sociological, scholarly conceptual category and not a common sense category; it exists in the world of the observers, not in the world of the observed.

In the second typology, the subjective meaning of social relationships can be changeable or permanent. When, for instance, a political relationship based on solidarity transforms into a conflict of interests, the problem arises of whether a new relation has appeared or whether the old relation has perhaps acquired a new meaning. It is also possible that the sense of one dynamic relation is partly constant and partly changing (1978, p. 28). According to Weber, the answer to the above question (regarding the potential emergence of a new relation) depends only on terminological convenience and the continuity of change. Therefore, we can conclude that the decision concerning the character of a particular relation is at least partly in hands of scholars.

Let us return to relations with a permanent subjective meaning. This permanent sense, or content, can be formulated "in terms of maxims which the parties concerned expect to be adhered to by their partners on the average or approximately" (1978, p. 28). Participants of a relation with a permanent sense can orient their actions on the belief in the existence of legitimate order. If one's behavior is oriented toward "determinable maxims", if the order is "in some appreciable way regarded by the actor as in some way obligatory or exemplary to him," the order will be called valid (1978, p. 31). Also in cases of evasion or disobedience, "the probability of their being recognized as valid norms may have an effect on action" (1978, p. 32). However, says Weber, when evasion or contravention of the meaning of an order becomes a rule, the order is valid only to a limited degree or even not valid at all.

In the third typology, the sense of a social relation can emerge in a spontaneous way or can be agreed upon by mutual consent. Weber is interested first and foremost in the second possibility. When we have to do with incidences of this mutual consent being agreed upon, the rational participants of the relationship count "in some degree on the fact that the other will orient his action to the meaning of the agreement as he (the first actor) understands it" (1978, p. 28).

The fourth typology divides relationships into communal and associative. A social relationship is communal when the orientation of social actions of its participants "is based on a subjective feeling of the parties, whether affectual or traditional, that they belong together" (1978, p. 40). A social relationship is associative, when the orientation of social actions of its partners "rests on a rationally motivated adjustment of interests or a similarly motivated agreement [...]. It is especially common [...] for the associative type of relationship to rest on a rational agreement by mutual consent. In that case the corresponding action is, at the pole of rationality, oriented either to a value-rational belief in one's own obligation, or to a rational (*zweckrationale*) expectation that the other party will live up to it" (1978, p. 41). Most real social relations have, according to Weber, a partly communal and partly associative character.

We can ask now what the place in this typology of social "conflict" and "hostility" is, to which Weber devoted much of his interest. Let us start with conflict.

> A social relationship will be referred to as "conflict" (*Kampf*) insofar as action is oriented intentionally to carrying out the actor's own will against the resistance of the other party or parties. The term "peaceful" conflict will be applied to cases in which actual physical violence is not employed. A peaceful conflict is "competition" insofar as it consists in a formally peaceful attempt to attain control over opportunities and advantages which are also desired by others. A competitive process is 'regulated' competition to the extent that its ends and means are oriented to an order (1978, p. 38).

Now we can answer the above posed question: "conflict and communal relationships are relative concepts" (1978, p. 42). This phrase was explained a little later by Weber:

> The communal type of relationship is [...] the most radical antithesis of conflict. This should not, however, be allowed to obscure the fact that coercion of all sorts is a very common thing in even the most intimate of such communal relationships if one party is weaker in character than the other. [...] Associative relationships, on the other hand, very often consist only in compromises between rival interests, where only a part of the occasion or means of conflict has been eliminated, or even an attempt has been made to do so. Hence, outside of the area of compromise, the conflict of interests, with its attendant competition for supremacy, remains unchanged (1978, p. 42).

According to Talcott Parsons, conflict is, in Max Weber, the third, beside communal and associative, basic type of social relationship (see 1949, p. 653).

The last, fifth, Weberian typology of social relations divides them into open and closed. It is valid in case of both communal and of associative relationships.

> A social relationship [...] will be spoken of as "open" to outsiders if and insofar as its system of order does not deny participation to anyone who whishes to join and is actually in a position to do so (1978, p. 43).

A social relation is

> "closed" against outsiders so far as, according to its subjective meaning, and its binding rules, participation of certain persons is excluded, limited or subjected to conditions. Whether a relationship is open or closed may be determined traditionally, affectually, or rationally in terms of values or of expediency. (1978, p. 43).

When discussing open and closed relationships, Max Weber presents another interesting characteristic of social relations. An "organization" (*Verband*) is such a closed relationship, in which

> regulations are enforced by specific individuals: a chief and, possibly, an administrative staff [...]. "Organized action" is (a) either the staff's action, which is legitimated by its executive or representative powers and oriented to realizing the organization's order, or (b) the members' action as directed by the staff (1978, p. 48).

It seems to me that in this sentence we have a presentation of one of the determinants of "collective action".

Znaniecki's Idea of Social Relations

Florian Znaniecki's sociological theory, as much as Max Weber's theoretical system in this discipline, is based on the concept of "social action" and the concept of "actor" (in Znaniecki – "agent"). Both scholars continued the German tradition of the anti-positivistic current in the humanities, as represented for instance by Wilhelm Dilthey, Wilhelm Windelband and Heinrich Rickert. Both worked, at least for a short time, in the same period, but relatively independently from each other. Weber (who died in 1920) hardly had a chance to know Znaniecki's theoretical work. Weber is not quoted in Znaniecki's *Wstęp do socjologii* [Introduction to Sociology] of 1922, in *Laws of Social Psychology* of 1925, in *Socjologia wychowania* [Sociology of Education] of 1929–1930, in *Ludzie teraźniejsi i cywilizacja przyszłości* [Contemporary People and the Civilization of the Future] of 1935, or in *Modern Nationalities* of 1952. He is quoted is *Social Actions* of 1936 twice, in relatively unimportant matters. He is mentioned (after Leopold von Wiese) in *Cultural Sciences* of 1952 as a scholar who introduced the concept of "ideal types", who stressed the importance of the analysis of social actions in sociology and who, unlike von Wiese, did not develop a general theory of interaction between social agents (see 1971, p. 658). He is quoted three times, again in unimportant contexts, in the posthumously published "Social Relations and Social Roles". It is interesting that in the context of social relations, Znaniecki quoted (not very often, though), Leopold von Wiese and Georg Simmel in *Wstęp do socjologii*, *Modern Nationalities*, *Cultural Sciences*, *Social Actions* and in *Social Relations and Social Roles*.

There is a very large number of books and articles devoted to Florian Znaniecki's sociological theory. This is particularly true in Poland (see, recently, Hałas 2000). These works on Znaniecki rarely refer to his sociology of social relations, but much more often to his sociology of social actions. Unlike Zbigniew Bokszański (see 1972) but like for instance Mariola Flis (see 1978), I will treat Znaniecki's theories of social actions and social relations here as a coherent whole, in spite of some evolution of the scholar's ideas. This evolution, and particularly the emergence of new ways of interpretation, will be taken into account in this article.

According to Mariola Flis, Znaniecki's model of action refers first and foremost to individual human acts. She says, however, that this model can also be used when studying group dynamics. Collective (group) action is not, for Znaniecki, a sum of individual actions, but it does consist of them (Flis 1978, p. 42). Flis writes that Znaniecki did not present anywhere a coherent and consistent idea of collective (group) actions but that he outlined a general sketch of theory. Whereas a social action of an individual is a system which ties together a subjective attitude and an objective social value, a social group action is a synthesis of individual actions which ties together two objective elements of culture: an ideological model of attitudes and social value (Flis 1978, pp. 42–44; see also Znaniecki 1971, pp. 463–478). In the present article, I would like to stress that Znaniecki refers to collective (group) actions *expressis verbis*, that he shows clearly what their basis is and suggests what the nature is of the above mentioned synthesis of individual actions in collective (group) action.

The concept of action as presented in *Wstęp do socjologii* is "idealistic" in character in the sense that every human act has a mental character, and "every ideal act brings, in a direct way, some real consequences" (Znaniecki 1988, p. 83). Since every action is ideal and brings real consequences, it is always creative (1988, p. 90). The human action is of a social character when its object is another human individual or a social group and if the purpose of this action of the subject (agent) is to exert influence on the action's object. Every subject of action (agent) can also be an object of somebody else's action and the other way around (1988, pp. 278, 280; see also Znaniecki 1973, pp. 322–373). Therefore, there is no reason to question the possibility of a social group's being an agent. Moreover, a social group cannot be reduced to an aggregate of individual actions because, the scholar writes, "more complex systems cannot be reduced to less complex systems [...], since every higher system is something more than a combination of lower systems" (1988, p. 285). One of the proofs of the subjective reality (and it is only the subjective reality that is interesting for Znaniecki) of social groups is the fact that people give them names (see, Znaniecki 1971, pp. 256–257). A group can "act and experience in a collective way"; being a "collective will", it is a social agent (1988, pp. 310–311). In *Cultural Sciences* of 1952 we read that what can be said about individual social actions, can also be said, although with some modifications, about collective actions which are "cultural forces". In the posthumous *Social Relations and Social Roles* the author says that "a social action can be performed individually or collectively" (1965, p. 16).

If we look at the problem from the point of view of the agent, we see, says Znaniecki, that

> a general determining characteristic of a social group [...] is its very ability to act collectively, to present itself as one entity, to focus individual actions into one common action — whether performed by all group members at the same time or divided among them or, finally, performed by a representative of the group, whose action is supported by potential actions of the remaining members (1988, p. 315).

Therefore, group action can have various shapes. It can be (a) an action of all members at the same time, (b) a common action being a result of the division of labor among the members, or (c) an action performed by authorized representatives. Regarding the first point we can ask, however, if it is necessary that ALL members of the group agent act; perhaps it is enough if only a part (though not authorized representatives), acts. In sum, the author's conclusion is that the total "activity of a group is a permanent synthesis of the social actions of its members" (1988, pp. 316, 317). In *Social Actions* of 1936 we read that a

> collective action, an action performed together by two or more individuals who have a common purpose, is a social action only if the purpose itself is social, i.e., if the action bears upon some other individual or collectivity as a social object and tries to provoke a social reaction by the use of social instruments and methods (1967, p. 168).

In the course of its action, a social group undergoes some transformations. A "collective task" is an element determining the group from the point of view of its future (and the pursuit of some ideal introduces a "temporal aspect" to its nature). "In the conscious collective attempt to perform its task a social group adapts to it its own way of functioning, normative demands directed to its members, and eventually, sometimes, its own composition and its form" (1988, p. 321).

After this introductory presentation of social actions, and particularly group actions, we can move to the subject matter of this article — social relations. As I have already mentioned, according to Znaniecki, in social life the acting subject can be also an object of social action:

> social value to which the action refers, i.e., individual or group who is the action's object, in this character in which this object is given to the agent. [...] object of action is at the same time an agent (1988, p. 280).

This idea leads us directly to the problematic of social relations.

> A relation [...] is a combination of two series of dynamic social acts of two individuals or groups, or a group and an individual, as partners of the relation; each partner is a subject of its own series of acts and an object of a series of acts of the second partner (1988, pp. 280–281).

Let us pay attention to the fact that the author has in mind the social relations between one individual and another individual, a group and another group, and finally between an individual and a group. In my opinion it is an important observation. It should also be mentioned that according to Znaniecki we have to do with a social relation when three additional conditions are met. They are as follows: (a) normative regulation of both series of social acts, (b) relative permanence, and (c) such a connection between these acts that "activity of an individual or group A is a *ratio* of activity of an individual or group B, and the other way around" (1988, p. 281; see also, for the analysis of normative regulation and permanence, 1971, pp. 660–665). In *Socjologia wychowania* of 1928–1930, the definition of a social relation is only slightly different. This relation is a "closed system" consisting of four elements. They are as follows: (a) two partners of relation (individuals or groups), (b) the partners' obligations, (c) a link or social contact between them (Znaniecki means first of all some common values which are, for both partners, a necessary and sufficient condition of their mutual interaction), and (d) the partners' duties. The difference between obligations and duties is interesting to me. A partner's obligation is his conduct towards the second partner according to the norms, whereas duty is

> the producing of some schematically regulated situations which would enable the second partner the performance of his obligation, or prevention or removing of some situations making this performance impossible (1973, p. 135).

Let us return to *Wstęp do socjologii*, though.

Independently of other similarities, Znaniecki's theory resembles the Weberian model in the fact that a social relation is considered to be a system of (first and foremost) two acting partners and the fact that this relation is basically a part of a larger concrete social collectivity. In Znaniecki's opinion it does not have to be so, but it usually is so, which has important consequences.

> A particular social relation can be unisolated and belong to a more general system of organization of a social group; in fact, majority of permanent and recurring relations is in this way included in a larger entity. In this case, to the above mentioned elements of this social system we must add the sanction which is to induce an individual (or smaller group) to unconditional realization of conditions that are to be a real base of the activities of the second partner; the same demand is posed by the sanction on the second partner (1988, p. 282).

It should be stressed that the common base of a relation, which is made up of the larger collectivity can, but does not have to, exist. When it does not exist, the norms regulating social relation are "empty", which means that there are no sanctions to support them. I will return to this issue. What is also important (but does not appear in the above citation), Florian Znaniecki clearly distinguishes between a social and a legal relation. The difference lies in the fact that a legal relation can be based only on agreement between the partners, whereas the social relation demands a real interaction between individuals or groups.

I have already written that an important feature of social relations and of the uniformity of many (meaning not all) of them are norms. They determine the activity of partners. Social norms have a reciprocal character. As we remember, they can be enforced by sanctions. Norms, the observation of which is demanded from a concrete partner, are social obligations. When the observation of norms is not enforced, they stop being "actual". Obligations, however, last even when an agent does not follow them, "if other elements of social relation exist. On the other hand, obligation ends when the social relation ends, while the norm, although it ceases to be actual, never does end and always can be resumed" (1988, p. 293). Every social relation contains at least two obligations which are different for different partners. They are, however, always obligations of all the parties.

As it turns out, according to the author of *Wstęp do socjologii*, a social relation does not have to imply a real interaction between the partners. Social contact is necessary, however, and it is this element of the platform (base) common to both partners which enables and determines the interaction and which makes their mutual obligation actually observed. It is social contact which creates a permanent, purely social unity between individuals and/or between groups which enter a social relation (1988, pp. 297–299).

In *Socjologia wychowania*, published a few years after *Wstęp do socjologii*, Znaniecki's ideas are formulated in a slightly different way. Now he stresses the problematic of the educational relation, a relation between an individual and a group to which he belongs. In this book, the concept of relation is introduced not so much by the concept of social action, as by the concept of contact. The latter concept has a slightly different meaning than in *Wstęp do socjologii*. In the chapter devoted to social groups we read that social contact (direct or indirect) occurs when one group

> performs a collective action (either in the form of action of all its members or by its representatives) which influences the social content of the second group, thereby modifying its value system and causing its reaction (1973, p. 62).

If this kind of situation takes place, we have to do with "co-existing groups". Sometimes, between such groups there is not a loose and fleeting contact but a permanent social relation regulated by social norms. In this latter case, each group has some obligations toward the other group,

> which means that it accepts the fact that it should perform toward it some collective actions and refrain itself (or make its members refrain) from some other actions and that it expects that the second group observes other or similar obligations toward the first group. In this situation, we are saying that the groups are *related to each other*; the social content of each of them is under a permanent influence of the social content of the other group and it, itself, influences that other group (1973, p. 62).

The permanent influence of the groups on each other and their obligations toward each other can be symmetric or asymmetric. When they are symmetric, we have to do with a relation of equality. The scholar gives the examples of nation-states united by a treaty, relations of a state and the Church (when regulated by the concordat). It seems to me interesting that in this model of the group there is a relation of equality between groups only when they are "positively" tied with each other. Znaniecki does not take into account a potential "equal" relation of conflict. In the second of the cases presented above (asymmetry), "when one group fully or partly determines both its own obligations and obligations of the second group", we have to do with relations of "domination/subordination" (1973, pp. 62–63). Relations between a state and the public schools, between the state and its constituent communities, between the Church and a religious order are examples given here by Znaniecki. As we can see, subordination is discussed in this context when one group can be considered a part of another group.

Like Max Weber, in the 1920s Florian Znaniecki was interested first and foremost in intra-group relations. However, when he gives examples of social relations, he lists, in addition to the marital relation, the relations between employer and employees, relations between a group and its members — also a "treaty between two nation-states" (1973, p. 134). In the latter case, he does not say that this relation takes place within a larger social arrangement, like a union, federation, or other superior entity consisting of these nation-states. This idea was not developed by the author, though. In a short essay *Potrzeby socjologii w Polsce* (Needs of Sociology in Poland), published simultaneously with *Socjologia wychowania*, intergroup relations are closed within a larger collectivity. Here, the author lists "marital relations, relations between the older and the younger generations, relations between employer and employees, between leaders and followers, etc." (1984, p. 138).

As I have already mentioned, what *Socjologia wychowania* is mainly concerned about is a relation between a group and its member(s). The author stresses that in this situation the group is the dominating partner and it "naturally presents much higher demands to the member than to itself" (1973, p. 213). The group member is obliged to do for the group everything that this group considers, "traditionally or rationally", necessary for its own functioning, while the group is obliged to do toward its member only as much as in the group's own view is necessary "to enable the member to perform dutifully his obligations toward the group" (1973, p. 214).

In *Socjologia wychowania*, the author suggests a possibility of a typology of social relations according to the common intentions of the partners (1973, p. 136). This suggestion was never realized, though.

In the above mentioned book, *Social Action*, the concept of a social relation is introduced first and foremost by the concept of action, but also by the concept of social contact. The content is, however, different than in the contexts analyzed earlier in the present article. The social contact is characterized in the following way. One agent, performing his actions, intentionally or unintentionally modifies a certain value which has been, or becomes, the object of the attitude of another agent. The first agent, though, does not intend to influence the second agent. Simultaneously or subsequently, the first agent, the second agent, or both agents experience the fact that the activity of the first agent modifies a value which is significant for the second agent.

> We call this value the *vehicle of social contact*. If, however, the purpose of the activity is to influence the other agent by modifying his values, we should then speak no longer merely of social contact, but of social action (1967, pp. 56–57).

Social contact can occur not only between individuals, but also between collectivities. There is no contact between collectivities, though, when what one of them does with its own values, does not influence the values of the second collectivity, meaning when the values of the first are not the values of the second. According to the author of *Social Action*, for a social contact to exist, there must be values common to both collectivities, but the significance of those values is different for both collectivities. Each of them tends to do something different with them, if one of them is to become interested in another, as a social object. This difference in the significance of the value is most obvious when it appears as opposition. Therefore, the collective interest in other collectivities appears most easily in cases of conflict. The opposition must assume, however that there is a social contact on the basis of common values. On the other hand, a difference between the collective attitudes toward the same value is not an opposition if these attitudes are considered not as mutually exclusive but as mutually supplementary (1967, pp. 63–64). War is an example of collective opposition toward a certain collectivity. It is interesting to note that for Znaniecki the "collective acts of opposition against collectivities do not always and necessarily express antagonistic tendencies of the individuals participating in them" (1967, pp. 450–451). Therefore, a collectivity is a superindividual body.

Earlier, in the context of the sociology of education, Znaniecki wrote about such inter-group relations as domination and subordination. That context suggested groups independent of each other. These groups do not have to be independent, though. Sometimes we have to do with cases of voluntary, collective, participative submission of a community or group to a small dominant body composed of its own members. Znaniecki is of the opinion that this kind of situation is difficult to study and has not been adequately analyzed yet. Individual active submission to a collectivity is a phenomenon easier to study (1967, pp. 255–156). As we remember, the latter issue was analyzed in detail in *Socjologia wychowania*.

In the previous paragraphs, I have mentioned the relation of opposition. Now, I would like to discuss the problem of hostility. This system of actions is not called by him a social relation. I think, though, that it meets the conditions presented earlier. The scholar does not say if hostility occurs only between individuals or also between collectivities, but the examples suggest that the latter was also taken into account by him. Moreover, these examples are important also from the point of view of the aims of the study of social relations, as suggested at the beginning of this article.

Znaniecki distinguishes four kinds of hostility and discusses some of them. First, there are actions tending to prevent social contacts between the agent and the social object, by *avoiding common experiences*. This case is called "avoidance". Avoidance of communication, as the scholar explains, is a mark of hostility when one agent does not wish to share experiences with somebody whom he values negatively despite the fact that he does not expect any definite harm from him and even knows that he could draw some benefits from the communication. Avoidance of communication can coexist with practical interaction in impersonal matters. This happens, according to Znaniecki, in cases of race and class relations. He points particularly to relations between Jews and Christians. The second kind of hostility is actions which tend to *eliminate the attitudes* of the social object from the agent's sphere of valuation. This kind of hostility is called "aversion". Aversion implies that the agent, aware of some sentiments of the social object, evaluates them negatively as a psychological phenomenon. Sometimes both the agent and the social object appreciate the same value positively or negatively, but the sentiments on the part of social object provoke an aversion in the agent and make the value undesirable to him. Thirdly, there are actions which tend to *frustrate activities* of the social object by counteracting them. It is here, according to Znaniecki, where the term "hostile fighting" is the most appropriate. Fourthly, there are actions deliberately *destroying* the social object's system of values not because of what he can do to the agent but because of what he has already done. Such a situation can be called "revenge" (1967, pp. 569–573).

The problem of conflict was interesting for Znaniecki also in his *Cultural Sciences* of 1952. As in Georg Simmel, we read here, for instance, about situations when a collective conflict against a hostile group influences the group's solidarity. The most interesting conflicts, from this point of view, are religious ones (1971, p. 605). The problem of conflicts between cultural groups was a subject of Znaniecki's studies even earlier. It is visible not only in *Cultural Sciences* and in *Modern Nationalities* (of the same year) but also in his *Studia nad antagonizmem do obcych* (Studies of Antagonism to Strangers) of 1931. Let us look at the problem chronologically, without going into unnecessary details.

In *Studia* ..., we find an interesting (and important in a somewhat different context) analysis of the nature of the relation between a social group and the "stranger". This is the problematic of relations between the group and the individual which is well known to us. In my opinion, we do not find here, unfortunately, significant, new, or more general ideas concerning this type of relation, but we read about unilateral and bilateral contacts, intentional and unintentional influences, about a contact between the agent and the social object and about the crossing of their own spheres of activity, and about contacts which appear based on common or divisible systems of values. We learn here that at least part of the value system must be common to the partners in order to have a contact. We find the important observation by Znaniecki that the human object, "who does not belong to the same social group to which the agent belongs, is experienced by the agent as a 'stranger', whenever he presents himself to him as actually or potentially entering by his action (intentionally or unintentionally) the space of this group, which is a closed system of social values" (1990, p. 303), but on the other hand, when the contact is based on common value systems, the same social object will not be treated as a "stranger" (1990, p. 309).

I have already written about antagonism to strangers as a relation between the individual and the social group. Znaniecki pays attention, however, to the fact that both the agent and the social object in this relation can be not only individuals but also social groups. In his opinion, the negative "intentions, the object of which is a strange group, differ from those, which refer to the individual considered to be strange, since they manifest in indifferent actions" (1990, pp. 327–328). In the same place we read, however, that the difference, even if real, "became by many sociologists exaggerated and at the same time obscured" (1990, p. 328). We learn about these differences that they do not refer to the nature of action, that they can appear or not, and that the collective antagonism is sometimes less reflexive and more spontaneous than the individual antagonism, but on the other hand it "sometimes contains in the eyes of participants the character of objective ethical justification, a specific 'loftiness' " (1990, pp. 328–329).

There is also an intermediate type between collective antagonism to an individual and antagonism to a collectivity. It is represented by situations in which

> although antagonistic activity refers to one or a few individuals, but rather to them not so as to individual persons, but as representatives of a certain (ethnic, class) group (1990, p. 328).

When we take into account the fact that antagonism can have an individual or a collective character, we can point out its four types, distinguishable only analytically. Firstly, we have antagonism in which the agent and social object are of a collective character. An example is a relation between a tribe, a sect, or a nation and another tribe, sect, or nation. Secondly, we have antagonism in which the agent is an individual and the object a collectivity. An example is hatred of an individual for a strange group by which the individual was, in his opinion, harmed. Thirdly, there is antagonism in which the agent is a collectivity and the object is an individual. Examples are the relation of villagers to a strange newcomer, of the ancient Greek aristocracy to the populist tyrant, and of Englishmen to Napoleon Bonaparte. The last, fifth type of antagonism is that between an individual agent and an individual object (1990, pp. 330–331). We can note that antagonisms, as presented by Znaniecki, are not strictly speaking social relations, because we do not have here bilateral courses of actions. However, it was worthwhile to discuss them here, since in the scholar's opinion they are important bases of social relations. When he writes, for instance, about "collective defensive antagonism," he presents it as a non-reflexive antipathy to the meeting of strangers, as "an attempt to react against any trial of contact on the part of others" (1990, p. 337). Therefore, we have here a kind of bilateral and practical "collective intention". Similarly, we read about group prejudices (which cannot, obviously, be considered social relations) that they are not false judgements and unjustified evaluations resulting from inadequate knowledge of a strange group, but are active defensive intentions, directed against potential spiritual ties with the strange group. Group defensive antagonism, we read later, passes into "aggressive" antagonism, when the strange group acts in such a way as to suggest its attempt to break such defensiveness down (1990, pp. 349–351).

In the 1931 essay *Siły społeczne w walce o Pomorze* (Social Forces in the Conflict over Pomerania) we find an important thesis regarding concrete social relations. The author writes that with other conditions stable the expansive group always wins in its conflicts with the exclusive, closed group (1990, p. 380).

If we look, from the point of view of the theory of social relations, at the book quoted above, *Modern Nationalities*, we find equally interesting ideas. Let me begin with an "editorial" issue — the arrangement of the material. Znaniecki discusses conflicts between nationalities earlier than the cooperation of societies with national cultures, even if the substantive importance seems to be the opposite in his theoretical system. He devotes nearly identical space to these issues. In the following analysis, I will briefly begin with the latter issue. Cooperation within nations, says Znaniecki, results from the necessity of defense against an enemy (1990, p. 206). The scholar stresses cooperation between various nations in different areas of culture, but he links this cooperation with the universal intention of cultural expansion (1990, pp. 208–209).

As in other works, in *Modern Nationalities* Znaniecki also does not present conflicts between nations as social relations, which is difficult to understand in the light of his most general theory of actions and of relations. The scholar says that sociologists should study "relations" occurring between nations in the times preceding the outbreak of mutual hostilities (1990, p. 166). He is of the opinion that all nations intend to expand (which differentiates them from, for instance, tribes or rural communities). Expansion can be of a creative character, by means of the development or enriching of national culture, or of a populational character, by means of gaining new members out of those people who, according to ethnic criteria, belong to the national group but were not aware of this fact. Expansion itself does not have to raise conflicts as understood (but barely mentioned) by Znaniecki. Conflict occurs only in situations when a given nation intentionally expands at the cost of another nation or when expansion of one nation is an obstacle to the expansion of another nation (1990, pp. 166–168). This kind of expansionist intention is called "aggression". It is not clear if both the above conditions formulated by Znaniecki must be met; it is not clear if conflict can be said to occur when a nation expands unintentionally; it is not clear if it occurs when expansion of one nation is not an obstacle to the expansion of another. It seems to me that conflict occurs, for Znaniecki, only when both groups undertake overt actions directed against each other. Only in this sense, in my opinion, does expansion or aggression not have to be considered a conflict.

There are, in Znaniecki, four types of aggressive expansion. It has, firstly, a geographical character, when group members migrate to other territories on which the native population still lives. The resistance of the latter will be a source of conflict. Separatism of both groups is not conflict, but it makes its occurrence possible. Aggressive expansion can be, secondly, described as economic, when it depends on the exploitation of the property and work of another group. Znaniecki discusses examples of such exploitation with the help of coercion, but he is of the opinion that it can also occur without coercion. What is called in other literature colonialism and neo-colonialism are examples given by Znaniecki. Conflict occurs when the nation subordinated by expansion realizes this and becomes united in its active intention to alter this state of affairs. Aggressive expansion can be, thirdly, of an assimilationist character. This is actually a continuation of population expansion. The author means here assimilation by a nation of a population which, according to ethnic criteria, does not belong to this nation. As a result of this expansion, resistance and conflict can occur. Conflict is also possible when two different nations wish to assimilate a population which does not belong to either of them. Very important, in my opinion, is Znaniecki's idea that serious conflicts occur when a group culturally assimilates another group but does not want to allow it to take advantage of all the privileges to which the members of the assimilating group are entitled. Aggressive expansion can be, finally, of an ideological character, when one nation wants to draw to its side the cultural leaders of other nations to win them over to its own ideas. Here, too, conflicts can occur (1990, pp. 168–195).

I will move now to those Znaniecki's ideas regarding social relations, which were presented in the large book *Social Relations and Social Roles* posthumously published in 1965. From the point of view presented at the beginning of this article, that is from the point of view of the analysis of inter-group relations of various characters, *Social Relations and Social Roles* seems to me much less useful than the books discussed above. Although in the opening fragments we read a suggestion that "sociology is a science of human relations or social relations, in the sense of relations between interacting human individuals or groups" and that there are two basic kind of interaction — cooperation and conflict (1965, pp. 15–17; see also pp. 19, 85) — soon after these sentences we read that social relations are first and foremost interactions between individuals and that they are of a cooperative character. Znaniecki is interested mostly in relations within social collectivities and much less in relations between collectivities.

As the above citation shows, the concept of social relation is introduced here by means of the concept of interaction. Therefore, we have to look at the discussion of interaction.

> When social interaction between two individuals occur, usually one of them, A, initiates it by performing a social action intended to influence another, B; then B 'reacts', i.e., performs an action in consequence of A's action. Sometimes each of them independently starts a social action bearing upon the other, and each reacts to the other's action [...]. In any case, the connection between a particular social action and the reaction to it is an elementary fact of social interaction, though it may also be a component of a more or less complex system (1965, p. 86).

It seems to me that for Znaniecki of this period, as in earlier periods, interaction is a social relation when it is permanent, organized, and normatively regulated.

Znaniecki is of the opinion that social thinkers have devoted more attention to social conflict than to social cooperation. In modern times, however, they might find objective reasons for the priority of cooperation, since the latter is essential for the very existence of human collectivity.

> Moreover, investigation of the origin of interindividual conflicts indicates that many of them are due to external influences which disturb such cooperation as already existed, while intergroup conflict presupposes close cooperation within each conflicting group. This means that, without adequate knowledge of social cooperation, no scientific generalizations about social conflict are possible (1965, pp. 17–18).

Therefore investigations of social cooperation is the primary task of sociology, and investigation of social conflict is its secondary task.

Let us return to the characteristics of social relations, of cooperation, and of conflicts. As I have already mentioned, Znaniecki pays much attention in *Social Relations and Social Roles* to the cooperative relations in communities. He is aware of the fact that in each community there is a lot of conflicts which interfere with the normatively regulated course of social relations. “Such conflicts have a disorganizing influence [...] on social relations as organized systems [...]” (1965, pp. 91–92). Again, we can see that conflicts are not social relations. The author devotes to them much attention in the chapter on social relations in communities, and in the chapter on the impact of social conflicts on social relations. He says that

> within certain relations some conflicting interaction is not only considered compatible with positive valuation of the individuals involved, but is required by the cultural pattern of the relation, as in competitive companionate play — sports, games of chance, and intellectual contests (1965, p. 92).

Conflict itself can be based on some standards and norms which are common to its parties. Conflicts do not have to engender disorganization. They can be anticipated by leaders, who cooperate to check or counteract them. As a result, they can bring the development of new standards and norms regulating social interactions (1965, pp. 93–94). This idea is similar to the well known discussion by Georg Simmel at the turn of the 20th century about the positive functions of social conflict (see, e.g., Simmel 1964).

In *Social Relations and Social Roles*, Znaniecki intended (which is very clear in Chapter Four) to present a general typology of all possible social relations, but he had no time to do it. He presents, however, his own conclusions of his discussion. He says that social relations within communities are in fact permanent social processes. All the time new social relations emerge, since newcomers enter communities, and individuals already there become partners in relations in which they earlier did not participate; former social relations come to an end. Moreover, in the course of this “evolution” new patterns of relations cross the communities’ borders and reach neighboring groups.

Summary

The classic, analytical concepts of the social relation which were presented in this article seem to me only partly useful for the analysis of ethnic, racial, industrial, and international problems. These concepts are first and foremost systems of ideas referring to interactions within social collectivities, and not between them. To the extent to which the problems mentioned above occur within collectivities (and, obviously, we usually can find a social system that is an umbrella “covering” them), these theoretical models can be useful. To the extent to which we have to do with issues between collectivities and when it is not reasonable to look for an “umbrella” system, it is more difficult to take advantage of them.

In the declarations of the authors Max Weber and Florian Znaniecki the most important partners of social relations are individuals and not social groups. The concept of the social relation was introduced by both scholars first of all by means of the concept of individual social action. What is common to Weber and Znaniecki (I mentioned the differences earlier in this text) is the concept of intentionality and the subjective sense of action. The actor/agent is mostly an individual person. However, in Max Weber the actor can be also a plurality of similar (from a given point of view) individuals or even collective bodies like states, communities, etc., consisting of (in a way difficult to reconstruct) individual actions having a subjective sense. In Florian Znaniecki, social actions are always directed toward individuals or groups and are intended to modify the attitudes of their social objects. The social object can be the subject of another action and the other way around. Collective action can also be conceptualized, and it is a synthesis of individual social actions which rests, as it seems to me, either on the simultaneous action of all members of the group or on the division of labor between the members or, finally, on the action of the authorized representatives of the groups. Collective action in the third case is similar to Weber's idea of action performed by the "administrative staff" of a social group.

In Max Weber, a social relation takes place when the social actions of a number of people are oriented toward each other in such a way that there is a chance that the social action of one actor will be met by the reaction of the other. There can be more than two partners in a social relation (although the author writes most of the time about two partners and does not explain how to conceptualize the relation if there are more partners). According to Weber, social relations do not have to consist in cooperation. Potentially we have here the base for investigation of various kinds of relations (cooperation but also conflict; within a social entity or between social entities) between various partners, but the German scholar limits himself to suggestions regarding social relations within communities or associations. Social relations are enacted, in Weber's model, first and foremost between individual partners, but he allows the possibility of relations between collective bodies like states, political parties, economic organizations or churches. It is not clear, however, how he would like to conceptualize these relations. Partners in a social relation do not have to accept the same definition of situation; that would be only a limiting case of the relation. Social relations in Weber's model can be typologized in various ways. In the present article the following typologies were presented: permanent or fleeting; organized or unorganized; keeping their particular sense all the time or changing the subjective sense; emerging spontaneously or agreed upon; communal, associative, or conflictual; open or closed.

In Florian Znaniecki, social relations are systems of actions in which each partner is a subject of one course of actions and object of the second course of actions, and each of these actions is a *ratio* for the second. Relations can occur between individual and individual, group and group, or between an individual and a group. Relations do not have to be symmetrical in the sense that the influences of partners upon each other do not have to be equal. Social relations are always relatively stable and subject to normative regulation. Social relations are, in this conceptualization, subsystems of a larger whole, even if they are inter-group relations. We find here, however, the important idea of relations between one group and a strange group, and between a group and a strange individual. The opposition between individuals, between groups, or between individuals and groups is always based on a common value system. Znaniecki analyzes in detail various situations of conflict and hostility between partners and devotes a lot of space to them, but in his conceptualization they do not constitute social relations. Social relations seem to be here always of a "positive" character. Florian Znaniecki discusses four types of antagonism, but again it does not seem to me that, literally speaking, they are social relations in this conceptualization. They seem to be rather mental aspects of not very clear ties between individual and collective partners. When the author writes about aggressive expansion, which is typologized by him and elaborated in detail, he does not present it as a social relation.

Despite the initial stress on the individual action as the basic social system, in Znaniecki's sociology of nationalities and in his sociology of culture we find an analysis of collective action, and, what follows, collective relations. In spite of the fact that the Polish-American scholar declared that social relations are only "positive" systems of actions, in fact he investigated in an interesting way many aspects of conflicts between cultural groups.

We can ask here where to locate the ideas presented above on the continuum "individualism-holism" and continuum "micro-macro" sociology. Weber's model, particularly in his own declarations, is very individualistic, although he did not deny the real existence of collective entities. Znaniecki's theory is rather holistic. Despite the stress on the significance of individual social action, the scholar showed clearly that this was itself a social system. The ideas of both scholars overcome, in my opinion (or perhaps ignore), the distinction between the macro and micro spheres of social life. The stress is put on individual relations, but examples are given most of the time from such social entities as the family, churches, local communities, and nation-states. I did not see here any methodological discussion of the problem of distinction between micro and macro. Both models have, in my opinion, a relational character (see Ritzer and Gindoff 1992; Emirbayer 1997).

If we follow *expressis verbis* the individualistic aspects of Weber's model and the analytical aspects of Znaniecki's theory, it would be difficult to base empirical investigations of social relations as presented at the beginning of this article on them. The situation changes when we depart from the classics' ways of construction of the concepts of social relation and concentrate on the implication of the concept presented by the same authors. Fortunately, the research practices of both scholars encourage us to abandon a literal interpretation of their models. They present a variety of interesting typologies of social relations and social objects between which these relations can exist.

AGH University of Science and Technology
Kraków
Poland
e-mail: jmucha@post.pl

REFERENCES

Bendix, R. (1962). *Max Weber: An Intellectual Portrait*. New York: Doubleday.

Bokszański, Z. (1972). Pojęcie działania społecznego w teorii socjologicznej Floriana Znanieckiego [The Concept of Social Action in Florian Znaniecki's Sociological Theory]. *Studia Socjologiczne* **2**, 241–258.

Emirbayer, M. (1997). Manifesto for a Relational Sociology. *American Journal of Sociology* **103** (2), 281–317.

Flis, M. (1978). Wartość, działanie, osobowość – w systemie socjologicznym Floriana Znanieckiego [Value, Action, Personality – in Florian Znaniecki's Sociological System]. *Studia Socjologiczne* **1**, 31–58.

Gerth, H. and C. Wright Mills, ed. (1991). *From Max Weber: Essays in Sociology*. London: Routledge.

Hałas, E., ed. (2000). *Florian Znaniecki's Sociological Theory and the Challenges of the 21st Century*. Frankfurt am Main: Peter Lang.

Kaesler, D. (1998). *Max Weber: Eine Einführung in Leben, Werk und Wirkung*. Frankfurt/New York: Campus Verlag.

Parsons, T. (1949). *The Structure of Social Action: A Study in Social Theory with Special Reference to a Group of Recent European Writers*. New York and London: The Free Press and Collier Macmillan Ltd.

Ritzer, G. and P. Gindoff (1992). Methodological Relationism: Lessons for and from Social Psychology. *Social Psychology Quarterly* **55** (2), 128–140.

Seligman, E. and A. Johnson, eds. (1930). *Encyclopaedia of the Social Sciences*. New York: Macmillan.

Sills, D. L., ed. (1968). *International Encyclopedia of the Social Sciences*. The Macmillan Company and The Free Press.

Simmel, G. (1908). *Soziologie*. Leipzig: Drunker und Humbolt.

Weber, M. (1978). *Economy and Society: An Outline of Interpretive Sociology*. Edited by G. Roth and C. Wittich. Berkeley *et al.*. University of California Press.

Znaniecki, F. (1965). *Social Relations and Social Roles: The Unfinished Systematic Sociology*. San Francisco: Chandler.

Znaniecki, F. (1967). *Social Actions*. New York: Russell and Russell.

Znaniecki, F. (1971). *Nauki o kulturze: Narodziny i rozwój* [Cultural Sciences]. Warszawa: PWN.

Znaniecki, F. (1973). *Socjologia wychowania* [Sociology of Education], vol. I and II. Warszawa: PWN.

Znaniecki, F. (1984). Potrzeby socjologii w Polsce [Needs of Sociology in Poland]. In: *Społeczne role uczonych* [Social Roles of Scholars], pp. 134–151. Warszawa: PWN.

Znaniecki, F. (1988). *Wstęp do socjologii* [Introduction to Sociology]. Warszawa: PWN.

Znaniecki, F. (1990). *Współczesne narody* [Modern Nationalities]. Warszawa: PWN.

Andrzej Pietruszczak

ON APPLICATIONS OF TRUTH-VALUE CONNECTIVES FOR TESTING ARGUMENTS WITH NATURAL CONNECTIVES

ABSTRACT. In introductory logic courses the authors often limit their considerations to the truth-value operators. Then they write that conditionals and biconditionals of natural language ("if" and "if and only if") may be represented as material implications and equivalences ("$\supset$" and "$\equiv$"), respectively. Yet material implications are not suitable for conditionals.[1] Lewis' strict implications ("$\strictif$") are much better for this purpose.[2] Similarly, strict equivalences ("$\strictfromto$") are better for representing biconditionals (than material equivalences). In this paper we prove that the methods from standard first courses in logic can be used for testing arguments with strict implications, strict equivalences and other operators which may represent connectives from natural language.[3]

1. Used Notions and Facts from Modal Logic

1.1. *Modal and Non-modal Formulæ*

Formulæ (modal and non-modal) are formed in a standard way from propositional letters ("p", "q", "r", "p_1", "p_2", "p_3", ...), brackets, and

---

[1]See, for example, Czerwiński (1958) and Priest (2001). The first author gives the following true material implication: "Copernicus had a son $\supset$ Copernicus was not a father" (it is true since Copernicus had no children). Priest writes (2001, 4.5.2, p. 64):

> Consider a true material conditional "The sun is shining $\supset$ Canberra is federal capital of Australia". One is inclined to reject this as a true conditional just because the truth of material conditional is too contingent an affair. Things could have been quite otherwise, in which case the material conditional would have been false. This suggests defining the conditional, "if A then B", as $\Box(A \supset B)$, where $\Box$ expresses an appropriate notion of necessity.

See also the so called "paradoxes of material implication": $q \models p \supset q$, $\neg p \models p \supset q$, $\neg(p \supset q) \models p$, $\neg(q \supset p) \models p \supset q$, $(p \wedge q) \supset r \models (p \supset r) \vee (q \supset r)$, $(p \supset r) \wedge (q \supset s) \models (p \supset s) \vee (q \supset r)$.

[2]Priest writes (2001, 4.5.3, p. 64):

> When Lewis created modern modal logic, he was not, in fact, concerned with modality as such. He was dissatisfied with the material conditional. He defined $A \strictif B$ as $\Box(A \supset B)$, and suggested this as a correct account of the conditional.

For "$\strictif$" we have not the paradoxes from Footnote 1, i.e.: $q \not\models p \strictif q$, $\neg p \not\models p \strictif q$, $\neg(p \strictif q) \not\models p$, $\neg(q \strictif p) \not\models p \strictif q$, $(p \wedge q) \strictif r \not\models (p \strictif r) \vee (q \strictif r)$, $(p \strictif r) \wedge (q \strictif s) \not\models (p \strictif s) \vee (q \strictif r)$. Yet we have the paradoxes of strict implication: $\Box p \models q \strictif p$ and $\Box\neg p \models p \strictif q$.

[3]These other connectives can be, for example, intensional disjunctions. In the inclusive case, "p and/or q" has the same meaning as "if not-p then q" (in other words, "if p then q" says that not-p and/or q). In the exclusive case, "p or q" has the same meaning as "q if and only if not-p".

In: J. Malinowski and A. Pietruszczak (eds.), *Essays in Logic and Ontology* (*Poznań Studies in the Philosophy of the Sciences and the Humanities*, vol. 91), 143–156. Amsterdam/New York, NY: Rodopi, 2006.

- truth-value operators: "$\neg$", "$\vee$", "$\wedge$", "$\supset$", and "$\equiv$" (negation, disjunction, conjunction, material implication, material equivalence, respectively);
- modal operators: the necessity sign "$\Box$", the possibility sign "$\Diamond$", and connectives of strict implication "$\prec$" and strict equivalence "$\mathrel{\ooalign{$\prec$\cr$\succ$}}$".

A formula is called *modal* if it contains one or more modal operators; otherwise, it is called *non-modal*. Let F be the set of all formulæ and F^0 be the set of all non-modal formulæ.

1.2. *PL-consequences. Tautologies*

Definition 1.1. (i) A *PL-valuation* on F is any assignment v of truth values, truth (1) and falsity (0), to all formulæ such that for all φ, ψ in F

($v_\neg$) $v(\neg\varphi) = 1$ iff $v(\varphi) = 0$,

($v_\wedge$) $v(\varphi \wedge \psi) = 1$ iff $v(\varphi) = 1$ and $v(\psi) = 1$,

($v_\vee$) $v(\varphi \vee \psi) = 1$ iff $v(\varphi) = 1$ or $v(\psi) = 1$,

($v_\supset$) $v(\varphi \supset \psi) = 1$ iff $v(\varphi) = 0$ or $v(\psi) = 1$,

($v_\equiv$) $v(\varphi \equiv \psi) = 1$ iff $v(\varphi) = v(\psi)$.

Thus formulæ $\ulcorner\Box\varphi\urcorner$, $\ulcorner\Diamond\varphi\urcorner$, $\ulcorner\varphi \prec \psi\urcorner$ and $\ulcorner\varphi \mathrel{\ooalign{$\prec$\cr$\succ$}} \psi\urcorner$ have arbitrary values.

(ii) A formula is *propositionally atomic* iff it is either a propositional letter or it has one of the following form: $\ulcorner\Box\varphi\urcorner$, $\ulcorner\Diamond\varphi\urcorner$, $\ulcorner\varphi \prec \psi\urcorner$, or $\ulcorner\varphi \mathrel{\ooalign{$\prec$\cr$\succ$}} \psi\urcorner$.

Thus, any PL-valuation is the extension (satisfying the truth conditions for the truth-value operators) of an assignment of truth values to the propositional atomic formulæ.

Definition 1.2. Let $\Gamma \subseteq \mathrm{F}$ and $\varphi \in \mathrm{F}$. We say that φ is a *PL-consequence* of Γ (written $\Gamma \models_{\mathrm{PL}} \varphi$) iff for every PL-valuation v such that $v(\psi) = 1$, for all $\psi \in \Gamma$, we have $v(\varphi) = 1$.

Definition 1.3. (1) A formula φ from F is a *tautology* iff for any PL-valuation v, $v(\varphi) = 1$, i.e. $\emptyset \models_{\mathrm{PL}} \varphi$. Let PL be the set of all tautologies.

(2) All tautologies from F^0 are called *non-modal tautologies*. Let PL^0 be the set of all non-modal tautologies, i.e., we put $\mathrm{PL}^0 := \mathrm{F}^0 \cap \mathrm{PL}$.

Obviously, $\varphi \in \mathrm{F}$ is a tautology iff φ is an instance of a non-modal tautology. By compactness of $\models_{\mathrm{PL}}$, for any $\Gamma \subseteq \mathrm{F}$ and any $\varphi \in \mathrm{F}$ we have

$$\Gamma \models_{\mathrm{PL}} \varphi \quad \text{iff} \quad \text{for some } \{\psi_1, \ldots, \psi_n\} \subseteq \Gamma \ (n \geqslant 0)^4 \quad \ulcorner(\psi_1 \wedge \cdots \wedge \psi_n) \supset \varphi\urcorner \in \mathrm{PL}$$

[4]For $n = 0$ we have $\{\psi_1, \ldots, \psi_n\} = \emptyset$ and we put $\ulcorner(\psi_1 \wedge \cdots \wedge \psi_n) \supset \varphi\urcorner = \varphi$

1.3. *Modal Systems and Logics*

Definition 1.4. A set Σ of formulæ is a *modal system* iff Σ is closed under the following rule

(RPL) if $\Gamma \models_{\mathrm{PL}} \varphi$ and $\Gamma \subseteq \Sigma$, then $\varphi \in \Sigma$,

for any $\Gamma \subseteq \mathrm{F}$ and $\varphi \in \mathrm{F}$. The *theses* of a system are just the formulæ in it.

Fact 1.1. A set Σ is a modal system iff $\mathrm{PL} \subseteq \Sigma$ and Σ is closed under the following rule of detachment for material implication for any $\varphi, \psi \in \mathrm{F}$

(MP) if φ and $\ulcorner\varphi \supset \psi\urcorner$ are members of Σ, so is ψ.

Definition 1.5. A set Σ of modal formulæ is a *logic* iff Σ is a modal system and is closed under the following rule of uniform substitution for any $\varphi \in \mathrm{F}$

(US) if $\varphi \in \Sigma$ then $\mathrm{s}\,\varphi \in \Sigma$,

where $\mathrm{s}\,\varphi$ is the result of uniform substitution of formulæ for propositional letters in φ.

The set PL is the smallest modal system. It is a logic.

We say that a system Σ' is an Σ-system iff $\Sigma \subseteq \Sigma'$. All modal systems are PL-systems.

For the next definition see e.g. (Bull & Segerberg 1984; Chellas 1980; Lemmon 1977).

Definition 1.6. Let $\Gamma \subseteq \mathrm{F}$ and $\varphi \in \mathrm{F}$. A formula φ is *deducible* from a set of formulæ Γ in a modal system Σ (written $\Gamma \vdash_{\Sigma} \varphi$) iff for some $\{\psi_1, \ldots, \psi_n\} \subseteq \Gamma$ ($n \geqslant 0$) we have $\ulcorner(\psi_1 \wedge \cdots \wedge \psi_n) \supset \varphi\urcorner \in \Sigma$.

Of course, for any modal system Σ: $\Sigma \vdash_{\Sigma} \varphi$ iff $\varphi \in \Sigma$ iff $\emptyset \vdash_{\Sigma} \varphi$. Moreover, given that $\mathrm{PL} \subseteq \Sigma$, $\models_{\mathrm{PL}} = \vdash_{\mathrm{PL}} \subseteq \vdash_{\Sigma}$ and we obtain the following fact (see e.g. Lemmon 1977, p. 16).

Fact 1.2. $\Gamma \vdash_{\Sigma} \varphi$ iff there is a finite sequence $\varphi_1, \ldots, \varphi_m$ such that $\varphi_m = \varphi$ and for each i, $1 \leqslant i \leqslant m$, either $\varphi_i \in \Gamma$, or $\varphi_i \in \Sigma$, or there are $j, k < i$ such that $\varphi_j = \ulcorner\varphi_k \supset \varphi_i\urcorner$.

1.4. *Correct Forms of Arguments*

Formulæ are forms for sentences of natural language. For arguments we put

Definition 1.7. A *form of argument* is any pair $\langle\Gamma, \varphi\rangle$, where Γ is a finite non-empty subset of F and $\varphi \in \mathrm{F}$. If $\Gamma = \{\psi_1, \ldots, \psi_n\}$ ($n > 0$), the pair $\langle\Gamma, \varphi\rangle$ is written $\ulcorner\psi_1, \ldots, \psi_n / \therefore \varphi\urcorner$; the members of Γ are called its *premises* and φ is called its *conclusion*.

DEFINITION 1.8. Let Λ be any modal logic. We say that a form of argument $\langle \Gamma, \varphi \rangle$ is *Λ-correct* iff $\Gamma \vdash_\Lambda \varphi$, i.e., iff $\ulcorner(\psi_1 \wedge \cdots \wedge \psi_n) \supset \varphi\urcorner \in \Lambda$, where $\psi_1, \ldots, \psi_n$ are all of the members of Γ.

Of course, we say that an argument is Λ-correct iff its form is Λ-correct.

1.5. *PL-regular Modal Logics*

In this paper we will consider only such modal logics which are logics and moreover contain the following formulæ

(df $\Diamond$) $\quad \Diamond p \equiv \neg\Box\neg p$

(df $\prec$) $\quad (p \prec q) \equiv \Box(p \supset q)$

(df $\succ\!\prec$) $\quad (p \succ\!\prec q) \equiv \Box(p \equiv q)$

This formulæ we will call the *basic definitions*.

DEFINITION 1.9. A modal logic Λ is *PL-regular* iff Λ contains the basic definitions and is closed under the following rule

($\mathrm{RR_{PL}}$) $\quad$ if $\ulcorner(\varphi \wedge \psi) \supset \chi\urcorner \in \mathrm{PL}$ then $\ulcorner(\Box\varphi \wedge \Box\psi) \supset \Box\chi\urcorner \in \Lambda$,

for any $\varphi, \psi, \chi \in \mathrm{F}$.

Fact 1.3. A modal logic Λ is PL-regular iff Λ contains the basic definitions and is closed under the following PL-Scott's rule for any $n > 0$ and for any $\chi_1, \ldots, \chi_n, \chi \in \mathrm{F}$

($\mathrm{RS_{PL}}$) $\quad$ if $\ulcorner(\chi_1 \wedge \cdots \wedge \chi_n) \supset \chi\urcorner \in \mathrm{PL}$ then $\ulcorner(\Box\chi_1 \wedge \cdots \wedge \Box\chi_n) \supset \Box\chi\urcorner \in \Lambda$.

Fact 1.4. Let Λ by any PL-regular modal logic. Then Λ is closed under the following rules

($\mathrm{RM_{PL}}$) $\quad$ if $\ulcorner\varphi \supset \psi\urcorner \in \mathrm{PL}$ then $\ulcorner\Box\varphi \supset \Box\psi\urcorner \in \Lambda$,

($\mathrm{RE_{PL}}$) $\quad$ if $\ulcorner\varphi \equiv \psi\urcorner \in \mathrm{PL}$ then $\ulcorner\Box\varphi \equiv \Box\psi\urcorner \in \Lambda$,

($\mathrm{RR_{PL}}\Diamond$) $\quad$ if $\ulcorner\varphi \supset (\psi \vee \chi)\urcorner \in \mathrm{PL}$ then $\ulcorner\Diamond\varphi \supset (\Diamond\psi \vee \Diamond\chi)\urcorner \in \Lambda$,

($\mathrm{RM_{PL}}\Diamond$) $\quad$ if $\ulcorner\varphi \supset \psi\urcorner \in \mathrm{PL}$ then $\ulcorner\Diamond\varphi \supset \Diamond\psi\urcorner \in \Lambda$,

($\mathrm{RE_{PL}}\Diamond$) $\quad$ if $\ulcorner\varphi \equiv \psi\urcorner \in \mathrm{PL}$ then $\ulcorner\Diamond\varphi \equiv \Diamond\psi\urcorner \in \Lambda$,

for any $\varphi, \psi \in \mathrm{F}$. Moreover, Λ contains the following formulæ

(df $\Box$) $\quad \Box p \equiv \neg\Diamond\neg p$

(df′ $\prec$) $\quad p \prec q \equiv \neg\Diamond(p \wedge \neg q)$

(df′ $\succ\!\prec$) $\quad p \succ\!\prec q \equiv ((p \prec q) \wedge (q \prec p))$

(K) $\quad \Box(p \supset q) \supset (\Box p \supset \Box q)$

(C) $\quad (\Box p \wedge \Box q) \supset \Box(p \wedge q)$

(M) $\quad \Box(p \wedge q) \supset (\Box p \wedge \Box q)$

(R) $\Box(p \wedge q) \equiv (\Box p \wedge \Box q)$

(R$\Diamond$) $\Diamond(p \vee q) \equiv (\Diamond p \vee \Diamond q)$

(1) $\Diamond(p \supset q) \equiv (\Box p \supset \Diamond q)$

Fact 1.5. If a modal logic Λ contains the basic definitions and either (K) or (C), and is closed under (RM_{PL}), then Λ is PL-regular.

Fact 1.6. If a modal logic Λ contains the basic definitions, (C) and (M), and Λ is closed under (RE_{PL}), then Λ is PL-regular.

By C1 Routley (1968) denoted the smallest sets of modal formulæ which includes PL and (K), and is closed under (MP), (US), and (RM_{PL}). Since in (Routley 1968) formulæ have only one modal operator "$\Box$", in our formalization we must add the basic definitions to Routley's axioms of C1. By Facts 1.4 and 1.5, C1 is the smallest PL-regular modal logic. Thus, for any modal logic Λ: Λ is PL-regular iff $C1 \subseteq \Lambda$.

Lemmon (1957) sets out the logic S0.5 and two groups of non-normal modal logics called the "D" and "E" logics, respectively. The "logics E1, E2, E3, E4, and E5 are intended as possible epistemic counterparts to the five logics S0.5, S2, S3, S4, and S5". The E-logics, just like the S-logics, contain the following formula

(T) $\Box p \supset p$

E1 is the smallest PL-regular modal logic which contain (T).

The five D-logics, D1, D2, D3, D4, and D5, were associated with the five E-logics. The distinguishing feature of D-logics is that the axioms (T) of the corresponding E-logics is weakened to the following formula

(D) $\Box p \supset \Diamond p$

D1 is the smallest PL-regular modal logic which contain (D).[5]

The Feys' S1° is Lewis' S1 without (T) and (D). Similarly, the logic S0.5° is obtained from Lemmon's S0.5. Moreover, in this paper by $S0.5^{D}$ we denote the logic S0.5° together with the axiom (D).

In (Lemmon 1966) the logic C2 is examined. It contains neither (T) nor (D), i.e., C2 is E2 without (T) and (D).[6]

Fig. 1 (see p. 150) depicts some well-known PL-regular logics (except PL and $PLT^{\dashv}$) and inclusions that hold among them.[7]

---

[5]By (1) in all PL-regular modal logics the formulæ (D) and "$\Diamond(p \supset p)$" are equivalent. So, by ($RE_{PL}\Diamond$), (D) is equivalent to $\ulcorner \Diamond\tau \urcorner$, where τ is an arbitrary tautology.

"These D-logics are intended for possible deontic interpretation". The logics D3, D4, D5 have also the formula $\ulcorner \Box\varphi \supset \varphi \urcorner$ for any fully modalized formula φ.

[6]C2 is the smallest regular modal logic. In (Chellas 1980) this logic is denoted by *R*.

[7]See e.g. diagrams in (Bowen 1979, p. 3; Chellas 1980, p. 132; Chellas & Segerberg 1996, p. 21; Lemmon 1957, p. 186; Lemmon 1966, p. 48; Lemmon 1977, p. 58).

1.6. *Models for Logics* S0.5°, S0.5^{D}, *and* S0.5

For the logics S0.5°, S0.5^{D}, and S0.5 we can use the following semantics.

DEFINITION 1.10. (i) An S0.5°-*model* is any quadruple $\langle W, w, A, V\rangle$ in which

- W is a non-empty set of objects (worlds);
- $w \in W$, w is a «distinguished» world;
- $A \subseteq W$, A is a (possibly empty) set of alternatives to the world w;
- $V\colon \mathrm{F} \times W \longrightarrow \{0, 1\}$ is a function such that
  - for every $x \in W$ the assignment $V(\cdot, x)\colon \mathrm{F} \longrightarrow \{0, 1\}$ satisfies $(v_{\neg})$, $(v_{\wedge})$, $(v_{\vee})$, $(v_{\supset})$, and $(v_{\equiv})$ from Definition 1.1,
  - for the world w and for any $\varphi, \psi \in \mathrm{F}$ the assignment V satisfies the following conditions

    $(V^A_{\Box})$ $\quad V(\Box\varphi, w) = 1$ iff $\forall_{y\in A}\ V(\varphi, y) = 1$,

    $(V^A_{\Diamond})$ $\quad V(\Diamond\varphi, w) = 1$ iff $\exists_{y\in A}\ V(\varphi, y) = 1$,

    $(V^A_{\dashv})$ $\quad V(\varphi \dashv \psi, w) = 1$ iff $\forall_{y\in A}$ either $V(\varphi, y) = 0$ or $V(\psi, y) = 1$,

    $(V^A_{\vDash\dashv})$ $\quad V(\varphi \vDash\dashv \psi, w) = 1$ iff $\forall_{y\in A}\ V(\varphi, y) = V(\psi, y)$.

    For every other world formulæ $\ulcorner\Box\varphi\urcorner$, $\ulcorner\Diamond\varphi\urcorner$, $\ulcorner\varphi \dashv \psi\urcorner$, and $\ulcorner\varphi \vDash\dashv \psi\urcorner$ have arbitrary values.

(ii) An S0.5^{D}-*model* is any S0.5°-model in which $A \neq \emptyset$.

(iii) An S0.5-*model* is any S0.5°-model in which $A = W$. So any S0.5-model may by identified with a triple $\langle W, w, V\rangle$ (of course in this case in the conditions $(V^A_{\Box})$, $(V^A_{\Diamond})$, $(V^A_{\dashv})$, and $(V^A_{\vDash\dashv})$ we put $A := W$).

For any φ from F we say that it is S0.5°-valid (resp. S0.5^{D}-valid, S0.5-valid) iff for every S0.5°-model (resp. S0.5^{D}-model, S0.5-model) $\langle W, w, A, V\rangle$ we have $V(\varphi, w) = 1$.

The following theorem says that S0.5° (resp. S0.5^{D} and S0.5) is sound with respect to the class of all S0.5°-models (resp. S0.5^{D}-models and S0.5-models). The proof of this theorem is immediate.

Theorem 1.1. If φ is a thesis of S0.5° (resp. S0.5^{D}, S0.5), them φ is S0.5°-valid (resp. S0.5^{D}-valid, S0.5-valid).

From the above theorem we obtain the following lemma which we use in the next section (see Theorem 2.5).

Lemma 1.1. For any $\varphi, \psi \in \mathrm{F}$, if $\ulcorner\varphi \supset \psi\urcorner \notin \mathrm{PL}$ then $\ulcorner\Box\varphi \supset \Box\psi\urcorner \notin \mathrm{S0.5^D}$.

PROOF. Suppose that $\ulcorner\varphi \supset \psi\urcorner \notin \mathrm{PL}$. Then there is a PL-valuation v_0 such that $v_0(\varphi) = 1$ and $v_0(\psi) = 0$. We construct a falsifying S0.5^{D}-model for $\ulcorner\Box\varphi \supset \Box\psi\urcorner$. Let w be an arbitrary object different from v_0. We put $A := \{v_0\}$ and $W := \{w, v_0\}$.

For any propositional atomic formula π and for any $x \in \{w, v_0\}$ we put $V(\pi, x) := v_0(\pi)$. Next we extend V on $\mathrm{F} \times \{w, v_0\}$ by the conditions $(v_\neg)$, $(v_\wedge)$, $(v_\vee)$, $(v_\supset)$, $(v_\equiv)$, $(V^A_\Box)$, $(V^A_\Diamond)$, $(V^A_\strictif)$, and $(V^A_\strictif)$ (the four last only for the world w). We obtain an $\mathrm{S0.5^D}$-model $\langle\{w, v_0\}, V\rangle$.

By $(v_\neg)$, $(v_\wedge)$, $(v_\vee)$, $(v_\supset)$, $(v_\equiv)$, we have $V(\varphi, v_0) = v_0(\varphi) = 1$ and $V(\psi, v_0) = v_0(\psi) = 0$, and so, $V(\Box\varphi, w) = 1$ and $V(\Box\psi, w) = 0$, by $(V^A_\Box)$. Thus $V(\Box\varphi \supset \Box\psi, w) = 0$, i.e. $\ulcorner\Box\varphi \supset \Box\psi\urcorner \notin \mathrm{S0.5^D}$, by soundness of $\mathrm{S0.5^D}$. □

The logics S0.5°, $\mathrm{S0.5^D}$, and S0.5 are also complete with respect to the class of all S0.5°-models, $\mathrm{S0.5^D}$-model, and S0.5-model, respectively. That is, for every $\varphi \in \mathrm{F}$: if φ is S5-valid (resp. $\mathrm{S0.5^D}$-valid, S0.5-valid), then φ is a thesis of S0.5° (resp. $\mathrm{S0.5^D}$, S0.5). For this fact see, for example, (Pietruszczak 2005).

1.7. *The Logic* $\mathrm{PLT}^{\strictif}$

Every modal logic which contains (df $\strictif$), (df $\strictiff$) and (T) also contains the following formulæ

(T$\strictif$) $\quad (p \strictif q) \supset (p \supset q)$

(T$\strictiff$) $\quad (p \strictiff q) \supset (p \equiv q)$

Let $\mathrm{PLT}^{\strictif}$ be the smallest modal logic that contains (T$\strictif$) and (T$\strictiff$). This logic neither closed under ($\mathrm{RE_{PL}}$) nor contains (T). But we have the following fact.

Fact 1.7. If Λ is a modal logic which contains (df $\strictif$) and (T$\strictif$), and is closed under ($\mathrm{RE_{PL}}$), then Λ contains (T).

Proof. By (T$\strictif$) and (US), "$\Box((p \supset p) \supset p) \supset ((p \supset p) \supset p)$" is in Λ. Since "$((p \supset p) \supset p) \equiv p$" is in PL, by ($\mathrm{RE_{PL}}$), "$\Box((p \supset p) \supset p) \equiv \Box p$" is in Λ. □

1.8. *The Logic* S5

So far we have presented the «bottom» of the diagram from Fig. 1. Now we consider its «top», i.e. the logic S5.

S5 is the smallest set of modal formulæ which includes PL, the basic definitions, (T), and the following formula

(E) $\quad \Diamond p \supset \Box\Diamond p$

and moreover is closed under (MP), (US) and the following rule

(RN) $\quad$ if $\varphi \in \Lambda$ then $\ulcorner\Box\varphi\urcorner \in \Lambda$,

for any $\varphi \in \mathrm{F}$.

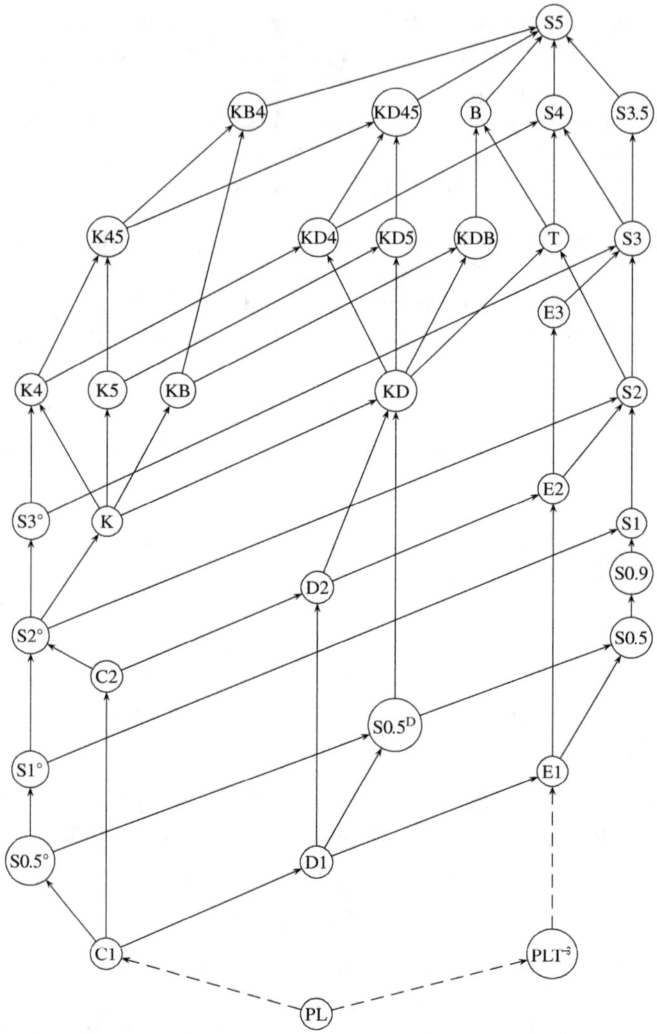

T = KT B = KTB

S4 = KT4 S5 = KT5 = KTB4 = KDB4 = KDB5

KB4 = KB5 S0.5 = S0.5° ⊕ (T)

S0.5° = PL ⊕ □PL, (K), (df ◇), (MP) $\quad$ S0.5^{D} = S0.5° ⊕ (D)

PL = the set of modal formulæ which are instances of classical tautologies

C1 = the smallest PL-regular modal logic

C2 = the smallest regular modal logic = *R* (in Chellas 1980)

Fig. 1. Some well-known PL-regular modal logics

S5 is a modal logic and it is closed under the following Scott's rule for any $n \geqslant 0$ and for any $\chi_1, \ldots, \chi_n, \chi \in \mathrm{F}$

(RS) if $\ulcorner(\chi_1 \wedge \cdots \wedge \chi_n) \supset \chi\urcorner \in \mathrm{S5}$ then $\ulcorner(\Box\chi_1 \wedge \cdots \wedge \Box\chi_n) \supset \Box\chi\urcorner \in \mathrm{S5}$.

Thus, given that PL $\subseteq$ S5, the logic S5 is PL-regular.

Definition 1.11. An S5-*model* is any pair $\langle W, V\rangle$ in which

- W is a non-empty set of objects (*worlds*);
- $V\colon \mathrm{F} \times W \longrightarrow \{0, 1\}$ is a function such that for every $x \in W$:
  - the assignment $V(\cdot, x)\colon \mathrm{F} \longrightarrow \{0, 1\}$ satisfies the conditions $(v_\neg)$, $(v_\wedge)$, $(v_\vee)$, $(v_\supset)$, and $(v_\equiv)$ from Definition 1.1,
  - for any $\varphi, \psi \in \mathrm{F}$ the assignment V satisfies four conditions:
    - $(V_\Box)$ $V(\Box\varphi, x) = 1$ iff $\forall_{y\in W}\ V(\varphi, y) = 1$,
    - $(V_\Diamond)$ $V(\Diamond\varphi, x) = 1$ iff $\exists_{y\in W}\ V(\varphi, y) = 1$,
    - $(V_{-\!3})$ $V(\varphi \mathrel{-\!3} \psi, x) = 1$ iff $\forall_{y\in W}$ either $V(\varphi, y) = 0$ or $V(\psi, y) = 1$,
    - $(V_{\varepsilon\!-\!3})$ $V(\varphi \mathrel{\varepsilon\!-\!3} \psi, x) = 1$ iff $\forall_{y\in W}\ V(\varphi, y) = V(\psi, y)$.

For any φ from F we say that it is S5-valid iff for every S5-model $\langle W, V\rangle$ and for any $x \in W$ we have $V(\varphi, x) = 1$.

The following theorem says that S5 is sound with respect to the class of all S5-models. The proof of this theorem is immediate.

Theorem 1.2. If $\varphi \in \mathrm{S5}$ then φ is S5-valid.

From the above theorem we obtain the following lemma[8] which we use in the next section (Theorems 2.2 and 2.4).

Lemma 1.2. For any $\alpha, \beta \in \mathrm{F}^0$, if $\ulcorner\alpha \supset \beta\urcorner \notin \mathrm{PL}^0$, then $\ulcorner\Box\alpha \supset \Box\beta\urcorner$ and $\ulcorner\Box\alpha \supset \beta\urcorner$ are not theses of S5.[9]

Proof. Suppose that $\ulcorner\alpha \supset \beta\urcorner \notin \mathrm{PL}^0$. Then there is a PL-valuation v_0 such that $v_0(\alpha) = 1$ and $v_0(\beta) = 0$. We construct a falsifying S5-model for $\ulcorner\Box\alpha \supset \Box\beta\urcorner$ and $\ulcorner\Box\alpha \supset \beta\urcorner$. Let W consist of exactly one world v_0. For any propositional letter λ we put $V(\lambda, v_0) := v_0(\lambda)$. Next we extend V on $\mathrm{F} \times \{v_0\}$ by the conditions $(v_\neg)$, $(v_\wedge)$, $(v_\vee)$, $(v_\supset)$, $(v_\equiv)$, $(V_\Box)$, and $(V_\Diamond)$. We obtain an S5-model $\langle\{v_0\}, V\rangle$.

By $(v_\neg)$, $(v_\wedge)$, $(v_\vee)$, $(v_\supset)$, $(v_\equiv)$, for every $\delta \in \mathrm{F}^0$, we have $V(\delta, v_0) = v_0(\delta)$. Hence $V(\alpha, v_0) = 1$, $V(\beta, v_0) = 0$, and so, $V(\Box\alpha, v_0) = 1$ and $V(\Box\beta, v_0) = 0$, by $(V_\Box)$. Therefore $V(\Box\alpha \supset \Box\beta, v_0) = 0$ and $V(\Box\alpha \supset \beta, v_0) = 0$. Thus $\ulcorner\Box\alpha \supset \Box\beta\urcorner$ and $\ulcorner\Box\alpha \supset \beta\urcorner$ are not theses of S5. □

[8] This lemma is a special case of some theorem which is presented e.g. in (Pollock 1967, p. 361; Hughes & Cresswell 1996, pp. 105–107).

[9] Notice that for all $\varphi, \psi \in \mathrm{F}$: if $\ulcorner\varphi \supset \psi\urcorner \in \mathrm{PL}$, then $\ulcorner\Box\varphi \supset \Box\psi\urcorner \in \mathrm{C1} \subseteq \mathrm{S5}$, by $(\mathrm{RM_{PL}})$, and $\ulcorner\Box\varphi \supset \psi\urcorner \in \mathrm{E1} \subseteq \mathrm{S5}$, by (T) and PL.

The logic S5 is also complete with respect to the class of all S5-models, i.e. for every $\varphi \in$ F: if φ is S5-valid then $\varphi \in$ S5. For this theorem see e.g. (Chellas 1980, p. 178; Hughes & Cresswell 1996, pp. 61 and 105–108).

2. Testing Arguments

We consider two problems concerning forms of arguments.

2.1. *Test 1*

The first problem concerns arguments of the following forms:

$$(\dagger)\quad \begin{array}{l} \varphi_1 \supset \psi_1 \\ \vdots \\ \varphi_n \supset \psi_n \\ \chi_1 \\ \vdots \\ \underline{\chi_m} \\ \therefore \chi \end{array} \qquad\qquad (\ddagger)\quad \begin{array}{l} \varphi_1 \dashv\!\!3\, \psi_1 \\ \vdots \\ \varphi_n \dashv\!\!3\, \psi_n \\ \chi_1 \\ \vdots \\ \underline{\chi_m} \\ \therefore \chi \end{array}$$

where $n > 0$, $m \geqslant 0$ and $\varphi_i, \psi_i, \chi_i, \chi$ may be modal formulæ. Notice that we have the following theorem.

Theorem 2.1. For any $\varphi_i, \psi_i, \chi_i, \chi \in$ F: if (†) is PL-correct, then (‡) is PLT$^{⥽}$-correct, so (‡) is correct in all modal logics which include the formula (T-⥽).

PROOF. Let (†) be PL-correct, i.e. $\ulcorner(\bigwedge_{i=1}^{n}(\varphi_i \supset \psi_i) \wedge \bigwedge_{i=1}^{m}\chi_i) \supset \chi\urcorner \in$ PL, where $n > 0$ and $m \geqslant 0$. Then $\ulcorner(\bigwedge_{i=1}^{n}(\varphi_i \dashv\!\!3\, \psi_i) \wedge \bigwedge_{i=1}^{m}\chi_i) \supset \chi\urcorner \in$ PLT$^{⥽}$, by (T-⥽) and PL. So (‡) is PLT$^{⥽}$-correct. □

Remark 2.1. Theorem 2.1 holds also for every meaning of "if ... then" for which "if p then q" entails "$p \supset q$". It is, more or less, what we can learn from (Czerwiński 1958), where conditionals are interpreted in a popular way. □

If formulæ φ_i, ψ_i, χ_i, and χ in (†) are non-modal, then the foregoing considerations are invertible in the following sense. If an argument of the form (†) is not PL-correct then the argument (‡) is not S5-correct. So then we may say that the argument (‡) is not correct in the sense of modal logic. More precisely, we have the following theorem[10]:

Theorem 2.2. For any $\varphi_i, \psi_i, \chi_i, \chi \in$ F^0: if (‡) is S5-correct, then (†) is PL-correct.

[10] In this theorem the assumption that all formulæ φ_i, ψ_i, χ_i, and χ are non-modal is essential. In Remark 2.2(ii) we prove that the version of Theorem 2.2 without this assumption does not hold even for C1 (in place of S5). Thus, this version would be false for all PL-regular logics.

PROOF. Suppose that (‡) is S5-correct, i.e. $\ulcorner(\bigwedge_{i=1}^{n}(\varphi_i \dashv\!\!\!\! \ni \psi_i) \wedge \bigwedge_{i=1}^{m}\chi_i) \supset \chi\urcorner \in$ S5, where $\varphi_i, \psi_i, \chi_i, \chi \in \mathrm{F}^0$. Then $\ulcorner(\bigwedge_{i=1}^{n}\Box(\varphi_i \supset \psi_i)) \supset (\bigwedge_{i=1}^{m}\chi_i \supset \chi)\urcorner \in$ S5, by (df ⥽) and PL. Hence we have $\ulcorner\Box(\bigwedge_{i=1}^{n}(\varphi_i \supset \psi_i)) \supset (\bigwedge_{i=1}^{m}\chi_i \supset \chi)\urcorner \in$ S5, by (M) and PL. Thus, by Lemma 1.2, $\ulcorner(\bigwedge_{i=1}^{n}(\varphi_i \supset \psi_i)) \supset (\bigwedge_{i=1}^{m}\chi_i \supset \chi)\urcorner \in \mathrm{PL}^0$, and so by PL, $\ulcorner(\bigwedge_{i=1}^{n}(\varphi_i \supset \psi_i) \wedge \bigwedge_{i=1}^{m}\chi_i) \supset \chi\urcorner \in \mathrm{PL}^0$. Thus, (†) is PL-correct. □

We obtain similar results to Theorems 2.1 and 2.2 and Remark 2.1, if in (†) one or more occurrences of "⊃" is replaced by "≡", and according to this in (‡) "⥽" is replaced by "⥽⥼".

Remark 2.2. (i) In Theorem 2.2 we need the assumption that all formulæ φ_i, ψ_i, χ_i, and χ are non-modal. For example,

(†$_a$)

$$\frac{\Diamond p \supset p \quad \Diamond p}{\therefore\ \Box p} \qquad \frac{p \supset \Box p \quad \Diamond p}{\therefore\ \Box p} \qquad \frac{p \supset \Box p \quad \Diamond p}{\therefore\ p}$$

(‡$_a$)

$$\frac{\Diamond p \dashv\!\!\!\! \ni p \quad \Diamond p}{\therefore\ \Box p} \qquad \frac{p \dashv\!\!\!\! \ni \Box p \quad \Diamond p}{\therefore\ \Box p} \qquad \frac{p \dashv\!\!\!\! \ni \Box p \quad \Diamond p}{\therefore\ p}$$

the three forms from (†$_a$) are not PL-correct, but those from (‡$_a$) are S5-correct[11].

(ii) Furthermore, if the assumption is omitted, the Theorem 2.2 does not hold even for C1. For example,

$$(\dagger_b)\ \frac{p \supset q \quad \Box p}{\therefore\ \Box q} \qquad\qquad (\ddagger_b)\ \frac{p \dashv\!\!\!\! \ni q \quad \Box p}{\therefore\ \Box q}$$

(†$_b$) is not PL-correct, but (‡$_b$) is C1-correct. □

2.2. *Test 2*

The second problem concerns the following forms of arguments:

$$(*)\ \frac{\varphi_1 \supset \psi_1 \quad \vdots \quad \varphi_n \supset \psi_n}{\therefore\ \varphi \supset \psi} \qquad (*')\ \frac{\varphi_1 \supset \psi_1 \quad \vdots \quad \varphi_n \supset \psi_n \quad \varphi}{\therefore\ \psi} \qquad (\star)\ \frac{\varphi_1 \dashv\!\!\!\! \ni \psi_1 \quad \vdots \quad \varphi_n \dashv\!\!\!\! \ni \psi_n}{\therefore\ \varphi \dashv\!\!\!\! \ni \psi}$$

where $n > 0$ and φ, φ_i, ψ and ψ_i may be modal formulæ. Of course, (∗) is PL-correct iff (∗′) is PL-correct. Firstly, we prove the following theorem.

[11] Given Fact 1.2, we can prove the following facts. Firstly, the formula "$((\Diamond p \dashv\!\!\!\! \ni p) \wedge \Diamond p) \supset \Box p$" is a thesis of any logic which contains (df ⥽), (K) and (E), and is closed under (RE$_{\mathrm{PL}}$). Secondly, the formula "$((p \dashv\!\!\!\! \ni \Box p) \wedge \Diamond p) \supset \Box p$" is a thesis of an arbitrary logic which contains (df ◇), (df ⥽), (K), and "$\Diamond\Box p \supset \Box p$", and is closed under (RE$_{\mathrm{PL}}$) (this logic may not contain (E), if it is not closed under the rule (RE): if $\ulcorner\varphi \equiv \psi\urcorner \in \Lambda$, then $\ulcorner\Box\varphi \equiv \Box\psi\urcorner \in \Lambda$). Thirdly, if this logic also contains (T), then we obtain "$((p \dashv\!\!\!\! \ni \Box p) \wedge \Diamond p) \supset p$" as its thesis. Thus, the last formula is, for example, a thesis of Åqvist's logic S3.5, which is obtained by adding (E) or "$\Diamond\Box p \supset \Box p$" to Lewis' logic S3. This formula is also a thesis of Lemmon's logic E5.

Theorem 2.3. For any $\varphi_i, \psi_i, \chi_i, \chi \in \mathrm{F}$: if (∗) is PL-correct, then (★) is C1-correct, i.e. (★) is correct in all PL-regular logics.

Proof. If (∗) is PL-correct, then $\ulcorner(\bigwedge_{i=1}^{n}(\varphi_i \supset \psi_i)) \supset (\varphi \supset \psi)\urcorner \in \mathrm{PL}$, where $n > 0$. Thus by the rule ($\mathrm{RS_{PL}}$), $\ulcorner(\bigwedge_{i=1}^{n} \Box(\varphi_i \supset \psi_i)) \supset \Box(\varphi \supset \psi)\urcorner \in \mathrm{C1}$. So $\ulcorner(\bigwedge_{i=1}^{n}(\varphi_i \dashv\!3\, \psi_i)) \supset (\varphi \dashv\!3\, \psi)\urcorner \in \mathrm{C1}$, by (df$\dashv\!3$) and PL. Thus (★) is C1-correct, and so it is correct for all PL-regular logics. □

If formulæ φ_i, ψ_i, φ, and ψ in (∗) and (∗∗) are non-modal, then the foregoing considerations are invertible. In other words, if an argument of the form (∗) or (∗∗) is not PL-correct, then the argument (★) is not S5-correct. Thus we may say that the argument (★) is not correct in the sense of modal logic. More precisely, we have the following theorem[12]:

Theorem 2.4. For any $\varphi_i, \psi_i, \chi_i, \chi \in \mathrm{F^0}$: if (★) is S5-correct, then (∗) is PL-correct.

Proof. Suppose that (★) is S5-correct, i.e. $\ulcorner(\bigwedge_{i=1}^{n}(\varphi_i \dashv\!3\, \psi_i)) \supset (\varphi \dashv\!3\, \psi)\urcorner \in \mathrm{S5}$, where $\varphi_i, \psi_i, \varphi, \psi \in \mathrm{F^0}$. Then $\ulcorner(\bigwedge_{i=1}^{n} \Box(\varphi_i \supset \psi_i)) \supset \Box(\varphi \supset \psi)\urcorner \in \mathrm{S5}$, by (df$\dashv\!3$) and PL. Hence $\ulcorner\Box(\bigwedge_{i=1}^{n}(\varphi_i \supset \psi_i)) \supset \Box(\varphi \supset \psi)\urcorner \in \mathrm{S5}$, by (R) and PL. Thus, by Lemma 1.2, $\ulcorner(\bigwedge_{i=1}^{n}(\varphi_i \supset \psi_i)) \supset (\varphi \supset \psi)\urcorner \in \mathrm{PL^0}$. Thus, (∗) is PL-correct. □

We obtain similar results to Theorems 2.3 and 2.4, if in (∗) and (∗∗) one or more occurrences of "⊃" is replaced by "≡", and according to this in (★) "$\dashv\!3$" is replaced by "$\varepsilon\!\dashv\!3$".

Remark 2.3. (i) In Theorem 2.4 we need the assumption that all formulæ φ_i, ψ_i, χ_i, and χ are non-modal. For example,

$$(\ast_{\mathrm{a}})\quad \frac{\Diamond p \supset p}{\therefore\ \Diamond p \supset \Box p} \qquad\qquad (\star_{\mathrm{a}})\quad \frac{\Diamond p \dashv\!3\, p}{\therefore\ \Diamond p \dashv\!3\, \Box p}$$

the form $(\ast_{\mathrm{a}})$ is not PL-correct, but $(\star_{\mathrm{a}})$ is S5-correct.

(ii) Furthermore, when the assumption is omitted, then even for S1° and E1 the theorem does not hold. Firstly, for S1° we have the following example

$$(\ast_{\mathrm{b}})\quad \frac{(\Box p \supset \Box q) \supset r}{\therefore\ \Box(p \supset q) \supset r} \qquad\qquad (\star_{\mathrm{b}})\quad \frac{(\Box p \supset \Box q) \dashv\!3\, r}{\therefore\ \Box(p \supset q) \dashv\!3\, r}$$

The form $(\ast_{\mathrm{b}})$ is not PL-correct, but $(\star_{\mathrm{b}})$ is S1°-correct, since all formulæ of the forms $\ulcorner((\varphi \dashv\!3\, \psi) \wedge (\psi \dashv\!3\, \chi)) \supset (\varphi \dashv\!3\, \chi)\urcorner$ and $\ulcorner\Box(\varphi \supset \psi) \dashv\!3\, (\Box\varphi \supset \Box\psi)\urcorner$ are thesis of S1°.

---

[12] Also in this theorem the assumption that all formulæ $\varphi_i, \psi_i, \chi_i$, and χ are non-modal is essential. In Remark 2.3(ii) we prove that when this assumption is omitted, then even for S1° and E1 the theorem does not hold. Moreover, in Theorem 2.5 we prove the version of Theorem 2.4 without this assumption holds for the logic $\mathrm{S0.5^D}$. Thus, this version would be true also for C1, D1, and S0.5°.

Secondly, for E1 we have the following example for any $\tau \in$ PL

$$(*_c)\quad \frac{\tau \supset \Box q}{\therefore\ \tau \supset q} \qquad\qquad (\star_c)\quad \frac{\tau \strictif \Box q}{\therefore\ \tau \strictif q}$$

The form $(*_c)$ is not PL-correct, but $(\star_c)$ is E1-correct. Indeed, for any $\varphi \in$ F, $\ulcorner(\tau \supset \varphi) \equiv \varphi\urcorner \in$ PL, and so $\ulcorner(\tau \strictif \varphi) \equiv \Box\varphi\urcorner \in$ E1, by (RE$_{\mathrm{PL}}$), (df $\strictif$), and PL. Hence, by (T), we obtain $\ulcorner(\tau \strictif \Box q) \supset (\tau \strictif q)\urcorner \in$ E1.

For the logic S0.9 we have the following more interesting examples

$$(*_d)\quad \frac{p \supset \Box q}{\therefore\ p \supset q} \qquad \frac{p \supset q}{\therefore\ \Box p \supset q} \qquad\qquad (\star_d)\quad \frac{p \strictif \Box q}{\therefore\ p \strictif q} \qquad \frac{p \strictif q}{\therefore\ \Box p \strictif q}$$

Both forms from $(*_d)$ are not PL-correct, but those from $(\star_d)$ are S0.9-correct, since $\ulcorner((\varphi \strictif \psi) \wedge (\psi \strictif \chi)) \supset (\varphi \strictif \chi)\urcorner$ and $\ulcorner\Box\varphi \strictif \varphi\urcorner$ are in S0.9. □

As we have already mentioned in Footnote 12, the version of Theorem 2.4 without this assumption holds for the logic S0.5^{D} (in place of S5). Thus, this version would be true also for C1, D1, and S0.5°.

Theorem 2.5. For any $\varphi_i, \psi_i, \chi_i, \chi \in$ F: if $(\star)$ is S0.5^{D}-correct, then $(*)$ is PL-correct.

Proof. Similarly to the proof of Theorem 2.4; we use Lemma 1.1 instead of Lemma 1.2. □

Remark 2.4. Our above considerations (Theorems 2.1–2.5) can be also applied to intensional disjunctions (see Footnote 3). It is sufficient assumption that for *inclusive* disjunction we have the following theses: $\ulcorner(\varphi \boxed{\vee} \psi) \equiv \Box(\varphi \vee \psi)\urcorner$, $\ulcorner(\varphi \boxed{\vee} \psi) \equiv \Box(\neg\varphi \supset \psi)\urcorner$ and $\ulcorner(\varphi \boxed{\vee} \psi) \equiv (\neg\varphi \strictif \psi)\urcorner$, where "$\boxed{\vee}$" is the sing of intensional inclusive disjunction. Moreover, for *exclusive* disjunction we have the following theses: $\ulcorner(\varphi \boxed{\underline{\vee}} \psi) \equiv \Box(\varphi \underline{\vee} \psi)\urcorner$, $\ulcorner(\varphi \boxed{\underline{\vee}} \psi) \equiv \Box(\neg\varphi \equiv \psi)\urcorner$ and $\ulcorner(\varphi \boxed{\underline{\vee}} \psi) \equiv (\neg\varphi \strictiff \psi)\urcorner$, where "$\underline{\vee}$" and "$\boxed{\underline{\vee}}$" are the sings of exclusive truth-value disjunction and intensional disjunction, respectively. □

Nicolaus Copernicus University
Department of Logic
ul. Asnyka 2
87-100 Toruń
Poland
e-mail: pietrusz@uni.torun.pl

REFERENCES

Bowen, K. A. (1979). *Model Theory for Modal Logic. Kripke Models for Modal Predicate Calculi*, Dordrecht–Boston: D. Reidel Publishing Company.

Bull, R. A., and K. Segerberg (1984). Basic Modal Logic. In: D. M. Gabbay and F. Guenthner, *Handbook of Pholosophical Logic*, vol. II, pp. 1–88. Dordrecht: D. Reidel Publishing Company.

Chellas, B. (1980). *Modal Logic. An Introduction*, Cambridge: Cambridge University Press.

Chellas, B. F., and K. Segerberg (1996). Modal Logics in the Vicinity of S1. *Notre Dame Journal of Formal Logic* **37**, 1–24.

Czerwiński, Z. (1958). O paradoksie implikacji [On the paradox of the implication]. *Studia Logica* **7**, 263–271.

Lemmon, E. J. (1957). New Foundations for Lewis Modal Systems. *The Journal of Symbolic Logic* **22**, 176–186.

Lemmon, E. J. (1966). Algebraic Semantics for Modal Logics I. *The Journal of Symbolic Logic* **31**, 46–56.

Lemmon, E. J., in collaboration with D. Scott (1977). *"Lemmon Notes": An Introduction to Modal Logic*. Edited by K. Segerberg, no. 11 in the American Philosophical Quarterly Monograph Series, Oxford: Basil Blackwell.

Hughes, G E., and M. J. Cresswell (1996). *A New Introduction to Modal Logic*. London: Routledge.

Pietruszczak, A. (2005). Relational Semantics for some Very Weak Lemmon's Systems. In preparation.

Pollock, J. L. (1967). Basic Modal Logic. *The Journal of Symbolic Logic* **32**, 355–365.

Priest, G. (2001). *An Introduction to Non-Classical Logic*. Cambridge: Cambridge University Press.

Routley, R. (1968). Decision Procedure and Semantics for C1, E1 and S0.5°. *Logique et Analyse* **44**, 468–469.

Itala M. Loffredo D'Ottaviano
Hércules Araujo Feitosa

TRANSLATING FROM ŁUKASIEWICZ'S LOGICS INTO CLASSICAL LOGIC: IS IT POSSIBLE?

ABSTRACT. This work presents some basic results on a theory of translations between logics and a short revision about Łukasiewicz's logics. Then, it is shown, using facts about algebraic semantics, that there is a conservative translation from every finite Łukasiewicz's logic into classical logic. However, this is not a constructive result.

Introduction

The use of special functions between logics in order to study relations between them has been developed since the beginning of the 20th century. But the idea of constructing a theory of such functions is not so old. A theory of translations between logics intends to analyze general logical properties, having these functions as tools.

In general, it is not too difficult to get a translation or a conservative translation from classical calculus into some other calculi, but the inverse is always hard to reach.

In several previous works we have initiated the development of a general theory of translations between logics and have introduced various translations involving classical logic and several non-classical logics (see D'Ottaviano & Feitosa 1999a; D'Ottaviano & Feitosa 1999b, 2000; and Feitosa & D'Ottaviano 2000).

In this work we consider the problem of constructing conservative translations associated to Łukasiewicz's calculi. In D'Ottaviano & Feitosa (1999a) it is presented a family of conservative translations from classical propositional calculus (**CPC**) into the n-valued Łukasiewicz's calculi $\mathbf{L}_n$, for every $n \in \mathbb{N}$, but it was not considered if there are conservative translations in the other direction.

In Section 3, by using known algebraic semantics results, we show that there are conservative translations in both directions. But, in the case here studied, from Łukasiewicz's calculi $\mathbf{L}_n$, $n \geqslant 3$, into classical calculus, the result is non-constructive.

1. A Little About Logical Translations

In this section we present some elements of a general theory of translations between logics, necessary for the central result of Section 3.

In: J. Malinowski and A. Pietruszczak (eds.), *Essays in Logic and Ontology* (*Poznań Studies in the Philosophy of the Sciences and the Humanities*, vol. 91), 157–168. Amsterdam/New York, NY: Rodopi, 2006.

Given a set E, $\mathcal{P}(E)$ denotes the set of all subsets of E. A *consequence operator* in E is a mapping $C: \mathcal{P}(E) \to \mathcal{P}(E)$, such that, for every $A, B \subseteq E$:

(i) $A \subseteq C(A)$

(ii) $A \subseteq B \Longrightarrow C(A) \subseteq C(B)$

(iii) $C(C(A)) \subseteq C(A)$.

For every consequence operator C, it is immediate the equality $C(C(A)) = C(A)$. Sometimes the consequence operator is named *closure operator*.

A consequence operator C in E is *finitary* when, for every $A \subseteq E$: $C(A) = \bigcup\{ C(A_0) \mid A_0$ is a finite subset of $A \}$.

Let C be a consequence operator in E. A set A is *closed* in E if $C(A) = A$, and A is *open* if the complement of A is closed. An element $x \in E$ is *dense* in E if $C(\{x\}) = E$.

A *logic*, or a *deductive system*, is a pair (E, C), where E is a set (the domain) and C is a consequence operator in E.

Well, besides this very general notion about logics, in general, we want to work with formal languages that include connectives and other specific logical objects. Thus, let us consider, even in a general way, some of these elements.

For a formal language L, let For(L) be the set of formulas of L. A *consequence operator in* L has as its domain the set For(L).

Given a formal language L, the free algebra For(L) of formulas of L is generated by the set of atomic formulas of L. A *substitution* is an endomorphism s on For(L), that is, $s \in \text{Hom}(\text{For}(L), \text{For}(L))$.

Let L be a formal language, $s \in \text{Hom}(\text{For}(L), \text{For}(L))$ and C a consequence operator in L. The consequence operator C is *structural* if, for every $\Gamma \subseteq \text{For}(L)$, $s(C(\Gamma)) \subseteq C(s(\Gamma))$. The operator C is *standard* if C is structural and finitary.

The concept of logical system permits us to characterize particular cases of logics, about which we can claim that the operator is finitary, structural or standard.

A *logical system* defined on L is a pair $\boldsymbol{L} = (L, C)$, where L is a formal language and C is a consequence operator in L.

If $\boldsymbol{L}$ is a logical system, the set For(L) is also denoted by For($\boldsymbol{L}$). If $\boldsymbol{L} = (L, C)$ is a logical system and $\Delta \subseteq \text{For}(\boldsymbol{L})$, a *theory* Δ of $\boldsymbol{L}$ is a closed set in $\boldsymbol{L}$. An element of the theory Δ of $\boldsymbol{L}$ is named a *theorem* of Δ. As a theorem of $\boldsymbol{L}$ we mean a member of $C(\emptyset)$, in this case $\Delta = \emptyset$, and the set of all theorems of $\boldsymbol{L}$ is denoted by $\text{Thm}(\boldsymbol{L}) = \{\alpha \mid \alpha \in C(\emptyset)\}$.

$C(\emptyset)$ and For($\boldsymbol{L}$) are, for sure, the smallest and biggest theories respectively associated to the consequence operator C.

After these initial concepts about logical systems we can give a definition of translation between such systems, which have been named Tarski's logical systems.

Let $\boldsymbol{L}_1$ and $\boldsymbol{L}_2$ be two logical systems. A *translation* from $\boldsymbol{L}_1$ into $\boldsymbol{L}_2$ is a mapping $T\colon \boldsymbol{L}_1 \to \boldsymbol{L}_2$ such that, for all $\alpha \in \text{For}(\boldsymbol{L}_1)$ and $\Gamma \subseteq \text{For}(\boldsymbol{L}_1)$:

$$\alpha \in C_1(\Gamma) \Longrightarrow T(\alpha) \in C_2(T(\Gamma)).$$

For logical systems with correct deductibility, that is, $\alpha \in C(\Gamma)$ in $\boldsymbol{L}$ if and only if $\Gamma \vdash_{\boldsymbol{L}} \alpha$, the function T is a translation if, and only if:

$$\Gamma \vdash_{\boldsymbol{L}_1} \alpha \Longrightarrow T(\Gamma) \vdash_{\boldsymbol{L}_2} T(\alpha).$$

As, from set theory, we have that $T(\emptyset) = \emptyset$, considering $\Gamma = \emptyset$, every translation takes theorems of $\boldsymbol{L}_1$ into theorems of $\boldsymbol{L}_2$, that is:

$$\vdash_{\boldsymbol{L}_1} \alpha \Longrightarrow \ \vdash_{\boldsymbol{L}_2} T(\alpha).$$

Proposition 1.1. A mapping $T\colon \boldsymbol{L}_1 \to \boldsymbol{L}_2$ is a translation if, and only if, for every $\Gamma \subseteq \text{For}(\boldsymbol{L}_1)$, $T(C_1(\Gamma)) \subseteq C_2(T(\Gamma))$, with $T(\Gamma) = \{T(\alpha) \mid \alpha \in \Gamma\}$.

Theorem 1.2. Let $T\colon \boldsymbol{L}_1 \to \boldsymbol{L}_2$ be a mapping between logical systems. The following sentences are equivalent

(i) T is a translation;
(ii) the inverse image of a closed set is a closed set;
(iii) the inverse image of an open set is an open set.

A mapping is *closed* if the image of every closed set is still a closed set. A *conservative mapping* between the logical systems $\boldsymbol{L}_1$ and $\boldsymbol{L}_2$ is a function $f\colon \boldsymbol{L}_1 \to \boldsymbol{L}_2$ such that, for every formula α of $\boldsymbol{L}_1$:

$$\vdash_{\boldsymbol{L}_1} \alpha \iff \ \vdash_{\boldsymbol{L}_2} f(\alpha).$$

A *conservative translation* is a mapping T from $\boldsymbol{L}_1$ into $\boldsymbol{L}_2$ such that, for every set $\Gamma \cup \{\alpha\} \subseteq \text{For}(\boldsymbol{L}_1)$:

$$\alpha \in C_1(\Gamma) \iff T(\alpha) \in C_2(T(\Gamma)).$$

In logical systems with correct deductibility, a function T is a conservative translation if, and only if:

$$\Gamma \vdash_{\boldsymbol{L}_1} \alpha \iff T(\Gamma) \vdash_{\boldsymbol{L}_2} T(\alpha).$$

Two logical systems $\boldsymbol{L}_1$ and $\boldsymbol{L}_2$ are *L-homeomorphic* if there is a bijective function $T\colon \boldsymbol{L}_1 \to \boldsymbol{L}_2$, such that T and T^{-1} are translations. In this case, the function T is named a *L-homeomorphism*.

Proposition 1.3. Let $T\colon \boldsymbol{L}_1 \to \boldsymbol{L}_2$ be a bijection. Then, T is a L-homeomorphism if, and only if, for every $\Gamma \subseteq \text{For}(\boldsymbol{L}_1)$, $T(C_1(\Gamma)) = C_2(T(\Gamma))$.

Proposition 1.4. Let $T\colon \boldsymbol{L}_1 \to \boldsymbol{L}_2$ be a function between logical systems. Then, T is an L-homeomorphism if, and only if, T is a conservative and bijective translation.

Let us consider that the language L_1 has only unary and binary connectives. Now, if L_2 is another language, a mapping $f\colon L_1 \to L_2$ is *schematic* if for every unary connective § and for every binary connective # of L_1 there exist schemes[1] $A, B_{\S}, C_{\#}$ of L_2 such that:

(i) $f(\sigma) = A(\sigma)$, for every atomic formula of L_1;

(ii) $f(\S\alpha) = B_{\S}(f(\alpha))$, for every formula α of L_1;

(iii) $f(\alpha \# \beta) = C_{\#}(f(\alpha), f(\beta))$, for all formulas α, β of L_1.

A schematic mapping is a homomorphism between languages, because it preserves the algebraic structures of the algebras of formulas associated to the respective languages.

A schematic mapping is *literal* if it translates every connective into itself, that is, for any atomic formula of L_1: $f(\sigma) = \sigma$, and for all formulas α, β of L_1: $f(\S\alpha) = \S(\alpha)$ and $f(\alpha \# \beta) = f(\alpha) \# f(\beta)$.

The function $T\colon \boldsymbol{L}_1 \to \boldsymbol{L}_2$ is a *schematic translation* if T is a translation and a schematic mapping.

2. On Łukasiewicz's Logics

As it is well known, Łukasiewicz's logics were initially introduced in 1920 and 1922, by semantical matrices.

Let $\mathbb{N}$ be the set of natural numbers, $\mathbb{Q}$ be the set of rational nubers, and $[0, 1]$ be the real interval. Let us consider the sets:

$$\mathrm{L}_n := \begin{cases} \{0, \frac{1}{n-1}, \frac{2}{n-1}, \frac{3}{n-1}, \ldots, 1\} & \text{if } n \in \mathbb{N} \text{ and } n \geqslant 2, \\ \mathbb{Q} \cap [0, 1] & \text{if } n = \aleph_0, \\ [0, 1] & \text{if } n = \aleph_1. \end{cases}$$

A *Łukasiewicz n-valued matrix* has the form:

$$\mathfrak{M}_n = ((\mathrm{L}_n, \neg, \to, \wedge, \vee, \leftrightarrow), \{1\}),$$

where $\neg$ is an unary symbol for the negation and $\to$, $\wedge$, $\vee$ and $\leftrightarrow$ are binary symbols for the conditional, conjunction, disjunction and biconditional, respectively, defined on the set L_n as below:

$$\neg x := 1 - x,$$

$$x \to y := \min\{1, 1 - x + y\},$$

[1] The schemes preserve the form or figure of formulas, such that, for every substitution in any subformula the form keeps the same.

$$x \vee y := \max\{x, y\} = (x \rightarrow y) \rightarrow y,$$
$$x \wedge y := \min\{x, y\} = \neg(\neg x \vee \neg y),$$
$$x \leftrightarrow y := 1 - |x - y| = (x \rightarrow y) \wedge (y \rightarrow x).$$

This way, we can see that $\mathfrak{M}_2$ coincides with the classical matrix.

In 1920 Łukasiewicz introduced his 3-valued logic $\mathbf{L}_3$ through the following tables for the negation and the conditional, extending the classical interpretation for these connectives:

x	$\neg x$
0	1
½	½
1	0

$x \rightarrow y$	0	½	1
0	1	1	1
½	½	1	1
1	0	½	1

The other connectives were defined as above.

Let us indicate the set of tautologies of $\mathfrak{M}_n$ by Taut($\mathfrak{M}_n$). Then, it is possible to demonstrate the following results (see Malinowski 1993).

PROPOSITION 2.1. Taut($\mathfrak{M}_n$) $\subseteq$ Taut($\mathfrak{M}_2$), for every $n \geqslant 2$.

PROPOSITION 2.2 (Lindenbaum, Łukasiewicz and Tarski). Taut($\mathfrak{M}_n$) $\subseteq$ Taut($\mathfrak{M}_m$) if, and only if, $m - 1 \mid n - 1$.

PROPOSITION 2.3 (Lindenbaum). Taut($\mathfrak{M}_{\aleph_0}$) = Taut($\mathfrak{M}_{\aleph_1}$).

PROPOSITION 2.4. Taut($\mathfrak{M}_{\aleph_0}$) = $\bigcap\{$ Taut($\mathfrak{M}_n$) $\mid n \geqslant 2$ and $n \in \mathbb{N}\}$.

Concerning the axiomatization of Łukasiewicz's logics we can see in Malinowski (1993, p. 39):

> It can be easily verified that the content of any Łukasiewicz matrix is closed under the detachment rule (for the original implication of the matrix in question). A proof that finite matrices are axiomatizable was given in Łukasiewicz and Tarski (1930). Nevertheless, the problem of formulation of a concrete finite axiomatization of $E(\mathfrak{M}_n)$ [E($\mathfrak{M}_n$) coincides with Taut($\mathfrak{M}_n$)] (with the exception of the case $n = 3$) remained open till 1952; see Rosser and Turquette (1952). Wajsberg (1931) showed that the ($\neg$, $\rightarrow$)-fragment of Łukasiewicz's three-valued propositional calculus may be axiomatized, accepting the rules MP [*modus ponens*] and SUB [substitution], in the following way:
>
> W1. $p \rightarrow (q \rightarrow p)$
> W2. $(p \rightarrow q) \rightarrow ((q \rightarrow r) \rightarrow (p \rightarrow r))$
> W3. $(\neg p \rightarrow \neg q) \rightarrow (q \rightarrow p)$
> W4. $((p \rightarrow \neg p) \rightarrow p) \rightarrow p$.
>
> Łukasiewicz also proposed an axiomatization for his infinite valued system, using five axioms, and conjectured that the axiomatization was complete for the semantical matrix.

Cignoli, D'Ottaviano and Mundici (1994) say:

> Wajsberg (1935, p. 240), mentioned that he had proved Łukasiewicz's conjecture, however his proof never appeared published (see Tarski 1956, p. 51; or Tarski 1983, p. 51). The first proof published about the Łukasiewicz's conjecture was that by Rose & Rosser 1958. Chang 1959 presented another proof of algebraic character. An exposition of these works is in Rosser 1960.

All that suggests us how difficult it was to clean the bridge between the semantical and syntactical aspects of Łukasiewicz's logics.

3. On the Conservative Translations from $\mathbf{L}_n$ to CPC

In D'Ottaviano & Feitosa (1999a) and Feitosa (1997), we have got conservative translations from **CPC** into $\mathbf{L}_n$, for every $n \geqslant 3$; however, we did not obtain such translations from $\mathbf{L}_n$ into **CPC**. Thus, it seems interesting to ask if there are such functions.

In this section, using results about algebraic semantics, we demonstrate that there exists a conservative translation from $\mathbf{L}_n$ into **CPC**, for every $n \geqslant 2$.

As we are going to use algebraic semantics, let us consider the classical propositional calculus (**CPC**) as presented in Rasiowa & Sikorski (1968), and the n-valued Łukasiewicz's calculi ($\mathbf{L}_n$) as presented in Bolc & Borowiki (1993). It is well known in the literature that each Boolean algebra is an algebraic model for **CPC** and, in particular, that the Lindenbaum algebra of **CPC** is a denumerable and without atoms Boolean algebra $\mathbf{B_2}$. The algebraic models for $\mathbf{L}_n$ are the Cignoli algebras and, as they are in general less familiar, we are introducing them here.

Let $n \in \mathbb{N} - \{0, 1\}$ and let us consider the following sets:

$$\mathrm{S}_n := \begin{cases} \emptyset & \text{if } n < 5, \\ \{(i, j) \in \mathbb{N}^2 \mid 3 \leqslant i \leqslant n-2, 1 \leqslant j \leqslant n-4 \text{ and } j < i\} & \text{if } n \geqslant 5; \end{cases}$$

$$\mathrm{T}_n := \begin{cases} \emptyset & \text{if } n < 4, \\ \{(i, j) \in \mathbb{N}^2 \mid 2 \leqslant i \leqslant n-2, 1 \leqslant j \leqslant n-3 \text{ and } j < i\} & \text{if } n \geqslant 4. \end{cases}$$

We can easily see that $\mathrm{S}_n \subseteq \mathrm{T}_n$ and, for $n = 5$, $|\mathrm{T}_n| = |\mathrm{S}_n| + 2$.

These sets are necessary for the concept of Cignoli algebra, that in the literature on algebraic semantics for many-valued logics was introduced as the concept of proper n-valued Łukasiewicz's algebra, or proper n-valued Moisil-Łukasiewicz's algebra.

Let $n \geqslant 2$. Then *n-valued Cignoli algebra* is a structure

$$\boldsymbol{C}_n = (C, \wedge, \vee, \Rightarrow, -, 0, 1, \{s_i\}_{1 \leqslant i \leqslant n-1}, \{F_{ij}\}_{(i,j) \in \mathrm{S}_n}),$$

such that:

(L_1) $(C, 0, 1, \wedge, \vee, -)$ is a distributive De Morgan lattice,

(L_2) $(C, 0, 1, \wedge, \vee, \Rightarrow)$ is a Heyting algebra,

and for all $a, b \in C$ and $1 \leqslant i, j, k \leqslant n-1$:

(L_3) $s_i(a \vee b) = s_i(a) \vee s_i(b)$,

(L_4) $s_i(s_j(a)) = s_j(a)$,

(L_5) $s_i(-a) = -s_{n-i}(a)$,

(L_6) $s_1(a) \vee -s_1(a) = 1$;

(L_7) $s_1(a) \vee a = a$;

(L_8) $s_i(a \Rightarrow b) = \bigwedge_{j=1}^{i}(s_j(a) \Rightarrow s_j(b))$,

(L_9) $s_k(F_{ij}(a, b)) = \begin{cases} 0 & \text{if } k \leqslant i-j \\ J_i(a) \wedge J_i(b) & \text{if } k > i-j, \end{cases}$
where $J_i(x) = s_{n-i}(x) \wedge -s_{n-i-1}(x)$, for every $x \in C$.

Every n-valued Cignoli algebra, according to Cignoli (1980) and Boicescu *et al.* (1991), is an algebraic model adequated for $\mathbf{L}_n$, and according to Rasiowa (1974, VIII, 9.5), it is a strongly adequated algebraic model for $\mathbf{L}_n$, $n \geqslant 2$, with a valuation v_C, such that:

$$v_C(p) = [p]$$
$$v_C(\neg\alpha) = -v_C(\alpha)$$
$$v_C(s_i(\alpha)) = s_i(v_C(\alpha))$$
$$v_C(\alpha \wedge \beta) = v_C(\alpha) \wedge v_C(\beta)$$
$$v_C(\alpha \vee \beta) = v_C(\alpha) \vee v_C(\beta)$$
$$v_C(\alpha \to \beta) = v_C(\alpha) \Rightarrow v_C(\beta)$$
$$v_C(F_{ij}(\alpha,\beta)) = F_{ij}(v_C(\alpha), v_C(\beta)).$$

According to Bolc & Borowiki (1993), if $\mathbf{L}_n$ is a Łukasiewicz n-valued system, then $(\text{For}(\mathbf{L}_n)/_{\leftrightarrow}, 1, 0, \to, \wedge, \vee, \neg, s_1, s_2, \ldots, s_{n-1}, \{F_{ij}\}_{(i,j)\in S_n})$ is the Lindenbaum algebra of $\mathbf{L}_n$, which is a n-valued freely generated Cignoli algebra.

Given a logic $\boldsymbol{L} = (L, \boldsymbol{C})$, let us consider the following equivalence relation in For($\boldsymbol{L}$):

$$\alpha \sim \beta \iff C(\alpha) = C(\beta).$$

The quotient mapping $Q\colon \boldsymbol{L} \to \boldsymbol{L}_{\sim}$, defined by:

$$Q(\alpha) := [\alpha] := \{\beta \mid \alpha \sim \beta\},$$

is a translation and $\boldsymbol{L}_{\sim} = (L_{\sim}, \boldsymbol{C}_{\sim})$ is the co-induced logic by Q and $\boldsymbol{L}$ (see Feitosa & D'Ottaviano 2000).

PROPOSITION 3.1. The mapping $Q\colon \boldsymbol{L} \to \boldsymbol{L}_{\sim}$ is a conservative translation.

PROOF. Considering the above affirmation, Q is a translation. Now, let $Q(\Delta) = \{[\alpha] \mid \alpha \in \Delta\}$. If $Q(\alpha) \in C_{\sim}(Q(\Delta))$ then, for some $\beta \in [\alpha]$, we have that $\beta \in C(\Delta)$. As $C(\beta) = C(\alpha)$, it follows that $\alpha \in C(\beta)$ and, therefore, $\alpha \in C(\Delta)$. □

THEOREM 3.2 (Feitosa & D'Ottaviano 2000). Let $\boldsymbol{L}_1$ and $\boldsymbol{L}_2$ be logics, with the domain of $\boldsymbol{L}_2$ denumerable. Then there is a conservative translation $T\colon \boldsymbol{L}_1 \to \boldsymbol{L}_2$ if, and only if, there exists a conservative translation $\tau : \boldsymbol{L}_{1\sim} \to \boldsymbol{L}_{2\sim}$.

COROLLARY 3.3. If there is the function τ of the previous theorem, then it is injective.

PROOF. Considering $\tau(Q_1(\alpha)) = \tau(Q_1(\beta))$, we have that $Q_2(T(\alpha)) = Q_2(T(\beta))$ and $T(\alpha) \sim_2 T(\beta)$, that is, $C_2(T(\alpha)) = C_2(T(\beta))$. As T is a conservative translation, so $C_1(\alpha) = C_1(\beta)$, that is, $\alpha \sim_1 \beta$. Then $Q_1(\alpha) = Q_1(\beta)$ or yet $[\alpha] = [\beta]$. □

In our case, the quotient logics $\boldsymbol{L}_{\sim} = (L_{\sim}, C_{\sim})$, where $\alpha \sim \beta$ if and only if $C(\alpha) = C(\beta)$, are the Lindenbaum algebras of **CPC** and $\mathbf{L}_n$, for $n > 2$.

In the following we will need the Stone's morphisms, through which we will get the indicated results. Thus, we start by the concept of a filter in a lattice.

A *filter* in a lattice $\boldsymbol{A}$ is a non empty set $F \subseteq \boldsymbol{A}$ such that, for every $a, b \in \boldsymbol{A}$:

(i) $a, b \in F \Longrightarrow a \wedge b \in F$;

(ii) $a \in F$ and $a \leq b \Longrightarrow b \in F$.

The filter F is *proper* if $F \subset \boldsymbol{A}$, that is, F is properly included in $\boldsymbol{A}$. The filter F is *prime* if it is proper and, for $a, b \in \boldsymbol{A}$:

(iii) $a \vee b \in F \Longrightarrow a \in F$ or $b \in F$.

PROPOSITION 3.4 (Rasiowa & Sikorski 1968, Proposition I.9.2). Let $\boldsymbol{A}$ be a distributive lattice. Given $a, b \in \boldsymbol{A}$, if $b \nleq a$, then there is a prime filter F such that $a \notin F$ and $b \in F$.

Given a lattice $\boldsymbol{A}$, we denote by $\mathbb{F}(\boldsymbol{A})$ the set of prime filters in $\boldsymbol{A}$. Now, let us consider the function $h\colon \boldsymbol{A} \to \mathbb{F}(\boldsymbol{A})$, defined by $h(a) := \{F \in \mathbb{F}(\boldsymbol{A}) \mid a \in F\}$, and the set

$$\mathcal{H}(\boldsymbol{A}) := \{\, h(a) \mid a \in \boldsymbol{A} \,\} \subseteq \mathcal{P}(\mathbb{F}(\boldsymbol{A})).$$

PROPOSITION 3.5 (Rasiowa & Sikorski 1968). Let $\boldsymbol{A}$ be a distributive lattice. Then $\mathcal{H}(\boldsymbol{A})$ is a distributive lattice of sets and h is an isomorphism from $\boldsymbol{A}$ to $\mathcal{H}(\boldsymbol{A})$.

Two algebraic structures $\boldsymbol{A}$ and $\boldsymbol{B}$ are *partially isomorphic*, what is denoted by $\boldsymbol{A} \cong_{\mathrm{p}} \boldsymbol{B}$, if there is a non empty set $\boldsymbol{I}$ of partial isomorphisms from $\boldsymbol{A}$ into $\boldsymbol{B}$ for which:

(a) for every $\imath \in \boldsymbol{I}$ and $a \in \boldsymbol{A}$, there is $\jmath \in \boldsymbol{I}$, with $\imath \subseteq \jmath$ and $a \in \mathrm{dom}(\jmath)$;

(b) for every $\imath \in \boldsymbol{I}$ and $b \in \boldsymbol{B}$, there is $\jmath \in \boldsymbol{I}$, with $\imath \subseteq \jmath$ and $b \in \mathrm{im}(\jmath)$.

THEOREM 3.6. Let $\boldsymbol{A}$ and $\boldsymbol{B}$ be Boolean algebras partially isomorphic. If the domain of $\boldsymbol{A}$ and $\boldsymbol{B}$ are denumerable, then they are isomorphic, that is $\boldsymbol{A} \cong \boldsymbol{B}$.

PROOF. See Ebbinghaus *et al.* (1984, p. 183) or Cifuentes (1992, p. 25). □

LEMMA 3.7 (Feitosa 1997). Let $\boldsymbol{A}$ and $\boldsymbol{B}$ be denumerable and dense Boolean algebras. If $\imath$ is a partial isomorphism from $\boldsymbol{A}$ into $\boldsymbol{B}$ and $\mathrm{dom}(\imath)$ is finite then for any $a \in \boldsymbol{A}$ and $b \in \boldsymbol{B}$ there is a partial isomorphism $\imath_{a.b}$ such that $\imath \subseteq \imath_{a.b}$ and still $a \in \mathrm{dom}(\imath_{a.b})$ and $b \in \mathrm{im}(\imath_{a.b})$.

THEOREM 3.8. If $\boldsymbol{A}$ and $\boldsymbol{B}$ are denumerable and dense Boolean algebras then they are isomorphic.

PROOF. Let $\{a_1, a_2, \ldots, a_n, \ldots\}$ and $\{b_1, b_2, \ldots, b_n, \ldots\}$ be enumerations of $\boldsymbol{A}$ and $\boldsymbol{B}$ respectively. Now let us construct the following sequence of partial isomorphisms, by recursion:

$$\jmath_0 \colon \boldsymbol{A}_0 = \{0, 1\} \to \boldsymbol{B}_0 = \{0, 1\},$$
$$\jmath_{n+1} \colon [\boldsymbol{A}_n \cup \{a\}] \to [\boldsymbol{B}_n \cup \{b\}], \qquad \jmath_{n+1} = (\jmath_n)_{a.b},$$

where $(\jmath_n)_{a.b}$ is the extension of $\jmath_n$, in which $a \in \mathrm{dom}((\jmath_n)_{a.b})$ and $b \in \mathrm{im}((\jmath_n)_{a.b})$ according to Lemma 3.7. From Theorem 3.6, we have that $\boldsymbol{A} \cong \boldsymbol{B}$. □

COROLLARY 3.9. Every denumerable Boolean algebra can be immerged into a denumerable and dense Boolean algebra $\boldsymbol{B}$.

PROOF. Let $\boldsymbol{A}$ be a denumerable Boolean algebra. So, $\boldsymbol{A} = \bigcup_{n \in \mathbb{N}} \boldsymbol{A}_n$ such that $\boldsymbol{A}_0 = \{0, 1\}$ and $\boldsymbol{A}_{n+1} = [\boldsymbol{A}_n \cup \{a\}]$, with $a \notin \boldsymbol{A}_n$ and $a \in \boldsymbol{A}$. For every partial isomorphisms $\imath \colon \boldsymbol{A} \cong_{\mathrm{p}} \boldsymbol{B}$ and for every $a \in \boldsymbol{A}$ we define — as in the Lemma 3.7 — a partial isomorphism $\imath_a$ such that $\imath \subseteq \imath_a$ and $a \in \mathrm{dom}(\imath_a)$.

Once more, by recursion:

$$\jmath_0 \colon \boldsymbol{A}_0 \to \boldsymbol{B}_0,$$
$$\jmath_{n+1} \colon [\boldsymbol{A}_n \cup \{a\}] \to [\boldsymbol{B}_n \cup \{b\}], \qquad \jmath_{n+1} = (\jmath_n)_a,$$

where $(\jmath_n)_a$ is the extension of $\jmath_n$, such that $a \in \mathrm{dom}((\jmath_n)_a)$ and $b \in \mathrm{im}((\jmath_n)_a)$. Then $\jmath = \bigcup_{n \in \mathbb{N}} \jmath_n$ is an isomorphism from $\boldsymbol{A}$ in $\mathrm{im}(\jmath) \subseteq \boldsymbol{B}$. □

A Boolean algebra A is *without atoms* when A does not have any atom, that is, for every $a \in A$, $a \neq 0$, there exists $b \in A$ such that $0 < b < a$.

PROPOSITION 3.10. A Boolean algebra A is dense if, and only if, A is without atoms.

PROOF. ($\Rightarrow$) If A is dense then, then for every $a \in A$, $a \neq 0$ there is a b such that $0 < b < a$. So, a can not be an atom.

($\Leftarrow$) Let A be without atoms, and $a, b \in A$ such that $a < b$: (i) If $a = 0$, since b is not an atom, then there is a $x \in A$ such that $0 < x < b$.

(ii) If $a \neq 0$, then $b \wedge -a \neq 0$. As A does not have atoms, then there is $x \in A$, such that $0 < x < b \wedge -a$. Let us verify that $a < a \vee x < b$.

For sure, $a \leq a \vee x$. However, $a \neq a \vee x$, because otherwise $x \leq a$ and, as $x < -a$, we would have $x \leq a \wedge -a = 0$, what contradicts the choice of x. Besides, as $a < b$ and $x < b$, so $a \vee x \leq b$; if $a \vee x = b$, then $(a \vee x) \wedge -x = b \wedge -x$; hence, $(a \wedge -x) \vee (x \wedge -x) = b \wedge -x$ and so, $a \wedge -x = b \wedge -x$; therefore $a = b$, what contradicts the hypothesis.

In any case, there is a $x \in A$, such that $a < x < b$. □

PROPOSITION 3.11. For every freely generated Cignoli algebra C_n, $n \geqslant 2$, there is an injective homomorphism (immersion) $\tau\colon C_n \to \mathbf{B_2}$, the Lindenbaum algebra of **CPC**.

PROOF. If C_n is a freely generated Cignoli algebra, then it is a distributive and denumerable lattice. By Proposition 3.5, $C_n \cong \mathcal{H}(C_n)$ and as $|C_n| = \aleph_0$, then $|\mathcal{H}(C_n)| = \aleph_0$. Now, as $\mathcal{H}(C_n)$ is a lattice of sets we will consider the Boolean algebra A generated by its elements. Thus $A = [h(a) \mid a \in C_n]$ is a denumerable Boolean algebra. By Corollary 3.9, there is an immersion of A into a denumerable and dense Boolean algebra B. And now, by Proposition 3.10, B is isomorphic to $\mathbf{B_2}$. Taking τ as the composition of these four functions, we have the intended immersion. □

COROLLARY 3.12. For every $n \geqslant 2$, let $\mathbf{C}_n$ be the Lindenbaum algebra of $\mathbf{L}_n$. In this case, there is an injective homomorphism $\tau\colon C_n \to \mathbf{B_2}$.

PROOF. It follows from Proposition 3.11 and from the fact that $\mathbf{C}_n$ is a freely generated Cignoli algebra. □

THEOREM 3.13. For every $n \geqslant 2$ there is a conservative translation from $\mathbf{L}_n$ into **CPC**.

PROOF. Let $n \geqslant 2$ and τ be the function from Corollary 3.12. Since τ is an injective homomorphism (immersion), so $\mathbf{C}_n \cong \mathrm{im}(\tau)$.

Therefore, $[\Gamma] \vdash_{\mathbf{L}_n} [\alpha]$ if, and only if, $\tau[\Gamma] \vdash_{\mathbf{CPC}} \tau[\alpha]$.

Then τ is a conservative translation from the Lindenbaum algebra of $\mathbf{L}_n$ into the Lindenbaum algebra of **CPC**.

Hence, by Theorem 3.2, it follows that there is a conservative translation from $\mathbf{L}_n$ into **CPC**. □

Universidade Estadual de Campinas (Unicamp)
Group of Theoretical and Applied Logic
Centre for Logic, Epistemology and the History of Science
Department of Philosophy
e-mail: itala@cle.unicamp.br

Universidade Estadual Paulista (Unesp)
Department of Mathematics
e-mail: haf@fc.unesp.br

REFERENCES

Ackermann, R. (1967). *Introduction to Many-Valued Logics*. New York: Dover.

Boicescu, V. and A. Filipoiu, G. Georgescu and S. Rudeanu (1991). *Łukasiewicz-Moisil Algebras. Annals of Discrete Mathematics*, vol. 49. Amsterdam: North-Holland.

Bolc, L. and P. Borowik (1992). *Many-Valued Logics: 1 Theoretical Foundations*. Berlin: Springer-Verlag.

Cifuentes, J. C. (1992). *O método dos isomorfismos parciais. Coleção CLE*, vol. 10. Campinas: UNICAMP/CLE.

Cignoli, R. (1980). Some Algebraic Aspects of Many-Valued Logics. In: A. I. Arruda, N. C. A. da Costa and A. M. Sette (eds.), *Proceedings of the Third Brazilian Conference on Mathematical Logic*, pp. 49–69. São Paulo: Sociedade Brasileira de Lógica.

Cignoli, R. L. O., I. M. L. D'Ottaviano and D. Mundici (1994). *Álgebras das lógicas de Łukasiewicz. Coleção CLE*, vol. 12. Campinas: UNICAMP/CLE.

Cignoli, R. L. O., I. M. L. D'Ottaviano and D. Mundici (2000). *Algebraic Fundations of Many-Valued Reasoning. Trends in Logic*, vol. 2. Dordrecht: Kluwer Academic Publishers.

D'Ottaviano, I. M. L. and H. A. Feitosa (1999a). Many-Valued Logics and Translations. *Journal of Applied Non-Classical Logics* **9** (1), 121–140.

D'Ottaviano, I. M. L. and H. A. Feitosa (1999b). Conservative Translations and Model-Theoretic Translations. *Manuscrito* XXII (2), 117–132.

D'Ottaviano, I. M. L. and H. A. Feitosa (2000). Paraconsistent Logics and Translations. *Synthese* **125** (1/2), 77–95.

Da Silva, J. J., I. M. L. D'Ottaviano and , A. M. Sette, (1999). Translations Between Logics. In: X. Caicedo and C. H. Montenegro (eds.), *Models, Algebras and Proofs*, pp. 435–448. *Lectures Notes in Pure and Applied Mathematics*, vol. 203. New York: Marcel Dekker.

Ebbinghaus, H. D., J. Flum and W. Thomas (1984). *Mathematical Logic*. New York: Springer-Verlag.

Epstein, R. L. (1990). *The Semantic Foundations of Logic*. Vol. 1: *Propositional Logics*. Dordrecht: Kluwer Academic Publishers.

Feitosa, H. A. (1997). *Traduções conservativas* [Conservative tranlations]. Tese de Doutorado em Lógica. Campinas: UNICAMP.

Feitosa, H. A., I. M. L. D'Ottaviano (2001). Conservative Translations. *Annals of Pure and Applied Logic*, 108, 205–227.

Hamilton, A. G. (1978). *Logic for Mathematicians*. Cambridge: Cambridge University Press.

Malinowski, G. (1993). *Many-Valued Logics*. Oxford: Clarendon Press.

Mendelson, E. (1964). *Introduction to Mathematical Logic*. Princeton: D. Van Nostrand.

Rasiowa, H. (1974). *An Algebraic Approach to Non-Classical Logics*. Amsterdam: North-Holland.

Rasiowa, H. and R. Sikorski (1968). *The Mathematics of Metamathematics*. 2nd edition. Warszawa: PWN (Polish Scientific Publishers).

Wójcicki, R. (1988). *Theory of Logical Calculi: Basic Theory of Consequence Operations*. *Synthese Library*, vol. 199. Dordrecht: Kluwer Academic Publishers.

Luis Fariñas del Cerro
Olivier Gasquet

MODAL TABLEAUX FOR REASONING ABOUT DIAGRAMS

ABSTRACT. This paper, we propose a modal logic satisfying minimal requirements for reasoning about diagrams via collection of sets and relations between them, following Harel's proposal. We first give an axiomatics of such a theory and then provide its Kripke semantics. Then we extend previous works of ours in order to obtain a decision procedure based on tableaux for this logic. Beside soundness and completeness of our tableaux, we manage to define a strategy of rule application ensuring termination by extending the usual loop test of modal logic S4 to whole sub-structures of the model being computed.

Introduction

Diagrammatic reasoning is of high interest in the knowledge representation community because diagrams are widely used by humans in various activities. It is a simple and clear way of representing complex systems like operating systems or graphical interfaces. Of course, the possibility to reason on such diagrammatic specification is also of high interest.

It is well-know that the representation of such systems require collections of sets together with specific relations between them, in Harel *et al.* (1987), Harel has proposed a kind of diagrams that enable to manipulate both collections of sets and relations between these collections, by marrying in a very elegant way Euler's circles — that represent collections of sets — and graphs — that represent relations. Also, conceptual graphs and terminological logics are other examples of formalisms in which collections of sets and relations between them are the basic notions.

In this paper we propose a bi-modal logic that follows Harel's approach: one modality expresses the subset inclusion between elements of the collection of sets, and the others express specific relations between these collections of sets, for example actions that lead from one state to another one. The aim of this paper is to provide a decision procedure for such a modal logic.

For this logic, we present its language, its semantics in terms of Kripke frames, an axiomatisation of it, and then we address the automated deduction problem for it: we present a naive deduction calculus which is both sound and complete but may not terminate. This deduction calculus is based on previous works of ours (e.g. Fariñas del Cerro & Gasquet 2002) and is a tableau calculus where rules are divided between propagation rules and structural rules.

In: J. Malinowski and A. Pietruszczak (eds.), *Essays in Logic and Ontology* (*Poznań Studies in the Philosophy of the Sciences and the Humanities*, vol. 91), 169–184. Amsterdam/New York, NY: Rodopi, 2006.

Then we define a strategy which ensures that the tableau for this logic is terminating while being still sound and complete. This strategy is based on: the order the rules are applied and a "loop test" that detects repetition of a substructure of the tableau. Our tableaux, supplied with this strategy, yield a decision procedure.

Our general framework for modal tableaux has been developed in Castilho *et al.* (1997) and in Fariñas del Cerro & Gasquet (2000, 2002). It is characterized by two ideas:

1. the propagation of formulas need not to be top-down,
2. the underlying structure need not to be a tree.

In Massacci (2000, 1994), Goré (1998), De Giacomo & Massacci (1996), the first idea has been investigated. Their work is intimately mixed with another feature: the so called single-step rules. Such rules allow to propagate formulas only from one node to one of its successors or predecessor in the tree. We generalize this notion to "pattern-driven rules": rules apply if some elementary pattern in the mathematical structure has been matched.

While trees are a good basis for tableaux for many usual modal logics (tableaux for systems K(D)(T)(4) are based on trees), they fail to support in a comprehensive way confluent or dense relations (corresponding, respectively, to the axioms $\Diamond\Box p \rightarrow \Box\Diamond p$ and $\Diamond p \rightarrow \Diamond\Diamond p$), and bimodal confluent systems too, i.e. systems which semantics is such that there may be several paths from the root of the structure (the tableau under computation) to some given world/node. This feature requires that the reasoning about the accessibility relation should be separated from the reasoning about formulas. At least two main approaches may be first considered:

- The use of labels (see Massacci 2000, 1994; Goré 1998; De Giacomo & Massacci 1996; Fitting 1983) where each possible world is associated with a label (its path from the root, the set of all path may be considered as a spanning tree of a graph) and an equationnal theory is needed in order to specify which paths denote the same world.
- The explicite use of the accessibility relation, as explored in Baldoni (1998a, 1998b); this turns out to be a model-construction approach. But this work focuses on "grammar logics" (multi-modal logics with axioms of the form $[a_1]\ldots[a_n]p \rightarrow [b_1]\ldots[b_m]p$) and mainly study their connections with formal languages, thus excluding properties like symmetry, euclideanness, confluence; moreover, it does not seem possible to extend decidability results and decision procedures for a property like density in their framework.

Our position (partially-explicit use of the accessibility relation), while being close to that of Baldoni *et al.* (1998), is different at least because as much as possible, we do not completely compute the accessibility relation: for example, we do not compute its transitive closure but *simulate* it by propagation of formulas. In fact, only existential properties require to explicitely compute the accessibility relation. One will find other examples of complete tableaux in Castilho *et al.* (1997).

We argue that rooted directed acyclic graphs (RDAG for short), which are DAGS having a distinguished node called the root, are better suited than trees. They allow to naturally handle some properties that do not marry easily with tree structures (like confluence, density, and also, permutation in the multi-modal case), while other properties (like transitivity, symmetry, ...) can still be handled by the propagation of formulas.

This leads us to identify two kinds of tableaux rules:

1. propagation rules
2. structural rules.

The former are formulated as "if in some node of such pattern there is such formula, then propagate such formula (the same or another one)", while the latter are "if there is such pattern then add some new node(s) and edge(s)".

They respectively correspond to two different families of relational properties:

- Propagation rules correspond to properties like reflexivity, transitivity, symmetry and euclideanness (i.e. the well-known axioms T, 4, B and 5 respectively), which are closure properties;
- Structural rules correspond to properties like seriality, density, confluence, ... which are existential properties.

Generally speaking, a tableau is a structure (usually a tree, in our case it will be an RDAG) whose nodes are labelled by sets of formulas. The completeness proof of a tableau method is in two main steps: the construction of a model from this structure, and the verification that this model satisfies the formulas of the nodes (the so-called Fundamental Lemma).

The first step is usually achieved by adding new arrows to the structure, according to the particular property of the accessibility relation of the logic under concern. For example, for the system S4 the accessibility relation is reflexive and transitive. Hence, given a tree (the underlying structure for S4), we must close it under reflexivity and transitivity in order to make an S4-model of it. In other terms, we must characterize when two nodes are related in the resulting closure. Then we can say that for a given node x another node y will be accessible from x if there is an $n \geqslant 0$ and $x_0, \ldots, x_i, x_{i+1}, \ldots, x_n$ such that x_0 is x, x_n is y and x_{i+1} is a child of x_i in the original tree. From this characterization of the closure of the initial tree under the additional properties of the logic under concern, we can "read off" the rules to be designed. Thus the rules will ensure the correct propagation of formulas, the proof being very easy.[1] This gives naturality and simplicity.

In addition, the rules that we have obtained fit closely to the intuition.

Thus we obtain tableaux calculi whose completeness proofs are neatly separated into three components:

1. the Relational Closure Lemma where the properties of the closure of an RDAG under some relational properties are expressed in terms of the initial RDAG,
2. the Structural Lemma where we check that the closure of an RDAG under some relational properties preserves some of its initial features (e.g. the transitive closure of a confluent RDAG yields a confluent relation, but not necessarily an RDAG),
3. the Box Lemma (Lemma 1) where we check that whenever x and y are related in the closure, and $\Box A \in x$ then the set of associated rules ensures that A was transported into y.

The completeness proof is completely factorized and soundness is straightforward to prove.

Our general framework first lead to the design of "naive" tableaux that may be non terminating. For addressing decidability and complexity we have to adapt these naive tableaux by providing a *strategy* that guides the computation and allow to conclude to the satisfiability of the initial formula even in those cases where the computation may run forever. As examples, in Fariñas del Cerro & Gasquet (2002) we designed such strategies for monomodal logics like KD4 plus density, or KD4 plus confluence, and obtain tight complexity bounds.

For backgrounds about tableaux for modal logic, the reader may look at de Swart (1980), Fitting (1983, 1993), Goré (1998) and our Fariñas del Cerro & Gasquet (2002).

The Logic

We formally here presents the language, syntax and semantics of our logic of diagrams that will henceforth denoted by `Diag`.

1. Language

The language is defined by the following: let P be a set of propositional symbols, and as usual let $\bot$ denotes falsity. The set Φ of formulas is given by the BNF:

$$\Phi ::= \bot \mid p \mid \neg\Phi \mid (\Phi \wedge \Phi) \mid \Box_{\subseteq}\Phi \mid \Box_{\alpha}\Phi \qquad (\text{where } p \in P)$$

[1]The correctness proof mainly consists of results of relational calculus.

and as usual, $\Diamond_{\subseteq}$ and $\Diamond_{\alpha}$ respectively abbreviate $\neg\Box_{\subseteq}\neg$ and $\neg\Box_{\alpha}\neg$, while $(\_ \vee \_)$ abbreviates $\neg(\_ \wedge \_)$, and $(\_ \rightarrow \_)$ abbreviates $(\neg\_ \vee \_)$.

Thus, $\Box_{\subseteq}\phi$ reads "ϕ holds for all subsets of the current set", while $\Box_{\alpha}\phi$ reads "ϕ holds for all set α-associated with the current one"[2].

2. Models and Frames

All along this paper, we will use the following well-known notations: let R and S be relations over a set W (we will indistinctly use $(x, y) \in R$ or xRy for denoting that x and y are R-related),

- I is the diagonal relation defined by: $I = \{(x, x) : x \in W\}$;
- $R \circ S$ denote the relation consisting of pairs $(x, y) \in W^2$ such that: $\exists z \in W$: $(x, z) \in R$ and $(z, y) \in S$;
- R^n for some integer n is defined by: $R^0 = I$ and $R^{n+1} = R \circ R^n$ if $n \geqslant 0$;
- R^+ is the *transitive closure* of R defined by: $(x, y) \in R^+$ iff there exists some integer $n \geqslant 1$ such that $(x, y) \in R^n$; hence $R^+ = \bigcup_{i<\infty} R^i$;
- R^* is the *reflexive and transitive closure* of R, defined by: $R^* = R^+ \cup I$;
- the *converse relation* $\overline{R}$ is defined by: $\overline{R} = \{(y, x) : (x, y) \in R\}$, i.e. $(x, y) \in R$ iff $(y, x) \in \overline{R}$.

The models we aim at capturing are described as follows:

A frame is a triple $(W, R_{\subseteq}, R_{\alpha})$ where W is a set of names of sets, and $R_{\subseteq}$, R_{α} are binary relations over W. Now, let C be the class of such frames verifying in addition:

- $R_{\subseteq}$ is a reflexive and transitive relation,
- R_{α} is a serial relation,[3]
- $\overline{R_{\subseteq}} \circ R_{\alpha} = R_{\alpha} \circ \overline{R_{\subseteq}}$ which concisely express the condition:

$$\forall x, y, z : ((x, y) \in R_{\alpha} \& (x, z) \in R_{\subseteq}) \rightarrow (\exists u \colon (y, u) \in R_{\subseteq} \; \& \; (z, u) \in R_{\alpha})).$$

And let C_0 be the class of frames of C which verifies in addition:

- $R_{\subseteq}$ is an irreflexive anti-symmetric tree
- R_{α} is also an irreflexive anti-symmetric tree

[2]For instance, in a dynamic interpretation, $\Box_{\alpha}\phi$ reads "After any execution of the action α the formula ϕ holds"

[3]I.e. each element of W has a successor by R_{α}.

A model M based on a frame $F = (W, R_{\subseteq}, R_{\alpha})$ is given by a triple (w, F, v) where:

- $w \in W$ is the so-called actual world,
- F is a frame
- v is a valuation function, $v(p)$ being the set of worlds where p is true (hence $v(p) \subseteq W$).

Validity and satisfiability of a formula are defined as usual.

Here is a very simple example of the kind of situation we aim at formalizing:

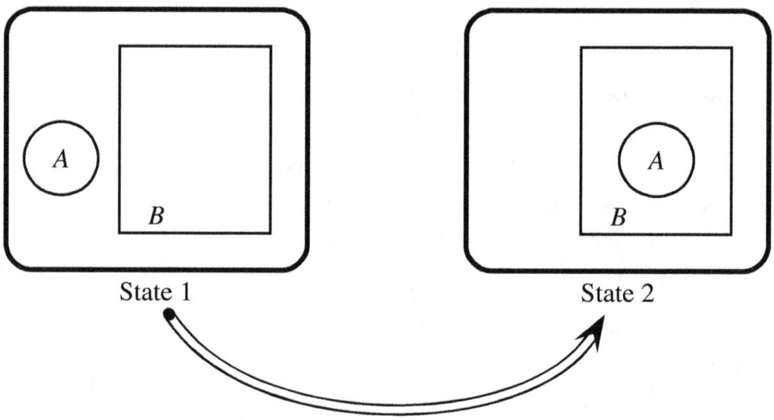

Action: put A into B

In this figure, State 1 represents two icons A and B, each of them representing a set of objects. State 2 represents the result of the action "put A into B": by confluence every object initially in A will still be in A after moving it. It is well-known that interpreting atomic actions within such a semantics is satisfactory, while complex actions (containing loops, ...) need a more complex semantics like that of PDL.

3. Axiomatics

From Salhlqvist's theorem of Sahlqvist (1975) — or its multimodal version (Catach 1988) — we know that the set Th of theorems defined below is sound and complete w.r.t. the class C of frames (where $R_{\subseteq}$ and R_{α} are not supposed to be anti-symmetrical trees). The set Th is defined by:

1. Th contains all classical propositional tautologies
2. if $\phi \in$ Th then $\Box_{\subseteq}\phi$ and $\Box_{\alpha}\phi$ are also in Th (Necessitation rule)

3. For any formulas ϕ_1 and ϕ_2, the following formulas are in $\mathtt{Th}$:

   (a) $(\Box(\phi_1 \rightarrow \phi_2) \wedge \Box\phi_1) \rightarrow \Box\phi_2$ where $\Box$ is either $\Box_{\subseteq}$ or $\Box_\alpha$ (K axiom)

   (b) $\Box_\alpha\phi_1 \rightarrow \Diamond_\alpha\phi_1$ (seriality axiom)

   (c) $\Box_{\subseteq}\phi_1 \rightarrow \phi_1$ (reflexivity axiom)

   (d) $\Box_{\subseteq}\phi_1 \rightarrow \Box_{\subseteq}\Box_{\subseteq}\phi_1$ (transitivity axiom)

   (e) $\Diamond_{\subseteq}\Box_\alpha\phi_1 \rightarrow \Box_\alpha\Diamond_{\subseteq}\phi_1$ (confluence axiom denoted by $\mathtt{Conf}$)

   Axioms and inference rules 1, 2 and 3a defines the basic bimodal logic denoted $K(\subseteq, \alpha)$.

In the sequel, we will design a decision procedure for testing the satisfiability of a formula w.r.t. the class C_0 of frames (which by the way proves that C and C_0 characterizes the same set of satisfiable formulas).

4. Rewriting Tableaux with Rules

Usually, tableaux calculi consist in rewriting a structure by using some appropriate set of rewriting rules (or simply rules). This structure is a triple $(\mathcal{N}, \Sigma, \Phi)$ where: $\mathcal{N}$ is a set of nodes (possible worlds), Σ is a pair $(\Sigma_{\subseteq}, \Sigma_\alpha)$ of binary relations over $\mathcal{N}$ (the accessibility relation) and Φ is a function that maps each element of $\mathcal{N}$ to its associated set of formula (by abuse of notation, we will sometimes write $A \in x$ instead of $A \in \Phi(x)$). Here are the rules we need for first designing a naive tableau calculus for our logic. These rules have the obvious operational meaning: *if some part of the current tableau under process matches the left-hand side of the rule then rewrite this part into the right-hand side of the rule*. Note that edges are implicitely left-to-right oriented.

We will use the following for graphical denotations for rules and tableaux: (edges are implicitely left-to-right directed):

S •	denotes a node associated with a set S of formulas
$S0$ •—$\subseteq$—• $S1$	denotes that two nodes respectively associated with $S0$ and $S1$ are $R_{\subseteq}$-related
$S0$ •—α—• $S1$	denotes that two nodes respectively associated with $S0$ and $S1$ are R_α-related

Here are the rules we need:

- Classical rules, T-rule and $\Diamond$ rules:

Rule $\bot$: $A, \neg A, S \implies A, \neg A, \bot, S$

Rule $\neg$: $\neg\neg A, S \implies \neg\neg A, A, S$

Rule $\wedge$: $A \wedge B, S \implies A \wedge B, A, B, S$

Rule $\vee$: $\neg(A \wedge B), S \implies \neg(A \wedge B), C, S$

where C is non-deterministically choosen among $\neg A$ and $\neg B$

Rule $T_{\subseteq}$: $S, \Box_{\subseteq} A \implies S, A, \Box_{\subseteq} A$

Usually, treatment of disjunction is presented by so-called *tableau branching*: both possibilities are computed in parallel (width-first computation), while in our presentation, we use a depth-first computation. These presentations are in fact equivalent, in practice only depth-first implementations are realized.

Rule $\Diamond_{\subseteq}$: $S, \Diamond_{\subseteq} A \implies \Diamond_{\subseteq} A, S \xrightarrow{\subseteq} A$

Rule $\Diamond_{\alpha}$: $S, \Diamond_{\alpha} A \implies \Diamond_{\alpha} A, S \xrightarrow{\alpha} A$

- Propagation rules:

Rule $\Box_{\subseteq}$: $\Box_{\subseteq} A, S \xrightarrow{\subseteq} S1 \implies \Box_{\subseteq} A, S \xrightarrow{\subseteq} A, S1$

Rule $\Box_{\alpha}$: $\Box_{\alpha} A, S \xrightarrow{\alpha} S1 \implies \Box_{\alpha} A, S \xrightarrow{\alpha} A, S1$

Rule $4_{\subseteq}$: $S, \Box_{\subseteq} A \xrightarrow{\subseteq} S1 \implies S, \Box_{\subseteq} A \xrightarrow{\subseteq} S1, \Box_{\subseteq} A$

In fact, rule $\Box_{\subseteq}$ is useless as it is subsumed by rule $4_{\subseteq}$ followed by rule $T_{\subseteq}$.

- Structural rules:

Rule D_{α}: $S \implies S \xrightarrow{\alpha} \emptyset$

Rule Cf:

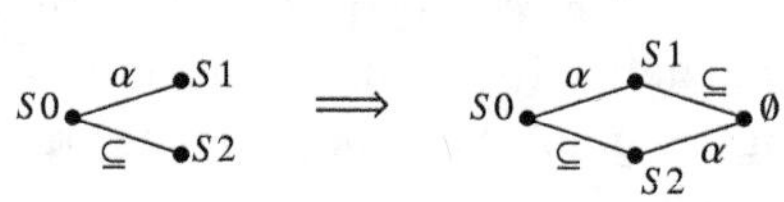

Of course, it may be discussed why not to consider $\Diamond$ rules as a structural rules. This decision is quite arbitrary, we consider that structural rules are those describing an existential property of the accessibility relation, while $\Diamond$ rules correspond to the expression *in the object language* of the existence of objects.

We define what we call *naive tableaux*, i.e. tableaux computed with no strategy at all, particularly strategies ensuring termination.

DEFINITION 1. A (naive) tableau for a formula ϕ is the limit of a sequence $\Upsilon_0, \ldots, \Upsilon_i, \Upsilon_{i+1}, \ldots$ where $\Upsilon_i = (\mathcal{N}_i, \Sigma_i = \Sigma_{\subseteq_i} \cup \Sigma_{\alpha_i}, \Phi_i)$. It is the limit in the sense that $\mathcal{N} = \bigcup_{i\geqslant 0} \mathcal{N}_i$, $\Sigma = \bigcup_{i\geqslant 0} \Sigma_i$, $\Phi(x \in \mathcal{N}) = \bigcup_{i\geqslant 0, x\in \mathcal{N}_i} \Phi_i(x)$, where $\Phi_i(x)$ denotes the set of formulas associated with node x at iteration i.

In addition we must have:

- Υ_0 only consists of one node whose associated set of formulas is $\{\phi\}$,
- Υ_{i+1} is obtained from Υ_i by applying one of the above rules,
- every applicable rule has been applied once.

DEFINITION 2. A tableau is closed if some node in it contains $\perp$; it is open otherwise. A formula is closed iff all its tableaux are closed.[4]

Of course, as implicitly stated above, we make the usual assumption of *fairness*: if at some iteration i, some rule is applicable then for some $j \geqslant i$ the rule has been applied. Thus, saying that completeness and soundness rely on the fairness assumption consists in saying that *they only hold for fair algorithms*. For example, an algorithm that would apply rule (D_α) first (hence with a priority greater than that of other rules) would not be fair since this rule may be applied infinitely giving no chance to apply any other rule: from a node x_0, create a successor x_1, then a successor x_2 of x_1, and so on. Note that since rules are monotonic (in the sense that any applicable rule remains so until it has been applied), since order of application does not change anything and since at each step only finitely many rules may be applied, there always exist a fair algorithm: at step n compute the set S of all applicable left-hand side of rules, and rewrite them all before going on with the computation.

DEFINITION 3. With the above rules, the tableaux that we obtain have a particular structure: If $\Upsilon = (\mathcal{N}, \Sigma_\subseteq, \Sigma_\alpha, \Phi)$ is a naive tableau, it is clear that the relation Σ_α is a set of trees. Each of these trees will be called *α-tree*.

This feature will be exploited in the next section to prove the decidability of the logic `Diag`.

In Fariñas del Cerro & Gasquet (2002) we proved that such a tableau calculus is both sound and complete for our logic (which is a particular case of those investigated). But, of course, we did not provide a terminating tableau.[5] This is what the sequel is devoted to after a brief sketch about completeness of naive tableaux.

5. Completeness

Soundness is immediate in this case. Completeness proof has the following structure: we show how, from a given open tableau for A we can construct a model for A.

---

[4]Due to the rule $\vee$, a formula may have several distinct tableaux.

[5]The well-known example of the formula $\Box_\subseteq \Diamond_\subseteq p$ leads to a non-terminating naive tableau.

Let $\Upsilon = (\mathcal{N}, \Sigma, \Phi)$ be an open tableau for ϕ with root r.
Now let $\mu = (W, R, \tau)$ be the Kripke model defined as follows:

DEFINITION 4. • $W = \mathcal{N}$

- $R_{\subseteq} = \Sigma_{\subseteq}^{\star}$ – reflexive and transitive closure of $\Sigma_{\subseteq}$
- $R_{\alpha} = \Sigma_{\alpha}$
- for all $w \in W$, $w \in \tau(p)$ iff $p \in w$

By construction, μ is a member of C. What remains is to prove that it satisfies the formula ϕ. It is easy to check that:

LEMMA 1 (Box Lemma). Let $\Upsilon = (\mathcal{N}, \Sigma, \Phi)$ be a tableau with root r. Let x, y be such that $(x, y) \in \Sigma_{\subseteq}^{\star}$ and $\Box_{\subseteq}\psi \in x$; then $\psi \in y$. As well, if x, y are such that $(x, y) \in \Sigma_{\alpha}$ and $\Box_{\alpha}\psi \in x$; then $\psi \in y$.

The following fundamental lemma brings us to the desired conclusion:

LEMMA 2 (Fundamental Lemma). Let Υ be an open tableau for A, let μ be the model defined as in Definition 4 w.r.t. Υ and let $\psi \in$ Subformulas(ϕ) then: if $\psi \in x$ then $\mu, x \models \psi$. Thus in particular $\mu, r \models \phi$.

PROOF. Straightforward induction on the structure of ψ. □

As a direct consequence of the previous lemma, we have:

COROLLARY 1. If ϕ has a fair open tableau then ϕ is satisfiable. Hence our tableaux calculus is complete under the fairness assumption.

Decision procedures for systems based on K($\subseteq, \alpha$) plus *non-interaction* axioms are treated in Catach (1991), and those based on K($\subseteq, \alpha$) plus *interaction* axioms of the form $\Diamond_{\alpha} p \rightarrow \Diamond_{\subseteq} p$ in (Demri 2000). With $\mathtt{Per} = \Box_{\subseteq}\Box_{\alpha} p \leftrightarrow \Box_{\alpha}\Box_{\subseteq} p$, the system K($\subseteq, \alpha$) + Per + Conf, also known as the weakest product modal logic $K \times K$ is investigated in Gabbay & Shehtman (1998). Its precise complexity is still not known. As well, the decidability of the system K($\subseteq, \alpha$)+Per+Cf+$4_{\subseteq}$+4_{α} is an open question.

6. Decidability

In the following lines, we present a terminating strategy for tableaux for the logic Diag. It mainly consists in paying attention to the order the rules are applied and in detecting when the calculus may run forever without closure. But first, we need some additional specific definitions:

DEFINITION 5. The $\subseteq$-depth of a node x in a tableau $(\mathcal{N}, \Sigma_{\subseteq}, \Sigma_{\alpha}, \Phi)$ is denoted by $d_{\subseteq}(x)$ and is defined by: if $\overline{\Sigma_{\subseteq}}(x) = \emptyset$ then 0 else $d_{\subseteq}(\overline{\Sigma_{\subseteq}}(x)) + 1$ — note that $\overline{\Sigma_{\subseteq}}(x)$ is either empty or is a singleton. Informally, $d_{\subseteq}(x)$ is the length of the shortest $\Sigma_{\subseteq}$-path from the root to some Σ_{α}-ancestor of x.

DEFINITION 6. The α-depth of a node x in a tableau $(\mathcal{N}, \Sigma_{\subseteq}, \Sigma_{\alpha}, \Phi)$ is denoted by $d_{\alpha}(x)$ and is defined by: if $\overline{\Sigma_{\alpha}}(x) = \emptyset$ then 0 else $d_{\alpha}(\overline{\Sigma_{\alpha}}(x)) + 1$ — as well $\overline{\Sigma_{\alpha}}(x)$ is either empty or is a singleton.

DEFINITION 7. The rules we use ensure that the converse relations $\overline{\Sigma_{\subseteq}}$ and $\overline{\Sigma_{\alpha}}$ are (partial) functions. Hence, we denote the unique $\Sigma_{\subseteq}$-predecessor (resp. Σ_{α} predecessor) of some node x by $\overline{\Sigma_{\subseteq}}(x)$ (resp. by $\overline{\Sigma_{\alpha}}(x)$).

DEFINITION 8. Given a node x in a tableau containing a formula $\Diamond_{\subseteq}\psi$ (resp. $\Diamond_{\alpha}\psi$), we say that this formula is *inactive* if there exists a $\Sigma_{\subseteq}$-sibling[6] (resp. Σ_{α}-sibling) y of x which contains ψ, it is *active* otherwise (This feature is useful to avoid multiple identical siblings).

DEFINITION 9. Let x, y, z be three nodes of a tableau $(\mathcal{N}, \Sigma_{\subseteq}, \Sigma_{\alpha}, \Phi)$ and ψ be a formula:

- the unique node created by applying rule $\Diamond_{\subseteq}$ (resp. $\Diamond_{\alpha}$) to node x and active formula $\Diamond_{\subseteq}\psi$ of x (resp. $\Diamond_{\alpha}\psi$) will be denoted by $\Diamond_{\subseteq}(x, \psi)$ (resp. $\Diamond_{\subseteq}(x, \psi)$).
- the unique node created by applying rule $\mathtt{Cf}$ to nodes x, y and z (with $(x, y) \in \Sigma_{\alpha}$ and $(x, z) \in \Sigma_{\subseteq}$) will be denoted by $\mathtt{Cf}(x, y, z)$.

DEFINITION 10. Let σ and τ, be two α-trees (Definition 3) in a tableau $T = (\mathcal{N}, \Sigma_{\subseteq}, \Sigma_{\alpha}, \Phi)$ whose respective roots are s and t and have the same α-depths.

Then $\sigma\Sigma_{\subseteq}\tau$ iff

- $t = \Diamond_{\subseteq}(s, \psi)$ (see Definition 9) for some active formula $\Diamond_{\subseteq}\psi$ of s; this implies that $(s, t) \in \Sigma_{\subseteq}$.
- $\forall x \in \tau : (x \neq t \wedge \overline{\Sigma_{\subseteq}}(x) \neq \emptyset) \Rightarrow x = \mathtt{Cf}(\overline{\Sigma_{\subseteq}}(\overline{\Sigma_{\alpha}}(x)), \overline{\Sigma_{\subseteq}}(x), \overline{\Sigma_{\alpha}}(x))$
- $\forall x \in \tau : (x \neq t \wedge \overline{\Sigma_{\subseteq}}(x) = \emptyset) \Rightarrow x = \Diamond_{\alpha}(\overline{\Sigma_{\alpha}}(x))$

Informally, $\sigma\Sigma_{\subseteq}\tau$ iff:

- t has been obtained by applying rule $\Diamond_{\subseteq}$ to s,

  and for other nodes than t:
- those for which $\overline{\Sigma_{\subseteq}}(x) \neq \emptyset$, have been obtained by applying rule $\mathtt{Cf}$

[6]A node y is a *sibling* of x iff they have a common predecessor, which in our case is unique.

- the remaining nodes, for which $\overline{\Sigma_{\subseteq}}(x) = \emptyset$, have been obtained by applying rule $\Diamond_{\alpha}$ to some node of τ.

Then we generalize: $\sigma(\Sigma_{\subseteq})^n\tau$ for some integer $n > 1$ iff there is some α-tree ρ of T such that $\sigma\Sigma_{\subseteq}\rho$ and $\rho(\Sigma_{\subseteq})^{n-1}\tau$.

DEFINITION 11. In addition, we note $\sigma|\Sigma_{\subseteq}|\tau$ to denote that $\sigma\Sigma_{\subseteq}\tau$ and $\forall x \in \tau$: $\overline{\Sigma_{\subseteq}}(x) \neq \emptyset$. Informally, the second condition express that in τ no node has been created by applying rule $\Diamond_{\alpha}$, in this case, it is easy to see that σ and τ are isomorphic (up to their labels or sets of formulas).

DEFINITION 12. Last, we note $\tau \subseteq \sigma$ iff $\exists n \geqslant 1$: $\sigma|\Sigma_{\subseteq}|\tau$ and $\forall x \in \tau$: $\Phi(x) \subseteq \Phi(\overline{\Sigma_{\subseteq}}^n(x))$. Informally, $\tau \subseteq \sigma$ iff σ and τ are pointwise $\Sigma_{\subseteq}{}^n$-related and each node of τ is included in its ancestor node in σ.

The algorithm

We first recall a possible PSPACE algorithm for the well-known logic S4 (with only one modality $\Box_{\subseteq}$) of reflexivity and transitivity:

Starting from $T_0 = (\mathcal{N}_0, (\Sigma_{\subseteq})_0, \Phi_0)$ where $\mathcal{N}_0$ only consists of one node r (the root), $(\Sigma_{\subseteq})_0$ is empty and Φ_0 associate the formula ϕ with r. Initially, the *new node* is r. Then compute $T_{i+1} = (\mathcal{N}_{i+1}, (\Sigma_{\subseteq})_{i+1}, \Phi_{i+1})$ from $T_i = (\mathcal{N}_i, (\Sigma_{\subseteq})_i, \Phi_i)$ by applying successively each of the following steps:

1. Loop step: consider the current *new node* $x \in \mathcal{N}_i$: if $\exists y \in \mathcal{N}_i$ and y is an ancestor of x (i.e. $(y, x) \in (\Sigma_{\subseteq})_i^*$) and $\Phi_i(x) \subseteq \Phi_i(y)$ then mark x as `loop-node`; (nodes that are "contained" in one already present in the branch needs not to be further developed)
2. Classical step: apply classical rules $((\bot), (\neg), (\vee), (\wedge)$ and $(T_{\subseteq}))$ on x as much as possible (also known as classical saturation) but only if it is not marked as `loop-node`;
3. Structural step: apply once one rule $(\Diamond_{\subseteq})$ on *active* formulas (cf. Definition 8) x but only if it is not marked as `loop-node` — this yields the next new node;
4. Propagation step: apply rules $(4_{\subseteq})$ as much as possible.
5. If rule is applicable, then backtrack to the last choice (at step 3 with rule $\Diamond_{\subseteq}$, or at step 2 with rule $\vee$)

This algorithm is depth-first, it explorates each tableau tree (there are several because of rule $\vee$) and each branch of each tree (there are several because of rule $\Diamond_{\subseteq}$) independently, and alternative choices are left to backtracking. If no closure is found, whatever choice is done, the initial formula has an open tableau.

In Ladner (1977), it was proved that the length of each branch is in fact of polynomial size w.r.t. the length of the initial formula, and since *each branch may be explored independently*[7], this yields a PSPACE algorithm.

Let us introduce the following conventions:

- `ClSat` $= (\bot + \neg + \vee + \wedge + T_{\subseteq})^*$ – Classical saturation, note that `ClSat` does always terminate.
- `Loop`? consists in the following: if at iteration i, the current new node x is such that $\Phi_i(x) \subseteq \Phi_i(y)$ for some ancestor of x (i.e. $(x, y) \subseteq (\Sigma_{\subseteq})_i^*$) then *mark x as* `loop-node`.
- the rule $\diamond_\alpha^\circ$ is the same as rule $\diamond_\alpha$ but applies only on *active* formulas.
- the rule $\diamond_\subseteq^\circ$ is the same as rule $\diamond_\subseteq$ but it only applies on *active* formulas of nodes which, in addition, are not marked as `loop-node`.
- Let R and S be any rules: R denotes a unique application of rule R, R^* denotes the repeated application, as much as possible of rule R, $(R + S)$ denotes the non-deterministic application of either R or S and $R; S$ denotes the application of R and then of S.

With these conventions, the above algorithm (for S4) may be described by: $(\texttt{ClSat}; \diamond_\subseteq^\circ; 4_\subseteq^*; \texttt{Loop?})^*$.

With the same conventions, a possible algorithm for the monomodal logic K_α may be: $(\texttt{ClSat}; \diamond_\alpha; \Box_\alpha^*)^*$.

Here the strict degree decrease of formulas (the highest is the α-degree of a node, the lowest is the modal degree of its formulas) ensures the termination of this algorithm whose complexity is again PSPACE since each branch of the tableau is of polynomial size w.r.t. the length of the initial formula (precisely, the length of each branch is bounded by the modal degree of the initial formula), and *each branch may be explored independently*.

Of course, in the case of the logic `Diag`, no such simple loop test will do it, our proof of termination is based on the repetition of a whole substructure, more precisely on the repetition of α-trees at distinct $\subseteq$-depths.

Now we come to the algorithm for `Diag`:

$$((\diamond_\alpha; \Box_\alpha^*; \texttt{ClSat})^*; ((\diamond_\subseteq^\circ; 4_\subseteq^*; \texttt{ClSat}); (\texttt{Cf}; (\Box_\alpha + 4_\subseteq)^*; \texttt{ClSat})^*); \texttt{Loop?})^*$$

Comments:

1. The part $(\diamond_\alpha; \Box_\alpha^*; \texttt{ClSat})^*$ consists in computing an initial α-tree;
2. The part $(\diamond_\subseteq^\circ; 4_\subseteq^*; \texttt{ClSat})$ consists in computing the root of the next α-tree; (this will be done only if rule $\diamond_\subseteq^\circ$ may be applied on a node of a non loop-α-tree — see step 4 below)

[7] This is crucial.

3. The part ($\texttt{Cf}; (\Box_\alpha + 4_\subseteq)^*; \texttt{ClSat})^*$ consists in completing the computation of the next α-tree, this part of the computation being immediately followed by the Loop? test.

4. The test Loop? consists in the following:

   - Let τ be the α-tree which has just been completed, if there exists an α-tree σ such that $\tau \subseteq \sigma$ then mark τ as loop-α-tree.

LEMMA 3. The above algorithm terminates.

SKETCH OF PROOF. The condition that rule $\Diamond_\alpha$ is applied only on active formulas ensures that the branching factor w.r.t. Σ_α is finite (and roughly bounded by the — linear — number $|\phi|$ (the length of ϕ) of $\Diamond_\alpha$ subformulas of the initial formula). Then the number of α-trees such that none of them is included in some other is finite too, though very high. Because of the rules, nodes of α-depth equal to the modal degree $d(\phi)$ of the initial formula are empty, hence, α-trees are of height bounded by $d(\phi)$, and since their branching factor is bounded by $|\phi|$: there are finitely many distinct trees of bounded height and branching factor. Each of their nodes may be associated with one of the exponentially many subsets of subformulas of the initial formula ϕ. This yields a very high but still finite number of possible α-trees. But, this is very rough and the precise complexity of the logic `Diag` is still under investigation. □

LEMMA 4. The strategy given above is sound for `Diag`.

PROOF. Straightforward since the resulting algorithm is a restriction (on the order of application of the rules) of the naive one which is sound. Hence, if there is an open naive tableau for ϕ, there is a fortiori an open (non-naive) tableau for ϕ. □

LEMMA 5. The strategy given above is complete for `Diag`.

SKETCH OF PROOF. The completeness of this strategy, relies on the fact that once a loop-α-tree ($\tau \subseteq \sigma$ and hence for some n: $\sigma|\Sigma_\subseteq{}^n|\tau$) has been detected, there is no need to further develop any of its nodes: potentially, it suffices to duplicate the structure "under" σ, i.e. act for x in τ as for $\overline{\Sigma_\subseteq}^n(x)$. □

THEOREM 1. The above algorithm is sound and complete for testing the satisfiability problem for the logic `Diag` and does terminate.

Conclusion

In this paper, we proposed a bi-modal logic to formalize simple situations where set inclusion is needed together with some logic of actions. We prove its decidability by means of a tableau calculus provided with a strategy which ensures its termination. It is a matter of on going work to extend our tableau method to star-free PDL for addressing more complex actions instead of the logic K for just one atomic action.

Université Paul Sabatier
IRIT
118 route de Narbonne
Toulouse Cedex 04
France
e-mail: {farinas, gasquet}@irit.fr
internet: http://www.irit.fr/ACTIVITES/LILaC

REFERENCES

Baldoni, M. (1998). *Normal Multimodal Logics: Automatic Deduction and Logic Programming Extension*. PhD thesis. Univ. Torino, Italy.

Baldoni, M., L. Giordano, and A. Martelli (1998). A Tableau Calculus for Multimodal Logics and Some (Un)Decidability Results. In: H. de Swart (ed.), *TABLEAUX'98*, LNAI 1397. Springer-Verlag.

Castilho, M. A., L. Fariñas del Cerro, O. Gasquet and A. Herzig (1997). Modal Tableaux with Propagation Rules and Structural Rules. *Fundamenta Informaticæ* **32** (3/4), 281–302.

Catach, L. (1988). Normal Multi-Modal Logics. In: *Proceedings National Conference on Artificial Inteligentce (AAAI-88).*

Catach, L. (1991). TABLEAUX: A General Theorem Prover for Modal Logics. *Journal of Automated Reasoning* **7**, 489–510.

De Giacomo, G. and F. Massacci (1996). Tableau and Algorithm for Propositional Dynamic Logic with Converse. In: M. A. McRobbie and J. K. Slaney (eds.), *Proc. CADE-13, LNAI 1104*. Springer.

Demri, S. (2000). Complexity of Simple Dependent Bimodal Logics. In: *Proc. Int. Conf. Automated Reasoning with Analytic Tableaux and Related Methods*, (TABLEAUX'2000), LNCS 1847. Springer.

de Swart, H. C. M. (1980). Gentzen-Type Systems for C, K and Several Extensions of C and K: Constructive Completeness Proofs and Effective Decision Procedure for these Systems. *Logique et Analyse* **90–91**, 263–284.

Fariñas del Cerro, L. and O. Gasquet (2000). Tableaux Based Decision Procedures for Modal Logics of Confluence and Density. *Fundamenta Informaticæ* **41** (1), 1–17.

Fariñas del Cerro, L. and O. Gasquet (2002). A General Framework for Pattern-Driven Modal Tableaux. *Logic Journal of the IGPL* **10** (1). 51–84.

Fitting, M. (1983). *Proof Methods for Modal and Intuitionistic Logics*. Dordrecht: D. Reidel.

Fitting, M. (1993). Basic Modal Logic. In: D. Gabbay *et al.* (eds.), *Handbook of Logic in Artificial Intelligence and Logic Programming: Logical Foundations*, vol. 1. Oxford University Press.

Gabbay, D. M. and V. Shehtman (1998). Products of Modal Logics: Part 1. *Logic Journal of the IGPL* **6**, 73–146.

Goré, R. (1998). Methods for Modal and Temporal Logics. In: D. Gabbay *et al.* (eds.), *Handbook of Tableau Methods*. Dordrecht: Kluwer.

Harel, D. *et al.* (1987). On the Formal Semantics of Statechart. *Proc. of the Second IEEE Symp. on Logic in Computer Science*. Ithaca.

Kripke, S. (1959). A Completeness Theorem in Modal Logic. *Journal of Symbolic Logic* **24**.

Kripke, S. (1963). Semantical Analysis of Modal Logic I: Normal Modal Propositional Calculi. *Zeitschrift für Mathematische Logik und Grundlagen der Mathematik* **9**.

Ladner, R. (1977). The Computational Complexity of Provability in Systems of Modal Logic. *SIAM Journal on Computing* **6**, 466–480.

Marx, M. and S. Mikulas (2002). An Elementary Constrution for a Non-Elementary Procedure. *Studia Logica* **72** (2), 253–263.

Massacci, F. (1994). Strongly Analytic Tableaux for Normal Modal Logics. In: A. Bundy (ed.), *Proc. CADE-12, LNAI 814*. Springer.

Massacci, F. (2000). Single Step Tableaux for Modal Logics: Computational Properties, Complexity and Methodology. *Journal of Automated Reasoning* **24** (3), 319–364.

Sahlqvist, H. (1975). Completeness and Correspondence in the First and Second Order Semantics for Modal Logics. In: S. Kanger (ed.), *Proc. 3rd Scandinavian Logic Symposium 1973*. Studies in Logic 82. North-Holland.

Diderik Batens

NARROWING DOWN SUSPICION IN INCONSISTENT PREMISE SETS*

ABSTRACT. Inconsistency-adaptive logics isolate the inconsistencies that are derivable from a premise set, and restrict the rules of Classical Logic only where inconsistencies are involved. From many inconsistent premise sets, disjunctions of contradictions are derivable no disjunct of which is itself derivable. Given such a disjunction, it is often justified to introduce new premises that state, with a certain degree of confidence, that some of the disjuncts are false. This is an important first step on the road to consistency: it narrows down suspicion in inconsistent premise sets and hence locates the real problems among the possible ones. In this paper I present two approaches for handling such new premises in the context of the original premises. The first approach may apparently be combined with all paraconsistent logics. The second approach does not have the same generality, but is decidedly more elegant.

Preludium

I met Jerzy Perzanowski in the early seventies in Cracow on an exchange between our universities. Finding out, upon arrival, that my lecture was scheduled in a Conference on the History of Logic, I apologized that I could only talk on logic. I was comforted right away, "Maybe your lecture will be historical", after which the Kraków chairman of Solidarność — the Kraków division encompassed Nowa Huta — brought me to a huge Russian-style hotel including a dollar shop. This initiated a long friendship, many meetings in Poland, Belgium and conferences abroad, and later a research cooperation. Logical ontology, modal and para(in)consistent logics, and political disagreements served as a reinforcement.

On my next visit to Cracow, Jerzy had left full-time for Jaśkowski's Toruń, the birthplace of Mikołaj Kopernik, and hence was part-time in Cracow — every Pole is actually more than one Pole. Then and during next stays in Poland my lodgings were of Middle-European comfort and definitely more cosy. But Jerzy retained his typical smile, suggesting he knows better. Usually he does.

My contribution will be on handling inconsistency, Ghent style with a strong Toruń flavor.

*The research for this paper was supported by the Fund for Scientific Research – Flanders, by the Research Fund of Ghent University, by the Royal Flemish Academy of Belgium for Science and the Arts, and indirectly by the Flemish Minister responsible for Science and Technology (contract BIL01/80). I am indebted to Leen De Vreese for careful comments on a former draft.

In: J. Malinowski and A. Pietruszczak (eds.), *Essays in Logic and Ontology* (*Poznań Studies in the Philosophy of the Sciences and the Humanities*, vol. 91), 185–209. Amsterdam/New York, NY: Rodopi, 2006.

1. The Problem

Although the scientific enterprise is directed at obtaining a body of consistent knowledge, it is well known that inconsistencies occurred in several central episodes of the development of scientific disciplines — see, for example, Norton (1987, 1993), Smith (1988), Nersessian (2002), Meheus (1993, 2002). In some cases an accepted theory — one that was superior to alternatives in view of its problem solving capacity — turned out to be inconsistent. In other cases progress in a discipline was realized by reasoning from several accepted theories that are mutually inconsistent. In still other cases, progress in a discipline was realized by reasoning from a theory and a set of data that are mutually inconsistent.

The reasoning that occurs in such situations is explicated by inconsistency-adaptive logics — see, for example, Batens (1989, 1985, 1999, 2000b, 2001a), Batens and Meheus (2006), Priest (1991), Meheus (2000), Vanackere (2000), De Clercq (2000). These logics isolate the involved inconsistencies and in this way provide an interpretation of the premises (theories and/or data) that is as consistent as possible. Precisely this type of interpretation is needed in order to reason from the inconsistent premises and in order to regain consistency.

Inconsistency-adaptive logics provide a maximal consistent interpretation of the premises, but do not themselves resolve the inconsistencies. I have argued on several occasions that the elimination of inconsistencies is not a task of logic but has to rely on empirical data, if the theory is an empirical one, or on conceptual analysis, in sum on extra-logical information. Although I do not want to change this position, I admit that inconsistency-adaptive logics leave the process that leads to consistency a mystery. The aim of the present paper is to clarify part of the mystery: on the road to consistency, logic has to play a role. To be more precise, it has to perform two different tasks. First, the logical analysis of the old theory and, where appropriate, data should indicate which specific extra-logical information should be sought. Next, once the information is obtained, logic should handle it, which includes combining it in the suitable way with the old theory. The present paper concerns the first step beyond interpreting the inconsistent premise set as consistently as possible. This will require a bit of explanation.

Suppose that one derived

(1) $(p \wedge {\sim} p) \vee (q \wedge {\sim} q) \vee (r \wedge {\sim} r)$

from the premises, but failed to derive $(p \wedge {\sim} p) \vee (q \wedge {\sim} q)$, $(p \wedge {\sim} p) \vee (r \wedge {\sim} r)$, or $(q \wedge {\sim} q) \vee (r \wedge {\sim} r)$. So according to the available insights, the premises require that either $p \wedge {\sim} p$ or $q \wedge {\sim} q$ or $r \wedge {\sim} r$ holds true, but do not specify which of them does. This will have consequences for interpreting the premises as consistently as possible. On the Reliability strategy, for example, $p \wedge {\sim} p$, $q \wedge {\sim} q$ and $r \wedge {\sim} r$

are all considered as unreliable, which means that no inference can be made that presupposes one of them to be false.

Of course, only the *minimal* disjunctions of inconsistencies have an impact. If $q \wedge \sim q$ is derived, then $(p \wedge \sim p) \vee (q \wedge \sim q)$ is obviously derivable but this does not provide a reason to consider $p \wedge \sim p$ as possibly true. Hence, once a disjunction of abnormalities has been derived, say (1), it is advisable to try to derive a "shorter" disjunction, for example $(p \wedge \sim p) \vee (q \wedge \sim q)$, etc. This is a sound heuristic advise, and the ensuing search process obviously does not involve new extra-logical information.

This brings us to our problem: when deriving a shorter disjunction of contradictions fails, one may attempt to narrow down the suspect formulas by means of new information. There may be reasons to consider either p or q or r as *not* suspect. One or more empirical criteria may be considered as extremely reliable. That they univocally lead to p may provide a good reason to consider both $\sim p$ and $p \wedge \sim p$ as false. A different reason may be that p is considered as well entrenched and unproblematic in the scientific discipline, or even in a set of disciplines. Similarly, the "personal constraints" of a researcher or research group may provide this researcher or group with a reason to consider p as not suspect, and hence to consider $p \wedge \sim p$ as false — as was argued in Meheus & Batens (1996), there is nothing irrational in such situations, even if a different researcher or research group may take the opposite decision. Especially in mathematical contexts but not only there, the fact that p corresponds to a clear and simple concept, whereas q and r do not, may provide the good reason. And of course, a researcher may just tinker, trying out what becomes of the premises if one of the contradictions is considered as false.

If (1) is derived and one fails to derive a shorter disjunction of contradictions, one will try to obtain new information of the aforementioned type. As (1) is derivable from the premises, these evoke (in the sense of Wiśniewski 1995) the question $?\{p \wedge \sim p, q \wedge \sim q, r \wedge \sim r\}$. Each member of the set is a direct answer to the question. The statement that, for example, $p \wedge \sim p$ is false is an indirect answer in view of (1) — it narrows down $\{p \wedge \sim p, q \wedge \sim q, r \wedge \sim r\}$ to $\{q \wedge \sim q, r \wedge \sim r\}$. Even if the information was available all along, it may have been considered irrelevant, or not have been included in the original premise set because this was supposed to be consistent anyway. So the derivation of a *Dab*-formula — a disjunction of abnormalities, here contradictions — provokes a heuristic advice (concerning inference) and, if this does not lead to a useful result, provokes the introduction of new information or the attempt to obtain new information.

If the new information offers good reasons to consider both $p \wedge \sim p$ and $q \wedge \sim q$ as false, $r \wedge \sim r$ is identified as the "real" inconsistency involved in the premises. This move does not render the premise set consistent. It *narrows down* the set of formulas that might be inconsistent. This is a sensible step towards restoring consistency. By this step, one locates the real problems, the real inconsistencies, that have to be resolved in order to restore consistency. Moreover, the step has an immediate effect on the maximal consistent interpretation of the premises.

I remarked before that logic has two tasks on the path towards consistency. And we have seen that inconsistency-adaptive logics themselves perform the first task: they indicate which specific extra-logical information should be sought by locating the minimal disjunctions of contradictions or, in undecidable waters, the best candidates in view of present insights in the premises. The rest of the paper is devoted to the the second task: handling the newly gathered information together with the old theory.

A straightforward but naive approach would proceed as follows. Given a minimal disjunction of contradictions, the problem solver might be allowed to state that some of the disjuncts are false. It is possible to do so by means of classical negation, $\neg$. Given that the original negation of the language, viz. $\sim$, is paraconsistent, it is advisable to extend the language with classical negation anyway: this does not harm the original premises or their logical treatment and greatly simplifies the metatheoretic stuff. In the present situation, classical negation seems to come in handy at the object level. Indeed, it is sufficient to add $\neg(p \wedge \sim p)$ and $\neg(q \wedge \sim q)$ to the premises of the example in order to derive $r \wedge \sim r$ and hence to locate it as the real inconsistency.

This naive approach is not suitable. Suppose that one allows a problem solver to add the new premises $\neg(p \wedge \sim p)$ and $\neg(q \wedge \sim q)$, or to add the new premise $r \wedge \sim r$. It may turn out later that $(p \wedge \sim p) \vee (q \wedge \sim q)$ is derivable from the original premises. If $\neg(p \wedge \sim p)$ and $\neg(q \wedge \sim q)$ were added as new premises, one ends up with a trivial consequence set. If $r \wedge \sim r$ was added as a new premise, one has introduced an inconsistency that was not derivable from the original premises; in other words one has created a problem rather than resolving one. Of course all this has only a serious effect at the predicative level, where there is no negative test for derivability.

Some people might propose to allow the problem solver to withdraw the new premises in such cases. This results in tinkering, not in a logic. That personal constraints of the problem solver play a role in the introduction of new premises is all right: the problem solver is building his or her or its approach to the problem. However, the logic should define derivability. If some premises are defeasible, *the logic* should eliminate their effects where this is suitable; no intervention of the problem solver should be required.[1]

The conclusion is that the new premises have to be introduced in such a way that they never lead to triviality and never introduce new inconsistencies.

[1]Conservative logicians claim that adaptive logics are not logics. (I have no objection to calling them giclos.) However, adaptive logics are and should be formally precise characterizations of correct reasoning mechanisms.

I shall present a short characterization of inconsistency-adaptive logics in Section 2. This is required for making the paper self-contained. Next I spell out the most obvious approach in Section 3. It results from combining the inconsistency-adaptive logic with a specific prioritized adaptive logic — see for example Batens *et al.* (2003) and Batens (2001a) for the latter. Given that the new premises are conjectures about the "safety" of certain formulas, viz. the falsehood of certain inconsistencies (and possibly the truth of other inconsistencies), they are introduced (by the standard prioritized adaptive mechanism) with a lower degree of confidence than the original premises — this come to saying that they are defeasible premises with a certain degree of defeasibility.

A brief digression is in place here. One way to look at the inconsistency-adaptive logics that are presented in Section 2 is to see $\sim A$ as a defeasible classical negation. "Normally" $\sim A$ is taken to mean that A is false, but where this turns out impossible, $\sim A$ is downgraded to the paraconsistent negation of A — $\sim A$ may then be true even if A is true. The new premises are defeasible in a different way. $\neg(A \wedge \sim A)$ states that $A \wedge \sim A$ is *false*, but is introduced with a degree of confidence which will be expressed by prefixing the statement by a sequence of possibility operators: $\Diamond\neg(A \wedge \sim A)$ or $\Diamond\Diamond\neg(A \wedge \sim A)$, etc. To simplify the notation, $\Diamond^i\neg(A \wedge \sim A)$ will abbreviate i occurrences of $\Diamond$ followed by $\neg(A \wedge \sim A)$. $\Diamond^i\neg(A \wedge \sim A)$ expresses a lower degree of confidence as i is larger. The prioritized adaptive logic enables one to derive $\neg(A \wedge \sim A)$ from $\Diamond^i\neg(A \wedge \sim A)$ unless (and until) this is proven false, which roughly means — a precise definition follows in Section 3 — that $A \wedge \sim A$ is derived from the original premises and from new premises that have a higher degree of confidence than i. This approach is related to Jaśkowski's, except that Jaśkowski did not consider degrees of confidence and did not come up with a non-monotonic (let alone adaptive) logic — but see Meheus (2006).

When I presented the result in a Flemish-Polish logic workshop in Ghent, Wojciech Sady complained about the use of two negations for formalizing premises from natural language. This remark triggered an approach which I present in Section 4. The idea is to handle the new premises in terms of a single negation, priorities being coupled to the degree of complexity of the negated contradiction.

Section 5 contains some comments in conclusion, mainly on the relations between both approaches.

2. The Environment

I shall briefly introduce two (flat) inconsistency-adaptive logics, and restrict attention to them in the rest of the paper. However, the results from the subsequent sections may be easily adapted to other inconsistency-adaptive logics. A flat adaptive logic **AL** is characterized by a triple:

(i) A *lower limit logic*: a monotonic logic.
(ii) A *set of abnormalities*: a set of formulas characterized by a logical form.
(iii) An *adaptive strategy*: this specifies what it means to interpret the premises "as normally as possible".

Let us consider the inconsistency-adaptive logics **ACLuN1** and **ACLuN2**. Their predicative versions were studied in Batens (1999). The most recent survey of adaptive logics may be found in Batens (2006). The subsequent description relies on these insights and hence depart from Batens (1999) in some respects.

Let $\mathcal{L}$ be the standard predicative language, with $\sim$ as the standard negation, extended (for reasons mentioned before) with a second negation, viz. $\neg$. Let $\mathcal{S}, \mathcal{C}, \mathcal{V}, \mathcal{P}^r, \mathcal{F}$ and $\mathcal{W}$ respectively be the sets of sentential letters, individual constants, individual variables, predicate letters of rank r, formulas, and wffs. Let $\mathcal{F}$ be the set of (open and closed) formulas of $\mathcal{L}$, and $\mathcal{W}$ the set of closed formulas. **CL** will denote classical logic.

The *lower limit logic* will be **CLuN**,[2] obtained by extending full positive **CL** with $A \vee \sim A$ and with suitable axioms for $\neg$.[3] This system has some properties that might come unexpected. Replacement of Equivalents does not hold in general, viz. does not hold within the scope of $\sim$ — for example $\vdash_{\mathbf{CLuN}} p \equiv (p \vee p)$ but $\nvdash_{\mathbf{CLuN}} \sim p \equiv \sim(p \vee p)$. The same holds for Replacement of Identicals — for example $a = b, Pa \vdash_{\mathbf{CLuN}} Pb$ but $a = b, \sim Pa \nvdash_{\mathbf{CLuN}} \sim Pb$.

I now spell out a provably adequate semantics for **CLuN**. To describe it in a simple way, I extend the language $\mathcal{L}$ to the pseudo-language $\mathcal{L}^+$ by a set of pseudo-constants $\mathcal{O}$ which has at least the cardinality of the largest model one wants to consider. Let $\mathcal{F}^+$ be the set of formulas of $\mathcal{L}^+$, $\mathcal{W}^+$ the set of closed formulas, and ${}^{\sim}\mathcal{W}^+ = \{\sim A \mid A \in \mathcal{W}^+\}$.

A model $M = \langle D, v\rangle$, in which D is a set and v an assignment function defined by:

C1.1 $v\colon \mathcal{S} \mapsto \{0, 1\}$
C1.2 $v\colon \mathcal{C} \cup \mathcal{O} \mapsto D$ (where $D = \{v(\alpha) \mid \alpha \in \mathcal{C} \cup \mathcal{O}\}$)
C1.3 $v\colon \mathcal{P}^r \mapsto \wp(D^r)$ (the power set of the r-th Cartesian product of D)
C1.4 $v\colon {}^{\sim}\mathcal{W}^+ \mapsto \{0, 1\}$

The valuation function $v_M\colon \mathcal{W}^+ \mapsto \{0, 1\}$ determined by M is defined by:

C2.1 where $A \in \mathcal{S}$, $v_M(A) = v(A)$
C2.2 $v_M(\pi^r\alpha_1 \ldots \alpha_r) = 1$ iff $\langle v(\alpha_1), \ldots, v(\alpha_r)\rangle \in v(\pi^r)$
C2.3 $v_M(\alpha = \beta) = 1$ iff $v(\alpha) = v(\beta)$

[2] It is as **CL**, except that it allows for gl*u*ts with respect to *n*egation, viz. $\sim$.

[3] If bottom, $\perp$, is present, one may define $\neg A =_{\mathrm{df}} A \supset \perp$.

C2.4 $v_M({\sim} A) = 1$ iff $v_M(A) = 0$ or $v({\sim} A) = 1$
C2.5 $v_M(A \supset B) = 1$ iff $v_M(A) = 0$ or $v_M(B) = 1$
C2.6 $v_M(A\&B) = 1$ iff $v_M(A) = 1$ and $v_M(B) = 1$
C2.7 $v_M(A \vee B) = 1$ iff $v_M(A) = 1$ or $v_M(B) = 1$
C2.8 $v_M(A \equiv B) = 1$ iff $v_M(A) = v_M(B)$
C2.9 $v_M((\forall\alpha)A(\alpha)) = 1$ iff $v_M(A(\beta)) = 1$ for all $\beta \in \mathcal{C} \cup \mathcal{O}$
C2.10 $v_M((\exists\alpha)A(\alpha)) = 1$ iff $v_M(A(\beta)) = 1$ for at least one $\beta \in \mathcal{C} \cup \mathcal{O}$
C2.11 $v_M(\neg A) = 1$ iff $v_M(A) = 0$

The valuation v_M is an interpretation of $\mathcal{W}^+$, and hence of $\mathcal{W}$, which is what we are interested in. The following definitions are standard: $M \models A$ (A is true in M, M verifies A) iff $v_M(A) = 1$; $M \models \Gamma$ (M is a model of Γ) iff $M \models A$ for all $A \in \Gamma$; $\Gamma \vDash A$ iff all models of Γ verify A; $\vDash A$ (A is valid) iff all models verify A.

Where $\exists A$ denotes the existential closure of A, the *set of abnormalities* is defined as follows: $\Omega = \{\exists(A \wedge {\sim} A) \mid A \in \mathcal{F}\}$. The upper limit logic is obtained by extending **CLuN** with an axiom that rules out abnormalities, for example $(A \wedge {\sim} A) \supset B$. This logic is **CL**, as expected.

The abnormal part of a model M is defined as

$$Ab(M) = \{A \wedge {\sim} A \mid M \models A \wedge {\sim} A\}.$$

CL is provably sound and complete with respect to the set of normal models of **CLuN**, in other words the **CLuN**-models M for which $Ab(M) = \emptyset$.

The two inconsistency-adaptive logics considered here differ from each other only with respect to the *adaptive strategy*: Reliability for **ACLuN1** and Minimal Abnormality for **ACLuN1**. We shall see the strategies at work in the dynamic proofs as well as in the semantics.

The lines of an annotated proof consist of five elements: (i) a line number, (ii) a formula, (iii) the line numbers of the formulas from which the formula is derived, and (iv) the rule by which the formula is derived, and (v) a *condition*, which is a subset of Ω. Lines are added to a proof in view of the rules of inference. With the addition of a new line, the proof is brought to its next *stage* — think of a stage as a sequence of lines. At each stage, some lines may be marked whereas the others are unmarked. The formulas derived at a stage are those derived in the lines that are unmarked at the stage. A line that is marked at a stage may be unmarked at the next, and *vice versa*. Marking is governed by the Marking definition.

The rules are common for both logics and are presented here in generic form. As only the conditions of the lines are unusual, let

$$A \quad \Delta$$

abbreviate that A occurs in the proof on the condition Δ, in other words that there is a line that has A as its second element and Δ as its fifth element. Let Γ be the set of premises. Where Δ is a finite subset of Ω, $Dab(\Delta)$ denotes the disjunction (in some preferred order) of the members of Δ.

PREM If $A \in \Gamma$:

$$\frac{\ldots \quad \ldots}{A \quad \emptyset}$$

RU If $A_1, \ldots, A_n \vdash_{\mathbf{CLuN}} B$:

$$\begin{array}{ll} A_1 & \Delta_1 \\ \ldots & \ldots \\ A_n & \Delta_n \\ \hline B & \Delta_1 \cup \cdots \cup \Delta_n \end{array}$$

RC If $A_1, \ldots, A_n \vdash_{\mathbf{CLuN}} B \vee Dab(\Theta)$

$$\begin{array}{ll} A_1 & \Delta_1 \\ \ldots & \ldots \\ A_n & \Delta_n \\ \hline B & \Delta_1 \cup \cdots \cup \Delta_n \cup \Theta \end{array}$$

There is a striking correspondence between dynamic proofs and **CLuN**-proofs: A is derivable on the condition Δ in a dynamic proof iff $A \vee Dab(\Delta)$ is derivable in a (standard) **CLuN**-proof (define "$\vee Dab(\emptyset)$" as the empty string).

Let us now turn to the *marking definitions*. Given a dynamic proof, we shall say that $Dab(\Delta)$ is a *minimal Dab-formula* at stage s of the proof iff, at that stage, $Dab(\Delta)$ occurs in the proof on the empty condition and, for any $\Delta' \subset \Delta$, $Dab(\Delta')$ does not occur in the proof on the empty condition. Where $Dab(\Delta_1)$, ..., $Dab(\Delta_n)$ are the minimal *Dab*-formulas at stage s of the proof, $U_s(\Gamma) = \Delta_1 \cup \cdots \cup \Delta_n$ is the set of unreliable formulas at stage s.

DEFINITION 1. Marking for Reliability (**ACLuN1**): A line is marked at stage s iff, where Δ is its condition, $\Delta \cap U_s(\Gamma) \neq \emptyset$.

Marking for the Minimal Abnormality strategy is slightly more tiresome. Let $\Phi^{\circ}_s(\Gamma)$ be the set of all sets that contain one disjunct out of each minimal *Dab*-formula at stage s. Let $\Phi_s(\Gamma)$ contain those members of $\Phi^{\circ}_s(\Gamma)$ that are not proper supersets of other members of $\Phi^{\circ}_s(\Gamma)$.

DEFINITION 2. Marking for Minimal Abnormality (**ACLuN2**): A line is marked at stage s iff, where A is derived on the condition Δ in the line, (i) there is no $\varphi \in \Phi_s(\Gamma)$ such that $\varphi \cap \Delta = \emptyset$, or (ii) for some $\varphi \in \Phi_s(\Gamma)$, there is no line at which A is derived on a condition Θ for which $\varphi \cap \Theta = \emptyset$.

In other words, a line at which A is derived on the condition Δ is *not* marked iff (i) $\varphi \cap \Delta = \emptyset$ for some $\varphi \in \Phi_s(\Gamma)$ and (ii) for any $\varphi \in \Phi_s(\Gamma)$, there is a line at which A has been derived on a condition Θ for which $\varphi \cap \Theta = \emptyset$.[4]

[4]Definition 2 will become transparent when we consider its semantic counterpart.

The formulas *derived* from Γ *at a stage* of the proof are those that are derived at a line that is unmarked at that stage. As the proof proceeds, unmarked lines may be marked and vice versa. So, it is important that one defines a different, stable, kind of derivability:

DEFINITION 3. A is *finally derived* from Γ on line i of a proof at stage s iff (i) A is the second element of line i, (ii) line i is not marked at stage s, and (iii) any extension of the proof in which line i is marked may be further extended in such a way that line i is unmarked.

DEFINITION 4. $\Gamma \vdash_{\mathbf{ACLuN}} A$ (A is *finally* **ACLuN**-*derivable* from Γ) iff A is finally derived on a line of a proof from Γ.

Note that **ACLuN** is not a logic but a variable name for the logics **ACLuN1** and **ACLuN2**, and that the only difference between these stems from Definitions 1 and 2.

Here is a very simple propositional example of a dynamic proof. Let Γ be $\{p, {\sim} q, t, r \supset q, {\sim} p \vee s, {\sim} p \vee q, {\sim} t \vee u\}$.

1	p	–	Prem	$\emptyset$
2	${\sim} q$	–	Prem	$\emptyset$
3	t	–	Prem	$\emptyset$
4	$r \supset q$	–	Prem	$\emptyset$
5	${\sim} p \vee s$	–	Prem	$\emptyset$
6	${\sim} p \vee q$	–	Prem	$\emptyset$
7	${\sim} t \vee u$	–	Prem	$\emptyset$
8	${\sim} r$	2, 4	RC	$\{q \wedge {\sim} q\}$
9	s	1, 5	RC	$\{p \wedge {\sim} p\}$
10	u	3, 7	RC	$\{t \wedge {\sim} t\}$

As lines 8–10 have a non-empty condition, one will try to find out whether the members of the conditions are indeed reliable. It turns out that some are not, as is seen below. At stage 11 (after line 11 is derived), lines 8 and 9 are marked while the other lines are not.

8	${\sim} r$	2, 4	RC	$\{q \wedge {\sim} q\}$	$\surd$
9	s	1, 5	RC	$\{p \wedge {\sim} p\}$	$\surd$
10	u	3, 6	RC	$\{t \wedge {\sim} t\}$	
11	$(p \wedge {\sim} p) \vee (q \wedge {\sim} q)$	1, 2, 7	RU	$\emptyset$	

It is easily seen that no other interesting consequences are derivable, and that the marks will not change if the proof is extended (without extending the premise set). So u is **ACLuN1**-derivable from the premises whereas ${\sim} r$ and s are not. The proof is utterly simple: all lines marked at stage 11 remain marked at all later stages.

If $\sim r \vee u$ were derived from $\sim r$ or from u, the line at which it is derived would either have the condition $\{q \wedge \sim q\}$ or the condition $\{p \wedge \sim p\}$, and hence will be marked. So $\sim r \vee u$ is not **ACLuN1**-derivable from the premises.[5]

I now turn to the semantics for **ACLuN1** and **ACLuN2**. A minimal *Dab*-consequence of the premise set Γ is, as one expects, a formula $Dab(\Delta)$ such that $\Gamma \vdash_{\mathbf{CLuN}} Dab(\Delta)$ and $\Gamma \nvdash_{\mathbf{CLuN}} Dab(\Delta')$ for any $\Delta' \subset \Delta$ — there is no reference to a stage of a proof here. Where $Dab(\Delta_1)$, $Dab(\Delta_2)$, ... are the minimal *Dab*-consequences of Γ,[6] $U(\Gamma) = \Delta_1 \cup \Delta_2 \cup \cdots$ is the set of formulas that are unreliable with respect to Γ.

Definition 5. A **CLuN**-model M of Γ is *reliable* iff $Ab(M) \subseteq U(\Gamma)$.

Definition 6. $\Gamma \vDash_{\mathbf{ACLuN1}} A$ iff A is verified by all reliable models of Γ.

Definition 7. A **CLuN**-model M of Γ is *minimally abnormal* iff there is no **LLL**-model M' of Γ such that $Ab(M') \subset Ab(M)$.

Definition 8. $\Gamma \vDash_{\mathbf{ACLuN2}} A$ iff A is verified by all minimally abnormal models of Γ.

It does not make sense to say that a model M is reliable or minimally abnormal, but only to say that M is a reliable model (or a minimally abnormal model) *of* Γ.

Both **ACLuN1** and **ACLuN2** have been shown (in Batens 1999) to be sound and complete with respect to their semantics. It has moreover been shown that a proof at a stage (for either logic) offers the best available insight in final derivability in view of the insights in the premises provided by the proof at the stage. In other words, derivability at a stage may be taken as a (defeasible) basis for rational decision — see especially Batens (1995).

In subsequent sections, I shall often write **ACLuN** to refer to either of **ACLuN1** and **ACLuN2**. Statements in terms of **ACLuN** also hold true for most other inconsistency-adaptive logics.

3. The First Way

The approach considered here combines the adaptive logics **ACLuN1** and **ACLuN2** with an adaptive logic, respectively $\mathbf{TuN}^r$ and $\mathbf{TuN}^m$, that handles

[5]However, it is **ACLuN2**-derivable from them. If $\sim r \vee u$ is derived on the condition $\{q \wedge \sim q\}$ as well as on the condition $\{p \wedge \sim p\}$, say at stage 13 of the proof, then $\Phi_{14}(\Gamma) = \{\{p \wedge \sim p\}, \{q \wedge \sim q\}\}$. Intuitively: all minimal abnormal models of Γ verify either $p \wedge \sim p$ or $q \wedge \sim q$, but none of them verifies both, and hence all verify $\sim r \vee u$.

[6]The minimal *Dab*-consequences of Γ may be semantically defined in view of the soundness and completeness of **CLuN** with respect to its semantics.

the premises of the form $\diamond^i(A \wedge {\sim} A)$. **TuN**^{r} may be seen as an infinite sequence of adaptive logics that all have the same structure — the first member suitably handles premises of the form $\diamond A$, the second premises of the form $\diamond^2 A$, etc. This is why the prioritized adaptive logic, if phrased as a single system, has a sequence of sets of abnormalities. Similarly for **TuN**^{m}.

The lower limit logic will be called **TuN** — it is like **T** but allows for gl*u*ts with respect to *n*egation. Put differently, it is **CLuN** extended with the **T**-modalities. Let $\mathcal{L}^M$ be the result of extending $\mathcal{L}$ with the modalities in the standard way. The pseudo-language $\mathcal{L}^{M+}$ is obtained by extending $\mathcal{L}^M$ with the set of pseudo-constants $\mathcal{O}$.

A **TuN**-model M is a quintuple $\langle W, w_0, R, D, v\rangle$ in which W is a set of worlds, $w_0 \in W$ the real world, R a binary relation on W, D a non-empty set and v an assignment function. The accessability relation R is reflexive. The assignment function is defined by:

C1.1 $v\colon \mathcal{S} \times W \mapsto \{0, 1\}$
C1.2 $v\colon (\mathcal{C} \cup \mathcal{O}) \times W \mapsto D$
C1.3 $v\colon \mathcal{P}^r \times W \mapsto \wp(D^r)$
C1.4 $v\colon {}^{\sim}\mathcal{W}^{+} \times W \mapsto \{0, 1\}$

The valuation function $v_M\colon \mathcal{W}^{M+} \times W \mapsto \{0, 1\}$, determined by the model M is defined by:

C2.1 where $A \in \mathcal{S}$, $v_M(A, w) = v(A, w)$
C2.2 $v_M(\pi^r\alpha_1 \ldots \alpha_r, w) = 1$ iff $\langle v(\alpha_1, w), \ldots, v(\alpha_r, w)\rangle \in v(\pi^r, w)$
C2.3 $v_M(\alpha = \beta, w) = 1$ iff $v(\alpha, w) = v(\beta, w)$
C2.4 $v_M({\sim} A, w) = 1$ iff $v_M(A, w) = 0$ or $v({\sim} A, w) = 1$
C2.5 $v_M(A \supset B, w) = 1$ iff $v_M(A, w) = 0$ or $v_M(B, w) = 1$
C2.6 $v_M(A \vee B, w) = 1$ iff $v_M(A, w) = 1$ or $v_M(B, w) = 1$
C2.7 $v_M(A \wedge B, w) = 1$ iff $v_M(A, w) = 1$ and $v_M(B, w) = 1$
C2.8 $v_M(A \equiv B, w) = 1$ iff $v_M(A, w) = v_M(B, w)$
C2.9 $v_M((\forall\alpha)A(\alpha), w) = 1$ iff $v_M(A(\beta), w) = 1$ for all $\beta \in \mathcal{C} \cup \mathcal{O}$
C2.10 $v_M((\exists\alpha)A(\alpha), w) = 1$ iff $v_M(A(\beta), w) = 1$ for at least one $\beta \in \mathcal{C} \cup \mathcal{O}$
C2.11 $v_M(\diamond A, w) = 1$ iff $v_M(A, w') = 1$ for at least one w' such that Rww'.
C2.12 $v_M(\neg A, w) = 1$ iff $v_M(A, w) = 0$

We define $M \models A$ (A is true in M, M verifies A) iff $v_M(A, w_0) = 1$; $M \models \Gamma$, etc. is defined as for **CLuN**.[7]

As the construction is a bit unusual, some explanation is useful. One may define a function d that assigns to each $w \in W$ its domain $d(w) = \{v(\alpha, w) \mid \alpha \in \mathcal{C} \cup \mathcal{O}\}$. If an element of an r-tuple of $v(\pi^r, w)$ does not belong to $d(w)$, then

[7] I shall not need $\Box A$; one of course may define $\Box A =_{\rm df} \neg\diamond\neg A$.

the r-tuple does not have any effect on the valuation.[8] If Rww', the question whether $v(\alpha, w)$ is or is not a member of $d(w')$ is immaterial for $v(A(\alpha), w)$. For example, the value of $v_M(\Diamond Pa, w)$ depends on the values of $v_M(Pa, w')$ for Rww', and $v_M(Pa, w')$ depends only on $v(a, w')$ and $v(P, w')$. The semantics may be rephrased as a counterpart semantics: $\mathsf{a} \in d(w)$ is a counterpart of $\mathsf{b} \in d(w')$ just in case there is an $\alpha \in C \cup O$ such that $v(\alpha, w) = \mathsf{a}$ and $v(\alpha, w') = \mathsf{b}$. An $\alpha \in C \cup O$ may be seen as picking a specific counterpart "path" on W.

Axiomatically **TuN** is characterized by **CLuN** plus the Neccessitation rule, the **T**-axioms, and the Barcan-formula.

The sets of (modal) abnormalities are $\Omega^i = \{\exists((A \wedge \sim A) \wedge \Diamond^i\neg(A \wedge \sim A)) \mid A \in \mathcal{F}\}$. That, for example, $\exists((A \wedge \sim A) \wedge \Diamond\neg(A \wedge \sim A))$ is an abnormality (of level 1) means that, if $A \wedge \sim A$ is possibly false, then the normal situation is that it is false. In other words, from $\Diamond^i\neg(A \wedge \sim A)$ one may derive $\neg(A \wedge \sim A)$ "unless and until proven otherwise". The abnormal parts of a model M are defined as follows:

DEFINITION 9. $Ab^i(M) =_{\text{df}} \{A \in \Omega^i \mid M \models A\}$

The upper limit logic is obtained by adding to **TuN** the axiom schema $\Diamond^i\neg(A \wedge \sim A) \supset \neg(A \wedge \sim A)$. If, as for the present application, all modal premises are of the form $\Diamond^i\neg(A \wedge \sim A)$, then the upper limit logic boils down to the **CLuN**-variant of **Triv** (**CLuN** with modalities devoid of meaning).[9]

From **TuN** and the Ω^i one builds the prioritized adaptive logics $\mathbf{TuN}^r$ by the Reliability strategy and $\mathbf{TuN}^m$ by the Minimal Abnormality strategy. Remember that the set of premises Γ is a set of formulas containing modal-free formulas (the original premises) as well as formulas of the form $\Diamond^i\neg(A \wedge \sim A)$ (the new premises). Suppose that, for some n, $i \leq n$ for all premises of the form $\Diamond^i\neg(A \wedge \sim A)$—the case where all $i \in \mathbb{N}$ occur is briefly discussed at the end of this section. Where $\mathbf{TuN}^a$ is used as a general name (the "a" varying over "r" and "m"), we want to have

(2) $$Cn_{\mathbf{TuN}^a}(\Gamma) = Cn_{\mathbf{TuN}^a_n}(Cn_{\mathbf{TuN}^a_{n-1}}(\ldots(Cn_{\mathbf{TuN}^a_1}(\Gamma))\ldots))$$

in which each $\mathbf{TuN}^a_i$ is a flat adaptive logic. This means that we want that first all premises of the form $\Diamond^1\neg(A \wedge \sim A)$ are taken into account, next those of the form $\Diamond^2\neg(A \wedge \sim A)$, etc.

Let $Dab^i(\Delta)$ abbreviate the disjunction (in some preferred order) of the members of $\Delta \in \Omega^i$ — note that $Dab^i(\Delta)$ is a meaningless expression if $\Delta \notin \Omega^i$.

---

[8] In other words, one obtains an equivalent semantics (one verifying the same formulas) if C1.3 is replaced by "$v\colon \mathcal{P}^r \times W \mapsto \wp((d(w))^r)$".

[9] Precisely this upper limit logic is obtained if one defines $\Omega^i = \{\exists(A \wedge \Diamond^i\neg A) \mid A \in \mathcal{F}\}$. The upper limit logic of the combined adaptive logic, which also rules out that members of $\Omega = \{\exists(A \wedge \sim A) \mid A \in \mathcal{F}\}$ be true, is then **Triv**.

Although (2) is all right as a definition, it may cause trouble from a computational point of view. This is why we shall circumvent it for the dynamic proofs: in a stage of a proof we shall "approach" $Cn_{\mathbf{TuN}^a}(\Gamma)$ by the order in which marking proceeds. Let us at once move to these dynamic proofs.

$Dab^i(\Delta)$ is a minimal Dab^i-formula at a stage s of a proof iff, at stage s, (i) there is an unmarked line that has $Dab^i(\Delta)$ as its formula and a $\Theta \subseteq \Omega^1 \cup \cdots \cup \Omega^{i-1}$ as its condition, and (ii) there is no $\Delta' \subset \Delta$ such that $Dab^i(\Delta')$ is the formula of an unmarked line that has a $\Theta \subseteq \Omega^1 \cup \cdots \cup \Omega^{i-1}$ as its condition. The inference rules are exactly as in Section 2 except that **CLuN** is replaced by **TuN** and that $Dab(\Delta)$ has to be replaced by $Dab^M(\Delta)$.

Let $U^i_s(\Gamma)$ be defined as $U_s(\Gamma)$, but in terms of minimal Dab^i-formulas at stage s of a proof. Let $\Phi^i_s(\Gamma)$ be defined as $\Phi_s(\Gamma)$, but in terms of minimal Dab^i-formulas at stage s of a proof.

The Marking definitions are as in Section 2 except that they should be applied starting from level 1 abnormalities. To avoid confusion, I spell them out:

Definition 10. Marking for Reliability ($\mathbf{TuN}^r$) starting from $i = 1$: A line is marked at stage s iff, where Δ is its condition, $\Delta \cap U^i_s(\Gamma) \neq \emptyset$.

Definition 11. Marking for Minimal Abnormality ($\mathbf{TuN}^m$) starting from $i = 1$: A line is marked at stage s iff, where A is the second element and Δ the fifth element of the line, (i) there is no $\varphi \in \Phi^i_s(\Gamma)$ such that $\varphi \cap \Delta = \emptyset$, or (ii) for some $\varphi \in \Phi^i_s(\Gamma)$, there is no line at which A is derived on a condition Θ for which $\varphi \cap \Theta = \emptyset$.

Final derivability is defined as before.

Turning to the semantics, I first recall that n is the largest i for which $\Diamond^i$ occurs in the premises. Let M vary over **TuN**-models of Γ. We define:

- $\Sigma^0 = \{M \mid M \models \Gamma\}$.
- Where $0 < i \leq n$, a minimal Dab^i-consequence of Γ is a (set-theoretically) shortest Dab^i-formula that is true in all $M \in \Sigma^{i-1}$.
- Where $0 < i \leq n$, and $Dab^i(\Delta_1)$, $Dab^i(\Delta_2)$, ... are the minimal Dab^i-consequences of Γ, $U^i(\Gamma) = \Delta_1 \cup \Delta_2 \cup \cdots$.
- Where $0 < i \leq n$, $\Sigma^i = \{M \in \Sigma^{i-1} \mid Ab^i(M) \subseteq U^i(\Gamma)\}$.
- The Reliable models ($\mathbf{TuN}^r$-models) of Γ are the members of Σ^n.

We proceed analogously for Minimal Abnormality, but here the matter is simpler. Where Σ^0 is as before:

- Where $0 < i \leq n$, $\Sigma^i = \{M \in \Sigma^{i-1} \mid \text{for no } M' \in \Sigma^{i-1}, Ab^i(M') \subset Ab^i(M)\}$.
- The Minimal Abnormal models ($\mathbf{CLuN}^m$-models) of Γ are the members of Σ^n.

$\Gamma \models_{\mathbf{TuN}^r} A$ iff A is verified by all reliable models of Γ. $\Gamma \models_{\mathbf{TuN}^m} A$ iff A is verified by all minimally abnormal models of Γ. It is provable that both logics are sound and complete with respect to their semantics.

To get home, we have to combine the inconsistency-adaptive logics from Section 2 with the prioritized adaptive logics from the present section. As few applications will justify that one combines Reliability for one logic with Minimal Abnormality for the other, two combinations seem most attractive.

As before, the set of (original and new) premises Γ will consist solely of non-modal formulas and of formulas of the form $\Diamond^i \neg(A \wedge {\sim} A)$, and I suppose that all $i \leq n$ for some n. The logic we are after is obtained by combining flat adaptive logics. Let the combination for Reliability be called $\mathbf{N}^r$. What we want is

$$(3) \qquad Cn_{\mathbf{N}^r}(\Gamma) = Cn_{\mathbf{ACLuN1}}(Cn_{\mathbf{TuN}^r}(\Gamma)).$$

This may be spelled out as follows:

$$(4) \qquad Cn_{\mathbf{N}^r}(\Gamma) = Cn_{\mathbf{ACLuN1}}(Cn_{\mathbf{TuN}^r_n}(Cn_{\mathbf{TuN}^r_{n-1}}(\ldots(Cn_{\mathbf{TuN}^r_1}(\Gamma))\ldots)))$$

which makes it transparent that first all premises of the form $\Diamond^i(A \wedge {\sim} A)$ are taken into account, starting with $i = 1$, and next the result is interpreted as consistently as possible.[10] The situation is similar for the Minimal Abnormality combination, $\mathbf{N}^m$. We want

$$(5) \qquad Cn_{\mathbf{N}^m}(\Gamma) = Cn_{\mathbf{ACLuN2}}(Cn_{\mathbf{TuN}^m}(\Gamma)).$$

In order to avoid confusion, let us rename the Ω from Section 2 to Ω^0 — so $\Omega^0 = \{\exists(A \wedge {\sim} A) \mid A \in \mathcal{F}\}$.

We move to the dynamic proofs. Let $Dab(\Theta)$ refer to the disjunction of a $\Theta \subset \Omega^0 \cup \Omega^1 \cup \Omega^2 \cup \cdots$. The inference rules for the combined adaptive logic are exactly like those of **ACLuN**, except that **CLuN** should be replaced by **TuN**.

The marking definitions too are easily obtained in view of (4). Where $i \geq 1$, let $Dab^i(\Delta)$ be as before in this section. Let $Dab^0(\Delta)$ denote the disjunction of the members of $\Delta \subseteq \Omega^0$. $Dab^0(\Delta)$ is a minimal Dab^0-formula at stage s of a proof iff, at stage s, $Dab^0(\Delta)$ is the second element of an unmarked line that has some $\Theta \subseteq \Omega^1 \cup \Omega^2 \cup \cdots$ as its fifth element — again, the order in which lines are marked will be essential. $U^0_s(\Gamma)$ is defined from the minimal Dab^0-formulas at stage s just as $U_s(\Gamma)$ was defined in Section 2 from (what was there called) the minimal Dab-formulas at stage s. $\Phi^0_s(\Gamma)$ is defined from the minimal Dab^0-formulas at stage s just as $\Phi_s(\Gamma)$ is defined in Section 2 from (what is there called) the minimal Dab-formulas at stage s.

Consider a proof at a stage. Depending on the chosen strategy, one first marks the lines in view of level 1 (see Definitions 10 and 11), next in view of level 2, and so on up to the highest level n — I supposed that there was one. Finally one marks the lines in view of level 0.

---

[10]There is some notational abuse as $Cn_{\mathbf{ACLuN1}}(\Gamma)$ is not defined for $\mathcal{L}^M$. But the idea is clear and $\mathbf{N}^r$ is correctly defined below.

Consider the simple example of a dynamic **ACLuN1**-proof from Section 2, and suppose that one introduces the new premise $\Diamond\Diamond\neg(p \wedge \sim p)$. The proof may then be continued as follows:

8	$\sim r$	2, 4	RC	$\{q \wedge \sim q\}$	$\surd$
9	s	1, 5	RC	$\{p \wedge \sim p\}$	
10	u	3, 7	RC	$\{t \wedge \sim t\}$	
11	$(p \wedge \sim p) \vee (q \wedge \sim q)$	1, 2, 6	RU	$\emptyset$	
12	$\Diamond\Diamond\neg(p \wedge \sim p)$	–	Prem	$\emptyset$	
13	$\neg(p \wedge \sim p)$	12	RC	$\{(p \wedge \sim p) \wedge \Diamond\Diamond\neg(p \wedge \sim p)\}$	
14	$q \wedge \sim q$	11, 13	RU	$\{(p \wedge \sim p) \wedge \Diamond\Diamond\neg(p \wedge \sim p)\}$	

For all $i > 0$, $U^i_{14}(\Gamma) = \emptyset$. (And of course $U^i(\Gamma) = \emptyset$ for all $i > 0$.) So no line is marked in view of any level higher than 0. $U^0_{14}(\Gamma) = \{q \wedge \sim q\}$, whence only line 8 is marked in view of level 0. So s is a $\mathbf{N}^r$-consequence of the present premise set and $\sim r$ is not — similarly for $\mathbf{N}^m$.

If not only $\Diamond\Diamond\neg(p \wedge \sim p)$ but also $\Diamond\Diamond\neg(q \wedge \sim q)$ is added as a new premise, then it is possible to continue as follows:

8	$\sim r$	2, 4	RC	$\{q \wedge \sim q\}$	$\surd$
9	s	1, 5	RC	$\{p \wedge \sim p\}$	$\surd$
10	u	3, 7	RC	$\{t \wedge \sim t\}$	
11	$(p \wedge \sim p) \vee (q \wedge \sim q)$	1, 2, 6	RU	$\emptyset$	
12	$\Diamond\Diamond\neg(p \wedge \sim p)$	–	Prem	$\emptyset$	
13	$\neg(p \wedge \sim p)$	12	RC	$\{(p \wedge \sim p) \wedge \Diamond\Diamond\neg(p \wedge \sim p)\}$	$\surd$
14	$q \wedge \sim q$	11, 13	RU	$\{(p \wedge \sim p) \wedge \Diamond\Diamond\neg(p \wedge \sim p)\}$	$\surd$
15	$\Diamond\Diamond\neg(q \wedge \sim q)$	–	Prem	$\emptyset$	
16	$((q \wedge \sim q) \wedge \Diamond\Diamond\neg(q \wedge \sim q)) \vee ((p \wedge \sim p) \wedge \Diamond\Diamond\neg(p \wedge \sim p))$	11, 12, 15	RU	$\emptyset$	

It follows that $U^1_{16}(\Gamma) = \emptyset$ and $U^2_{16}(\Gamma) = \{(q \wedge \sim q) \wedge \Diamond\Diamond\neg(q \wedge \sim q), (p \wedge \sim p) \wedge \Diamond\Diamond\neg(p \wedge \sim p)\}$. As a result line 14 is marked, and the formula of line 11 is again a minimal Dab^0 formula. It follows that $U^0_{16}(\Gamma) = \{p \wedge \sim p, q \wedge \sim q\}$ and hence that both lines 8 and 9 are marked (in view of level 0) at stage 16. The present premises state that both disjuncts of 11 are false, but, as they attach the same priority to both $\neg(p \wedge \sim p)$ and $\neg(q \wedge \sim q)$, both $p \wedge \sim p$ and $q \wedge \sim q$ are unreliable. It follows that neither s nor $\sim r$ is a $\mathbf{N}^r$-consequence of the present premise set.

Suppose that the researcher or research group finds a better reason to consider $p \wedge \sim p$ as not suspect, and adds $\Diamond\neg(p \wedge \sim p)$ as a third new premise. If this is the case, the proof would be continued as follows

8	$\sim r$	2, 4	RC	$\{q \wedge \sim q\}$	$\surd$
9	s	1, 5	RC	$\{p \wedge \sim p\}$	
10	u	3, 7	RC	$\{t \wedge \sim t\}$	

11	$(p \wedge \sim p) \vee (q \wedge \sim q)$	1, 2, 6	RU	$\emptyset$	
12	$\Diamond\Diamond\neg(p \wedge \sim p)$	–	Prem	$\emptyset$	
13	$\neg(p \wedge \sim p)$	12	RC	$\{(p \wedge \sim p) \wedge \Diamond\Diamond\neg(p \wedge \sim p)\}$	$\surd$
14	$q \wedge \sim q$	11, 13	RU	$\{(p \wedge \sim p) \wedge \Diamond\Diamond\neg(p \wedge \sim p)\}$	$\surd$
15	$\Diamond\Diamond\neg(q \wedge \sim q)$	–	Prem	$\emptyset$	
16	$((q \wedge \sim q) \wedge \Diamond\Diamond\neg(q \wedge \sim q)) \vee ((p \wedge \sim p) \wedge \Diamond\Diamond\neg(p \wedge \sim p))$	11, 12, 15	RU	$\emptyset$	
17	$\Diamond\neg(p \wedge \sim p)$	–	Prem	$\emptyset$	
18	$\neg(p \wedge \sim p)$	17	RC	$\{(p \wedge \sim p) \wedge \Diamond\neg(p \wedge \sim p)\}$	
19	$q \wedge \sim q$	11, 18	RU	$\{(p \wedge \sim p) \wedge \Diamond\neg(p \wedge \sim p)\}$	

As $U^1_{19}(\Gamma) = \emptyset$, line 19 is unmarked, $U^0_{19}(\Gamma) = \{q \wedge \sim q\}$, and hence line 9 is unmarked. So s is a $\mathbf{N}^r$-consequence (as well as a $\mathbf{N}^m$-consequence) of the present premise set and $\sim r$ is not.

Incidentally, it is possible to continue the proof, for example by

20	$(q \wedge \sim q) \wedge \Diamond\Diamond\neg(q \wedge \sim q)$	16, 18	RU	$\{(p \wedge \sim p) \wedge \Diamond\neg(p \wedge \sim p)\}$

This causes lines 13 and 14 to be unmarked, but changes nothing to the marks of the other lines.

Let us briefly turn to the semantics. First we define minimal Dab^0-consequences of Γ as the minimal $Dab^0(\Delta)$ that are $\mathbf{TuN}^r$-consequences (respectively $\mathbf{TuN}^r$-consequences) of Γ. $Ab^0(M) = \{A \in \Omega^0 \mid M \models A\}$ as expected. The Reliable models ($\mathbf{N}^r$-models) of Γ are the $\mathbf{TuN}^r$-models M of Γ for which $Ab^0(M) \subseteq U^0(\Gamma)$. The Minimal Abnormal models ($\mathbf{N}^m$-models) of Γ are the $\mathbf{TuN}^r$-models M of Γ for which there is no $\mathbf{TuN}^r$-model M' of Γ such that $Ab^0(M') \subset Ab^0(M)$. $\Gamma \vDash A$ is defined as before.

It can be shown for both systems that final derivability (which is defined by Definition 4 for all adaptive logics) is sound and complete with respect to the semantics. I do not even sketch the proofs because they are by now standard in view of the fact that the prioritized adaptive logics are themselves combinations of flat adaptive logics — see (2) and (4).

I promised to return to the case in which the premise set is infinite and moreover contains formulas of the form $\Diamond^i\neg(A \wedge \sim A)$ for all $i \in \mathbb{N}$. Let us consider $\mathbf{N}^r$. In this case, one can still define $Cn_{\mathbf{TuN}^r}(\Gamma)$ as the limit of (2) and settle $Cn_{\mathbf{N}^r}(\Gamma)$ by (3). Similarly for the semantic consequence relation. So the only trouble is with the dynamic proofs. This does not come unexpected. Compactness with respect to the derivability relation fails for nearly all adaptive logics. In some cases the metatheory enables one to get around this. In the other cases, it can still be shown that the dynamic proofs offer a sensible estimate of final derivability — the best possible one in view of the insight in the premises that is provided by the proof. With respect to applications, for example in the philosophy of science, all this is rather esoteric. Thus the kind of application considered in this paper will never require an infinite set of new premises. And it clearly does not make sense to introduce consistency requirements with infinitesimally small priorities, except as a manner of speech.

4. The Second Way

The approach presented here is attractive in view of its sparing use of logical symbols in the premises — classical negation does not occur in the premises. Moreover, the approach is rather fascinating as it does not lead to a combination of known adaptive logics but opens up a formerly unexplored path.

Suppose that one derived

(6) $(p \wedge \sim p) \vee (q \wedge \sim q)$

and that this actually is a minimal *Dab*-formula — we are of course back at the terminology of Section 2. Suppose that one introduces $\sim(q \wedge \sim q)$ as a new premise. (6) states that either $p \wedge \sim p$ or $q \wedge \sim q$ is true but does not specify which of them is true. The new premise denies that $q \wedge \sim q$ is true. Can this be taken as a reason to consider $p \wedge \sim p$ as "the real problem"? If this question is answered in the positive, a second question arises. Having sat at Egypt's meat pots in Section 3, we know the advantages of prioritized denials of inconsistency. So the second question is whether one is able to express defeasible denials of contradictions within the standard predicative language (not recurring to classical negation $\neg$). Unlikely as it may appear — it appeared impossible to me when I first tried out the approach — both questions may be answered positively. Let us start with the first problem, and more specifically with a philosophical point.

The inconsistency-adaptive logics from Section 2 *isolate* negations of formulas. For example, $p \wedge \sim p$ is not derivable from q and $\sim(p \vee q)$. The formula

(7) $(p \wedge \sim p) \vee ((q \wedge \sim q) \wedge \sim(q \wedge \sim q))$

is **CLuN**-derivable from (6) together with the new premise $\sim(q \wedge \sim q)$, and (6) is **CLuN**-derivable from (7). In other words, to consider the new premise as a good reason to defeasibly derive $p \wedge \sim p$ from (6) comes to considering contradictions of contradictions as more likely false than (simple) contradictions. But is this justifiable? What is so special about $\sim(A \wedge \sim A)$, or about $(A \wedge \sim A) \wedge \sim(A \wedge \sim A)$?

There is an answer to these questions, and a rather convincing one with respect to the intended applications. Given that the original theory was meant as consistent and that **CL** was taken as its underlying logic, the negation of a contradiction is a logical theorem. So it does not make sense that it is explicitly affirmed by the original premises. In other words, if one goes out of one's way to affirm the negation of a contradiction, then one is obviously affirming something special, in our case, the fact that the contradiction is false — defeasibly of course: unless and until proven otherwise. Some people might see trouble coming from the fact that $\sim(A \wedge \sim A)$ is **ACLuN1**-derivable from Γ whenever $A \wedge \sim A \notin U(\Gamma)$.[11] However, this reinforces my point rather than weakening it. Indeed, it shows that, even in the context of **ACLuN1**, it does not make sense to introduce $\sim(A \wedge \sim A)$ as a new premise unless $A \wedge \sim A$ is a disjunct of a minimal *Dab*-formula of the proof. Given all this, $\sim(A \wedge \sim A)$ can sensibly be taken as a defeasible rejection of $A \wedge \sim A$.

Of course, I still have to show that the approach is workable from a technical point of view. But let us first turn to the second problem.

The priorities from Section 3 are attractive with respect to the intended application. The conjecture that a certain inconsistency is false may rely on observational data, but the trustworthiness of observational data is affected by error margins and by theory-ladenness; it may rely on personal constraints, but these too may be tentative; and so on. In sum, degrees of confidence come in useful for the new premises. Here is a proposal. If both $\sim(q \wedge \sim q)$ and $\sim(p \wedge \sim p)$ are added as new premises, then the previous reasoning could be applied to let the further premise $\sim((q \wedge \sim q) \wedge \sim(q \wedge \sim q))$ locate the real problem with $p \wedge \sim p$. In the present context, this premise expresses a stronger denial of $q \wedge \sim q$.

But suppose that (6) is derived from the premises, and that $\sim((q \wedge \sim q) \wedge \sim(q \wedge \sim q))$ is the *only* new premise. Apparently this premise cannot be assigned a sensible effect on the derivability of disjunctions of contradictions. All that would follow is

$$(p \wedge \sim p) \vee ((q \wedge \sim q) \wedge \sim((q \wedge \sim q) \wedge \sim(q \wedge \sim q)))$$

and the second disjunct is not even a contradiction. Moreover, apart from (6) itself, no *Dab*-formula is derivable from (6) and *this* new premise.

However, there is a solution to this riddle. Although priorities cannot be expressed by negations of contradictions — $\sim(A \wedge \sim A)$, $\sim((A \wedge \sim A) \wedge \sim(A \wedge \sim A))$, etc. — they can be expressed by conjunctions of negations of contradictions thus:

$$\sim(A \wedge \sim A),\ \sim(A \wedge \sim A) \wedge \sim((A \wedge \sim A) \wedge \sim(A \wedge \sim A)), \ldots$$

Indeed, if (6) is derived from the premises, then $\sim(q \wedge \sim q) \wedge \sim((q \wedge \sim q) \wedge \sim(q \wedge \sim q))$ is sufficient to put the blame on $p \wedge \sim p$, for it warrants the derivability of

$$(p \wedge \sim p) \vee (((q \wedge \sim q) \wedge \sim(q \wedge \sim q)) \wedge \sim((q \wedge \sim q) \wedge \sim(q \wedge \sim q))).$$

Even if $\sim(p \wedge \sim p)$ were also added, it would still follow by **CLuN** that

$$((p \wedge \sim p) \wedge \sim(p \wedge \sim p)) \vee \\ (((q \wedge \sim q) \wedge \sim(q \wedge \sim q)) \wedge \sim((q \wedge \sim q) \wedge \sim(q \wedge \sim q)))$$

[11] Indeed, as $(A \wedge \sim A) \vee \sim(A \wedge \sim A)$ is a **CLuN**-theorem, $\sim(A \wedge \sim A)$ is derivable in any proof from any Γ on the condition $\{A \wedge \sim A\}$. So if $A \wedge \sim A \notin U(\Gamma)$, then $\sim(A \wedge \sim A)$ is finally **ACLuN1**-derivable from Γ.

and this would be sufficient to eliminate the second disjunct as the "more complex" contradiction, whence the blame would still be on $p \wedge \sim p$.

Lest I should need formulas that run out of the margins, let $!A$ abbreviate $\exists A \wedge \sim A$, whence $!!A$ abbreviates $(\exists A \wedge \sim A) \wedge \sim(\exists A \wedge \sim A)$, etc. Next let $!^i A$ abbreviate whatever is abbreviated by i exclamation marks followed by A. Finally, let $¡^i A$ abbreviate $\sim !^1 A \wedge \sim !^2 A \wedge \cdots \wedge \sim !^i A$.

A second conventional matter is that the complexity of a contradiction $!A$ will be said to be i if $!A$ is **CLuN**-equivalent to $!^i B$ in which B is not a contradiction.

We seem to have arrived at an approach that is justifiable. We have seen that introducing the new premise $¡^1 A$ is a sensible way of rejecting $!A$ if $!A$ is a disjunct of a minimal *Dab*-formula (at the stage of the proof). Now consider a minimal *Dab*-formula $Dab(\Delta)$. It is obvious that from $Dab(\Delta)$ a *Dab*-formula $Dab(\Theta)$ is **CLuN**-derivable (possibly $Dab(\Delta)$ itself) of which each disjunct (each member of Θ) is a contradiction of complexity 1. Let $A \wedge \sim A \in \Theta$.[12] Stating $¡^1 A$ comes to denying $A \wedge \sim A$ to degree 1; this means that $A \wedge \sim A$ is stated to be false unless all other members of Θ are also denied to degree 1 — that is, unless $Dab(\{!^2 B \mid B \in \Theta\})$ is **CLuN**-derivable from the present premises. Stating $¡^2 A$ comes to denying $A \wedge \sim A$ to degree 2; this means that $A \wedge \sim A$ is stated to be false unless all other members of Θ are also denied to degree 2 — that is, unless $Dab(\{!^3 B \mid B \in \Theta\})$ is **CLuN**-derivable from the present premises. And so on.

This approach has an important advantage over the one from Section 3. Suppose that one introduces the new premise $\Diamond^1 \neg(A \wedge \sim A)$ and that one later wants to introduce a new premise to deny $B \wedge \sim B$ even more strongly than $B \wedge \sim B$. Then clearly the only way out is to introduce as a new premise, for example, $\Diamond^1 \neg(B \wedge \sim B)$, to replace the premise $\Diamond^1 \neg(A \wedge \sim A)$ by $\Diamond^2 \neg(A \wedge \sim A)$, and to revise the proof in view of the replaced premise. Such a revision is never required by the present approach. Suppose that one introduced the premise $¡^i A$. If one wants to more strongly deny $B \wedge \sim B$, one introduces a premise $¡^j B$ with $j > i$. If one wants to deny $B \wedge \sim B$ less strongly than $A \wedge \sim A$, but stronger than a contradiction C denied by the premise $¡^{i-1} C$, one introduces the premise $¡^i B$ together with $¡^{i+1} A$ (and possibly some other 'upgrading premises' for other formulas). In sum, it is always possible to revise one's views by *adding* further premises.

So the approach is justifiable. All I have to show is that it is technically sound.

The lower limit logic will of course be **CLuN**. The strategies will be Reliability and Minimal Abnormality. The only change required concerns the set of abnormalities Ω.

---

[12] And, to keep things simple, suppose that $A \wedge \sim A$ is not a disjunct of another minimal *Dab*-formula all members of which have complexity 1.

We actually need a sequence of sets of abnormalities, let us call them Ω^1, Ω^2, ... They are defined as follows: $\Omega^i = \{!^i A \mid A \in \mathcal{F}\}$. Ω^1 is just the Ω from Section 2. Note also that $\cdots \subset \Omega^2 \subset \Omega^1$.[13] Let us again restrict attention to premise sets for which $i \leq n$ for all formulas of the form $¡^i A$. $Dab^i(\Delta)$ denotes the disjunction of the members of $\Delta \subset \Omega^i$ and will be called a Dab^i-formula—if $\Delta \not\subseteq \Omega^i$, $Dab^i(\Delta)$ is a meaningless expression as before. $Ab^i(M) = \{A \in \Omega^i \mid M \models A\}$ as expected.

The rules for the dynamic proofs are exactly those from Section 2. For the marking definitions we need some terminology first. $Dab^i(\Delta)$ is a minimal Dab^i-formula at a stage s of a proof iff, at stage s, (i) there is an unmarked line that has $Dab^i(\Delta)$ as its formula and a $\Theta \subseteq \Omega^{i+1} \cup \Omega^{i+2} \cup \cdots$ as its condition, and (ii) there is no $\Delta' \subset \Delta$ such that $Dab^i(\Delta')$ is the formula of an unmarked line that has a $\Theta \subseteq \Omega^{i+1} \cup \Omega^{i+2} \cup \cdots$ as its condition. For each $i \geq 0$, $U^i_s(\Gamma)$ and $\Phi^i_s(\Gamma)$ are defined from the minimal Dab^i-formulas at stage s in the same way as $U_s(\Gamma)$ and $\Phi_s(\Gamma)$ are defined in Section 2 from the minimal Dab-formulas at stage s. The marking rules are displayed below and final derivability is defined as before.

Definition 12. Marking for Reliability ($\mathbf{CLuN}^{rn}$) starting from the largest i: A line is marked at stage s iff, where Δ is its condition, $\Delta \cap U^i_s(\Gamma) \neq \emptyset$.

Definition 13. Marking for Minimal Abnormality ($\mathbf{CLuN}^{mn}$) starting from the largest i: A line is marked at stage s iff, where A is second element and Δ the fifth element of the line, (i) there is no $\varphi \in \Phi^i_s(\Gamma)$ such that $\varphi \cap \Delta = \emptyset$, or (ii) for some $\varphi \in \Phi^i_s(\Gamma)$, there is no line at which A is derived on a condition Θ for which $\varphi \cap \Theta = \emptyset$.

To exemplify the dynamic proofs, consider the example with the same premises as in the previous sections. Of course the three new premises that are consecutively introduced are now formalized respectively as (for example) $¡^1 p$, $¡^1 q$ and $¡^2 p$. The logic is $\mathbf{CLuN}^{rn}$. I do not repeat lines 1–7 from Section 2 and present at once stage 20 of the proof.

8	$\sim r$	2, 4	RC	$\{!^1 q\}$	√
9	s	1, 5	RC	$\{!^1 p\}$	
10	u	3, 7	RC	$\{!^1 t\}$	
11	$(!^1 p) \vee (!^1 q)$	1, 2, 6	RU	$\emptyset$	
12	$¡^1 p$	–	Prem	$\emptyset$	
13	$(!^2 p) \vee (!^1 q)$	11, 12	RU	$\emptyset$	
14	$!^1 q$	13	RC	$\{!^2 p\}$	
15	$¡^1 q$	–	Prem	$\emptyset$	

[13] If A is an open formula, then $\exists((A \wedge \sim A) \wedge \sim(A \wedge \sim A)) \in \Omega^1 - \Omega^2$. This is a theoretical remark; introducing such a premise (for an open formula A) is not sensible for the considered applications.

16	$(!^2p) \vee (!^2q)$	13, 15	RU	$\emptyset$
17	$¡^2p$	–	Prem	$\emptyset$
18	$(!^3p) \vee (!^2q)$	16, 17	RU	$\emptyset$
19	$!^2q$	18	RC	$\{!^3p\}$
20	$!^1q$	19	RU	$\{!^3p\}$

The marks at stage 20 of the proof are exactly as in the example of Section 3. The reader is prayed to run through the stages of the proof from stage 11 on to see that the marks come and go exactly as they did in Section 3.

We now come to the semantics. Let n be the largest i for which some $Dab^i(\Delta)$ is a **CLuN**-consequence of Γ and let M vary over **CLuN**-models of Γ. We define:

- $\Sigma^{n+1} = \{M \mid M \models \Gamma\}$.
- Where $1 \leq i \leq n$, a minimal Dab^i-consequence of Γ is a (set-theoretically) shortest Dab^i-formula that is true in all $M \in \Sigma^{i+1}$.
- Where $1 \leq i \leq n$, and $Dab^i(\Delta_1)$, $Dab^i(\Delta_2)$, ... are the minimal Dab^i-consequences of Γ, $U^i(\Gamma) = \Delta_1 \cup \Delta_2 \cup \cdots$.
- Where $1 \leq i \leq n$, $\Sigma^i = \{M \in \Sigma^{i+1} \mid Ab^i(M) \subseteq U^i(\Gamma)\}$.
- The Reliable models (**CLuN**^{rn}-models) of Γ are the members of Σ^1.

We proceed analogously for Minimal Abnormality, but here the matter is simpler. Where Σ^{n+1} is as before:

- Where $1 \leq i \leq n$, $\Sigma^i = \{M \in \Sigma^{i+1} \mid$ for no $M' \in \Sigma^{i+1}$, $Ab^i(M') \subset Ab^i(M)\}$.
- The Minimal Abnormal models (**CLuN**^{mn}-models) of Γ are the members of Σ^1.

It can be shown that, for both strategies, final derivability is sound and complete with respect to the corresponding semantics. Given that **CLuN**^{rn} can be seen as resulting from the consecutive application of a sequence of flat adaptive logics (**CLuN**$^{r(n)}$, **CLuN**$^{r(n-1)}$, ..., **CLuN**$^{r(1)}$, in which the number in parentheses refers to the set of abnormalities and hence to the minimal complexity of the involved abnormalities), the proofs are standard. Similarly for **CLuN**^{mn}.

5. In Conclusion

There can be little doubt that the approach from Section 3 is less elegant than the one from Section 4. However the former approach is more general. The reason for this is that the second approach cannot be applied to many paraconsistent logics. It cannot be applied to such logics as **CLuNs** (see for example Batens 2000a) or **LP** (see for example Priest 1987) in which all formulas of the form $\sim(A \wedge \sim A)$ are theorems. It cannot be applied to such logics as the one ascribed to Vasil'ev in Arruda (1977) or to da Costa's $\mathbf{C}_i$ systems (see for example da

Costa 1974) because there certain formulas of the form $\sim(A \wedge \sim A)$ are logically equivalent to $\neg(A \wedge \sim A)$ and hence express a non-defeasible denial of $A \wedge \sim A$. I tend to think that, with respect to certain applications, **CLuN** is superior to all such logics. **CLuN** enables one to deny a certain contradiction, and does not itself deny all contradictions on a par (and hence vacuously). Nevertheless this denial is defeasible in **CLuN**, and hence does not entail triviality if the contradiction turns out unavoidable. Precisely the combination of both features makes the approach from Section 4 possible. Of course, the adherents of systems that do not have these features might reply that they can realize the same effect by relying on the approach from Section 3. This requires a more sophisticated language, viz. a modal one, and sophistication need not be a disadvantage provided it remains within reasonable limits. So everyone is bound to like the present paper.

The approach from Section 3 is not only more general with respect to handling inconsistency. Consider any adaptive logic of the type characterized in Section 2. The dynamics of the proofs of this logic depends essentially on the minimal *Dab*-formulas at a stage. So the logic may be combined with $\mathbf{TuN}^r$ or $\mathbf{TuN}^m$ in order to handle statements expressing that some disjunct of a minimal *Dab*-formula is (to a certain degree) likely to be false. The matter was recently studied for the logic of inductive generalization of Batens & Haesaert (2001b). It there leads to the sensible and desirable result that a researcher may (with some degree of priority) select as a conjecture a member of a minimal set of generalizations that are jointly incompatible with the data.[14] The justification of such conjectures is of the same type as that of the "new information" mentioned in Section 1.

In Section 1, I mentioned the possibility to introduce a defeasible premise that selects a disjunct of a Dab^0-formula as the real inconsistency. This might be realized by extending the first approach to the effect that it can handle premises of the form $\Diamond^i(A \wedge \sim A)$. The second approach cannot be extended to this effect, which seems a disadvantage. However, some reflection on the situation in the second approach reveals that it is not wise to extend the first approach in the suggested way. Suppose indeed that (1) is derived, and that $\Diamond(r \wedge \sim r)$ is added as a premise. The effect of this is that, for all that we know of the premises, both $p \wedge \sim p$ and $q \wedge \sim q$ are reliable. In other words, the premise $\Diamond(r \wedge \sim r)$ does not express that, presumably, $r \wedge \sim r$ is true, but that, presumably, $r \wedge \sim r$ is true *whereas* both $p \wedge \sim p$ and $q \wedge \sim q$ are false. In other words, one needs to know that both $p \wedge \sim p$ and $q \wedge \sim q$ are presumably unproblematic. But then it is possible to express this by $\Diamond^i\neg(p \wedge \sim p)$ and $\Diamond^j\neg(q \wedge \sim q)$, for some (possibly

[14] The logic of inductive generalization determines these sets — each of them comprising the disjuncts of a minimal *Dab*-formula — and hence guides the conjectures. Blind conjectures are demonstrably useless and uninteresting.

arbitrary) i and j. In sum, introducing new premises of the form $\Diamond^i(A \wedge \sim A)$ is useless.

The relation between both approaches brings up an old problem: the modal foundation of paraconsistency. This is the kernel of the 'Polish approach to parainconsistency' — see Perzanowski (2001). Suppose that the premises are given as a sequence of sets of formulas of the language $\mathcal{L}$: Γ^0 containing the original premises, and each other Γ^i containing prioritized denials of contradictions, with the priority decreasing as i increases. These premises may be expressed *within* the language of both approaches (Sections 3 and 4). It is obvious that both approaches will lead to results that are in an obvious sense equivalent albeit formulated in a very different language. Now consider the denial of the contradiction $A \wedge \sim A$ that has priority n, and let the maximal priority in the premise set be m. On the modal approach this comes to $\Diamond^n(A \wedge \sim A)$ or, and taking **TuN**-derivability into account, to $\Diamond^n(A \wedge \sim A) \wedge \Diamond^{n+1}(A \wedge \sim A) \wedge \cdots$. On the **CLuN**-approach it comes to $¡^{m-n}A$ or, taking **CLuN**-derivability into account, to $¡^1 A \wedge \cdots \wedge ¡^{m-n}A$. Of course, the priorities are not absolute, and hence $\Diamond^n(A \wedge \sim A)$ may be replaced by $\Diamond^{r.(n+k)}(A \wedge \sim A)$ whereas $¡^{m-n}A$ may be replaced by $¡^{(m+s)-(r.(n+k))}A$ (for suitable values of s, r and k). At first sight, all this suggests that the **CLuN**-approach is not reducible to the modal one. It appears even less reducible if one realizes that $\Diamond\neg(A \wedge \sim A)$ **TuN**-entails $\Diamond(\neg A \vee \neg \sim A)$, which is equivalent to $\Diamond\neg A \vee \Diamond\neg \sim A$, whereas $\sim(A \wedge \sim A)$ **CLuN**-entails no suitable formula of $\mathcal{L}$ corresponding to $\Diamond(\neg A \vee \neg \sim A)$. On the other hand, precisely these insights might contain the clue to a reduction of **CLuN**.

My final comment concerns applications. I restricted the attention to a theory (possibly conjoined to a set of empirical data) that was meant to be consistent but turned out to be inconsistent. However, narrowing down suspicion in inconsistent premise sets is sensible in nearly any context in which inconsistency arises. In this connection too I refer the reader to Perzanowski (2001). No narrowing down is mentioned there, whence the source is unsuspect. But the reader of the present paper will readily see the use and need for narrowing down suspicion in all inconsistent premise sets mentioned there.

Universiteit Gent
Centre for Logic and Philosophy of Science
Blandijnberg 2
B-9000 Ghent
Belgium
e-mail: Diderik.Batens@UGent.be

REFERENCES

Arruda, A. I. (1977). On the Imaginary Logic of N.A. Vasil'ev. In: A. I. Arruda, N. C. A. da Costa, and R. Chuaqui (eds.), *Non-classical Logics, Model Theory and Computability*, pp. 3–24. Amsterdam: North-Holland.

Batens, D. (1985). Dynamic Dialectical Logics as a Tool to Deal with and Partly Eliminate Unexpected Inconsistencies. In: J. Hintikka and F. Vandamme (eds.), *The Logic of Discovery and the Logic of Discourse*, pp. 263–271. New York: Plenum Press.

Batens, D. (1989). Dynamic Dialectical Logics. In: G. Priest, R. Routley and J. Norman (eds.), *Paraconsistent Logic: Essays on the Inconsistent*, pp. 187–217. München: Philosophia Verlag.

Batens, D. (1995). The Clue to Dynamic Aspects of Logic. *Logique et Analyse* **150–152**, 285–328. Appeared 1997.

Batens, D. (1999). Inconsistency-Adaptive Logics. In: E. Orłowska (ed.), *Logic at Work: Essays Dedicated to the Memory of Helena Rasiowa*, pp. 445–472. Heidelberg, New York: Physica Verlag (Springer).

Batens, D. (2000a). Minimally Abnormal Models in Some Adaptive Logics. *Synthese* **125**, 5–18.

Batens, D. (2000b). A Survey of Inconsistency-Adaptive Logics. In: D. Batens *et al.* (2000), pp. 49–73.

Batens, D. (2001a). A General Characterization of Adaptive Logics. *Logique et Analyse* **173–175**, 45–68. Appeared 2003.

Batens, D. (2006). A Universal Logic Approach to Adaptive Logics. In: J.-Y. Béziau (ed.). *Directions in Universal Logic*. Birkhauser. In print.

Batens, D. and L. Haesaert (2001b). On Classical Adaptive Logics of Induction. *Logique et Analyse* **173–175**, 255–290. Appeared 2003.

Batens D. and J. Meheus (2006). Recent Results by the Inconsistency-adaptive Labourers. In: J.-Y. Béziau and W. A. Carnielli (eds.), *Paraconsistent Logic with no Frontiers*. Studies in Logic and Practical Reasoning. North-Holland/Elsevier. In print.

Batens, D., J. Meheus, D. Provijn and L. Verhoeven (2003). Some Adaptive Logics for Diagnosis. *Logic and Logical Philosophy* **11**, 36–65.

Batens, D., Ch. Mortensen, G. Priest and J. P. Van Bendegem, eds. (2000). *Frontiers of Paraconsistent Logic*. Baldock, UK: Research Studies Press.

da Costa, N. C. A. (1974). On the Theory of Inconsistent Formal Systems. *Notre Dame Journal of Formal Logic* **15**, 497–510.

De Clercq, K. (2000). Two New Strategies for Inconsistency-Adaptive Logics. *Logic and Logical Philosophy* **8**, 65–80. Appeared 2002.

Meheus, J. (1993). Adaptive Logic in Scientific Discovery: The Case of Clausius. *Logique et Analyse* **143–144**, 359–389. Appeared 1996.

Meheus, J. (2000). An Extremely Rich Paraconsistent Logic and the Adaptive Logic Based on it. In: D. Batens *et al.* (2000), pp. 189–201.

Meheus, J. (2002). Inconsistencies in Scientific Discovery: Clausius's Remarkable Derivation of Carnot's theorem. In: H. Krach, G. Vanpaemel and P. Marage (eds.), *History of Modern Physics*, pp. 143–154. Brepols: Brepols.

Meheus, J. (2006). An Adaptive Logic Based on Jaśkowski's **D2**. *Journal of Philosophical Logic*. In print.

Meheus, J. and D. Batens (1996). Steering Problem Solving Between Cliff Incoherence and Cliff Solitude. *Philosophica* **58**, 153–187.

Nersessian, N. (2002). Inconsistency, Generic Modeling, and Conceptual Change in Science. In: J. Meheus (ed.), *Inconsistency in Science*, pp. 197–211. Dordrecht: Kluwer.

Norton, J. (1987). The Logical Inconsistency of the Old Quantum Theory of Black Body Radiation. *Philosophy of Science* **54**, 327–350.

Norton, J. (1993). A Paradox in Newtonian Gravitation Theory. *PSA 1992* **2**, 421–420.

Perzanowski, J. (2001). Parainconsistency: Or Inconsistency Tamed, Investigated and Exploited. *Logic and Logical Philosophy* **9**, 5–24. Appeared 2002.

Priest, G. (1987). *In Contradiction: A Study of the Transconsistent*. Dordrecht: Nijhoff.

Priest, G. (1991). Minimally Inconsistent **LP**. *Studia Logica* **50**, 321–331.

Smith, J. (1988). Inconsistency and Scientific Reasoning. *Studies in History and Philosophy of Science* **19**, 429–445.

Vanackere, G. (2000). **HL2**: An Inconsistency-Adaptive and Inconsistency-Resolving Logic for General Statements that Might have Exceptions. *Journal of Applied Non-Classical Logics* **10**, 317–338.

Wiśniewski, A. (1995). *The Posing of Questions: Logical Foundations of Erotetic Inferences*. Dordrecht: Kluwer.

Joke Meheus

DISCUSSIVE ADAPTIVE LOGICS: HANDLING INTERNAL AND EXTERNAL INCONSISTENCIES†

ABSTRACT. In this paper, I present the discussive adaptive logic $\mathbf{DLI}^r$. As is the case for other discussive logics, the intended application context of $\mathbf{DLI}^r$ is the interpretation of discussions. What is new about the system is that it does not lead to explosion when some of the premises are self-contradictory. It is argued that this is important in view of the fact that human reasoners are not logically omniscient, and hence, that it may not be evident to discover the inconsistencies in one's beliefs. In addition to this, $\mathbf{DLI}^r$ can handle cases in which different participants contradict each other. It is shown that, in both kinds of cases, $\mathbf{DLI}^r$ leads to an interpretation of the discussion that is as rich as possible (even though no discussive connectives are introduced).

Dedication

The first time I saw Jerzy Perzanowski was at the *First World Congress on Paraconsistency* (Ghent 1997) where he was one of the invited speakers. Since then, partly because of a cooperation between the Logic Group in Toruń and my own research group, I had both the opportunity and the pleasure to attend several other lectures by him. Some of these were on paraconsistency — or parainconsistency as Jerzy prefers to call it. Others were on modal logic. Many were on both. To honour him, I can think of no better topic than one that explores relations between the two. Ontology would be an alternative — but I figure I better leave that to philosophers older and wiser than myself.

1. Introduction

The relations between modal logic and paraconsistency go back to 1948 when Stanisław Jaśkowski used a modal framework to introduce the notion of discussive logics (also called discursive logics) which constitute the first family of formal paraconsistent logics. As their name suggests, the inspiration for discussive logics originated from Jaśkowski's interest in the interpretation of discussions.

Suppose, with Jaśkowski, that one is trying to interpret a discussion in which some of the participants contradict each other. If one lumps together all the statements made, one risks arriving at triviality. So, how should one proceed?

†Research for this paper was supported by the Fund for Scientific Research – Flanders (Belgium), and indirectly by the Flemish Minister responsible for Science and Technology (contract BIL98/73 and BIL01/80).

In: J. Malinowski and A. Pietruszczak (eds.), *Essays in Logic and Ontology* (*Poznań Studies in the Philosophy of the Sciences and the Humanities*, vol. 91), 211–223. Amsterdam/New York, NY: Rodopi, 2006.

The most natural solution, according to Jaśkowski, is to interpret the conflicting statements in a modal way: "someone accepts A; someone accepts $\sim A$". From this, neither "someone accepts B" nor someone accepts "$A \wedge \sim A$" follows, which is exactly what we want.

Thus, where Γ is the set of all statements made in the debate, it seems that inferences should be made, not from Γ, but from $\Gamma^{\diamond} = \{\diamond A \mid A \in \Gamma\}$. This at once gives us the basic idea behind discussive logics. Where **L** is some modal logic and $\Gamma^{\diamond} = \{\diamond A \mid A \in \Gamma\}$, a discussive logic **DL**, associated with **L**, is obtained by specifying the language of **DL** and by stipulating that $\Gamma \vdash_{\mathbf{DL}} A$ iff $\Gamma^{\diamond} \vdash_{\mathbf{L}} \diamond A$.

There is a catch, however. If one simply uses some standard modal logic to make inferences from $\Gamma^{\diamond}$, one obtains an interpretation that is much too poor. For instance, if one would define **DL** as $\Gamma \vdash_{\mathbf{DL}} A$ iff $\Gamma^{\diamond} \vdash_{\mathbf{S5}} \diamond A$, then **DL** would not even validate *Modus Ponens*.

This is why Jaśkowski dismissed the idea to formulate discussive logics in terms of the classical connectives (see Jaśkowski [1948] 1969, pp. 149–150). Instead, he proposed to define, in **L**, a number of "discussive connectives" that satisfy some or all of the valid schemes and rules of the positive fragment of **CL**. For instance, Jaśkowski's favorite discussive logic $\mathbf{D_2}$ is obtained by defining a discussive conjunction, $A \wedge_{\mathrm{d}} B = \diamond A \wedge B$, a discussive implication, $A \supset_{\mathrm{d}} B = \diamond A \supset B$, and a discussive equivalence, $A \equiv_{\mathrm{d}} B = (A \supset_{\mathrm{d}} B) \wedge_{\mathrm{d}} (B \supset_{\mathrm{d}} A)$. The language of $\mathbf{D_2}$ is as that for **CL**, except that the classical conjunction as well as the material implication and equivalence are replaced by their discussive counterparts. The definition of $\mathbf{D_2}$ is completed by stipulating that, where $\mathbf{S5^d}$ is **S5** extended with the above definitions and where A and the members of Γ are well-formed formulas of $\mathbf{D_2}$, $\Gamma \vdash_{\mathbf{D_2}} A$ iff $\Gamma^{\diamond} \vdash_{\mathbf{S5^d}} \diamond A$.[1]

It is easily observed that, in $\mathbf{D_2}$, "$\supset_{\mathrm{d}}$" satisfies *Modus Ponens* (in view of $\diamond(A \supset_{\mathrm{d}} B), \diamond A \vdash_{\mathbf{S5^d}} \diamond B$) and that "$\wedge_{\mathrm{d}}$" satisfies Adjunction (in view of $\diamond A, \diamond B \vdash_{\mathbf{S5^d}} \diamond(A \wedge_{\mathrm{d}} B)$). This way out, however, has been criticized in a number of ways. Some have argued that there is no intuitive justification for the way in which modal operators sneak into the definitions of the implication and the conjunction; others have pointed out that the usual relations between the disjunction and the conjunction are lost (see the contributions in Priest *et al.* (1989) for both criticisms).

A criticism of a different kind is that Jaśkowski's solution is inadequate with respect to the intended application context. When interpreting a debate, we assume that a statement is accepted by *all* participants in a debate, unless and until this assumption turns out to be mistaken. This is justified in view of the

[1] In the original version, the language of $\mathbf{D_2}$ does not include a discussive conjunction. The latter was introduced in a later paper (see Jaśkowski [1949] 2001 for a translation). Axiomatizations of $\mathbf{D_2}$ are presented in, for instance, da Costa (1977), Kotas *et al.* (1979), and Dubikajtis *et al.* (1980); the axiomatization presented in Dubikajtis *et al.* (1980) can also be found in Arruda (1989, p. 117).

fact that participants are expected to object when they do not agree with some of the statements made. Hence, even if we are willing to "isolate" statements that contradict each other, we tend to conjoin them *unless and until* they are proven to be inconsistent. Thus, if in a discussion A is asserted by some participant and B by another, we assume that all participants accept "A and B", unless and until proven otherwise. However, the discussive conjunction is inadequate to capture this. This is easily seen from the fact that, in $\mathbf{D_2}$, $A \wedge_d \sim A$ follows from A and $\sim A$.

In Meheus (2006), an alternative solution is proposed that meets each of these criticisms. The idea is that, instead of relying on a series of discussive connectives, a discussive system should validate the multiple-inference rules of **CL** "as much as possible". The logic presented in Meheus (2006) is called $\mathbf{DL^r}$ and is meant as an alternative for $\mathbf{D_2}$. Like $\mathbf{D_2}$, $\mathbf{DL^r}$ incorporates the attractive properties of discussive logics: it is paraconsistent and does not allow to derive a (classical) contradiction ($A \wedge \sim A$) from an inconsistency (A, $\sim A$). However, unlike $\mathbf{D_2}$, $\mathbf{DL^r}$ is entirely formulated in the language of **CL**.

Like all discussive logics, $\mathbf{DL^r}$ is defined with respect to some modal logic, called **AJ**. So, where $\Gamma \cup \{A\}$ is a set of formulas of the predicative language of **CL**, $\Gamma \vDash_{\mathbf{DL^r}} A$ iff $\Gamma^{\Diamond} \vDash_{\mathbf{AJ}} \Diamond A$.

The logic **AJ** is formulated within the adaptive logics programme. The first adaptive logic was designed by Diderik Batens around 1980 and was meant to interpret inconsistent theories "as consistently as possible".[2] Meanwhile a whole variety of adaptive logics is available — see Batens (2004, 2001b) for recent surveys and introductions. Some of these handle logical abnormalities (such as inconsistency and negation-incompleteness). Others extend the usual deductive inference rules with some "ampliative" ones (for instance, rules for abduction, enumerative induction, compatibility, ...). What all adaptive logics have in common, however, is that they define consequence relations that are dynamical and/or non-monotonic.[3]

The idea behind **AJ** is to assume that the participants in a debate agree with each other "as much as possible". Thus, where Γ stands for the statements made in some debate, it is assumed, for every $A \in \Gamma$, that $\Box A$ (everybody accepts A) is derivable from $\Diamond A$ (someone accepts A), *unless and until proven otherwise*. Let us call this the *unanimity presupposition*. The resulting consequence relation is *non-monotonic*: conclusions drawn on the assumption that all participants agree with each other may be withdrawn when the latter is proven false.

As is the case for all adaptive logics, the semantics of **AJ** is obtained by

---

[2]Logics that satisfy this property are called inconsistency-adaptive logics.

[3]I say that a consequence relation is dynamical if the mere analysis of the premises may lead to a withdrawal of a previously accepted conclusion. Not all dynamical consequence relations are non-monotonic. In Batens (2001a), it is shown, for instance, that the pure logic of relevant implication may be characterized in a dynamical way.

selecting, for each set of premises, a *subset* of the models of some monotonic system (in this case the models of **S5**). Intuitively, those **S5**-models of $\Gamma^{\Diamond}$ are selected that interpret $\Gamma^{\Diamond}$ as much as possible in accordance with the unanimity presupposition. Thus, where $\Gamma^{\Diamond} = \{\Diamond p, \ \Diamond \sim p, \ \Diamond q\}$, some **S5**-models verify $\Diamond q \wedge \Diamond \sim q$, others verify $\Box q$. **AJ** selects the latter. As a consequence, $\Diamond(p \wedge q)$ is an **AJ**-consequence of $\Gamma^{\Diamond}$ (but not an **S5**-consequence of it). This is as it should be: as there is no (explicit) disagreement with respect to q, all participants are supposed to accept q. Hence, as one of the participants accepts p, some of them accept $p \wedge q$.

Let me also put this somewhat more generally. Where $\exists A$ stands for the existential closure of A, **AJ** selects those **S5**-models of $\Gamma^{\Diamond}$ that verify not more formulas of the form $\exists(\Diamond A \wedge \Diamond \sim A)$ than is necessary for the truth of the members of $\Gamma^{\Diamond}$.[4]

As is argued in Meheus (2006), $\mathbf{DL^r}$ leads to a more natural interpretation of discussions than $\mathbf{D_2}$. There is, however, one drawback that $\mathbf{DL^r}$ shares with $\mathbf{D_2}$. If some member of Γ is self-contradictory, the system leads to triviality. Thus, $\mathbf{DL^r}$ can handle cases in which different participants in a debate contradict each other, but is inadequate for cases in which some participants contradict themselves.

At first sight, this may seem an advantage rather than a disadvantage. After all, should a participant that contradicts him or herself not be considered as irrational (or at least non-rational)? And hence, should one not expect to obtain nonsensical results as long as self-contradictory sentences are included in the set of statements one considers?

There are, however, two important remarks to be made in this respect. The first is that humans are not logically omniscient. Consequently, discovering some inconsistency in one's own beliefs may be far from evident. The second is that, in a debate, it may be just as important to expose internal inconsistencies (one participant contradicting him or herself) as it is to expose external inconsistencies (different participants contradicting each other).

The aim of this paper is to generalize the logic from Meheus (2006) to handle both internal and external inconsistencies. The resulting logic will be called $\mathbf{DLI^r}$. Like $\mathbf{DL^r}$, $\mathbf{DLI^r}$ validates the multiple-premise rules of **CL** "as much as possible". It moreover warrants that self-contradictory premises do not lead to explosion. If no member of Γ is self-contradictory, then $\mathbf{DLI^r}$ leads to exactly the same results as $\mathbf{DL^r}$.

At first sight, the design of $\mathbf{DLI^r}$ may seem obvious. Like all discussive logics, it should be formulated with respect to some modal logic. And, if it should validate Adjunction as much as possible, this modal logic should be adaptive. Now, if one wants to allow for (internal) inconsistencies, the latter cannot be

[4]The selection mechanism will be defined in more exact terms in Section 6.

defined with respect to **S5**. It seems, however, that all one needs is an **S5**-like version of some (standard) paraconsistent logic (that allows for the truth of $A \wedge {\sim} A$).[5]

However, for reasons that will be explained in the next section, matters are not all that simple. If one simply allows for contradictions, one obtains an interpretation of the debate that is much too poor. The solution will be to interpret the individual statements *as consistently as possible*. As we shall see below, this also has the advantage that, if some statement is not self-contradictory, it is interpreted in exactly the same way as with **AJ**.

Technically, this will be realized by making a combination of two adaptive logics: an inconsistency-adaptive logic (that interprets the individual statements as consistently as possible) and one that is analogous to **AJ** (that interprets the different statements as "unanimously" as possible).

I shall proceed as follows. After explaining why we need an inconsistency-adaptive logic (Section 2), I briefly discuss the inconsistency-adaptive logic $\mathbf{P}^{\mathrm{r}}$ and the monotonic paraconsistent logic **P** on which it is based (Section 3). Next, I present an **S5**-like version of **P** that constitutes the basis for the two adaptive logics (Section 4). In Sections 5 and 6, I present the logics **S5PI** and **S5PE**. The first of these interprets the individual statements as consistently as possible and proceeds in terms of $\mathbf{P}^{\mathrm{r}}$; the second one is similar to **AJ**. The way in which both logics are combined to obtain $\mathbf{DLI}^{\mathrm{r}}$ is discussed in Section 7. I end with a short conclusion (Section 8).

For reasons of space, I shall only discuss the semantics of the different systems. In view of the results from Meheus (2006) and from Batens (2001b), the design of the corresponding proof theories is rather straightforward.

2. The Problem

As is explained in the previous section, the discussive logic $\mathbf{DL}^{\mathrm{r}}$ is defined with respect to the modal adaptive logic **AJ**, and the latter is obtained by selecting, for each set of premises $\Gamma^{\Diamond}$, a subset of the **S5**-models of $\Gamma^{\Diamond}$. As no **S5**-model verifies a formula of the form $\Diamond(A \wedge {\sim} A)$, **AJ** leads to triviality whenever one of the members of Γ is self-contradictory.

Now, consider some paraconsistent logic **PL** in which $A \wedge {\sim} A$ does not lead to explosion,[6] and suppose that we define an **S5**-like version for it. Some models of the resulting logic, call it **MPL**, will verify $\Diamond(A \wedge {\sim} A)$. Hence, in order

[5] I say that a paraconsistent logic is standard if the consequence relation defined by it is reflexive, transitive and monotonic.

[6] As an example, one may think of the full positive fragment of Classical Logic (possibly extended with $A \vee {\sim} A$).

to obtain an alternative for **AJ** that can handle both internal and external inconsistencies, it seems sufficient to define the selection, not with respect to the **S5**-models of $\Gamma^{\Diamond}$, but with respect to its **MPL**-models.

There is, however, an important problem with this approach. Standard monotonic logics invalidate *Ex Falso Quodlibet* by invalidating some of the inference rules of **CL**.[7] For instance, most standard paraconsistent logics invalidate Disjunctive Syllogism; those that do not necessarily invalidate Addition (A / $A \vee B$).[8] As a consequence, basing the selection on the **MPL**-models of $\Gamma^{\Diamond}$, would in many cases result in an interpretation of the individual statements that is much too poor.

Consider, for instance, $\Gamma^{\Diamond} = \{\Diamond(p \wedge ({\sim} p \vee q)), \Diamond(p \wedge {\sim} p)\}$. In that case, one of the participants is "disagreeing" with him or herself concerning p. This, however, is the only disagreement. As a consequence, an **MPL**-model of $\Gamma^{\Diamond}$ would be selected iff it verifies $\Diamond p \wedge \Diamond {\sim} p$, but falsifies any other formula of the form $\exists(\Diamond A \wedge \Diamond {\sim} A)$. But, if **MPL** is based on a paraconsistent logic that invalidates Disjunctive Syllogism, some **MPL**-models of $\Gamma^{\Diamond}$ will satisfy this requirement while *falsifying* $\Diamond q$.[9] So, $\Diamond q$ would not be an **MPL**-consequence of $\Gamma^{\Diamond}$. This does not seem justified, however. Even if one of the members of $\Gamma^{\Diamond}$ is self-contradictory, $\Diamond(p \wedge ({\sim} p \vee q))$ is not. Hence, there is no reason why some of the **S5**-consequences should not be derivable from it. Or, put into terms of the intended application context, the fact that one of the participants in the debate is maintaining a self-contradictory statement should not lead to an impoverished understanding of what the other participants are asserting.

The way out is to make sure that the individual statements are interpreted "as consistently as possible". As we shall see below, this may be realized by making a preliminary selection from the **MPL**-models (before the one in accordance with the unanimity presupposition). Intuitively, the preliminary selection warrants that, for every individual statement, there is a "possible world" in which this statement is interpreted as consistently as possible. For instance, where $\Gamma^{\Diamond}$ is as in the previous paragraph, only those **MPL**-models will be selected that include a world in which $p \wedge ({\sim} p \vee q)$ is verified, but $p \wedge {\sim} p$ is falsified, and hence, in which q is verified.

In this paper, I shall base the preliminary selection on the inconsistency-adaptive logic $\mathbf{P}^{\mathrm{r}}$. As we shall see in the next section, $\mathbf{P}^{\mathrm{r}}$ is based on a very poor

[7] This does not necessarily hold true for some non-standard paraconsistent systems. In Batens (2006), it is shown, for instance, that it is possible to design a proof procedure that *only* invalidates *Ex Falso Quodlibet* (while validating all other inference rules of **CL**). The price to be paid, however, is that the resulting consequence relation is not transitive.

[8] An example of a standard paraconsistent logic that validates Disjunctive Syllogism can be found in Meheus (2000).

[9] The example will be explained in more detail in Section 6. For the moment, note only that **MPL**-models that verify $\Box(p \wedge {\sim} p)$ verify all members of $\Gamma^{\Diamond}$, even if they falsify $\Diamond q$.

paraconsistent logic (one that invalidates Disjunctive Syllogism as well as all other "negative" inference rules of **CL**). However, for all sentences that behave consistently, $\mathbf{P}^r$ is just as rich as **CL**.

$\mathbf{P}^r$ is not the only inconsistency-adaptive logic currently available. It is chosen because, on the one hand, it is a paradigmatic example to understand the functioning of an inconsistency-adaptive logic and, on the other hand, it makes very little assumptions about the domain one is dealing with.[10]

3. The Logics P and $\mathbf{P}^r$

The logic $\mathbf{P}^r$ is an adaptive version of the standard paraconsistent logic **P**.[11] In a sense, **P** is the most straightforward paraconsistent logic that one can obtain from **CL**. Where the negation of **CL** is defined by (i) a consistency requirement (A and its negation cannot both be true) and (ii) a negation-completeness requirement (A and its negation cannot both be false), the paraconsistent negation of **P** is obtained by dropping (i) while retaining (ii).

The language $\mathcal{L}$ of **P** is as the standard predicative language for **CL**, except that it contains two symbols for negation: "$\sim$" and "$\neg$". The former stands for the paraconsistent negation, the latter is the standard negation of **CL**. The language also includes $\bot$. In the premises, negation is formalized by "$\sim$". The main function of "$\neg$" is to simplify the meta-theoretic proofs.

Let $\mathcal{S}$, $\mathcal{C}$, $\mathcal{P}^r$, $\mathcal{F}$ and $\mathcal{W}$ stand for, respectively, the sets of sentential letters, individual constants, predicate letters of rank r, (open and closed) formulas, and wffs (closed formulas) of $\mathcal{L}$.

To simplify the semantic handling of quantifiers, the language $\mathcal{L}$ is extended to the pseudo-language $\mathcal{L}^+$ by introducing a set of pseudo-constants $\mathcal{O}$ that has at least the cardinality of the largest model one wants to consider. Let $\mathcal{W}^+$ denote the set of wffs of $\mathcal{L}^+$ (in which $\mathcal{C} \cup \mathcal{O}$ plays the role played by $\mathcal{C}$ in $\mathcal{L}$) and let ${}^{\sim}\mathcal{W}^+ = \{\sim A \mid \sim A \in \mathcal{W}^+\}$.

A **P**-model M is a couple $\langle D, v\rangle$, in which D is a set and v an assignment function. Every such model is an interpretation of $\mathcal{W}^+$, and hence of $\mathcal{W}$, which is what we are interested in. The assignment function v is defined by:

C1.1 $v\colon \mathcal{S} \longrightarrow \{0, 1\}$

C1.2 $v\colon \mathcal{C} \cup \mathcal{O} \longrightarrow D$ (where $D = \{v(\alpha) \mid \alpha \in \mathcal{C} \cup \mathcal{O}\}$)

C1.3 $v\colon \mathcal{P}^r \longrightarrow \wp(D^r)$ (the power set of the r-th Cartesian product of D)

C1.4 $v\colon {}^{\sim}\mathcal{W}^+ \longrightarrow \{0, 1\}$

---

[10]Most inconsistency-adaptive logics lead to a somewhat richer consequence set than $\mathbf{P}^r$. They do so, however, by making several assumptions about the application context (for instance, that there are reasons to *minimize* the inconsistencies in some or other way).

[11]The most complete presentation of $\mathbf{P}^r$ may be found in Batens (1999).

The valuation function, $v_M\colon \mathcal{W}^+ \longrightarrow \{0, 1\}$, determined by M is defined by:

C2.1 where $A \in \mathcal{S}$, $v_M(A) = v(A)$
C2.2 $v_M(\bot) = 0$
C2.3 $v_M(\pi^r\alpha_1 \ldots \alpha_r) = 1$ iff $\langle v(\alpha_1), \ldots, v(\alpha_r)\rangle \in v(\pi^r)$
C2.4 $v_M(\alpha = \beta) = 1$ iff $v(\alpha) = v(\beta)$
C2.5 $v_M({\sim} A) = 1$ iff $v_M(A) = 0$ or $v({\sim} A) = 1$
C2.6 $v_M(\neg A) = 1$ iff $v_M(A) = 0$
C2.7 $v_M(A \wedge B) = 1$ iff $v_M(A) = v_M(B) = 1$
C2.8 $v_M(A \vee B) = 1$ iff $v_M(A) = 1$ or $v_M(B) = 1$
C2.9 $v_M(A \supset B) = 1$ iff $v_M(A) = 0$ or $v_M(B) = 1$
C2.10 $v_M(A \equiv B) = 1$ iff $v_M(A) = v_M(B)$
C2.11 $v_M((\forall\alpha)A(\alpha)) = 1$ iff $v_M(A(\beta)) = 1$ for all $\beta \in \mathcal{C} \cup \mathcal{O}$
C2.12 $v_M((\exists\alpha)A(\alpha)) = 1$ iff $v_M(A(\beta)) = 1$ for some $\beta \in \mathcal{C} \cup \mathcal{O}$.

A is true in a **P**-model M (M verifies A) iff $v_M(A) = 1$; $\Gamma \vDash_{\mathbf{P}} A$ iff all **P**-models that verify all members of Γ also verify A; $\vDash_{\mathbf{P}} A$ (A is valid) iff it is verified by all **P**-models.

The semantics for $\mathbf{P}^{\mathrm{r}}$ is obtained in a rather straightforward way from that for **P**. Let Ω be the set of all formulas of the form $\exists(A \wedge {\sim} A)$ (where $A \in \mathcal{F}$) and let $Dab(\Delta)$ denote the disjunction $\bigvee(\Delta)$, in which $\Delta \subset \Omega$. A formula is called a *Dab-consequence of* Γ iff it is of the form $Dab(\Delta)$ and it is a **P**-consequence of Γ; it is called a *minimal Dab*-consequence of Γ iff it is a *Dab*-consequence of Γ, and $Dab(\Delta')$ is not a *Dab*-consequence of Γ, for any $\Delta' \subset \Delta$.

If $Dab\{A_1, \ldots, A_n\}$ is a minimal *Dab*-consequence of Γ, Γ provides the information that one of the A_i behaves inconsistently without revealing which one does. So, it is sensible not to rely on the consistent behaviour of any of the A_i (or, put somewhat differently, to consider all A_i as "unreliable"). This is the strategy $\mathbf{P}^{\mathrm{r}}$ is based on.

The *abnormal part* of a **P**-model is defined by:

Definition 1. For any **P**-model M, $Ab(M) = \{A \in \Omega \mid v_M(A) = 1\}$.

The set $U(\Gamma)$ of formulas that are *unreliable* with respect to Γ is defined by:

Definition 2. $U(\Gamma) = \bigcup\{\Delta \mid Dab(\Delta)$ is a minimal *Dab*-consequence of $\Gamma\}$.

For any set of premises Γ, those **P**-models are selected that verify an inconsistency only if it is unreliable with respect to Γ:

Definition 3. A **P**-model M of Γ is *reliable* iff $Ab(M) \subseteq U(\Gamma)$.

Note that it does not make sense to say that a **P**-model is reliable, but only that it is a reliable model *of* some set of premises Γ. Below, I shall sometimes use the term "$\mathbf{P}^{\mathrm{r}}$-models of Γ" to refer to the reliable models of Γ. The semantic consequence relation is defined with respect to the selected models:

Definition 4. $\Gamma \vDash_{\mathbf{P}^{\mathrm{r}}} A$ iff A is true in all reliable models of Γ.

To see what the $\mathbf{P}^{\mathrm{r}}$-semantics comes to, consider a simple propositional example. Let Γ be $\{p, q, \sim p \vee \sim q, \sim p \vee r, s, \sim s \vee t\}$. In that case, there is only one minimal *Dab*-consequence of Γ, namely $(p \wedge \sim p) \vee (q \wedge \sim q)$. This teaches us that either $p \wedge \sim p$ or $q \wedge \sim q$ behaves inconsistently with respect to Γ. As it is not possible to establish which one, both are considered as unreliable. It follows that some selected models verify $p \wedge \sim p$ and falsify r (if both p and $\sim p$ are true, then, in view of C2.8, $\sim p \vee r$ is true, even if r is false). However, as $s \wedge \sim s$ is reliable with respect to Γ, all selected models falsify $s \wedge \sim s$, and hence, verify t. So, r is not a $\mathbf{P}^{\mathrm{r}}$-consequence of Γ, but t is. (I leave it to the reader to check that neither r nor t are **P**-consequences of Γ.)

4. The Modal Paraconsistent Logic S5P

In this section, I present a modal version of **P** which will form the basis for the adaptive logics discussed in the next sections. The relation between **S5P** and **P** is as that between **S5** and **CL**.

Let $\mathcal{L}^M$ be the standard modal language (including $\perp$) and let $\mathcal{L}^{M+}$ be obtained from $\mathcal{L}^M$ in the same way as $\mathcal{L}^+$ is obtained from $\mathcal{L}$. Let $\mathcal{W}^{p+}$ be the set of primitive wffs of $\mathcal{L}^+$ (those wffs of $\mathcal{L}^+$ that contain no other logical constants than identity).

An **S5P**-model is a couple $\mathcal{M} = \langle \Sigma, M_0 \rangle$ in which Σ is a *set* of **P**-models and $M_0 \in \Sigma$. The valuation determined by an **S5P**-model $\mathcal{M}$ is defined by the following clauses:

MC1 where $A \in \mathcal{W}^{p+}$ or $A \in {}^{\sim}\mathcal{W}^+$, $v_{\mathcal{M}}(A, M_i) = v_{M_i}(A)$
MC2 $v_{\mathcal{M}}(\neg A, M_i) = 1$ iff $v_{\mathcal{M}}(A, M_i) = 0$
MC3 $v_{\mathcal{M}}(\sim \Box A, M_i) = 1$ iff $v_{\mathcal{M}}(\neg \Box A, M_i) = 1$
MC4 $v_{\mathcal{M}}(\sim \Diamond A, M_i) = 1$ iff $v_{\mathcal{M}}(\neg \Diamond A, M_i) = 1$
MC5 $v_{\mathcal{M}}(A \wedge B, M_i) = 1$ iff $v_{\mathcal{M}}(A, M_i) = v_{\mathcal{M}}(B, M_i) = 1$
MC6 $v_{\mathcal{M}}(A \vee B, M_i) = 1$ iff $v_{\mathcal{M}}(A, M_i) = 1$ or $v_{\mathcal{M}}(B, M_i) = 1$
MC7 $v_{\mathcal{M}}(A \supset B, M_i) = 1$ iff $v_{\mathcal{M}}(A, M_i) = 0$ or $v_{\mathcal{M}}(B, M_i) = 1$
MC8 $v_{\mathcal{M}}(A \equiv B, M_i) = 1$ iff $v_{\mathcal{M}}(A, M_i) = v_{\mathcal{M}}(B, M_i)$
MC9 $v_{\mathcal{M}}((\forall \alpha)A(\alpha), M_i) = 1$ iff $v_{\mathcal{M}}(A(\beta), M_i) = 1$ for all $\beta \in \mathcal{C} \cup \mathcal{O}$
MC9 $v_{\mathcal{M}}((\exists \alpha)A(\alpha), M_i) = 1$ iff $v_{\mathcal{M}}(A(\beta), M_i) = 1$ for some $\beta \in \mathcal{C} \cup \mathcal{O}$
MC10 $v_{\mathcal{M}}(\Box A, M_i) = 1$ iff $v_{\mathcal{M}}(A, M_j) = 1$ for all $M_j \in \Sigma$
MC11 $v_{\mathcal{M}}(\Diamond A, M_i) = 1$ iff $v_{\mathcal{M}}(A, M_j) = 1$ for some $M_j \in \Sigma$.

A is true in an **S5P**-model $\mathcal{M} = \langle \Sigma, M_0 \rangle$ ($\mathcal{M}$ verifies A) iff $v_{M_0}(A) = 1$; $\Gamma \vDash_{\mathbf{S5P}} A$ iff all **S5P**-models that verify all members of Γ also verify A; $\vDash_{\mathbf{S5P}} A$ (A is valid) iff it is verified by all **S5P**-models.

Note that, in front of a modal operator, the paraconsistent negation behaves as the classical negation. As a consequence the usual relations between "□" and

"$\Diamond$" are not preserved. For instance, as both $\Box A$ and $\Diamond \sim A$ may be true in the same **S5P**-model, some **S5P**-models verify $\Box A$ while falsifying $\sim \Diamond \sim A$. Note also that, for any **S5P**-model $\mathcal{M} = \langle \Sigma, M_0 \rangle$, $\alpha = \beta$ may be true in some members of Σ, while false in others. Hence, $\alpha = \beta$ does not entail $\Box \alpha = \beta$.

5. The Modal Adaptive Logic S5PI

In view of the intended application context, the logic **S5PI** is only defined with respect to sets of premises of the form $\Gamma^{\Diamond}$ (where $\Gamma \subset \mathcal{W}$). Semantically, it is obtained by selecting a subset of the **S5P**-models of $\Gamma^{\Diamond}$. The idea is that only those models are selected that interpret each individual member of $\Gamma^{\Diamond}$ as consistently as possible.

Consider, for instance, $\Gamma^{\Diamond} = \{\Diamond((p \wedge \sim p) \wedge (q \wedge (\sim q \vee r))),\ \Diamond(p \wedge (\sim p \vee s))\}$. In that case, some **S5P**-models of $\Gamma^{\Diamond}$ falsify $\Diamond r$, others falsify $\Diamond s$. However, in any interpretation of $(p \wedge \sim p) \wedge (q \wedge (\sim q \vee r))$ that is as consistent as possible, r is true. Analogously, in any interpretation of $p \wedge (\sim p \vee s)$ that is as consistent as possible, s is true. This suggests that an **S5P**-model $\mathcal{M} = \langle \Sigma, M_0 \rangle$ of $\Gamma^{\Diamond}$ should be selected only if, for every $A \in \Gamma$, some $M \in \Sigma$ interprets A as consistently as possible. Translated into terms of the inconsistency-adaptive logic $\mathbf{P}^{\mathrm{r}}$, this brings us to the following selection criterion:

DEFINITION 5. An **S5P**-model $\mathcal{M} = \langle \Sigma, M_0 \rangle$ is an **S5PI**-model of $\Gamma^{\Diamond}$ iff, for every $\Diamond A \in \Gamma^{\Diamond}$, some $M \in \Sigma$ is a $\mathbf{P}^{\mathrm{r}}$-model of A.

As is the case for $\mathbf{P}^{\mathrm{r}}$, the semantic consequence relation is defined with respect to the selected models:

DEFINITION 6. $\Gamma^{\Diamond} \vDash_{\mathbf{S5PI}} A$ iff A is true in all **S5PI**-model of $\Gamma^{\Diamond}$.

I leave it to the reader to check that, where $\Gamma^{\Diamond}$ is as in the second paragraph of this section, both $\Diamond r$ and $\Diamond s$ are **S5PI**-consequences of $\Gamma^{\Diamond}$.

6. The Modal Adaptive Logic S5PE

Like **S5PI**, **S5PE** is defined with respect to sets of premises of the form $\Gamma^{\Diamond}$ and is obtained by making a selection of the **S5P**-models. The selection criterion is analogous to the one for **AJ** (see Meheus 2006).

Let $\Omega^{\Diamond}$ be the set of all formulas of the form $\exists(\Diamond A \wedge \Diamond \sim A)$ in which A is a *primitive* wff of $\mathcal{L}$, and let $Dab^{\Diamond}(\Delta)$ denote the disjunction $\bigvee(\Delta)$, in which $\Delta \subset \Omega^{\Diamond}$. Intuitively, the set $\Omega^{\Diamond}$ contains all (primitive) formulas that violate the unanimity presupposition.[12]

[12] The reason why only primitive formulas are considered is explained in Meheus (2006).

A formula will be called a *minimal* $Dab^{\Diamond}$*-consequence of* $\Gamma^{\Diamond}$ iff (i) it is of the form $Dab^{\Diamond}(\Delta)$, (ii) it is an **S5P**-consequence of $\Gamma^{\Diamond}$, and (iii) $Dab^{\Diamond}(\Delta')$ is not an **S5P**-consequence of $\Gamma^{\Diamond}$, for any $\Delta' \subset \Delta$.

As for $\mathbf{P}^{\mathrm{r}}$, the selection proceeds in terms of the formulas that are unreliable with respect to the set of premises and the semantic consequence relation is defined with respect to the selected models:

Definition 7. For any **S5P**-model $\mathcal{M}$, $Ab(\mathcal{M}) = \{A \in \Omega^{\Diamond} \mid v_{\mathcal{M}}(A) = 1\}$.

Definition 8. $U(\Gamma^{\Diamond}) = \bigcup\{\Delta \mid Dab^{\Diamond}(\Delta)$ is a minimal $Dab^{\Diamond}$-consequence of $\Gamma^{\Diamond}\}$.

Definition 9. An **S5P**-model $\mathcal{M}$ of $\Gamma^{\Diamond}$ is *reliable* iff $Ab(\mathcal{M}) \subseteq U(\Gamma^{\Diamond})$.

Definition 10. $\Gamma^{\Diamond} \vDash_{\mathbf{S5PE}} A$ iff A is true in all reliable models of $\Gamma^{\Diamond}$.

Let me illustrate the **S5PE**-semantics with a simple example. Suppose that $\Gamma^{\Diamond} = \{\Diamond(p \wedge {\sim} p), \Diamond(p \wedge ({\sim} p \vee {\sim} q)), \Diamond r, \Diamond s\}$. The only minimal $Dab^{\Diamond}$-consequence of $\Gamma^{\Diamond}$ is $\Diamond p \wedge \Diamond {\sim} p$. Hence, all **S5PE**-models of $\Gamma^{\Diamond}$ verify $\Diamond p \wedge \Diamond {\sim} p$, but falsify any other member of $\Omega^{\Diamond}$. One consequence of this is that all selected models verify $\Box r$ as well as $\Box s$, and hence, also $\Box(r \wedge s)$. Another consequence, however, is that some **S5PE**-models of $\Gamma^{\Diamond}$ *falsify* $\Diamond {\sim} q$. Consider, for instance, the **S5P**-model $\mathcal{M} = \langle\{M\}, M\rangle$, in which $v_M(p) = v_M({\sim} p) = 1$ and $v_M({\sim} A) = 0$, for any primitive wff A that differs from p. It is easily observed that $\mathcal{M}$ is an **S5P**-model of $\Gamma^{\Diamond}$ and that it verifies no other member of $\Omega^{\Diamond}$ than $\Diamond p \wedge \Diamond {\sim} p$. Hence, $\mathcal{M}$ is an **S5PE**-model of $\Gamma^{\Diamond}$ that falsifies $\Diamond {\sim} q$.

So, by itself, **S5PE** does not lead to the desired results. It does, however, when it is combined in the appropriate way with **S5PI**. Suppose, for instance, that **S5PE** is applied, not to $\Gamma^{\Diamond}$, but to the set $\Gamma^{\dagger} = \{\Diamond A \mid \Gamma^{\Diamond} \vDash_{\mathbf{S5PI}} \Diamond A\}$. As $\Diamond {\sim} q$ is an **S5PI**-consequence of $\Gamma^{\Diamond}$, all **S5PE**-models of $\Gamma^{\dagger}$ verify $\Diamond {\sim} q$. In line with the idea that all individual statements should be interpreted as consistently as possible, this is exactly what we want.

7. The Discussive Adaptive Logic $\mathbf{DLI}^{\mathrm{r}}$

Where $\Gamma \cup \{A\} \subset \mathcal{W}$ and $\Gamma^{\dagger} = \{\Diamond A \mid \Gamma^{\Diamond} \vDash_{\mathbf{S5PI}} \Diamond A\}$, the logic $\mathbf{DLI}^{\mathrm{r}}$ is obtained by the following definition:

Definition 11. $\Gamma \vDash_{\mathbf{DLI}^{\mathrm{r}}} A$ iff $\Gamma^{\dagger} \vDash_{\mathbf{S5PE}} \Diamond A$.

An important difference between $\mathbf{DLI}^{\mathrm{r}}$ and the logic $\mathbf{DL}^{\mathrm{r}}$ from Meheus (2006) is that the former allows for the (non-trivial) derivation of contradictions. It is easily observed, however, that contradictions are only derivable if some of the premises are self-contradictory. Thus, although $p \wedge {\sim} p$ is (non-trivially)

$\mathbf{DLI}^{\mathrm{r}}$-derivable from $\{p \wedge (\sim p \wedge q)\}$, it is not $\mathbf{DLI}^{\mathrm{r}}$-derivable from $\{p, \sim p \wedge q\}$. This is in line with the ideas underlying discussive logics: from "someone accepts p" and "someone accepts $\sim p$" one should not be able to derive "someone accepts $p \wedge \sim p$". The latter should only be derivable if some participant is contradicting him or herself.

8. Conclusion

In this paper, I presented a brief exposition of the discussive adaptive logic $\mathbf{DLI}^{\mathrm{r}}$. Unlike other discussive logics currently available, $\mathbf{DLI}^{\mathrm{r}}$ enables one to handle both internal and external inconsistencies. I have argued that, because of this property, $\mathbf{DLI}^{\mathrm{r}}$ is an adequate tool to interpret discussions in which some participants not only contradict each other, but also themselves. It seems, however, that the interpretation of discussions is not the only possible application context for $\mathbf{DLI}^{\mathrm{r}}$. In the sciences, for instance, one frequently has to deal with theories that are not only incompatible with some other theories in the domain, but also internally inconsistent. It would be interesting to check whether $\mathbf{DLI}^{\mathrm{r}}$ leads to satisfactory results also for this kind of situation.

Other topics for further research concern the proof theory and the metatheory of the adaptive logics used in the definition of $\mathbf{DLI}^{\mathrm{r}}$. An important open problem in this respect is whether it is possible to design a direct characterization for $\mathbf{DLI}^{\mathrm{r}}$ (that is, one that does not proceed in modal terms).

Ghent University
Centre for Logic and Philosophy of Science
Blandijnberg 2
B-9000 Ghent
Belgium
e-mail: Joke.Meheus@rug.ac.be

REFERENCES

Arruda, A. I. (1989). Aspects of the Historical Development of Paraconsistent Logic. In: Priest *et al.* (1989), pp. 99–130.

Batens, D. (1999). Inconsistency-Adaptive Logics. In: Ewa Orłowska (ed.), *Logic at Work: Essays Dedicated to the Memory of Helena Rasiowa*, pp. 445–472. Heidelberg, New York: Physica Verlag (Springer).

Batens, D (2001a). A Dynamic Characterization of the Pure Logic of Relevant Implication. *Journal of Philosophical Logic* **30**, 267–280.

Batens, D. (2001b). A General Characterization of Adaptive Logics. *Logique et Analyse* **173–175**, 45–68.

Batens, D. (2004). The Need for Adaptive Logics in Epistemology. In: S. Rahman, J. Symons, D. M. Gabbay and J. P. Van Bendegem (eds.), *Logic, Epistemology, and the Unity of Science*, pp. 459–485. Dordrecht: Kluwer.

Batens, D. (2006). A Paraconsistent Proof Procedure Based on Classical Logic. To appear.

da Costa, N. C. A. (1977). On Jaśkowski's Discussive Logic. In: A. I. Arruda, N. C. A. da Costa and R. Chuaqui (eds.), *Non-Classical Logic, Model Theory and Computability*, pp. 37–56. Amsterdam: North-Holland.

Dubikajtis, L., E. Dudek and J. Konior (1980). On Axiomatics of Jaśkowski's Discussive Propositional Calculus. In: A. I. Arruda, N. C. A. da Costa and A. M. Sette (eds.), *Proceedings of the Third Brazilian Conference on Mathematical Logic*, pp. 109–117. São Paulo: Sociedade Brasileira de Lógica.

Jaśkowski, S. ([1948] 1969). Propositional Calculus for Contradictory Deductive Systems. *Studia Logica* **24**, 143–157. Originally published in Polish in: *Studia Scientarium Torunensis*, Sec. A II (1948), 55–77.

Jaśkowski, S. ([1949] 2001). On the Discussive Conjunction in the Propositional Calculus for Inconsistent Deductive Systems. *Logic and Logical Philosophy* **7**, 57–59. Originally published in Polish in 1949.

Kotas, J. and N. C. A. da Costa (1979). A New Formulation of Discussive Logic. *Studia Logica* **38**, 429–445.

Meheus, J. (2000). An Extremely Rich Paraconsistent Logic and the Adaptive Logic Based on it. In: D. Batens, Ch. Mortensen, G. Priest and J. P. Van Bendegem (eds.), *Frontiers of Paraconsistent Logic*, pp. 189–201. Baldock, UK: Research Studies Press.

Meheus, J. (2006). An Adaptive Logic Based on Jaśkowski's D2. *Journal of Philosophical Logic* **35** (6), 539–567.

Priest, G., R. Routley and J. Norman, eds. (1989). *Paraconsistent Logic: Essays on the Inconsistent*. München: Philosophia Verlag.

Sergei P. Odintsov

ABSURDITY AS UNARY OPERATOR

ABSTRACT. It was shown in the previous work of the author that one can avoid the paradox of minimal logic $\{\varphi, \neg\varphi\} \vdash \neg\psi$ defining the negation operator via reduction not a constant of absurdity, but to a unary operator of absurdity. In the present article we study in details what does it mean that negation in a logical system can be represented via an absurdity or contradiction operator. We distinguish different sorts of such presentations. Finally, we consider the possibility to represent the negation via absurdity and contradiction operators in such well known systems of paraconsistent logic as D.Batens's logic **CLuN** and Sette's logic P^1.

1. Introduction

The present article is a continuation of Odintsov (2003) and it also contains results presented in author's lecture at the II Flemish-Polish workshop on Ontological Foundations of Paraconsistency. In the lecture presented at the First Flemish-Polish Workshop (see Odintsov 1999a) as well as in a series of other works (Odintsov 1998a, 1998b, 1999b, 2000, 2001) we studied the class **Jhn** of non-trivial extensions of minimal or Johansson's logic. Minimal logic (we will denote it **Lj**) was suggested for the fist time by Kolmogorov (1925) and received the name "Johansson's logic" after the work (1937) by Ingebrigt Johansson. It is a paraconsistent analog of Heyting's intuitionistic logic **Li** (a logic is *paraconsistent* if it has inconsistent but non-trivial theories). However, this logic was not popular among specialists in the field of paraconsistency due to the following undesirable property. In **Lj**, one can prove $\{\varphi, \neg\varphi\} \vdash \neg\psi$ for arbitrary formulæ φ and ψ. This condition means that although inconsistent **Lj**-theories may be non-trivial, they are trivial with respect to negation. Any negated formula is provable in any inconsistent **Lj**-theory.

Due to this "paraconsistent paradox" of minimal logic it looks natural to finish an investigation of the class of **Lj**-extension with an attempt to overcome it or, at least, to better understand this paradoxical property. In Odintsov (2003), we have tried to do it by emerging the class of **Lj**-extensions in a more general class of paraconsistent logics and pointing out some special property distinguishing extensions of minimal logic in the latter class. We assumed also that negation in logics from the above mentioned class should preserve the most essential property of intuitionistic negation, namely, that it can be defined as reduction to absurdity. In intuitionistic logic, negation is characterized by three

In: J. Malinowski and A. Pietruszczak (eds.), *Essays in Logic and Ontology* (*Poznań Studies in the Philosophy of the Sciences and the Humanities*, vol. 91), 225–242. Amsterdam/New York, NY: Rodopi, 2006.

important features: 1) we assert $\neg\varphi$ if supposing φ leads to an absurdity; 2) the absurdity may be explicated as a propositional constant $\bot$; 3) the absurdity implies everything, $\bot \supset p$, or $\neg p \supset (p \supset q)$. Note that Item 3 implies, in fact, Item 2. If any contradiction implies everything, all contradictions are equivalent, and so we may use a propositional constant do denote an arbitrary contradictory or absurd statement. In minimal logic **Lj**, Item 3 is rejected, which allows one to distinguish contradictions. But Item 2 is preserved in **Lj**, which results, in particular, in the above mentioned "paraconsistent paradox". In this situation, it looks natural to make the next step and to resign not only from Item 3, but also from its consequence, Item 2. This gives rise to a question, how in this case the absurdity should be explicated in a logical system?

The answer suggested in Odintsov (2003) was based on the interrelations between the logic of classical refutability (hereafter, **Le**) and the modal logic **Ł** of Łukasiewicz. The logic **Le** of classical refutability can be obtained from **Lj** by extending its positive fragment up to the classical positive logic. It was suggested by P. Bernays (1953) and received his name from H. B. Curry (1952). In Odintsov (1998b), it was proved that **Le** is the greatest paraconsistent extension of minimal logic and that the simplest characteristic matrix for **Le** has four elements, and so **Le** can be considered as a four-valued logic as well as Łukasiewicz's modal logic **Ł** (see Łukasiewicz 1953, 1957; and Font & Hajek 2000). The close interrelation between **Le** and **Ł** was established by J. Porte (1979, 1984), who proved that **Le** is definitially equivalent to the positive fragment of **Ł**. In Odintsov (2003), Porte's result was slightly modified and it was shown that the necessity operator $\Box$ of **Ł** may be identified with the contradiction operator $\mathsf{C}(\varphi) := \varphi \wedge \neg\varphi$ of the logic of classical refutability. This leads to the idea of considering the class of C-logics in the language $\langle \vee, \wedge, \supset, \mathsf{C} \rangle$ with a primary connective C for the contradiction operator and treating negation as the following abbreviation $\neg\varphi := \varphi \supset \mathsf{C}(\varphi)$.[1] We assume also that positive fragments of C-logics contain positive logic and that any C-logic has an axiom $\mathsf{C}(p) \supset p$. The last two condition and the definition of negation imply that the equivalence $\mathsf{C}(p) \equiv p \wedge \neg p$ holds in any C-logic, which justify the name "contradiction operator" for C. Moreover, it turns out that the class **Jhn** can be identified with the class of C-logics satisfying the following axioms:

$$\mathsf{C}(p \wedge q) \equiv \mathsf{C}(p) \wedge q \quad \text{and} \quad \mathsf{C}(p \wedge q) \equiv p \wedge \mathsf{C}(q),$$

which coincides, in turn, with the class of C-logics L satisfying the "paradox of minimal logic":

$$\{\varphi, \neg\varphi\} \vdash_L \psi.$$

[1]The results of Odintsov (2003) was obtained independently of Carnielli & Marcos (2002), but, in fact, the idea of presentation of negation through contradiction operator lies in the setting of the general approach to classification of paraconsistent logics presented in Carnielli & Marcos (2002).

If we rewrite the first of the equivalences above in terms modal operators, we obtain an Ł-theorem $\Box(p \wedge q) \equiv \Box p \wedge q$ equivalent modulo the underlying classical logic to $\Diamond(p \vee q) \equiv \Diamond p \vee q$. Both statements have an obvious paradoxical nature. Indeed, if we accept the conjunction of two conditions as necessary (whatever it means), it means from the intuitive point of view more then stating that the first of these conditions is necessary and the second simply takes place. In a similar way, assuming that it is possible (what ever it means) that one of the two conditions takes place should be weaker than the alternative of one of the conditions and the possibility of the other. So, the unusual, at first glance, analogy between the contradiction and necessity operators goes even further. A more natural modal properties of contradiction operator result in a more satisfiable consequence relation from the point of view of paraconsistent logic.

Further, it was shown in Odintsov (2003) that the same negation may be defined via different operators. There is a sense to distinguish, for example, between the contradiction operator $\mathtt{C}$ that should satisfy the axiom $\mathtt{C}(\varphi) \supset \varphi$ and the more general absurdity operator, for which similar axiom is not obeying.

In the present article, we study in details what does it mean that negation in one or another system can be represented via a contradiction or absurdity operator. We distinguish different forms of such presentations. Further, we take two well known paraconsistent logics, namely, the logic **CLuN** by D. Batens (Batens 1980b; Batens *et al.* 1999) and Sette's maximal paraconsistent logic P^1 (see Pynko 1995; Sette 1973), and investigate the possibility to define negations of these two systems via suitable contradiction or absurdity operators. It will be shown, in particular, that the negation of **CLuN** may be defined via an absurdity operator about which nothing is postulated. In this way **CLuN** may be considered as a counterpart of **Le**, where negation is defined via the constant $\bot$ about which nothing is postulated.

2. Preliminaries

Our considerations will be restricted to the propositional case. We will consider several propositional languages, all of which are extensions of the positive language $\mathcal{L}^+ := \{\supset, \vee, \wedge\}$. The equivalence $\equiv$ is definable as $\varphi \equiv \psi := (\varphi \supset \psi) \wedge (\psi \supset \varphi)$. By **Lp** we denote positive logic and by **Lk**$^+$ classical positive logic (the positive fragment of classical logic; see Rautenberg 1979).

We denote by $\mathcal{L}^\neg$ the language obtained by adding to $\mathcal{L}^+$ the symbol $\neg$ for negation, $\mathcal{L}^\neg = \{\supset, \vee, \wedge, \neg\}$. Two further languages

$$\mathcal{L}^{\mathtt{C}} = \{\supset, \vee, \wedge, \mathtt{C}\} \quad \text{and} \quad \mathcal{L}^{\mathtt{A}} = \{\supset, \vee, \wedge, \mathtt{A}\}$$

are obtained by adding to $\mathcal{L}^+$ symbols $\mathtt{C}$ for a unary contradiction operator and $\mathtt{A}$ for a unary absurdity operator, respectively.

Our consideration will be restricted by the following classes of logics. Let $\diamond \in \{\neg, \mathsf{C}, \mathsf{A}\}$. By a $\diamond$-logic we mean a set of formulæ in the language $\mathcal{L}^{\diamond}$ containing axiom schemes of positive logic and closed under the rules of substitution and *modus ponens*. In case of C-logics, we assume additionally that they have an axiom $\mathsf{C}(p) \supset p$. This distinction is connected with different intuitions behind operators of contradiction and absurdity. By $\mathsf{C}(\varphi)$ we mean a contradiction expressed in terms of φ, i.e., a simultaneous stating φ and its negation $\neg\varphi$, from which follows that $\mathsf{C}(\varphi)$ should imply φ. By $\mathsf{A}(\varphi)$ we mean such statement that reducing φ to it, i.e., proving the implication $\varphi \supset \mathsf{A}(\varphi)$, is enough to negate φ. This understanding of $\mathsf{A}(\varphi)$ do not assume that $\mathsf{A}(\varphi)$ implies φ. The formulæ $\mathsf{A}(\varphi)$ and φ may be incomparable, as it takes place in case of **Lj**, where negation can be represented as a reduction to a constant absurdity operator ($\mathsf{A}(\varphi) \equiv \bot$) and the formulæ φ and $\bot$ are incomparable in general case. In fact, the contradiction operator C can be considered as a special case of an absurdity operator satisfying an additional assumption that $\mathsf{C}(\varphi) \supset \varphi$ for any φ.

The following assertion was proved in Odintsov (2003):

Proposition 2.1. Let L be an A-logic. Define the contradiction operator C as $\mathsf{C}(\varphi) := \varphi \wedge \mathsf{A}(\varphi)$ and the negation $\neg$ as $\neg\varphi := \varphi \supset \mathsf{A}(\varphi)$. The following formulæ are provable in L:

$$\neg p \equiv p \supset \neg p \text{ and } \neg p \equiv p \supset \mathsf{C}(p),$$
$$\mathsf{C}(p) \equiv p \wedge \neg p \text{ and } \mathsf{C}(p) \supset p,$$
$$\mathsf{C}(p) \supset \mathsf{A}(p) \text{ and } \mathsf{A}(p) \supset \neg p.$$

This proposition shows that C defined in terms of the absurdity operator A can really be considered as a contradiction operator corresponding to the negation defined via A and that there is the whole "interval" of absurdity operators, from C to $\neg$, defining the same negation.

With every logic L one can associate in a standard way a consequence relation $\vdash_L$. The relation $X \vdash_L \varphi$ means that one can pass in a finite number of steps to a formula φ using elements of X, theorems of L and the rule of *modus ponens*. For a logic L and a set X of formulæ in the language of L, $L + \{X\}$ denotes the minimal extension of L containing X.

Let $\mathcal{L}_1$ and $\mathcal{L}_2$ be propositional languages, L_1 and L_2 logics in the languages $\mathcal{L}_1$ and $\mathcal{L}_2$, respectively, and let $\tau\colon \mathcal{L}_1 \to \mathcal{L}_2$ and $\iota\colon \mathcal{L}_2 \to \mathcal{L}_1$ be translations from the language $\mathcal{L}_1$ into the language $\mathcal{L}_2$ and visa versa, i.e., simply mappings from the set of $\mathcal{L}_1$-formulæ to the set $\mathcal{L}_2$-formulæ and visa versa. We will say that L_1 is *faithfully embedded into L_2 via τ* if for any $\mathcal{L}_1$-formula φ we have the equivalence

$$L_1 \vdash \varphi \Leftrightarrow L_2 \vdash \tau(\varphi).$$

Further, the logic L_1 is said to be *definitially equivalent to L_2 via translations τ and ι* if L_1 is faithfully embedded into L_2 via τ, L_2 is faithfully embedded into

L_1 via ι, and moreover, for any $\mathcal{L}_1$-formula φ and $\mathcal{L}_2$-formula ψ we have the equivalences

$$L_1 \vdash \varphi \equiv \iota\tau(\varphi) \quad \text{and} \quad L_2 \vdash \psi \equiv \tau\iota(\psi).$$

A translation τ from the language $\mathcal{L}_1$ to the language $\mathcal{L}_2$ *preserves propositional variables* if $\tau(p) = p$ for any propositional variable p and it *preserves an n-ary connective* $*$ if $* \in \mathcal{L}_1 \cap \mathcal{L}_2$ and

$$\tau(*(\varphi_1, \ldots, \varphi_n)) = *(\tau\varphi_1, \ldots \tau\varphi_n)$$

for all formulæ $\varphi_1, \ldots, \varphi_n$.

Now we want to define exactly what does it mean that negation in one or another logic can be defined via absurdity or contradiction operator. We restrict ourselves to the class of $\neg$-logics. To this end we need several translations between languages $\mathcal{L}^\neg$, $\mathcal{L}^{\mathsf{C}}$, and $\mathcal{L}^{\mathsf{A}}$.

Translations $\theta\colon \mathcal{L}^\neg \to \mathcal{L}^{\mathsf{C}}$, $\varrho\colon \mathcal{L}^{\mathsf{C}} \to \mathcal{L}^\neg$, $\xi\colon \mathcal{L}^\neg \to \mathcal{L}^{\mathsf{A}}$, and $\zeta\colon \mathcal{L}^{\mathsf{C}} \to \mathcal{L}^{\mathsf{A}}$ preserve propositional variables and all positive connectives and act on other connectives as follows:

$$\theta(\neg\varphi) = \theta\varphi \supset \mathsf{C}(\theta\varphi) \qquad \varrho(\mathsf{C}(\varphi)) = \varrho\varphi \wedge \neg\varrho\varphi;$$
$$\xi(\neg\varphi) = \xi\varphi \supset \mathsf{A}(\xi\varphi) \qquad \zeta(\mathsf{C}(\varphi)) = \zeta\varphi \wedge \mathsf{A}(\zeta\varphi).$$

Let L be a $\neg$-logic. An A-logic L' such that ξ defines a faithful embedding of L into L' will be called an A-*presentation* of L. A C-*presentation* of L is a C-logic L' such that θ defines a faithful embedding of L into L'.

By an *exact* A-(C-)*presentation* of an $\neg$-logic L we mean its A-(C-) presentation which is the least among all A-(C-)presentations of L.

A *strong* C-*presentation* of L is a C-logic L' definitially equivalent to L via θ and ϱ.

If L is a C-logic, then its A-*presentation* is an A-logic L' such that ζ embeds faithfully L into L', and its *exact* A-*presentation* is the least A-presentation.

First we state some simple properties of the notions introduced.

Proposition 2.2. 1. Let L be an A-(C-)logic. There exists a unique $\neg$-logic L' such that L is an A-(C-)presentation of L'.

2. Let L be an A-logic. There exists a unique C-logic L' such that L is an A-presentation of L'.

Proof. 1. It will be enough to consider the case of A-logics. We define by induction the subset of $\neg$-formulæ in the set of all formulæ in the language $\mathcal{L}^{\mathsf{A}}$:

1. a propositional variable is a $\neg$-formula;

2. if φ and ψ are $\neg$-formulæ then $\varphi \vee \psi$, $\varphi \wedge \psi$, and $\varphi \supset \psi$ are $\neg$-formulæ;
3. if φ is a $\neg$-formula, then $\varphi \supset \mathsf{A}(\varphi)$ is a $\neg$-formula.

For a $\neg$-formula φ, we define a formula $\varphi^{\neg}$ as a result of replacing all subformulæ of the form $\psi \supset \mathsf{A}(\psi)$ by the subformula $\neg\psi$. Note that $\varphi = \zeta(\varphi^{\neg})$ and consider the intersection $\mathcal{S}$ of the logic L with the set of all $\neg$-formulæ. It is clear that $\mathcal{S}$ is closed under the substitution of $\neg$-formulæ and the restricted version of *modus ponens*: if $\varphi, \varphi \supset \psi \in \mathcal{S}$ and ψ has not the form $\mathsf{A}(\varphi)$, then $\psi \in \mathcal{S}$. Therefore, the set

$$(1) \qquad L' := \{\varphi^{\neg} \mid \varphi \in \mathcal{S}\}$$

will be a $\neg$-logic. Check that L is an A-presentation of L'.

If $L' \vdash \varphi$, then $\varphi = \psi^{\neg}$ for a suitable formula $\psi \in \mathcal{S} \subseteq L$. Consequently, $\xi\varphi = \xi(\psi^{\neg}) = \psi \in L$. Conversely, if $L \vdash \xi\varphi$, then $\xi\varphi \in \mathcal{S}$ and $(\xi\varphi)^{\neg} = \varphi$, whence, $L' \vdash \varphi$.

The uniqueness of a $\neg$-logic L', for which L is an A-presentation easily follows from the above considerations. Indeed, every such logic L' must have form (1).

2. We define the set of C-formulæ by replacing Item 3 in the definition of $\neg$-formulæ by

3′. if φ is a C-formula, then $\varphi \wedge \mathsf{A}(\varphi)$ is a C-formula.

The proof of this item is similar to the previous one. The desired C-logic will have the form

$$L' := \{\varphi^{\mathsf{C}} \mid \varphi \in L \text{ and } \varphi \text{ is a } \mathsf{C}\text{-formula}\},$$

where φ^{C} is obtained from φ by replacing every subformula $\psi \wedge \mathsf{A}(\psi)$ by $\mathsf{C}(\psi)$. □

In view of this statement Proposition 4.2 of Odintsov (2003) can be reworded as

Corollary 2.3. Let L be a C-logic. Then

$$\mathsf{C}(p \wedge q) \equiv \mathsf{C}(p) \wedge q,\ \mathsf{C}(p \wedge q) \equiv p \wedge \mathsf{C}(q) \in L$$

iff L is a C-presentation for some extension of **Lj**.

Proposition 2.4. 1. If L is a $\neg$-logic and L has an A-(C-)presentation, then L has also an exact A-(C-)presentation.

2. If L is a C-logic having an A-presentation, it has also an exact A-presentation.

Proof. This statement follows from the next simple lemma.

LEMMA 2.5. Let $\mathcal{L}_0$ and $\mathcal{L}_1$ be propositional languages and $\tau\colon \mathcal{L}_0 \to \mathcal{L}_1$ a translation. Further, let L be a logic in the language $\mathcal{L}_0$ and $\{L_i \mid i \in I\}$ a family of logics in the language $\mathcal{L}_1$ such that τ faithfully embeds L into L_i for all $i \in I$. Then τ faithfully embeds L into $\bigcap_{i \in I} L_i$.

In this way, if the set of A-(C-)presentations of some logic is non-empty, the intersection of all such presentations will give an exact A-(C-)presentation of that logic. □

3. The Case of Extensional Logics

To prove further results we impose the restriction of extensionality on the considered logics. Let L be a $\neg$-(A-, or C-)logic. We say that negation $\neg$ (respectively, absurdity operator A, contradiction operator C) is *extensional* in L if the logic L is closed under the rule

$$\frac{p \equiv p}{\neg p \equiv \neg q}, \quad (\text{resp. } \frac{p \equiv q}{\mathtt{A}(p) \equiv \mathtt{A}(q)}, \quad \text{or} \quad \frac{p \equiv q}{\mathtt{C}(p) \equiv \mathtt{C}(q)}).$$

If L is a $\neg$-(C-, or A-)logic and the operator $\neg$ (C, or, respectively, A) is extensional in L, we will say that L is an *extensional* $\neg$-(C-, or, respectively, A-) logic.

REMARK. Note that in all logics under consideration, positive connectives $\vee$, $\wedge$, and $\supset$ are extensional in the natural sense, which easily follows from the fact that $\neg$-(C-, and A-)logics was defined so as to satisfy all axioms of positive logic **Lp**.

PROPOSITION 3.1. If a $\neg$-logic L has an A-presentation, then we have

$$L \vdash \neg p \equiv p \supset \neg p.$$

Conversely, if the equivalence $\neg p \equiv p \supset \neg p$ can be proved in a $\neg$-logic L and, additionally, L is an extensional $\neg$-logic, then L has an A-presentation.

PROOF. Let a $\neg$-logic L has an A-presentation L'. By the contraction law for $\supset$, we have

$$L' \vdash (p \supset \mathtt{A}(p)) \equiv (p \supset (p \supset \mathtt{A}(p))).$$

This equivalence graphically equals to $\theta(p) \equiv \theta(p \supset \neg p)$ which is equal in turn to $\theta(p \equiv p \supset \neg p)$. By assumption, θ is a faithful embedding of L into L', whence,

$$L \vdash \neg p \equiv p \supset \neg p.$$

Now we assume that L is an extensional $\neg$-logic and $L \vdash \neg p \equiv p \supset \neg p$.

For a formula φ in the language $\mathcal{L}^{\neg}$, we define a formula $\varphi^{\mathtt{A}}$ as a result of replacing every subformula of the form $\neg\psi$ by the subformula $\mathtt{A}(\psi)$. Put

$$L^{\mathtt{A}} := \{\varphi^{\mathtt{A}} \mid \varphi \in L\}.$$

It is clear that $L^{\mathtt{A}}$ is an $\mathtt{A}$-logic, and the extensionality of $\neg$ in L immediately implies that the operator $\mathtt{A}$ is extensional in $L^{\mathtt{A}}$. We check that ξ is a faithful embedding of L into $L^{\mathtt{A}}$. First of all, we note that for any formula φ,

$$L \vdash \varphi \text{ iff } L^{\mathtt{A}} \vdash \varphi^{\mathtt{A}}.$$

Therefore, it is enough to prove that $L^{\mathtt{A}} \vdash \varphi^{\mathtt{A}} \equiv \xi(\varphi)$ for any φ. We use the induction on the structure of formula.

This statement trivially holds for propositional variables. The case of positive connectives easily follows by the extensionality of them. It remains to consider the case $\varphi = \neg\psi$. By the induction hypothesis, we have $\psi^{\mathtt{A}} \equiv \xi(\psi)$ in $L^{\mathtt{A}}$. This fact and the extensionality of $\mathtt{A}$ imply $L^{\mathtt{A}} \vdash \mathtt{A}(\psi^{\mathtt{A}}) \equiv \mathtt{A}(\xi(\psi))$. On the other hand, we have $\mathtt{A}(\xi\psi) \equiv (\xi(\psi) \supset \mathtt{A}(\xi(\psi)))$ since the equivalence $\neg p \equiv (p \supset \neg p)$ holds in L by assumption. The last two fact together imply

$$L^{\mathtt{A}} \vdash \mathtt{A}(\psi^{\mathtt{A}}) \equiv (\xi(\psi) \supset \mathtt{A}(\xi(\psi))),$$

i.e., $L^{\mathtt{A}} \vdash (\neg\psi)^{\mathtt{A}} \equiv \xi(\neg\psi)$. □

PROPOSITION 3.2. 1. If an extensional $\neg$-logic L has an $\mathtt{A}$-presentation, it has also a $\mathtt{C}$-presentation.

2. If an extensional $\neg$-logic L has a $\mathtt{C}$-presentation, it is unique and it will be also an exact and strong $\mathtt{C}$-presentation.

PROOF. 1. The set C of formulæ in the language $\mathcal{L}^{\neg}$ we define as the least set of formulæ satisfying the following conditions:

a) a propositional variable p belongs to C;
b) if $\varphi \in C$ and $\psi \in C$, then the formulæ $\varphi \vee \psi$, $\varphi \wedge \psi$, and $\varphi \supset \psi$ also are in C;
c) if $\varphi \in C$, then $\varphi \wedge \neg\varphi \in C$.

Let $L^C := L \cap C$. The set L^C is closed under the rules of *modus ponens* and substitution of formulæ from C, moreover it contains all formulæ of the form $(\varphi \wedge \neg\varphi) \supset \varphi$ for φ in C. Therefore, the set

$$L' := \{\varphi^{\mathtt{C}} \mid \varphi \in L^C\},$$

where $\varphi^{\mathtt{C}}$ is obtained from φ by substituting a formula $\mathtt{C}(\psi)$ for any subformula of the form $\psi \wedge \neg\psi$, forms a $\mathtt{C}$-logic.

Check that L' is the desired C-presentation of L. First, we note that the equivalence

$$L \vdash \varphi \text{ iff } L' \vdash \varphi^{\mathsf{C}}$$

holds for any $\varphi \in C$. The logic L has an A-presentation by assumption, whence, by Proposition 3.1 we have in L the equivalences

$$\neg p \equiv (p \supset \neg p) \equiv (p \supset (p \wedge \neg p)).$$

Therefore, in L any formula φ is equivalent to a formula φ' obtained by replacing every subformula $\neg\psi$ of φ by $\psi \supset (\psi \wedge \neg\psi)$. Note that φ' belongs to C, moreover, $\theta(\varphi) = \varphi'^{\,\mathsf{C}}$. In this way, we arrive at the following chain of equivalences

$$L \vdash \varphi \Leftrightarrow L \vdash \varphi' \Leftrightarrow L' \vdash \varphi'^{\,\mathsf{C}} \Leftrightarrow L' \vdash \theta\varphi,$$

which completes the proof of this item.

2. It follows easily from the considerations of the previous item that any C-presentation of an extensional $\neg$-logic L coincides with the set $\{\varphi^{\mathsf{C}} \mid \varphi \in L \cap C\}$. Therefore, the logic L may have at most one C-presentation. The exact C-presentation equals to the intersection of all C-presentations, therefore it coincides with the unique C-presentation of L.

It remains to prove that a C-presentation L' of L will be also strong, i.e., that ϱ is a faithful embedding of L' into L and that the translations θ and ϱ are mutually inverse. It is clear that for any formula φ in the language $\mathcal{L}^{\mathsf{C}}$, the formula $\varrho(\varphi)$ belongs to the set C and, conversely, any formula in C has the form $\varrho(\varphi)$. Therefore, the above presentation of L' immediately implies that ϱ is a faithful embedding.

Let φ be a formula in the language $\mathcal{L}^{\neg}$. The equivalence $\varphi \equiv \varrho\theta(\varphi)$ follows from the fact that $\varrho\theta(\varphi)$ graphically equals to the formula φ' defined in the previous item. For a formula φ in the language $\mathcal{L}^{\mathsf{C}}$, the equivalence $\varphi \equiv \theta\varrho(\varphi)$ easily follows from the extensionality of L' and the equivalence $\mathsf{C}(p) \equiv (p \wedge (p \supset \mathsf{C}(p)))$ which holds in any C-logic. □

PROPOSITION 3.3. Let L be an extensional $\neg$-logic, L_1 an extensional A-presentation of L, and L_2 a C-presentation of L. Then L_1 is an A-presentation of L_2.

PROOF. By previous proposition L_2 is a strong C-presentation of L. Therefore, we have for any formula φ,

$$L_2 \vdash \varphi \Leftrightarrow L \vdash \varrho\varphi \Leftrightarrow L_1 \vdash \xi\varrho\varphi.$$

Moreover, $\xi\varrho\,\mathsf{C}(\varphi) = \xi(\varrho\varphi \wedge \neg\varrho\varphi) = \varrho\varphi \wedge (\varrho\varphi \supset \mathsf{A}(\varrho\varphi))$. The latter formula is equivalent in L_1 to $\varrho\varphi \wedge \mathsf{A}(\varrho\varphi)$. This fact and extensionality of L_1 may be combined to yield $L_1 \vdash \xi\varrho\varphi \equiv \zeta\varphi$ for any φ. □

The above results concerning existence of A- and C-presentations, uniqueness of C-presentations, etc., used essentially the extensionality property. Most of well-known systems of paraconsistent logic are not extensional and need, therefore, a special treatment. In the following sections, we take, only as the first examples, two simple systems of paraconsistent logic and consider for these systems the problem of presenting negation via absurdity and contradiction operators.

4. Logic CLuN

All logics considered in this and the following section will have the same positive fragment, namely, $\mathbf{Lk}^+$, and will admit semantical characterization in terms of $\{0, 1\}$-interpretations non-truth-functional for operators of negation, absurdity, or contradiction.

For a language $\mathcal{L}^\diamond$, where $\diamond \in \{\neg, \mathsf{A}, \mathsf{C}\}$, we define an interpretation ν as a mapping from the set of formulæ of $\mathcal{L}^\diamond$ to the set $\{0, 1\}$ satisfying for arbitrary formulæ φ and ψ the following properties:

1^ν. $\nu(\varphi \wedge \psi) = 1$ iff $\nu(\varphi) = 1$ and $\nu(\psi) = 1$;

2^ν. $\nu(\varphi \vee \psi) = 1$ iff $\nu(\varphi) = 1$ or $\nu(\psi) = 1$;

3^ν. $\nu(\varphi \supset \psi) = 1$ iff $\nu(\varphi) = 0$ or $\nu(\psi) = 1$.

The logic **CLuN** was introduced for the first time in Batens (1980b) (under the name *PI*). The name **CLuN** we can understand as classical logic admitting gl*u*ts (see also Batens *et al.* 1999). It is a quite weak paraconsistent logics which is in a sense basic for the class of paraconsistent extensional propositional logics considered by Batens (1980b). The logic **CLuN** can be axiomatized modulo classical positive logic via the law of excluded middle, i.e., we have

$$\mathbf{CLuN} = \mathbf{Lk}^{+\neg} + \{p \vee \neg p\},$$

where $\mathbf{Lk}^{+\neg}$ is an expansion of $\mathbf{Lk}^+$ to the language $\mathcal{L}^\neg$.

We denote by $\mathcal{U}$ the class of interpretations ν satisfying for any φ the property:

U. If $\nu(\varphi) = 0$, then $\nu(\neg\varphi) = 1$.

PROPOSITION 4.1 (Batens 1980b, p. 205). For any formula φ, $\mathbf{CLuN} \vdash \varphi$ iff $\nu(\varphi) = 1$ for all $\nu \in \mathcal{U}$.

The proofs of this statement as well as of all other completeness theorems in this section can be done via a simple modification of a general method developed in Batens (1980a) and, therefore, they will be omitted.

We start to consider A- and C-presentations of the logic **CLuN** with the following observation.

PROPOSITION 4.2. The logic **CLuN** has no strong C-presentation.

PROOF. If **CLuN** have a strong C-presentation, the equivalence $\varphi \equiv \varrho\theta\varphi$ holds in **CLuN** for any φ. Take a formula $\neg\neg p$ and compute

$$\varrho\theta(\neg\neg p) = (p \supset (p \wedge \neg p)) \supset ((p \supset (p \wedge \neg p)) \wedge \neg(p \supset (p \wedge \neg p))) \equiv$$
$$\equiv (p \supset \neg p) \supset \neg(p \supset (p \wedge \neg p)).$$

Let v be a valuation from $\mathcal{U}$ such that $v(p) = 0$, $v(\neg p) = v(\neg\neg p) = 1$, and $v(\neg(p \supset (p \wedge \neg p))) = 0$. There is such valuation in $\mathcal{U}$, because $v(p) = 0$ implies $v(p \supset (p \wedge \neg p)) = 1$. As we can see $v(\varrho\theta(\neg\neg p)) = 0$, and so $v(\neg\neg p \equiv \varrho\theta(\neg\neg p)) = 0$. By the previous proposition we have

$$\mathbf{CLuN} \nvdash \neg\neg p \equiv \varrho\theta(\neg\neg p). \qquad \square$$

However, one can find exact A- and C-presentations of **CLuN**.

Define an A-logic $\mathbf{CLuN}^{\mathsf{A}}$ as an expansion of $\mathbf{Lk}^{+}$ to the language $\mathcal{L}^{\mathsf{A}}$, and let $\mathcal{I}^{\mathsf{A}}$ be the set of all interpretations of the language $\mathcal{L}^{\mathsf{A}}$. We have the following completeness theorem.

PROPOSITION 4.3. For any φ, $\mathbf{CLuN}^{\mathsf{A}} \vdash \varphi$ if and only if $v(\varphi) = 1$ for all $v \in \mathcal{I}^{\mathsf{A}}$.

PROPOSITION 4.4. The logic $\mathbf{CLuN}^{\mathsf{A}}$ is an exact A-presentation of **CLuN**.

PROOF. First we check that for any φ, if $\mathbf{CLuN} \vdash \varphi$, then $\mathbf{CLuN}^{\mathsf{A}} \vdash \xi\varphi$. The logics **CLuN** and $\mathbf{CLuN}^{\mathsf{A}}$ have common classical fragment, therefore, it is enough to check that the formula $\xi(p \vee \neg p) = p \vee (p \supset \mathsf{A}(p))$ is provable in $\mathbf{CLuN}^{\mathsf{A}}$, which is obviously true.

Conversely, assume that for some φ_0, we have $\mathbf{CLuN} \nvdash \varphi_0$. In this case, $v(\varphi_0) = 0$ for a suitable $v \in \mathcal{U}$. Consider an interpretation v' of the language $\mathcal{L}^{\mathsf{A}}$ satisfying the following conditions: $v'(p) = v(p)$ for any propositional variable p; $v'(\mathsf{A}(\xi\psi)) = v(\psi \wedge \neg\psi)$ for any ψ in the language $L^{\neg}$. Obviously, we can construct such an interpretation, and it belongs to the class $\mathcal{I}^{\mathsf{A}}$, which contains all interpretations of the language $\mathcal{L}^{\mathsf{A}}$.

We prove by induction on the length of formula that for all ψ, $v(\psi) = v'(\xi\psi)$. This is obviously true for propositional variables. The case of positive connectives is trivial, and so it remains to consider the case of negation. Assume that $v(\psi) = v'(\xi\psi)$ and calculate

$$v'(\xi(\neg\psi)) = v'(\xi\psi \supset \mathsf{A}(\xi\psi)) = v'(\xi\psi) \supset v'(\mathsf{A}(\xi\psi)) =$$
$$= v\psi \supset v(\psi \wedge \neg\psi) = v(\psi \supset (\psi \wedge \neg\psi)) = v(\neg\psi).$$

The last equation is due to the fact that $\mathbf{CLuN} \vdash \neg p \equiv p \supset (p \wedge \neg p)$. According to the proven equality $v'(\xi\varphi_0)) = v(\varphi_0) = 0$, i.e., $\mathbf{CLuN} \nvdash \xi\varphi_0$. We have thus proved that $\mathbf{CLuN}^{\mathsf{A}}$ is an A-presentation of **CLuN**. Exactness of this embedding immediately follows from the fact that $\mathbf{CLuN}^{\mathsf{A}}$ is the least A-logic with classical positive fragment. Obviously, any A-presentation of **CLuN** must have $\mathbf{Lk}^+$ as its positive fragment. □

Similarly to the case of A-presentation we can take the least C-logic with classical positive fragment as a C-presentation of **CLuN**. Put

$$\mathbf{CLuN}^{\mathsf{C}} := \mathbf{Lk}^{+\mathsf{C}} + \{\mathsf{C}(p) \supset p\},$$

where $\mathbf{Lk}^{+\mathsf{C}}$ is an expansion of $\mathbf{Lk}^+$ to the language $\mathcal{L}^{\mathsf{C}}$. We denote by $\mathcal{U}^{\mathsf{C}}$ the class of all interpretations v of the language $\mathcal{L}^{\mathsf{C}}$ satisfying for any φ the condition:

U^{C}. If $v(\varphi) = 0$, then $v(\mathsf{C}(\varphi)) = 0$.

PROPOSITION 4.5. For any φ, $\mathbf{CLuN}^{\mathsf{C}} \vdash \varphi$ if and only if $v(\varphi) = 1$ for all $v \in \mathcal{U}^{\mathsf{C}}$.

PROPOSITION 4.6. The logic $\mathbf{CLuN}^{\mathsf{C}}$ is an exact C-presentation of **CLuN**.

The proof of this statement may be obtained, in fact, by replacing A by C and ξ by θ in the proof of Proposition 4.4.

Due to Proposition 4.2 the logic $\mathbf{CLuN}^{\mathsf{C}}$ can not be a strong C-presentation, but we can prove that the inverse translation also embeds faithfully $\mathbf{CLuN}^{\mathsf{C}}$ into **CLuN**.

PROPOSITION 4.7. The logic **CLuN** is an exact $\neg$-presentation of $\mathbf{CLuN}^{\mathsf{C}}$.

PROOF. The fact that $\mathbf{CLuN}^{\mathsf{C}} \vdash \varphi$ implies $\mathbf{CLuN} \vdash \varrho\varphi$ for any φ can be checked directly.

Assume that for some formula φ_0, we have $\mathbf{CLuN}^{\mathsf{C}} \nvdash \varphi_0$ and consider an interpretation $v \in \mathcal{U}^{\mathsf{C}}$ with $v(\varphi_0) = 0$. One can find an interpretation v' of the language $\mathcal{L}^{\neg}$ such that $v'(p) = v(p)$ for all propositional variables p, $v'\neg\varrho\psi = v(\psi \supset \mathsf{C}\psi)$ for any ψ, and if φ has not the form $\varrho\psi$ and $v'\varphi = 0$, then $v'(\neg\varphi) = 1$. Using induction on the structure of formula we check that $v'\varrho\psi = v\psi$ for any ψ. Clearly, it is enough to consider the case of C-operation. Let $v'\varrho\psi = v\psi$ for some ψ. Then

$$v'(\varrho\,\mathsf{C}\psi) = v'(\varrho\psi \wedge \neg\varrho\psi) = v'\varrho\psi \wedge v'(\neg\varrho\psi) = v\psi \wedge v(\psi \supset \mathsf{C}\psi) = v(\mathsf{C}\psi).$$

The equality $v'\varrho\psi = v\psi$ implies, on one hand, that $v'\varrho\varphi_0 = 0$. On the other hand, if $v'(\varrho\psi) = 0$, then $v\psi = 0$ and

$$v'\neg\varrho\psi = v(\psi \supset \mathsf{C}\psi) = 1,$$

i.e., $v' \in \mathcal{U}$ and $\mathbf{CLuN} \nvdash \varrho\varphi_0$. We have thus proved that **CLuN** is a C-presentation of $\mathbf{CLuN}^{\mathsf{C}}$.

Further, any C-presentation of $\mathbf{CLuN}^{\mathrm{C}}$ have positive fragment $\mathbf{Lk}^{+}$ and contains the formula $p \vee \neg p$, because this formula is equivalent modulo $\mathbf{Lk}^{+}$ to $p \vee (p \supset (p \wedge \neg p)) = \varrho(p \vee (p \supset \mathrm{C}p))$. Therefore, **CLuN** is the least C-presentation of $\mathbf{CLuN}^{\mathrm{C}}$, and so it is exact. □

5. Sette's Logic P^1

Now we turn to the first example of maximal paraconsistent logic, namely, to the system P^1 suggested by A. Sette in (1973). This is a subsystem of classical logic maximal in the sense that adjoining to P^1 any new classical tautology yields the classical logic. The negation of P^1 is not extensional, however, as we will see, P^1 has even strong C-presentation. Originally P^1 was defined as a three-valued logic with the set of truth-values $\{T^0, T^1, F\}$, where T^0 and F are, in fact, classical truth and falsehood, and T^1 is in a sense a second truth, whose negation is also true. Both truths T^0 and T^1 are distinguished. The matrix for P^1 is $\langle\{T^0, T^1, F\}, \{T^0, T^1\}, \supset, \neg\rangle$, where operations $\supset$ and $\neg$ are defined via the following tables:

$\supset$	T^0	T^1	F
T^0	T^0	T^0	F
T^1	T^0	T^0	F
F	T^0	T^0	T^0

$\neg$	
T^0	F
T^1	T^0
F	T^0

The connectives $\vee$ and $\wedge$ are introduced via the definitions:

$$\varphi \vee \psi := (\varphi \supset \neg\neg\varphi) \supset (\neg\varphi \supset \psi),$$
$$\varphi \wedge \psi := (((\varphi \supset \varphi) \supset \varphi) \supset \neg((\psi \supset \psi) \supset \psi)) \supset \neg(\varphi \supset \neg\psi).$$

Sette (1973) gives the following axiomatics for P^1:

1. $p \supset (q \supset p)$
2. $(p \supset (q \supset r)) \supset ((p \supset q) \supset (p \supset r))$
3. $(\neg p \supset \neg q) \supset ((\neg p \supset \neg\neg q) \supset p))$
4. $(p \supset p) \supset \neg\neg(p \supset q)$

The only inference rules are, as usual, substitution and *modus ponens*. It is clear that P^1 is a subsystem of **Lk**, moreover, the following theorem holds.

THEOREM 5.1 (Sette 1973). For any $\varphi \in \mathbf{Lk}$, if φ is not provable in P^1, then $P^1 + \{\varphi\} = \mathbf{Lk}$.

In Alves & Sette (1996), it was stated that P^1 is equivalent to the system β_2 introduced in Loparić & da Costa (1986). Axiomatics of β_2 includes traditional axioms of classical positive logic and the following axiom schema for negation:

1^β. $(\neg\varphi \supset \psi) \supset ((\neg\varphi \supset \neg\psi) \supset \varphi)$, where ψ is molecular.

From β_2 we can easily pass to an axiomatization of P^1 convenient for our goals.

PROPOSITION 5.2. The logic P^1 is axiomatizable modulo the classical positive logic via the following axiom schemes:

1^S. $\neg\neg\varphi \supset \varphi$

2^S. $(\varphi \supset \psi) \supset ((\varphi \supset \neg\psi) \supset \neg\varphi)$, where ψ is molecular.

PROOF. On one hand, we can use truth-tables to check that schemes 1^S and 2^S hold in P^1. On the other hand, the schema 1^β easily follows from $(\neg\varphi \supset \psi) \supset ((\neg\varphi \supset \neg\psi) \supset \neg\neg\varphi)$, a substitutional case of 2^S, and 1^S. □

Using the maximal property we can easily obtain a semantical characterization of P^1 in terms of $\{0, 1\}$-interpretations. Denote by $\mathcal{P}$ the class of all interpretations v of the language $\mathcal{L}^\neg$ satisfying the properties:

1^P. If $v\varphi = 0$, then $v\neg\varphi = 1$

2^P. If φ is molecular and $v\varphi = 1$, then $v\neg\varphi = 0$.

PROPOSITION 5.3. For any formula φ, $P^1 \vdash \varphi$ if and only if $v\varphi = 1$ for all $v \in \mathcal{P}$.

PROOF. Put $L^* := \{\varphi \mid v\varphi = 1 \text{ for all } v \in \mathcal{P}\}$. We have $P^1 \subseteq L^* \subseteq \mathbf{Lk}$. Indeed, the former inclusion can be verified directly, the latter follows from the fact that all classical interpretations belong to $\mathcal{P}$. Due to Theorem 5.1 one of these inclusions is not proper. In $\mathcal{P}$, there is an interpretation v such that $vp = v\neg p = 1$ and $v\neg\neg p = 0$, which means that $p \supset \neg\neg p \notin L^*$, i.e., $L^* \neq \mathbf{Lk}$. Therefore, $L^* = P^1$. □

We have convenient axiomatics and semantics for P^1 and we may turn to its presentations.

LEMMA 5.4. Let L be an $\mathtt{A}$-presentation of P^1. Then L contains the following schemes:

$1^{\mathtt{A}}$. $\mathtt{A}(\varphi \supset \mathtt{A}\varphi) \supset \varphi$

$2^{\mathtt{A}}$. $(\psi \wedge \mathtt{A}\psi) \supset \varphi$, where ψ is molecular.

PROOF. Consider a ξ-translation of 2^S,

$$(\varphi \supset \psi) \supset ((\varphi \supset (\psi \supset \mathtt{A}\psi)) \supset (\varphi \supset \mathtt{A}\varphi).$$

Modulo $\mathbf{Lk}^+$ it is equivalent to

$$(\varphi \supset (\psi \wedge \mathtt{A}\psi)) \supset (\varphi \supset \mathtt{A}\varphi) \equiv$$
$$\equiv (\varphi \wedge (\varphi \supset (\psi \wedge \mathtt{A}\psi))) \supset \mathtt{A}\varphi \equiv (\varphi \wedge (\psi \wedge \mathtt{A}\psi)) \supset \mathtt{A}\varphi \equiv$$
$$\equiv (\psi \wedge \mathtt{A}\psi) \supset (\varphi \supset \mathtt{A}\varphi).$$

In particular, we have

$$L \vdash (\psi \wedge \mathtt{A}\psi) \supset ((\varphi \supset \mathtt{A}\varphi) \supset \mathtt{A}(\varphi \supset \mathtt{A}\varphi)).$$

At the same time, the formula

$$(2) \qquad ((\varphi \supset \mathtt{A}\varphi) \supset \mathtt{A}(\varphi \supset \mathtt{A}\varphi)) \supset \varphi,$$

a ξ-translation of $\neg\neg\varphi \supset \varphi$, also is provable in L. From the last two formulæ we immediately have

$$L \vdash (\psi \wedge \mathtt{A}\psi) \supset \varphi,$$

where ψ is molecular.

Further, the positive axiom $\mathtt{A}(\varphi \supset \mathtt{A}\varphi) \supset ((\varphi \supset \mathtt{A}\varphi) \supset \mathtt{A}(\varphi \supset \mathtt{A}\varphi))$ together with (2) yields the scheme $1^{\mathtt{A}}$. □

Denote by $P^{\mathtt{A}}$ the $\mathtt{A}$-logic axiomatized modulo $\mathbf{Lk}^+$ by schemes $1^{\mathtt{A}}$ and $2^{\mathtt{A}}$ and by $\mathcal{P}^{\mathtt{A}}$ the class of interpretations ν of the language $\mathcal{L}^{\mathtt{A}}$ satisfying the following condition

$1^{\mathtt{A}\nu}$. If φ is molecular and $\nu\varphi = 1$, then $\nu(\mathtt{A}\varphi) = 0$.

Lemma 5.5. If $P^{\mathtt{A}} \vdash \varphi$, then $\nu\varphi = 1$ for all $\nu \in \mathcal{P}^{\mathtt{A}}$.

The *proof* is by direct verification.

Proposition 5.6. The logic $P^{\mathtt{A}}$ is an exact $\mathtt{A}$-presentation of P^1.

Proof. Let $L^* := \{\varphi \mid \xi\varphi \in P^{\mathtt{A}}\}$. As it was stated in the proof of Lemma 5.4, the ξ-translation of 2^S is equivalent in $P^{\mathtt{A}}$ to a partial case of $2^{\mathtt{A}}$. The formula $\xi(\neg\neg\varphi \supset \varphi)$ has the following quasi-proof in $P^{\mathtt{A}}$:

1) $\mathtt{A}(\varphi \supset \mathtt{A}\varphi) \supset \varphi$, an axiom;

2) $((\varphi \supset \mathtt{A}\varphi) \supset \mathtt{A}(\varphi \supset \mathtt{A}\varphi)) \supset ((\varphi \supset \mathtt{A}\varphi) \supset \varphi)$, by monotonicity of $\supset$ in the second argument from 1);

3) $((\varphi \supset \mathtt{A}\varphi) \supset \varphi) \supset \varphi$, the Peirce law;

4) $((\varphi \supset \mathtt{A}\varphi) \supset \mathtt{A}(\varphi \supset \mathtt{A}\varphi)) \supset \varphi$ $(= \xi(\neg\neg\varphi \supset \varphi))$, by transitivity from 2) and 3).

We have thus proved the inclusion $P^1 \subseteq L^*$.

Let $P^{\mathsf{A}} \vdash \xi\varphi_0$ and ν is an arbitrary classical interpretation. Define and an interpretation ν' of $\mathcal{L}^{\mathsf{A}}$ by the conditions $\nu' p = \nu p$ for any p and $\nu'(\mathsf{A}\varphi) = |1 - \nu\varphi|$ for any φ. It is not hard to see that $\nu' \in \mathcal{P}^{\mathsf{A}}$ and $\nu'\xi\varphi = \nu\varphi$ for any φ. By Lemma 5.5, $\nu\varphi = 1$. And we have in this way $L^* \subseteq \mathbf{Lk}$. This inclusion is proper, because the formula $\xi(p \supset \neg\neg p)$ is not in P^{A}. The last fact can be easily checked using Lemma 5.5. Only one of the inclusions $P^1 \subseteq L^* \subseteq \mathbf{Lk}$ is proper, whence, $P^1 = L^*$.

We have thus proved that P^{A} is an A-presentation of P^1, the exactness immediately follows from Lemma 5.4. □

We turn now to C-presentations of P^1.

LEMMA 5.7. If L is a C-presentation of P^1, then it contains the schema:

1^{C}. $\mathsf{C}\psi \supset \varphi$, where ψ is molecular.

PROOF. By Lemma 5.4 $L \vdash (\psi \wedge \mathsf{C}\psi) \supset \varphi$ for molecular ψ, which is equivalent to 1^{C} in view of the axiom $\mathsf{C}\,p \supset p$ of C-logics. □

Let P^{C} be a C-logic axiomatized modulo $\mathbf{Lk}^+$ by the axiom scheme 1^{C}. We denote by $\mathcal{P}^{\mathsf{C}}$ the class of all interpretations ν of $\mathcal{L}^{\mathsf{C}}$ satisfying the conditions:

$1^{\mathsf{C}\nu}$. If $\nu\varphi = 0$, then $\nu\,\mathsf{C}\varphi = 0$.

$2^{\mathsf{C}\nu}$. If φ is molecular and $\nu\varphi = 1$, then $\nu\,\mathsf{C}\varphi = 0$.

LEMMA 5.8. If $P^{\mathsf{C}} \vdash \varphi$, then $\nu\varphi = 1$ for all $\nu \in \mathcal{P}^{\mathsf{C}}$.

PROPOSITION 5.9. The logic P^{C} is an exact and strong C-presentation of P^1.

PROOF. The fact that P^{C} is an exact presentation can be proved similarly to Proposition 5.6. We can directly check that P^1 is a $\neg$-presentation of P^{C}. It remains to prove that any formula φ is equivalent to $\varrho\theta\varphi$ in P^1 and that ψ is equivalent to $\theta\varrho\psi$ in P^{C}. This can be easily stated using induction on the structure of formulæ and the facts that $P^1 \vdash (\varphi \equiv \psi) \supset (\neg\varphi \equiv \neg\psi)$ for molecular φ and ψ and that $P^{\mathsf{C}} \vdash \mathsf{C}(\varphi) \equiv \mathsf{C}(\psi)$ for molecular φ and ψ. □

REMARK. In conclusion of the section we note that P^1 is not an exact $\neg$-presentation of P^{C}. The least $\neg$-presentation of P^{C} can be defined via the class of interpretations ν satisfying the condition $\nu\varphi = 1 \Rightarrow \nu\neg\varphi = 0$ for molecular φ. And this logic have no C-presentation.

Institute of Mathematics
Department of Mathematical Logic
630090 Novosibirsk
Russia
e-mail: odintsov@math.nsc.ru

REFERENCES

Alves, E. H. and A. M. Sette (1996). On the Equivalence Between Some Systems of Non-classical Logic. *Bull. Sect. Logic* **25** (2), 68–72.

Batens, D. (1980a). A Completeness Proof Method for Extensions of The Implicational Fragment of the Propositional Calculus. *Notre Dame J. Form. Logic* **21**, 509–517.

Batens, D. (1980b). Paraconsistent Extensional Propositional Logics. *Logique et Analyse* **23**, 195–234.

Batens, D., K. De Clercq, N. Kurtonina (1999). Embedding and Interpolation for Some Paralogics: The Propositional Case. *Rep. on Math. Logic* **33**, 29–44.

Bernays, P. (1953). Review of Curry (1952a, 1952b). *J. Symb. Logic* **18**, 266–268 .

Carnielli, W. A. and J. Marcos (2002). A Taxonomy of **C**-systems. In: W. A. Carnielli, M. E. Cognilio and I. M. L. D'Ottaviano (eds.), *Paraconsistency – the Logical Way to the Inconsistent*, pp. 1–94. Lecture Notes in Pure and Applied Mthematics, vol. 228. New York: Marcel Dekker.

da Costa, N. C. A. (1974). On the Theory of Inconsistent Formal Systems. *Notre Dame J. Form. Logic* **15**, 497–510. Reprinted in: *Bull. Sect. Logic* **23** (2) (1994), 98–105.

Curry H. B. (1952a). On the Definition of Negation by a Fixed Proposition in the Inferential Calculus. *J. Symb. Logic* **17**, 98–104.

Curry H. B. (1952b). The System LD. *J. Symb. Logic* **17**, 35–42.

Curry H. B. (1963). *Foundations of Mathematical Logic*. New York: McGrow-Hill Book Company.

Font, J. M. and P. Hajek (2000). Łukasiewicz and Modal Logic. *Math. Preprint Ser.*, no. 278. Univ. de Barselona.

van Heijenoort, J. (1967). *From Frege to Gödel: A Source Book in Mathematical Logic, 1879–1931*. Cambridge: Harvard University Press.

Johansson, I. (1937). Der Minimalkalkül: Ein reduzierter intuitionistischer Formalismus. *Compositio Mathematika* **4**, 119–136.

Kolmogorov, A. N. (1925). On the Principle of Excluded Middle [in Russian]. *Matematicheskij Sbornik* **32**, 646–667. English translation in: Heijenoort (1967), pp. 414–437.

Loparić, A. and N. C. A. da Costa (1986). Paraconsistency, paracompleteness and induction. *Logique et Analyse* **113**, 73–80.

Łukasiewicz, J. (1953). A System of Modal Logic. *The Journal of Computing Systems* **1**, 111–149. Reprinted in: Łukasiewicz (1970), pp. 352–390.

Łukasiewicz, J. (1957). *Aristotle's Syllogistic from the Standpoint of Modern Formal Logic*. 2nd enlarged edition. Oxford: Clarendon Press.

Łukasiewicz, J. (1970). *Selected Works*. Edited by L. Borkowski. *Stud. in Logic and the Found. of Math.* Amsterdam: North-Holland.

Odintsov, S. P. (1998a). Isomorphs of the Logic of Classical Refutability and their Generalizations. *Proceedings of the Seminar of Logical Center*, 48–61. Moscow: Inst. of Philosophy of RAS.

Odintsov, S. P. (1998b). Maximal Paraconsistent Extension of Johansson Logic. *Logique et Analyse* **161–162–163**, 107–120.

Odintsov, S. P. (1999a). Representation of j-algebras and Segerberg's Logics. *Logique et Analyse* **165–166**, 81–106.

Odintsov, S. P. (1999b). Algebraic Semantics and Kripke Semantics for Extensions of Minimal Logic [in Russian]. *Logical Studies*, no. 2 (http://www.logic.ruLogStud/02/LS2.html).

Odintsov, S. P. (2000). Negatively Equivalent Extensions of Minimal Logic and their Logics of Contradictions. *Logical Investigations* **7**, 117–129.

Odintsov, S. P. (2001). Logic of Classical Refutability and Class of Extensions of Minimal Logic. *Logic and Logical Philosophy* **9**, 91–107.

Odintsov, S. P. (2003). "Reductio ad Absurdum" and Łukasiewicz Modalities. *Logic and Logical Philosophy* **11**, 223–240.

Porte, J. (1979). The Ω-system and the L-system of Modal Logic. *Notre Dame J. Formal Logic* **20**, 915–920.

Porte, J. (1984). Łukasiewicz's L-modal System and Classical Refutability. *Logique Anal. Now. Ser.* **27**, 87–92.

Pynko, A. P. (1995). Algebraic Study of Sette'S Maximal Paraconsistent Logic. *Studia Logica* **54**, 89–128.

Rautenberg, W. (1979). *Klassische und nichtclassische Aussagenlogik*. Vieweg: Braunschweig.

Sette, A. M. (1973). On the Propositional Calculus P^1. *Mathematica Japonicæ* **18** (3), 173–180.

Guido Vanackere

A WORLD OF EXPERIENCES, AN ADEQUATE LANGUAGE, AND SELF-REFERENCE REVISED

ABSTRACT. The paper presents a new, intuitive formal language, $\mathcal{L}^{\mathrm{E}}$, that fits in with a world view in which experiences are central entities. It is shown how classical logic and an "objective making" adaptive logic can be applied to formulas of $\mathcal{L}^{\mathrm{E}}$. The latter logic sheds an interesting light on the creation of theories about "the objective world". The paper also contains a small comment on sentences that are not translatable in $\mathcal{L}^{\mathrm{E}}$. In the last section, I revise self-referring sentences by means of their translations in $\mathcal{L}^{\mathrm{E}}$.

1. Introduction

This paper can be read as plea for the following philosophical attitude: "Let us pay more attention to our reality. Language is a medium that we need in order to communicate about our reality. If our best present-day language turns out to be imperfect or inadequate, we better try to create a new language".

1.1. *The Subject-Predicate Form*

Let me start this paper with two quotes from Whitehead's *Process and Reality*. Although I am not a "Whiteheadian", I fully agree with these observations.

> An old established metaphysical system gains a false air of adequate precision from the fact that its words and phrases have passed into current literature. Thus propositions expressed in its language are more easily correlated to our flitting intuitions into metaphysical truth (Whitehead 1978, p. 13).

> This dominance of Aristotelian logic from the late classical period onwards has imposed on metaphysical thought the categories naturally derivative from its phraseology. This dominance of his logic does not seem to have been characteristic of Aristotle's own metaphysical speculations. The divergencies, such as they are, in these lectures from other philosophical doctrines mostly depend upon the fact that many philosophers, who in their explicit statements criticize the Aristotelian notion of "substance", yet implicitly throughout their discussions presuppose that the "subject-predicate" form of proposition embodies the finally adequate mode of statement about the actual world (Whitehead 1978, p. 30).

The world according to classical predicative logic (henceforth **CL**) displays the same structure as the "Aristotelian" proposition. The semantics of **CL** suggests that the world is a domain of elements that meet or do not meet properties, and that stand or do not stand in relations with each other.

In: J. Malinowski and A. Pietruszczak (eds.), *Essays in Logic and Ontology* (*Poznań Studies in the Philosophy of the Sciences and the Humanities*, vol. 91), 243–256. Amsterdam/New York, NY: Rodopi, 2006.

It cannot be denied that the exact sciences — classical mechanics in the first place — strongly influence our world view(s). As it is a widespread assumption that **CL** is the underlying logic of classical mechanics (henceforth **CM**), one may conclude that the world according to **CM** (and presumably also the world according to many of us) is the same as the world according to **CL**. In **CM** and other scientific theories most formulas are of the following form:

$$(\forall x_1)\ldots(\forall x_n)A \qquad (n \geq 1)$$

For instance, if "Mxy", "Vxy" and "Kxy" respectively stand for "the mass of x is y", "the speed of x is y", and "the kinetic energy of x is y", then the definition of kinetic energy of x can be written as:

$$(\forall x_1)(\forall x_2)(\forall x_3)(\forall x_4)((Mx_1x_2 \;\&\; Vx_1x_3 \;\&\; x_4 = 1/2 \times x_2 \times x_3^2) \supset Kx_1x_4)$$

Nevertheless, if one takes a closer look at handbooks of **CM**, one finds out that those scholars do not talk about kinetic energy as a relation between elements of the domain (such as particles or bodies) and arithmetical values. One simply does not find any particle or body in the laws and definitions of **CM**. Instead, one finds expressions as:

$$K = 1/2 \times M \times V^2 \tag{1}$$

This observation suggests that **CM** does not presuppose an atomic world view; **CM**-scholars rather live in a world of forces and movement. Hence, **CM** offers a good argument against the opinion that the subject-predicate form of proposition is the most adequate one.

Admittedly, the subject-predicate form is very useful for people who want to communicate fluently, but this argument in favour of the subject-predicate form may simply be due to the fact that people around here have been using it for ages. If we take into account that the age old use of the subject-predicate form of proposition strongly influences our world view, whereas our world view has an important impact on the way we organize and value our life, we have good reasons to question the hegemony of the subject-predicate form of proposition. In this paper I want to introduce a new formal language, that does not presuppose an atomic world view. It is a language that fits in with a world in which our experiences are important real entities, and in which "objects", "experiences" and "subjects" are interdependent.

1.2. *Our Experiences are Important Real Entities*

How can we give birth to a new language that fits in with the structure of our reality? The first thing to do is to keep an open mind for what is real. We may wonder which objects of our experiences are real beyond doubt, but I think that our experiences themselves are our first real entities. It may be true that the world consists of particles, but our life consists of our experiences. Whenever one observes or conceives a particle, one may doubt whether the particle really exists or not, but one does not doubt whether one observes or conceives. On each of our waking moments, our actual experience is our first real entity. Nobody can deny the reality of the experience she is having right here right now. One's life, at each of her moments, consists of her actual experience. The statement "our experiences are very important real entities" is an understatement.

Hence, we better try to speak about our experiences in an adequate way. How do we talk about our experiences nowadays? We have strange habits. We say "I am happy", "I am sick", or "I don't like it", *i.e.* we talk about the subject of our experiences instead of our experiences. We say "I see a huge tree" or "that tree is huge", if it is our experience to see a huge tree; we say that the water is warm, if it is our experience that the water feels warm: we talk about the object of our experiences instead of our experiences. In the rare occasions in which we use the word "experience" when we are talking about our experiences, our vocabulary is very poor. We say "that was a good experience", or "such a bad experience never again". I think there is more to say about experiences than "good" or "bad", and I propose to develop a new language.

2. Talking About Experiences

Can we use the language of **CL** if we want to talk about our experiences? If so, the question rises whether our everyday experiences are things we refer to by means of individual constants or variables, or events that we refer to by means of formulas. If it is my experience that it is cold, do I mean that my present experience meets some property? Or do I mean that this experience is a fact? I do not think that there is a general answer to the question whether an experience should be referred to by means of an individual or a propositional constant. I think that the ontological distinction between "things" and "events" is not an adequate one when we are talking about our experiences. In my view, this ambivalence of the nature of our experiences is only apparent: it derives from our linguistic and metaphysical habits. If we change these habits, the ambivalence will disappear. The main purpose of this paper is to start a new linguistic habit. Keeping in mind Whitehead's observations, this new habit may be important in view of a revision of our metaphysical habits.

2.1. *Representation and Characterization of Experiences*

In this section, I introduce a new and natural way of talking about our experiences. In order to represent experiences, we need a set Σ of beings who "have experiences", and a set Θ of (very short) periods. Questions about the ontological status of the members of Σ are not important here. We are used to indicate each other with a name, and that is all we need. The elements of Θ can be ordered chronologically.[1] If $s \in \Sigma$ and $i \in \Theta$, then "E_s^i" is the representation of the experience of s on i. We obtain an equivalent notation if we write "$\langle s, i \rangle$". The latter notation shows that these representations of experiences function as coordinates over "the space of our experiences". The expression "E_s^i" (or "$\langle s, i \rangle$") does not contain any information about s's experience on i. Such information can be expressed by means of members of Ψ, the set of characterizations of experiences. The idea is as simple as this: any expression that can be used to characterize an experience is a member of Ψ. I give some examples. The reader immediately notices that this way of putting things, strongly influences our picture of ourselves and our world.

If "S" stands for "being sick", then "$E_s^i : S$" stands for "s's experience on i can be characterized as being sick". In everyday English, we say: "On i, s is sick". Hence, in the world according to the new language, the experience is the real entity, whereas in the world according to the traditional language, s is the real entity.

If "o", "$P(X)$" and "t" respectively stand for "now", "perceiving X", and "that tree over there", then, "$E_s^o : P(t)$" stands for "s's actual experience can be characterized as perceiving that tree over there". In everyday English, we say: "s is perceiving that tree over there". Hence, in the world according to the new language, the act of perceiving is the central real entity, whereas in the world according to the traditional language, s and the tree are the central real entities. Moreover, if s simply says "that tree over there is huge" it sounds as if she is saying something about the objective world, (and not about her experience of "the objective world").

If "$B(A)$" stands for "believing that A", then "$E_{s_2}^i : B(E_{s_1}^i) : s$" stands for: "$s$'s experience on i can be characterized as believing that s_1 is sick on that moment". In everyday English: "(On i, s_2 believes that) s_1 is sick (on that moment)". In the world according to the new language, s's belief has the nature of an experience, whereas in the traditional language, s's belief has the nature of a disposition[2].

If "G" stands for "good", then "$E_s^i : G$" stands for "s's experience on i can be characterized as good". In traditional English we say: "s has a good experience", which suggests that s is the central real entity.

If we extend our language with the classical connectives, we can write complex formulas as: $E_s^i : S \supset \sim E_s^i : G$. Please notice that the first occurrence of

[1] Θ may be infinite, and thus there is a bijection between Θ and the set of integers, that preserves the relation $<$.

[2] Confer Quine & Ullian ([1970] 1978, p. 10): "[B]elieving is a disposition that can linger latent and unobserved".

E^i_s seems to be a proposition (s is sick), whereas the second occurrence seems to be an individual (the experience is good). Hence, if we use the language of **CL**, the translation of $E^i_s : S \supset \sim E^i_s : G$ would be something like $(E^i_s \equiv S\,s) \supset \sim Ge^i_s$ in which "$S\,s$" stands for "s is sick (on i)", and "G" stands for "is good".[3]

We can also make use of quantifiers over the members of Σ and Θ. "$(\forall z)(\forall y)(E^y_z : A \supset E^y_z : B)$" may stand for "if one's experience can be characterized as A, it can also be characterized as B. "$(\forall y)(E^y_s : Z \supset E^y_s : P)$" may stand for: "if s's experience can be characterized as "swimming", it can also be characterized as pleasant".[4]

The language $\mathcal{L}^{\mathrm{E}}$ is defined as follows. The set of formulas of $\mathcal{L}^{\mathrm{E}}$ is the smallest set Δ that fulfills the following conditions:

1. Σ is a set of persons, $\mathcal{V}^\Sigma$ is a set of variables for persons.
2. Θ is a set of periods, $\mathcal{V}^\Theta$ is a set of variables for periods.
3. Ψ is a set of (schematic representations of) characterizations.
4. If $\alpha \in \Sigma \cup \mathcal{V}^\Sigma, \beta \in \Theta \cup \mathcal{V}^\Theta$, and $\psi \in \Psi$, then $E^\beta_\alpha : \psi \in \Delta$.
5. If $A \in \Delta$, then $\sim A \in \Delta$.
6. If $A, B \in \Delta$, then $A \supset B, A \& B, A \vee B, A \equiv B \in \Delta$.
7. If $\alpha \in \mathcal{V}^\Sigma \cup \mathcal{V}^\Theta$, and $A \in \Delta$, then $(\forall\alpha)A, (\exists\alpha)A \in \Delta$.

The set of well formed (closed) formulas is defined in the same way as for **CL**.

We can also extend the language with specifications of subsets of Σ or Θ. For instance, if "o" stands for "now", then $(\forall y)(y \leq o \supset (E^y_s : P(Ce) \supset E^y_s : P(Be)))$ can be read as: "till now s has only seen black crows".

2.2. *We are Always Talking About our Experiences*

I think it is a self-evident philosophical observation that, even if we mean to talk about the objective world, we always talk about our experiences (of the objective world). One does not need to be a solipsist to agree with this. We *conceive that* the world turns around the sun, and we say "the earth turns around the sun". Whenever we perceive a crow, we notice that it is black, and we say that crows are black. Whatever we say about the world is said about our experience of the world.

Sometimes it is not clear which experiences are expressed by our utterances. If one says "donkeys are mammals", one may mean to say "I know that donkeys

[3]In view of the fact that the two occurrences of E^i_s have different grammatical functions, **CL** would demand to use different symbols: E^i_s and e^i_s.

[4]"s likes swimming", for short.

are mammals", or "If your experience can be characterized as perceiving or conceiving a donkey, it can also be characterized as dealing with a mammal", or "whenever one has examined a donkey, one perceived it was mammal". This example indicates that $\mathcal{L}^{\mathrm{E}}$is more precise than the language of **CL**: the **CL**-formula $(\forall x)(Dx \supset Mx)$ can be translated in at least three ways:

(i) $E^i_s : K((\forall x)(Dx \supset Mx))$,

(ii) $(\forall z)(\forall y)(E^y_z : P(D) \supset E^y_z : P(M))$, or

(iii) $(\forall z)(\forall y)(y \leq o \supset ((E^y_z : P(D) \supset E^y_z : P(M)))$

Another interesting example. If "g" stands for "I", then the classical formula "Sg" (I am sick) can be translated as:

(i) $E^o_g : S$, or

(ii) $(\forall z)(E^o_z : O(g) \supset E^o_z : N(Sg))$.

(i) can be read as: "my actual experience can be characterized as being sick"; (ii) can be read as: "if someone observes me now, he will notice that I am sick".

Some of the above examples show that we can translate our beliefs about the world into formulas of the language $\mathcal{L}^{\mathrm{E}}$. We simply have to focus on experiences that can be characterized as "believing that ...". Of course, the expressive power of $\mathcal{L}^{\mathrm{E}}$ is much stronger than that. It allows to say whatever one wants to say about experiences.

Expressive power is one thing. Another thing is that $\mathcal{L}^{\mathrm{E}}$ is very useful in reasoning processes. In Section 3, I show how we can apply **CL** to formulas of $\mathcal{L}^{\mathrm{E}}$; this way we can make use of the full expressive power of $\mathcal{L}^{\mathrm{E}}$. In Section 4, I show how the language $\mathcal{L}^{\mathrm{E}}$ sheds a new light on the construction of collective theories about the ''objective" world; in this case, we need to restrict our characterizations to expressions of the form "knowing that A", "conceiving that A", or "perceiving that A".

In Section 5 "our utterances are expressions of experiences" is treated as a norm for our natural language. In the last section, this norm (the language razor) is applied to self-referring sentences.

3. CL and $\mathcal{L}^{\mathrm{E}}$

It is easily seen that any formula of the form "$E^i_s : A$" has a unique translation in the language of **CL**, *viz.* "Asi", in which A is a predicate of rank 2, and s and i are individual constants.[5] Given this translation of primitive formulas of $\mathcal{L}^{\mathrm{E}}$, the

[5] Of course, if "$E^i_s : A$" is the formalization in $\mathcal{L}^{\mathrm{E}}$ of some sentence α in natural language, then the formula "Asi" is not the standard **CL**-formalization of α; see, *e.g.*, the examples in Section 2.1.

reader can easily verify that every formula of $\mathcal{L}^{\mathrm{E}}$ has a unique translation in the language of **CL**. The translation of "$(\forall z)(\forall y)(E_z^y : A \supset E_z^y : B)$", for instance, is "$(\forall z)(\forall y)(Azy \supset Bzy)$". This translation suffices to show that we can apply **CL** to the formulas of $\mathcal{L}^{\mathrm{E}}$.

4. $\mathcal{L}^{\mathrm{E}}$, OAL, and Theories About the World

In the previous sections, it is sufficiently shown that the language $\mathcal{L}^{\mathrm{E}}$ is useful if we want to talk about our experiences. Still, our actual scientific theories are theories about the object of our experiences. It is not my purpose to throw away these theories. To the contrary. In this section I want to show that $\mathcal{L}^{\mathrm{E}}$, combined with a very intuitive ambiguity-adaptive logic, allows us to reconstruct any scientific theory. This reconstruction sheds an interesting light on the construction of new theories.[6]

To begin with, we restrict the set Ψ of characterizations to formulas of the form "knowing, conceiving or perceiving that A", formalized as "$X(A)$", in which A is a formula of **CL**. Next, we restrict the formulas of $\mathcal{L}^{\mathrm{E}}$ to the primitive formulas of the form $E_s^i : X(A)$. For our present purposes, we rewrite a formula of this form as follows: "$A^{\langle s,i \rangle}$". Finally, we interpret the latter formula as an equivalent for the formula A in which every non-logical constant C (henceforth NLC) is replaced by $C^{\langle s,i \rangle}$. E.g., $(\forall x)(Pxa\&q) \supset Rx)^{\langle s,i \rangle}$ stands for $(\forall x)(P^{\langle s,i \rangle}xa^{\langle s,i \rangle}\&q^{\langle s,i \rangle}) \supset R^{\langle s,i \rangle}x)$.

Furthermore, we need a simple adaptive logic, *viz.* the ''objective making" adaptive logic **OAL**, which is nothing but a simplified version of the ambiguity-adaptive logic **ACL2**, that was presented in Vanackere (1997). In this section, I briefly introduce the semantics and the proof-theory of **OAL**,[7] and I briefly illustrate how we obtain "objective" statements by means of $\mathcal{L}^{\mathrm{E}}$ and **OAL**.

In order to apply **OAL**, we need a set Γ of premises of the form $A^{\langle s,i \rangle}$. The superscript depends on the experience that was characterized as $X(A)$. **CL** applied to such a set Γ, is the lower-limit logic of **OAL**: if $\Gamma \vdash_{\mathbf{CL}} A$, then $\Gamma \vdash_{\mathbf{OAL}} A$. *Moreover*, for every NLC $C^{\langle s,i \rangle}$ occurring in some $A^{\langle s,i \rangle} \in \Gamma$, the logic **OAL** assumes that $C^{\langle s,i \rangle}$ can be replaced by C, *unless and until a classical inconsistency derives from this assumption.*

Let Γ^o be obtained from Γ by deleting all superscripts occurring in its formulas. The upper-limit logic of **OAL** is **CL** applied to Γ^o. If Γ^o is consistent,

[6]The philosophical idea behind this reconstruction is borrowed from Husserl. See *e.g.* his Husser (1936, 1977). Objects get their objective character when the expression of someone's experience ''of this object" is no longer read as a statement about an experience but as a statement about the objective world. Admittedly, what I am doing here is an oversimplification of what Husserl had in mind. Nevertheless — or thanks to this oversimplification — it is very useful.

[7]For more information about ambiguity-adaptive logics, see Vanackere (1997).

then $\Gamma \vdash_{\mathbf{OAL}} A$ if and only if $\Gamma^o \vdash_{\mathbf{CL}} A$. If Γ^o is inconsistent, then $Cn_{\mathbf{OAL}}(\Gamma)$ constitutes an interpretation of the premises that is as "objective" as possible.

Let $\langle \Sigma, \Theta \rangle = \{\langle z, y \rangle \mid z \in \Sigma, y \in \Theta\}$. The set of ambiguities $\mathcal{A}$ is the smallest set that fulfils these conditions, for all $l \in \langle \Sigma, \Theta \rangle$:

1. For every propositional constant P : $\sim(P \equiv P^l) \in \mathcal{A}$.
2. For every predicate of rank r π: $\sim(\forall \alpha_1)...(\forall \alpha_r)(\pi \alpha_1 ... \alpha_r \equiv \pi^l \alpha_1 ... \alpha_r) \in \mathcal{A}$.
3. For every individual constant β: $\sim(\beta = \beta^l) \in \mathcal{A}$.

Let $?P^l$ be $\sim(P \equiv P^l)$, let $?\pi^l$ be $\sim (\forall \alpha_1)...(\forall \alpha_r)(\pi \alpha_1 ... \alpha_r \equiv \pi^l \alpha_1 ... \alpha_r)$, and let $?\beta^l$ be $\sim(\beta = \beta^l)$. Where $?C_1, ..., ?C_n \in \mathcal{A}$, $\mathsf{DA}(C_1, ..., C_n)$ is called a DA-formula and refers to a disjunction of ambiguities. The NLCs $C_1, ..., C_n$ are called the factors of $\mathsf{DA}(C_1, ..., C_n)$. In view of the commutativity of the disjunction, a permutation of $\mathsf{DA}(C_1, ..., C_n)$ results in $n!$ equivalent disjunctions. Therefore we will consider sets of factors, and use the notation $\mathsf{DA}\{C_1, ..., C_n\}$.

Definition 1. A DA-formula A is a DA-*consequence of* Γ if and only if $\Gamma \vdash_{\mathbf{CL}} A$.

4.1. *Semantics of* **OAL**

Definition 2. Where $\mathfrak{M}$ is a **CL**-model, $\mathsf{AC}(\mathfrak{M}) = \{C \mid ?C \in \mathcal{A} \text{ and } v_{\mathfrak{M}}(?C) = 1\}$.

Definition 3. A **CL**-model $\mathfrak{M}$ is *minimally abnormal with respect to* Γ if and only if $\mathfrak{M}$ is a **CL**-model of Γ, and there is no **CL**-model $\mathfrak{M}'$ of Γ such that $\mathsf{AC}(\mathfrak{M}') \subset \mathsf{AC}(\mathfrak{M})$.

Definition 4. $\mathfrak{M}$ is *an* **OAL**-*model of* Γ if and only if $\mathfrak{M}$ is minimally abnormal with respect to Γ.

Definition 5. $\Gamma \models_{\mathbf{OAL}} A$ if and only if A is true in all **OAL**-models of Γ.

4.2. *Proof Theory*

The format of **OAL**-proofs is obtained from the format of **CL**-proofs, by adding a fifth element to each line. Each line of a proof consists of five elements:

(i) a line number,
(ii) the formula derived,
(iii) the line numbers of the formulas from which (ii) is derived,
(iv) the rule of inference that justifies the derivation, and
(v) the NLC on the non-ambiguous behaviour of which we rely in order for (ii) to be derivable by (iv) from the formulas of the lines enumerated in (iii).

Every formula that is **CL**-derivable from Γ, is also **OAL**-derivable from Γ; the fifth element of these lines remains empty. Whenever a formula of the form $A \vee \mathsf{DA}\{C_1, ..., C_n\}$ is **CL**-derived from Δ, we can derive a new line in an **OAL**-proof with A as second element and $\{C_1, ..., C_n\}$ as fifth element. When such line is used in the derivation of new lines, all NLCs occurring in its fifth element have to occur in the fifth element of the new line.

The formulas in the fifth element of a line function as conditions on which that line is derived. If the conditions are overruled, the line has to be marked. In actual proofs a line is marked by writing the symbol "†" before the line number. A marked line does not belong to the proof. As a result, formulas *derived* at some stage of a proof, are not in general *finally derivable*, because the line in which they occur can be marked at a later stage. After each step, the marks are updated, (removed or added). Of course each set of premises must (and will) have a unique set of final **OAL**-consequences.

The role of DA-formulas is crucial in **OAL**-proofs. Clearly, when a DA-formula occurs unconditionally at some line of a proof, at least one of the disjuncts of that DA-formula is true. Some DA-formulas occurring unconditionally in a proof may be disregarded. Where Σ and Θ are sets of NLC, let us stipulate that

DEFINITION 6. $\mathsf{DA}(\Sigma)$ is a *minimal* DA-*formula at a stage of a proof* if and only if it occurs unconditionally in the proof at that stage and there is no $\Theta \subset \Sigma$ for which $\mathsf{DA}(\Theta)$ occurs unconditionally in the proof at that stage.

DEFINITION 7. *A occurs unconditionally* at some line of a proof if and only if the fifth element of that line is empty.

Let Φ^*_s be the set of all sets that contain one factor out of each minimal DA-formula at stage s of the proof. Φ^*_s may contain redundant elements: the same factor may occur in different minimal DA-formulas. If $\mathsf{DA}\{C_1, C_2\}$ and $\mathsf{DA}\{C_1, C_3\}$ are minimal DA-formulas, then $\Phi^*_s = \{\{C_1\}, \{C_1, C_2\}, \{C_1, C_3\}, \{C_2, C_3\}\}$. Of these $\{C_1, C_2\}$ and $\{C_1, C_3\}$ are redundant. Both $\mathsf{DA}\{C_1, C_2\}$ and $\mathsf{DA}\{C_1, C_3\}$ are true if $?C_1$ is true; there is no need that also $?C_2$ and $?C_3$ be true. So, let Φ_s be obtained from Φ^*_s by eliminating elements from it that are proper supersets of other elements. Hence, the members of Φ_s are sets of formulas, such that, if all members of such a set are true, then all DA-formulas that occur unconditionally in the proof at stage s are true.

Where Φ_s is as defined above and A is the second element of line j, line j *fulfils the integrity criterion at stage s* if and only if (*i*) the intersection of some member of Φ_s and of the fifth element of line j is empty, and (*ii*) for each $\varphi \in \Phi_s$ there is a line k such that the intersection of φ and of the fifth element of line k is empty and A is the second element of line k.

These are the **OAL**-rules:

RU (*unconditional rules*) All inference rules valid in **CL**, are also valid in **OAL**. The fifth element of the new line is the union of the fifth elements of the lines mentioned in its third element.

RC (*conditional rule*) If $\mathsf{DA}\{C_1, ..., C_m\} \vee A$ occurs as the second element of a line of a proof, with Δ as fifth element, then add a new line with A as second element, and $\Delta \cup \{C_1, ..., C_m\}$ as fifth element.

RQ A line is marked (with †) at a stage of a proof if and only if it does not fulfil the integrity criterion.

DEFINITION 8. A is *finally derived* at some line in an **OAL**-proof if and only if it is the second element of that line and any (possibly infinite) extension of the proof can be further extended in such way that the line is unmarked.

DEFINITION 9. $\Gamma \vdash_{\mathbf{OAL}} A$ (A is **OAL**-*finally derivable* from Γ) if and only if A is finally derived at some line of an **OAL**-proof from Γ.

I give an example of a **OAL**-proof. Let $s_1, s_2, s_3, s_4 \in \Sigma$, $i_1, i_2, i_3, i_4 \in \Theta$. Suppose the experience of s_1 on i_1 can be characterized as conceiving that $(\forall x)(Ax \supset Bx)$. Suppose the experience of s_2 on i_2 can be characterized as perceiving that Ac. Suppose the experience of s_3 on i_3 can be characterized as conceiving that $(\forall x)(Ax \supset Dx)$. Suppose the experience of s_4 on i_4 can be characterized as perceiving that Ae. Let $\langle s_l, i_l\rangle = l$ for $l \in \{1, 2, 3, 4\}$. Hence we obtain the set of premises $\Gamma = \{(\forall x)(Ax \supset Bx)^1, Ac^2, (\forall x)(Ax \supset Dx)^3, Ae^4, \}$.

1.	$(\forall x)(A^1x \supset B^1x)$	Premise	$\emptyset$
2.	A^2c^2	Premise	$\emptyset$
3.	$(\forall x)(A^3x \supset D^3x)$	Premise	$\emptyset$
4.	A^4e^e	Premise	$\emptyset$
5.	$(\forall x)(Ax \supset Bx)$	1; RC	$\{A^1, B^1\}$
6.	Ac	2; RC	$\{A^2, c^2\}$
7.	$(\forall x)(Ax \supset Dx)$	3; RC	$\{A^3, D^3\}$
8.	Ae	4; RC	$\{A^4, e^4\}$
9.	$Ac \supset Bc$	5; RU	$\{A^1, B^1\}$
10.	Bc	9,6; RU	$\{A^1, B^1, A^2, c^2\}$
11.	$Ae \supset Be$	5; RU	$\{A^1, B^1\}$
12.	Be	11,8; RU	$\{A^1, B^1, A^4, e^2\}$
13.	$Ac \supset Dc$	7; RU	$\{A^3, D^3\}$
14.	Dc	13,6; RU	$\{A^3, D^3, A^2, c^2\}$
15.	$Ae \supset De$	7; RU	$\{A^3, D^1\}$
16.	De	15,8; RU	$\{A^3, D^3, A^4, e^4\}$

From these premises, Bc, Be, Dc and De are finally derived, respectively in lines 10, 12, 14 and 16.

Let also $s_5 \in \Sigma$, and $i_5 \in \Theta$. Suppose the experience of s_5 on i_5 can be characterized as perceiving that $\sim Bc$. Let $\langle s_5, i_5 \rangle = 5$. Hence we obtain the set of premises $\Gamma' = \Gamma \cup \{\sim Bc^5\}$. We revise the proof from line 5 on.

†	5.	$(\forall x)(Ax \supset Bx)$	1; RC	$\{A^1, B^1\}$
†	6.	Ac	2; RC	$\{A^2, c^2\}$
	7.	$(\forall x)(Ax \supset Dx)$	3; RC	$\{A^3, D^3\}$
	8.	Ae	4; RC	$\{A^4, e^4\}$
†	9.	$Ac \supset Bc$	5; RU	$\{A^1, B^1\}$
†	10.	Bc	9,6; RU	$\{A^1, B^1, A^2, c^2\}$
†	11.	$Ae \supset Be$	5; RU	$\{A^1, B^1\}$
†	12.	Be	11,8; RU	$\{A^1, B^1, A^4, e^4\}$
	13.	$Ac \supset Dc$	7; RU	$\{A^3, D^3\}$
†	14.	Dc	13,6; RU	$\{A^3, D^3, A^2, c^2\}$
	15.	$Ae \supset De$	7; RU	$\{A^3, D^1\}$
	16.	De	15,8; RU	$\{A^3, D^3, A^4, e^4\}$
	17.	$\sim B^5c^5$	Premise	$\emptyset$
†	18.	$\sim Bc$	17; RC	$\{B^5, c^5\}$
†	19.	$\perp$	10,17; RU	$\{A^1, B^1, A^2, c^2, B^5, c^5\}$
	20.	$\mathsf{DA}\{A^1, B^1, A^2, c^2, B^5, c^5\}$	11; EC[8]	$\emptyset$

At stage 20 of the proof, every line with a factor of $\mathsf{DA}\{A^1, B^1, A^2, c^2, B^4, c^4\}$ in its fifth element is marked. Hence *Bc*, *Be* and *Dc* are no longer derivable from Γ'. Still, *De* in line 16 is finally derived.

This simple example illustrates that the language $\mathcal{L}^{\mathrm{E}}$ is very useful for the construction of theories "about the world". If we are interested in statements about the world, we have to look at the **OAL**-consequences without superscripts. If the premises (without superscripts) are consistent, we obtain all classical results. If the premises (without superscripts) are inconsistent, the set of **OAL**-consequences is richer than (or in border cases, equal to) the set of **CL**-consequences.[9]

5. The Language Razor

The language $\mathcal{L}^{\mathrm{E}}$ generates an interesting exercise, *viz.*: for any sentence *A* (in natural language) try to discover the experiences of which *A* is an expression. I am strongly convinced that a sentence in natural language, that cannot be translated in $\mathcal{L}^{\mathrm{E}}$, *i.e.* that cannot be interpreted as the expression of someone's experiences, is an absurd sentence. I cannot see what would be the point of saying something that has nothing to do with any of anybody's experiences.

[8] EC, (elimination of conditions) is a derivable rule: from a line with $\perp$ as second element and Δ as fifth element, derive a line with $\mathsf{DA}(\Delta)$ as second element, and $\emptyset$ as fifth element.

[9] People who want to read more about the role of ambiguities in the construction of are referred to Vanackere (2001).

So let us clean up our languages, and weed up absurd sentences by means of the following "razor".

> *A sentence that cannot be interpreted as the expression of one or more experiences, is nonsense. (If a sentence is nonsense, we should not pay too much attention to it.)*

In other words: if a sentence cannot be translated into the language $\mathcal{L}^{\mathrm{E}}$, we do not have to pay attention to it. Please take into account that not a single statement, uttered by someone who wants to say something about our reality, will be weeded out.

6. Self-reference Revised

The razor presented in Section 5, sheds an interesting light on paradoxes of self-reference. I only consider statements of the form "this sentence meets predicate P". First, let us pay attention to their ambivalent grammatical function. Obviously, a self-referring sentence is a sentence. At the same time, a self-referring sentence is the subject of a sentence, at least if we identify the self-referring sentence with "this sentence".[10] A self-referring sentence can be formalized as "$S \equiv P(S)$" or "$S = P(S)$". Notice the similarities between these formalizations and, *e.g.*, $E_s^i : Aw(E_s^i)$, the formalization of "s's experience on i can be characterized as being aware of that experience". Let us examine the self-awareness of s's experience on i. Can an experience possibly be characterized as being aware of itself? I think so. If there is no awareness, there is no experience. And if there were no immediate awareness of our experience-right-here-right-now, we would never be able to think of any experience explicitly. I would even say that every experience is aware of itself "by definition". The important thing about this immediate awareness, is that it is not mediated by any medium; there are no words involved in it, it is *implicit* awareness. On $i \in \Theta$, I cannot express the experience that I have on that very moment i.

Let us now take a closer look at self-referring sentences, and the experiences they are expressing.

- "This sentence is written in black ink". This is not a problematic sentence. It says that whenever some person is perceiving this sentence, she will notice that it is written in black ink.
- "This sentence is written in red ink". This sentence expresses that whenever some person is perceiving this sentence, she will notice that it is written in red ink. It is a false sentence, but it is not a problematic one.
- "This sentence is true" is a problematic one, and so is "this sentence is false".

---

[10] If we did not make this identification, there would not be self-reference.

If one says that sentence A is true, one expresses that she is believing the sentence true. And if one says that sentence B is false, one expresses that she believes the negation of that sentence true. Whether one believes A true, depends on the meaning she dedicates to A. If we use the language razor, the meaning of A should have something to do with someone's experience(s). So, suppose this sentence is an expression of some experience(s) of some $g \in \Sigma$. A straightforward interpretation is: some experience of g can be characterized as uttering "this sentence is true". If g is just uttering this sentence, *i.e.* if g does not have the intention to express some experience by means of this sentence, our language razor tells us not to pay attention to it. The only no-nonsense interpretation I can think of, is the following.[11] The sentence "this sentence is true" is the written or spoken counterpart of the thought "this experience is true". Clearly, as we are dealing with a case of self-reference, "this experience is true" expresses something about this experience. How do we characterize this self-referring experience? At least, there is immediate awareness involved in this experience: the awareness of this experience, the awareness of this experience being true. We already observed that this kind of immediate awareness is implicit; hence the *utterance* "this experience is true" is not involved in it. From this observation, we have to conclude that either (*i*) this kind of self-reference is not about our experiences (and hence is non-sense), or (*ii*) the sentence "this sentence is true" is the explicit expression of an implicit experience.

Still, we are curious about what a person who utters "this experience is true" or "this experience is false" has in mind.

Suppose it is freezing cold, whereas s's actual experience can be characterized as feeling hot. In such a situation, we simply conclude that there is something wrong with the way s is perceiving the temperature. In such a situation, "s's experience is false" is a correct and meaningful statement.[12] In general, "s's experience (of X) is false" can be interpreted as "there is something wrong with the way s is experiencing (X). "s's experience (of X) is true" can be interpreted as "s's experience (of X) is alright". Hence, "this sentence is true", uttered by s, can be interpreted as "s's has an immediate awareness of her actual experience, which "implicitly tells her" that the way she is experiencing is alright".[13] Analogously, "this experience is false" can be interpreted as "s has an immediate awareness of her actual experience, which "implicitly tells her" that there is something wrong with the way she is experiencing".[14]

---

[11]If the reader thinks of some other no-nonsense interpretation, I will be glad to hear it.

[12]"This sentence is false" does not mean that it is not true that s is feeling hot. I think we should respect her experience as it is, because it is s's reality on that moment.

[13]Perhaps, we are confronted here with an especially good experience, or with a self-assuring experience.

[14]I am not sure what the deeper meaning of such an experience could be, but I think that we are dealing here with a personal problem of s, and not with some central philosophical issue. If one

A last remark. When philosophers utter the sentence "this sentence is false", they usually mean to show that the language in which "this sentence is false" is a grammatically correct sentence, does not meet our classical standards of perfectness. When this imperfectness becomes an obstacle, I think it is time to keep a new and open mind for the part of reality we want to talk about.

University of Ghent
Centre for Logic and Philosophy of Science
Blandijnberg 2
B-9000 Ghent
Belgium
e-mail: Guido.Vanackere@UGent.be

REFERENCES

Husserl, E. (1936). *Vom Ursprung der Geometrie*.

Husserl, E. (1977). *Over de oorsprong van de meetkunde*. Vertaald door J. Duytschaever. Ingeleid en geannoteerd door Rudolf Boehm. Dixit.

Quine, W. V. and J. S. Ullian ([1970] 1978). *The Web of Belief*. 2nd edition. New York: McGraw-Hill, Inc.

Vanackere, G. (1997). Ambiguity-Adaptive Logic. *Logique et Analyse* **159**, 261–280 (appeared 1999).

Vanackere, G. (2001). The Role of Ambiguities in the Construction of Collective Theories. *Logique et Analyse* **173–175**, 189-214. Appeared 2003.

Whitehead, A. N. (1978). *Process and Reality: An Essay in Cosmology*. Corrected edition by David Ray Griffin and Donald W. Sherburne. New York–London: The Free Press.

just means to say that "this sentence is false" is a grammatically correct sentence, then I think she is right, but please take into a account that there are a lot of grammatically correct sentences that do not express any experience, (*e.g.* "Seventeen is swimming across saturday's ashes"), unless some weird poet is uttering them.

Sonja Smets

FROM INTUITIONISTIC LOGIC TO DYNAMIC OPERATIONAL QUANTUM LOGIC*

ABSTRACT. Research within the operational approach to the logical foundations of physics has recently pointed out a new perspective in which quantum logic can be viewed as an intuitionistic logic with an additional operator to capture its essential, i.e., non-distributive, properties. In this paper we will offer an introduction to this approach. We will focus further on why quantum logic has an inherent dynamic nature which is captured in the meaning of "orthomodularity" and on how it motivates physically the introduction of dynamic implication operators, each for which a deduction theorem holds with respect to a dynamic conjunction. As such we can offer a positive answer to the many who pondered about whether quantum logic should really be called a logic. Doubts to answer the question positively were in first instance due to the former lack of an implication connective which satisfies the deduction theorem within quantum logic.

1. Introduction

In his *Grundlagen der Quantenmechanik* (1932), J. von Neumann introduces the notion of a logical calculus based on the relation between the properties of a physical system and projections defined on a Hilbert Space. The short passage in von Neumann's book and the paper he wrote together with G. Birkhoff in 1936 form the birth of quantum logic. In Birkhoff & von Neumann (1936), their aim was to discover the logical structure in physical theories like quantum mechanics, which do not conform to classical logic. The conclusion they obtain is that, "based on admittedly heuristic arguments, one can reasonably expect to find a calculus of propositions which is formally indistinguishable from the calculus of linear subspaces with respect to *set products*, *linear sums*, and *orthogonal complements* — and resembles the usual calculus of propositions with respect to *and*, *or*, and *not*" (Birkhoff & von Neumann 1936, p. 1). In particular they point out that by entering the field of logic through physical and technical reasoning, it is not negation but distributivity of conjunction and disjunction which forms the weakest link in the algebra of logic. The following way in which this failure of distributivity is expressed by Birkhoff and von Neumann can be seen as a starting point to understand the perspective on quantum logic which we present below:

> Our conclusion agrees perhaps more with those critiques of logic, which find most objectionable the assumption that $a' \cup b = \square$ implies $a \subset b$ (or dually, the assumption that

*We dedicate this paper to Jerzy Perzanowski in honor of his sixtieth anniversary.

In: J. Malinowski and A. Pietruszczak (eds.), *Essays in Logic and Ontology* (*Poznań Studies in the Philosophy of the Sciences and the Humanities*, vol. 91), 257–275. Amsterdam/New York, NY: Rodopi, 2006.

> $a \cap b' = \circ$ implies $b \supset a$ — the assumption that to deduce an absurdity from the conjunction of a and not b, justifies one in inferring that a implies b) (Birkhoff & von Neumann 1936, p. 17).

where they use $\circ$ to denote the "absurd proposition" and $\square$ for the "self-evident proposition". Since this implies immediately that a material implication $a \to b = a' \vee b$ satisfying the strengthened law of entailment ($a \leq b$ iff $a \to b = 1$) is lacking in quantum logic, we are right at the core of the so called implication problem. In a more general framework it is the implication adjunction, $a \wedge c \leq b$ iff $c \leq a \to b$, which fails and of which it is known that it goes hand in hand with the distributivity of conjunction and disjunction in classical and intuitionistic logic. The question then arises whether this implication adjunction can be replaced by another one, motivated on physical grounds. As pointed out in Coecke & Smets (2004), the solution lies in the nature of the orthomodular identity, which replaces the distributive identities, within quantum logic. Due to the occurrence of so-called superposition properties when dealing with non-classical physical systems we have to cope with situations in which "$a \vee b$" expresses a true proposition while this does not imply that a is true or b is true. It are these situations which are responsible for the failure of the distributive identities, i.e. while $(a \wedge b) \vee (a \wedge c) \leq a \wedge (b \vee c)$, the converse fails. The orthomodular identity, on which we come back in detail below, can be expressed as follows: $\mathcal{L}$, a complete and orthocomplemented lattice is orthomodular iff $\forall a, b \in \mathcal{L}$ and $a \leq b$ we have $b = a \vee (b \wedge a')$ or similarly as $\forall a, b \in \mathcal{L}$: if $a \leq b$ and $a' \wedge b = 0$ then $a = b$. As has been shown in Coecke & Smets (2004), the orthomodular identity is equivalent to the so called Sasaki adjunction, i.e. $\varphi^*_a(b) \leq c$ iff $b \leq \varphi_{a,*}(c)$ with $\varphi^*_a : \mathcal{L}_1 \to \mathcal{L}_2 : b \mapsto a \wedge (a' \vee b)$ and $\varphi_{a,*} : \mathcal{L}_2 \to \mathcal{L}_1 : b \mapsto a' \vee (a \wedge b)$, or in short $\varphi^*_a \dashv \varphi_{a,*}$ for every $a \in \mathcal{L}$. This means that the Sasaki adjunction captures a very essential feature of quantum logicality and deserves to be focused upon in detail. From a dynamic logical perspective we will show that the Sasaki adjunction defines new causal (consequence) relations, each labeled with φ_a, $\forall a \in \mathcal{L}$. These relations can be uses to introduce labeled dynamic implications, each for which a deduction theorem holds.

In Section 1 we offer a short overview of operational quantum logic (OQL), i.e. the Geneva approach to the foundations of physics which originated with the work of C. Piron and J. M. Jauch in the late sixties. We follow the recent surveys in Coecke *et al.* (2003a, 2003b), Moore (1999), Smets (2001a) and focuss in particular on the basic notion of a property lattice. We take the liberty to start our exposition with an analysis of the link between the properties of a physical system and observables, the physical quantities of a physical system. In Section 2, we elaborate on operational quantum logic as an intuitionistic logic with an additional quantum operator, following the main results presented in Coecke (2002). In Section 3 we come back to the Sasaki adjunction and the notion of orthomodularity which plays a fundamental role in quantum logic. Following Smets (2001a, Chapter 11), we first focus on the problem of implication in quantum logic and show that a new perspective is needed to solve some old problems. So we finish off with an analysis of how a dynamic approach towards operational quantum logic — we refer to Coecke (2003), Coecke *et al.* (2004), Smets (2001a) for recent surveys — offers such a new perspective.

2. From Observables to Properties

Following Piron ([1990] 1998), OQL is founded on the basic notion of a particular physical system which can be conceived as an idealization of some phenomenon, i.e. as a characterizable and separable part of the external phenomenal world. Moreover, a particular physical system is believed to be completely describable by means of its "physical properties" or, as we will explain below, by means of its "two-valued physical quantities", i.e. its two-valued observables. Another cornerstone in OQL is that at every moment, a physical system can be represented by means of a point in a given state space. This space exists of all the points which represent the possible realizations of the given system. Piron refers with respect to this matter in Piron ([1990] 1998) to E. Cartan who in 1920 has introduced the notion of an abstract state space. In order to describe a simple point particle, the state space was defined just as the space $\mathbb{R}^7$, of which the coordinates are besides time, the components of the momentum and position quantities Piron ([1990] 1998). This space, we denote it by Σ, allows us to characterize besides momentum and position also all other physical quantities as well as the properties of a physical system. Let us in this paragraph first concentrate on classical physical systems. Piron gives as an example a point particle of which the set $\{(\vec{p}, \vec{q}, t) \mid (\vec{p}, \vec{q}, t) \in \mathbb{R}^7 \text{ and } \vec{p} = \vec{p}_0\}$ characterizes the physical property: "the particle possesses the momentum quantity $\vec{p}_0$" where a state $\mathcal{E}$ of this particle is represented by $(\vec{p}, \vec{q}, t)$, Piron ([1990] 1998). An Observable or physical quantity X can in this picture be represented by the collection of maps $O_X : \Sigma \to O_X(\Sigma)$ which attach to each state $\mathcal{E} \in \Sigma$ an element of another space such as for example a real number when $\mathbb{R} = O_X(\Sigma)$. With respect to the given example where $\Sigma = \mathbb{R}^7$, the observable "position" can be captured by the mappings from Σ into the vector space $\mathbb{R}^3$. As stated in Piron (1976), instead of defining an observable X by means of O_X which maps each state into the value taken by X, "it is just as natural to define an observable by the correspondence which associates a subset of Σ with each measuring interval Δ" (p. 11). In particular we will shift our focus from $O_X : \Sigma \to O_X(\Sigma)$, to its inverse image O_X^{-1} taken on powerset level, Piron ([1990] 1998):

$$O_X^{-1} : P(O_X(\Sigma)) \to P(\Sigma) : \Delta \mapsto O_X^{-1}(\Delta) = \{\mathcal{E} \in \Sigma \mid O_X(\mathcal{E}) \in \Delta\}.$$

As proven in Piron (1976), O_X^{-1} preserves lower and upper bounds and orthocomplementation.[1] Properties on the other hand, correspond to subsets of the state space, which in this classical case implies that the structure of the properties of a physical system is the structure of the powerset of the state space which, equipped with the operations of inclusion, orthocomplementation, union and intersection, forms a boolean algebra. We now see that each observable Y represented by $O_Y^{-1} : P(\{0, 1\}) \to P(\Sigma)$, which takes only the two values 1 or 0 in every state of the system, is associated directly to a property (and its orthocomplement) of a physical system (see Piron 1976). In the latter case of a two-valued observable, the specification of the subspace of Σ which is the inverse image of $\{1\}$ characterizes a physical property completely, where the specification of the subspace which is the inverse image of $\{0\}$ will characterize the orthocomplemented physical property. For a given system in state $\mathcal{E} \in \Sigma$ we call the property corresponding to $T \in P(\Sigma)$ "actual" iff $\mathcal{E} \in T$ and "potential" otherwise. So, a property y corresponding to a two-valued observable Y is actual in a given state $\mathcal{E}$ when $\mathcal{E} \in O_Y^{-1}(\{1\})$ and potential when $\mathcal{E} \in O_Y^{-1}(\{0\})$ in which case the orthocomplement of y is actual. Indeed, for classical physical systems, every property is actual or its orthocomplement is actual. For quantum physical systems, the latter utterance can not be sustained. Even more, the description of a physical system by means of its properties as thus far presented can not be upheld, that is to say, for non-classical systems not every subset of the state space will correspond to a physical property. Indeed, the union (classical disjunction) of those states in which a physical property is actual and of those states in which its orthocomplement is actual does not give us the full set of states, what is lacking here are exactly the superpositions of states which express an inherent quantum nature. For more details on the notion of superposition within OQL we refer to Coecke (2002). It is for the latter reasons that OQL has introduced a new physical language based on the notion of "yes/no question", or similarly called "definite experimental project", to arrive in first instance at a more general characterization of a physical property. The description we have in mind holds for classical and non-classical physical systems. In particular, an observable can then be represented by the mappings:

(1) $\quad O_X^{-1} : P(E) \to \mathcal{L}$

where E is the space in which observable X takes its values and $\mathcal{L}$ is a system's property lattice. The notion of a property lattice will be analyzed in detail in the following paragraphs, we only want to stress here that in (1) $\mathcal{L}$ is indeed isomorphic to $P(\Sigma)$ when the system is classical (see Piron [1990] 1998, Theorem 1.5.15).

[1]For the mathematical details of "orthocomplementation", we refer to the following section.

OQL takes an operational point of view in which a physical property will now be viewed as the set of equivalent "definite experimental projects" which all yield the same positive (negative) result. A definite experimental project is an experimental procedure and can best be viewed as a list of concrete actions which one has to define a priori. This list of actions has to be accompanied with a rule that specifies "in advance" when we ought to call the result at the end, positive. The underlying idea is that physical properties exist in reality, can be described by physicists and are measurable. Measurability is an important factor since the notion of when a property is to be called actual, is closely linked to the notion of "element of reality" introduced by Einstein, Podolsky and Rosen in (1935),

> If, without in any way disturbing a system, we can predict with certainty [...] the value of a physical quantity, then there exists an element of physical reality corresponding to this physical quantity.

A property of a particular physical system is then called "actual" if and only if the definite experimental projects which test it are certain and is potential otherwise. We call a definite experimental project certain for a particular system if it is sure that the positive response would obtain should we perform the experiment, Moore (1999). To the extent that one may assign an ontological value to elements of reality, they correspond exactly to what we called actual properties. Clearly, when a property is actual or not, depends on the state in which one considers the system to be. The reader might wonder now if the whole setup does not reduce itself to the debate about the notion of "certainty" which at first sight seems to play a central role with which everything stands or falls. Indeed the notion of "certainty" has caused already much confusion. When following Moore (1999), "it is certain" does not mean "it is necessary" because "at the very least one would have to assume that no uncontrolled external agent could act on the system during the course of the experiment in question". So, that is why in Coecke & Smets (2004) we have given a slightly different description of an actual physical property to avoid the certainty-debate:

> A physical property a of a system in a particular realization $\mathcal{E}$, is actual if it is the case that: whenever the system (in realization $\mathcal{E}$) is within environment e_a then it causes phenomenon α_a to happen.

Placing the system within a specific environment e_a can be seen as the actions included in a definite experimental project while the specification of phenomenon α_a can be seen as deciding upon the positive result within a definite experimental project. Therefore we assume also in our description that for every a, phenomenon α_a is specified from "in advance" whenever environment e_a is taken into consideration. The advantage of our reformulation is that the role played by the environment, i.e. measurement-context, presents itself more explicitly within the OQL-framework.

Given a particular physical system, let us denote the set of all its specifiable possible realizations by Σ. Within this domain of specification, we denote the set of all its properties: $\mathcal{L}$. A partial order relation can be introduced on $\mathcal{L}$ as follows: $a \leq b$ if and only if b is an actual property every time a is an actual property. The poset $\mathcal{L}$ is closed under the operation of meet, i.e. every subset $A \subseteq \mathcal{L}$ has a greatest lower bound $\bigwedge A \in \mathcal{L}$. And thus, for every subset $A \subseteq \mathcal{L}$: $\bigwedge A$ is an actual property in a state $\mathcal{E}$ iff $\forall a \in A$ is an actual property in state $\mathcal{E}$. Operationally: $\bigwedge A$ is an actual property of a system in state $\mathcal{E}$ if it is the case that for every arbitrarily chosen environment e_a, from the collection $\{e_a\}_{a\in A}$, necessary to test the actuality of a property contained in A we have: "whenever the system is placed within an environment e_a then it causes phenomenon α_a to happen". Remark that we use a disjunction, i.e. arbitrary choice, of environments as an operational motivation to introduce conjunctions of properties. Since this aspect has provoked many misconceptions of the basic OQL machinery, we refer the careful reader to Foulis & Randall (1984) and Smets (2003) for a reply to published misunderstandings. Further, as already noted in Coecke & Smets (2004), we cannot operationally motivate the introduction of a join of properties since that would require a notion of simultaneity of measurements which causes problems in quantum theory. Due to the fact that for every subset A of properties in $\mathcal{L}$, $\bigwedge A$ is a property as well, $\mathcal{L}$ is a complete meet-semilattice. It is bounded with the trivial property 1 as top element and the absurd property 0 as the bottom element. Property 1 can be examined by the experimental project "do whatever you wish with the system and assign the positive response". In other words, there is an environment e_1 which always causes phenomenon α_1 to happen whenever the system is placed within e_1, independently from its specific realization. Property 0 can be examined by the experimental project: "do whatever you wish with the system and assign the negative response". Also here, there is an environment e_0 which always causes the related phenomenon α_0 not to happen whenever the system is placed within e_0, independently from the specific realization of the system. As such one easily verifies that 1 is an always actual property while 0 can never be actual. Coming back to the introduction of "joins", Birkhoff's theorem assures us that every complete meet-semilattice is a complete lattice where the "join" or least upper bound of any collection of properties is obtained as follows:

$$\bigvee A = \bigwedge \{b \in \mathcal{L} \mid \forall a \in A : a \leq b\}.$$

We will now operationally introduce an orthogonality relation on the set of possible states of a physical system. We translate the definition of orthogonal states given in Moore (1999) as follows:

> Call $\mathcal{E}_1$ and $\mathcal{E}_2$ orthogonal, written $\mathcal{E}_1 \perp \mathcal{E}_2$, if there exists an environment e_a which is such that: whenever the system in realization

$\mathcal{E}_1$ is within environment e_a then it causes phenomenon α_a to happen and whenever the system in realization $\mathcal{E}_2$ is within environment e_a then it causes phenomenon α_a not to happen.

The orthogonality relation $\perp \subseteq \Sigma \times \Sigma$ is symmetric and antireflexive, Moore (1999). Given the fact that actuality of properties is state-dependent, it is natural to represent states within $\mathcal{L}$. Piron introduced the notion of "state property" in Piron ([1990] 1998) to refer to the strongest property of a physical system which is actual in a given realization. We call the properties $(\forall \mathcal{E} \in \Sigma) : p_{\mathcal{E}} \in \mathcal{L}$ state properties when there exists the following map:

$$p_{-} : \Sigma \to \mathcal{L} : \mathcal{E} \mapsto p_{\mathcal{E}} = \bigwedge \{a \in \mathcal{L} \mid a \text{ is actual in } \mathcal{E}\}$$

and when they satisfy the following axiom:

$$\text{If } p_{\mathcal{E}_1} \leq p_{\mathcal{E}_2} \text{ then } p_{\mathcal{E}_1} = p_{\mathcal{E}_2} .$$

The latter axiom states that one state property can never imply the actuality of another state property. So, while several properties can be actual in two different states in which the system can be realized, there is always at least one property which is actual in one state but not in the other. This axiom implies that the state properties of a physical system are the atoms in a property lattice which, following Moore (1999), can briefly be characterized as the minimal non-zero elements. One can now prove the theorem stating that $\mathcal{L}$ is atomistic, i.e. $\forall a \in \mathcal{L}$: $a = \bigvee \{p \leq a \mid p \text{ is an atom}\}$, which means that every property can be written as the join of its underlying state properties — we drop the subscripts of state properties when no confusion is possible. In order to equip $\mathcal{L}$ with an orthocomplementation within OQL, one first introduces an axiom to state that for every given realization $\mathcal{E}$ there exists at least one environment e_a which causes phenomenon α_a to happen when the system is in a possible state $\mathcal{E}'$ if and only if $\mathcal{E} \perp \mathcal{E}'$. In other words, for every state property $p_{\mathcal{E}} \in \mathcal{L}$, there exists a property $p^{\#} \in \mathcal{L}$ which is actual in state $\mathcal{E}'$ if and only if $\mathcal{E} \perp \mathcal{E}'$. Following Moore (1999), this allows us to represent the opposite of a property by means of the following map:

$$' : \mathcal{L} \to \mathcal{L} : a \mapsto \bigwedge \{p^{\#} \mid p \leq a\}.$$

The axiom assuring us that this map is surjective allows us to say that each property is the opposite of another one. With the two last axioms given, we may call $' : \mathcal{L} \to \mathcal{L}$ an orthocomplementation, i.e. $a \leq b \Rightarrow b' \leq a'$, $a \wedge a' = 0$, $a \vee a' = 1$ and $a'' = a$. Within OQL we can now describe both classical and quantum systems. As noted before, when the system is classical then is a physical property $a \in \mathcal{L}$ or its orthocomplement $a' \in \mathcal{L}$ actual in every of its possible realizations $\mathcal{E}$. When on the other hand the system is quantum, we can encounter situations in which a physical property and its orthocomplement are both potential in the same realization. In case we want to describe pure quantum systems, we have to take into consideration that the operations of meet and join in $\mathcal{L}$ are not distributive. As such quantum logicians focus on orthomodular structures due to the fact that the lattice of closed subspaces of a Hilbert space is "orthomodular" — for a survey of this notion we refer to Coecke *et al.* (2003a). In terms of property lattices, when $\mathcal{L}$ is complete and orthocomplemented, then we call $\mathcal{L}$ weak modular if and only if for $a, b \in \mathcal{L}$ and $a \leq b$ we have $b = a \vee (b \wedge a')$. The notion of an orthomodular lattice refers as such to an orthocomplemented weak modular lattice.

3. Intuitionistic Logic with a Quantum Operator

Let us first look back at the beginning of the previous section, if not every collection of realizations of a particular physical system corresponds to a physical property, then what is the link between the property lattice $\mathcal{L}$ of a system and its set of subsets of states $P(\Sigma)$? It is explicit in Moore (1999) that, physically, the notions of state and property are dual, i.e. to each state $\mathcal{E}$ we can associate the collection $S(\mathcal{E}) = \{a \in \mathcal{L} \mid p_{\mathcal{E}} \leq a\}$ of properties which are actual if the system is realized in $\mathcal{E}$ and dually, to each property $a(\neq 0)$ we can associate the family $\mu(a) = \{\mathcal{E} \in \Sigma \mid p_{\mathcal{E}} \leq a\}$ of states in which a is actual. More explicitly, $\mathcal{E} \in \mu(a) \Leftrightarrow p_{\mathcal{E}} \leq a \Leftrightarrow a \in S(\mathcal{E})$. This shows that a property is actual in a certain state $\mathcal{E}$ if $\mathcal{E} \in \mu(a)$ and is potential otherwise Piron ([1990] 1998). Note that $\mu\colon \mathcal{L} \to P(\Sigma)$ is referred to as the Cartan map in Piron ([1990] 1998), where it is proven to be injective, order and meet preserving. The difference between classical and non-classical physical systems shows itself in the fact that the cartan map is injective in general but surjective for classical (sub)systems or in other words, when we restrict the co-domain of μ to the biorthogonal closed subets of Σ i.e. subsets $T \subseteq \Sigma$ which satisfy $T^{\perp\perp} = T$ with $T^{\perp} = \{\mathcal{E} \mid (\forall \mathcal{E}_t \in T)\mathcal{E} \perp \mathcal{E}_t\}$. So under the condition of the mentioned axioms in the previous section, the application of the Cartan map determines an isomorphism between $\mathcal{L}$ and $(\Sigma, \perp)$ the lattice of biorthogonally closed subsets of Σ, ordered by inclusion, with the singletons $\{\mathcal{E}\}$ as its atoms. The physical duality between states and properties can been given a categorical reformulation (see Coecke *et al.* 2003a; Moore 1999; Stubbe 1999). It is this reformulation which forms the core of the transition from the "property lattice"-perspective into "quantum logic as an intuitionistic logic with a quantum operator". The tool we need for this categorical reformulation is the notion of a Galois adjunction, also called residuation. Following Coecke *et al.* (2003a), Coecke & Smets (2004), a pair of maps $f\colon \mathcal{L} \to \mathcal{M}$ and $g\colon \mathcal{M} \to \mathcal{L}$ between posets is Galois adjoint, written as $f \dashv g$, if and only if the following adjunction condition is satisfied:

$$f(a) \leq b \Leftrightarrow a \leq g(b).$$

When two maps f and g are Galois adjoint then f preserves all existing joins while g preserves all existing meets. In case $\mathcal{L}$ and $\mathcal{M}$ are complete lattices, any meet preserving map:

$$g\colon b \mapsto \bigvee\{a \in \mathcal{L} \mid f(a) \leq b\}$$

has a unique join preserving left Galois adjoint and any join preserving map:

$$f\colon a \mapsto \bigwedge\{b \in \mathcal{M} \mid a \leq g(b)\}$$

has a unique meet preserving right Galois ajoint. Looking back at the meet-preserving Cartan map μ, it has as a join preserving Galois adjoint the so-called "operational resolution"-map, Coecke & Stubbe (1999a, 1999b):

$$R_c\colon P(\Sigma) \to \mathcal{L}\colon T \mapsto \bigvee(T).$$

In the specific case to which we restrict ourselves here, i.e. $\mathcal{L}$ is atomistic and $\Sigma \subset \mathcal{L}$, the Galois adjunction $R_c \dashv \mu$ is characterizable as follows:

$$\begin{aligned} &R_c\colon P(\Sigma) \to \mathcal{L} : T \mapsto \bigvee\{\mathcal{E} \in \Sigma \mid \mathcal{E} \in T\} = \bigwedge\{a \in \mathcal{L} \mid T \subseteq \mu(a)\},\\ &\mu\colon \mathcal{L} \to P(\Sigma) : a \mapsto \bigcup\{T \in P(\Sigma) \mid R_c(T) \leq a\}, \end{aligned}$$

where $\mathcal{E} \in \mu(a)$ if and only if $R_c(\{\mathcal{E}\}) \leq a$. The operational resolution $R_c(T)$ assigns to every subset T of the state space Σ, the strongest property in $\mathcal{L}$ which is definitely actual in case the system would be realized in an arbitrary state $\mathcal{E} \in T$. Operationally such a resolution is useful when we have a lack of knowledge about the precise realization of a given system, of course within the boundaries of possibilities contained in a set $T \subseteq \Sigma$. Categorically, the Galois adjunction $R_c \dashv \mu$ gives rise to a closure operator

$$R := \mu \circ R_c\colon P(\Sigma) \to P(\Sigma) : T \mapsto \bigcap\{F \in \mu([\mathcal{L}]) \mid T \subseteq F\},$$

where $\mu[\mathcal{L}] := \{\mu(a) \mid a \in \mathcal{L}\}$ are the μ-closed subsets of the state space Σ (Coecke 2003). It is important to note that we can represent an atomistic property lattice $\mathcal{L}$ in terms of its state space Σ equipped with a T_1-closure operator[2] C for which we have given two possibilities: $C :=\perp$ or $C := R$. What is the advantage of passing from the old OQL-theory on property lattices into such a closure space? As remarked in Coecke (2003), working with $(P(\Sigma), C)$ enables us for instance to encode non-deterministic state or property-transitions. The appearance of disjunctions or unions within $P(\Sigma)$ allows us to deal with situations

[2] C is a closure operator on Σ if for all $T, T' \in P(\Sigma)$: $T \subseteq C(T), T \subseteq T' \Rightarrow C(T) \subseteq C(T'), C(C(T)) = C(T)$. This closure is called T_1 if in addition it satisfies: $C(\emptyset) = \emptyset$ and $C(\{\mathcal{E}\}) = \{\mathcal{E}\}$.

where we have a lack of certainty on the exact realization of a physical system. Such situations arise when we consider transitions of systems due to quantum measurements. In the next paragraph we will focus in detail on the case (H, R_H) where, for an underlying atomistic property lattice, H is a complete boolean algebra when it is equivalent to $P(\Sigma)$ and R_H is to be conceived as a quantum operator.

When it is our aim to incorporate a lack of knowledge about the actual properties or the states in which a system is realized, the idea arises to work with sets of properties. Though not all elements of $P(\mathcal{L})$ will be useful to work with, a first condition one might think of is order-preservance, i.e. for $a, b \in \mathcal{L}$ if $a \leq b$ and $a \in A \subseteq \mathcal{L}$, $b \in B \subseteq \mathcal{L}$ then $A \subseteq B$. Another feature of the sets we want to work with is that especially the classically behaving joins within $\mathcal{L}$ are captured on set-theoretical level. Which means that for sets of properties A, B and $A \subseteq B$, we have $\bigvee A \in B$ whenever the following condition is satisfied Coecke (2002):

(2) $\forall b \in \mathcal{L}\colon b \wedge \bigvee A = \bigvee \{b \wedge a \mid a \in A\}.$

As the results in Coecke (2002) show, what is necessary is the inclusion of a property lattice $\mathcal{L}$ in its distributive hull H, i.e.

$$\xi\colon \mathcal{L} \to H : a \to \downarrow a$$

characterizing the elements of $H \subset P(\mathcal{L})$ as “distributive ideals”. Distributive ideals, i.e. $A, B, \ldots \in H$ (or $\downarrow a, \downarrow b, \ldots \in H$ when explicitly referring to the embedded elements from $\mathcal{L}$), are downsets which recapture those underlying joins which behave as classical disjunctions besides all the meets (classical conjuctions) including the trivial and absurd properties. More concretely, an ideal I is distributive if it is an order ideal, i.e.

$$a \leq b \in I \Rightarrow a \in I \quad \text{and} \quad I \neq \emptyset$$

and

$$A \subset I \Rightarrow \bigvee A \in I \quad \text{under condition (2).}$$

The above mentioned quantum operator can be introduced as follows (see also Coecke 2002):

$$\begin{aligned} R_H\colon H \to H : A &\mapsto \bigcap \{B \in \xi[\mathcal{L}] \mid A \subseteq B\} \\ &= \downarrow \bigwedge_{\mathcal{L}} \{b \in \mathcal{L} \mid \forall a \in A : a \leq b\} = \downarrow \bigvee_{\mathcal{L}} A\,. \end{aligned}$$

Since $(H, R_H) \cong \mathcal{L}$, this operator assures us that all quantum properties can be recovered on a higher level. With all the ingredients present, we can now look at the logical structure of (H, R_H).

The language from which we start is a quotient of H, for notational reasons we denote it just as H, existing of equivalent distributive ideals or propositions for short. The equivalence relation $\sim\ \subseteq P(\mathcal{L}) \times P(\mathcal{L})$ is given in Coecke (2002):

$$A \sim B \iff C(A) = C(B),$$

where $C\colon P(\mathcal{L}) \to P(\mathcal{L}) : A \mapsto \bigcap\{B \in H \mid A \subseteq B\}$ and the logical connectives are algebraically introduced:

$$\begin{aligned}
&\bigwedge_H : P(H) \to H : \mathcal{A} \mapsto \bigcap \mathcal{A},\\
&\bigvee_H : P(H) \to H : \mathcal{A} \mapsto C(\bigcup \mathcal{A}),\\
&R_H : H \to H : A \mapsto \downarrow \bigvee_{\mathcal{L}} A,\\
&\Rightarrow_H : H \times H \to H : (B, C) \mapsto \bigvee_H \{A \in H \mid A \cap B \subseteq C\},\\
&\neg_H : H \to H : A \mapsto (A \Rightarrow_H \downarrow 0).
\end{aligned}$$

Clearly we obtain a complete Heyting algebra equipped with a quantum operator, where only for classical (distributive) systems we obtain $\forall A \in H : R_H(A) = A$. As proven in Coecke (2002), in case $H \cong P(\Sigma)$ we end up with a complete Boolean algebra. In the next paragraph we concentrate on the implication problem in quantum logic, and show later that the above presented perspective removes a lot of old difficulties when we focus on the nature of orthomodularity.

4. The Problem of Implication and the Nature of Orthomodularity

There are certain problems with an implication-connective in $\mathcal{L}$. Note that we introduced an implication relation $\leq$ above, this should however not be confused with the introduction of an implication operation on $\mathcal{L}$. Although not stated in exactly those terms, this point has already been noted by G. Birkhoff and J. von Neumann in (1936):

> The following point should be mentioned in order to avoid misunderstanding: If a, b are two physical qualities, then $a \cup b$, $a \cap b$ and a' [...] are physical qualities too [...] But $a \subset b$ is not a physical quality; it is a relation between physical qualities[3] (p. 25).

Indeed, if a and b are elements of a property lattice $\mathcal{L}$; $a \leq b$ does not express a new property, hence $\leq$ is not a connective or an operation. But if we are to

[3]Note that in Birkhoff & von Neumann (1936), physical qualities stand in a one-one relation with the closed linear subspaces of a Hilbert space — which are, as we understand it, their mathematical representatives — and can be interpreted as well as sets of equivalent experimental propositions.

concentrate on an implication operation, which criteria ought it to satisfy? This question has been answered for instance by G. M. Hardegree in (1979) where he analyzes such criteria for the case of a classical (boolean) lattice and for the case of an ortho(modular) lattice. The implicative condition or implication adjunction as we called it above,

(3) $a \wedge c \leq b \Leftrightarrow c \leq a \to b$

is equivalent to the so-called law of exportation ($a \wedge c \leq b \Rightarrow c \leq a \to b$) taken together with modus ponens ($a \wedge (a \to b) \leq b$) and is used to characterize classical implicative lattices as they are confusingly called in Hardegree (1979) or implicative lattices as they are mostly known and called in for instance Curry ([1963] (1977)).[4] We are in favor of the latter term since one uses the implicative condition in classical and intuitionistic logic, in particular can we say that the implicative condition is present in classical implicative lattices and in relatively pseudo-complemented lattices. The given law of exportation, as part of the implicative condition, is equivalent to the so-called "deduction theorem" in the following form:

(4) $\mathcal{A}, A \vdash B$ only if $\mathcal{A} \vdash A \to B$

with a standard consequence relation $\vdash$ (see Dunn & Hardegree 2001, chapter 11). In Dunn & Hardegree (2001) it is noted that:

> [...] the deduction theorem in its form (4) is central to the classical and intuitionistic systems, particularly to the intuitionistic system wherein all the pure implicational theorems may be deduced from (4) and its converse (modus ponens). Accordingly, (3) is central to the algebraic treatment of the intuitionistic system [...] (p. 381–382).

However, condition (3) can not be encountered in the case of pure quantum logic due to the fact that it implies distributivity of $\wedge$ and $\vee$. More explicitly, we refer to a theorem proved by Skolem (see also Curry [1963] 1977; and Dunn & Hardegree 2001). Following Herman *el al.* (1975) Skolem's theorem states that (our insertion of brackets):

> Any lattice with a binary connective satisfying [the implicative condition] (3) is necessarily distributive. Moreover, $a \to b$ is the maximal solution for c in the inequality $c \wedge a \leq b$ (p. 308).

Returning to orthomodular lattices, a straightforward generalization of the implicative condition by means of residual (Galois adjoint) conditionals is given in Hardegree (1979, 1981). Let us elaborate on such kind of generalization. Note first that the implicative condition gives rise to the adjunction $f \dashv g$ with

$$f\colon \mathcal{L} \to \mathcal{M} : c \mapsto (a \wedge c),$$

[4]The law of exportation holds in the reverse direction by applying modus ponens $x \wedge a \leq a \wedge (a \to b) \leq b$.

$$g\colon \mathcal{M} \to \mathcal{L} : b \mapsto (a \to b).$$

In order to consider implications besides the classical and intuitionistic ones, there are two possibilities for a generalization: either we restrict the domain of residuation — we take some special subset — or we allow that the left adjoints are not necessarily defined by the meet. Concentrating on the latter, the most common example in an orthomodular lattice is given via the Sasaki adjunction where the Sasaki hook

$$\varphi_{a,*}(-) : \mathcal{L}_2 \to \mathcal{L}_1 : b \mapsto a' \vee (a \wedge b)$$

is the right adjoint of the Sasaki projection defined by

$$\varphi_a^*(-) : \mathcal{L}_1 \to \mathcal{L}_2 : c \mapsto a \wedge (a' \vee c).$$

More explicitly, one can prove:

(5) $\varphi_a^*(c) \leq b \Leftrightarrow c \leq \varphi_{a,*}(b).$

Similar as before we now see that an ortholattice satisfying condition (5); or in other words: for all $a \in \mathcal{L}$ that we have $\varphi_a^*(-) \dashv \varphi_{a,*}(-)$, is necessarily weak modular (see Coecke & Smets 2003). Many have defined the action of the Sasaki hook as an internal operation, i.e. as a two-variable lattice polynomial. Let us denote such two-variable polynomials as $p_i(a, b)$. A full proof is given in Kalmbach (1983) of the fact that among the distinct two-variable polynomials which emerge from the elements of the free orthomodular lattices generated by a two-element set, the internal Sasaki hook operation is one the following five polynomials:

$$\begin{aligned}
p_1(a, b) &= (a' \wedge b) \vee (a' \wedge b') \vee (a \wedge (a' \vee b)),\\
p_2(a, b) &= (a' \wedge b) \vee (a \wedge b) \vee ((a' \vee b) \wedge b'),\\
p_3(a, b) &= a' \vee (a \wedge b),\\
p_4(a, b) &= b \vee (a' \wedge b'),\\
p_5(a, b) &= (a' \wedge b) \vee (a \wedge b) \vee (a' \wedge b'),
\end{aligned}$$

which satisfy the strengthened law of entailment, i.e.

(6) $p_i(a, b) = 1 \iff a \leq b\,.$

The strengthened law of entailment is yielded by the law of entailment ($a \leq b \Rightarrow p_i(a, b) = 1$) together with modus ponens. Note that the classical material implication is not one of the mentioned five polynomials satisfying the strengthened law of entailment in the non-boolean orthomodular case. Indeed, under the definition of $p(a, b) = a' \vee b$ we obtain that $a \leq b$ implies $p(a, b) = 1$ but not vice versa due to the presence of counterexamples. For instance, we have $a' \vee b = 1$ but not $a \leq b$ in the orthomodular lattice called M02 or "Chinese Lantern" with atoms a, a', b, b', Kalmbach (1983). We want to stress as well that the five polynomials we mentioned, fail to satisfy the strengthened law of entailment in the non-weak modular case. A proof in Dalla Chiara (1986), Dalla Chiara & Giuntini (2003) shows that for the internal Sasaki hook operation, if $a \leq b$ then $a' \vee (a \wedge b) = 1$ iff $a \wedge (a \vee b)' = 0$ only holds in case we have at least orthomodularity, the proofs for the other four polynomials are similar. Note that the Sasaki hook operation attained more attention than the other candidates which satisfy the strengthened law of entailment due to the fact that it behaves locally-boolean, i.e. $p_4(a, b) = a' \vee b$ if and only if a, b are compatible properties (see for instance Hardegree 1979). But, although the Sasaki hook operation satisfies the strengthened law of entailment given an orthomodular lattice, it still in general does not come close enough to what one from an "implication" would expect. Malinowski focuses in (1990) on the problems with a deduction theorem while Hardegree shows in (1975, 1979) that the Sasaki hook operation — similar as the counterfactual or subjunctive conditional — in the non-boolean orthomodular case falls short of certain laws namely strong transitivity $p(a, b) \wedge p(b, c) \leq p(a, c)$, weakening $p(b, c) \leq p((a \wedge b), c)$ and contraposition $p(a, b) = p(b', a')$. The failure of strong transitivity and contraposition of the Sasaki hook operation is pointed out explicitly by means of an example of a non-boolean orthomodular lattice in Herman *et al.* (1975). And due to the fact that weakening $p(b, c) \leq p((a \wedge b), c)$ can be obtained from strong transitivity $p(a, b) \wedge p(b, c) \leq p(a, c)$ by exchanging the a for $a \wedge b$, in Herman, Marsden and Piziak's example, weakening fails as well for the Sasaki hook operation. There is a way to overcome these disadvantages by shifting to the new perspective on quantum logic given above.

According to us, the problem with defining the action of the Sasaki Hook as an internal implication connective in standard quantum logic is due to the fact that it is disconnected from its context, i.e. adjunction (5). For this reason and since eq. (5) is equivalent to "orthomodularity" (see Coecke & Smets 2002), it is the Sasaki adjunction itself from which one ought to start rather than from the strengthened law of entailment (6) with $i = 4$. The first question one could ask is how the introduction of this adjunction obtains a physical motivation. In Coecke & Smets (2004) we pointed out that the Sasaki adjunction incarnates a causal duality. The notion "causal duality" refers here to a causal relation between the physical system before and after a quantum measurement has been performed and the fact that, within a mathematical framework, this relation can be used to derive the adjunction between maps expressing propagation and causation of physical properties respectively, Coecke *et al.* (2001), Stubbe (2002). For the particular case of the Sasaki adjunction, $\varphi_a^* \dashv \varphi_{a,*}$, the underlying causal relation captures the possible state/property transitions of a physical system due to an ideal measurement of the first kind of property a with a positive result.[5] For the physical interpretation of the Sasaki projection and Sasaki hook we refer to for instance Smets (2001b) where it is pointed out that the action of a Sasaki projection captures the perturbation suffered by a quantum system due to an ideal measurement of the first kind while the Sasaki hook can be conceived as assigning causes for properties of physical systems to be actual. As we will show in the next paragraph, the mentioned causal relation underlying the Sasaki adjunction will be labeled with actions and is necessary to introduce labeled dynamic implications.

Explicitly we define the labeled causal relation $\overset{\varphi_a}{\leadsto}$ as follows:

$$\forall a, b, c \in \mathcal{L} : \varphi_a^*(c) \leq b \Leftrightarrow c \overset{\varphi_a}{\leadsto} b \Leftrightarrow c \leq \varphi_{a,*}(b).$$

The interpretation we attach to $c \overset{\varphi_a}{\leadsto} b$ is that the actuality of c guarantees that after positively performing the measurement φ_a, i.e. an ideal measurement of the first kind of property a, property b will be actual.[6] Note that, although we restricted to Sasaki adjunctions, every action-based adjunction on property lattices defines a causal relation and, vice versa one can start from an operationally motivated definition of a causal relation and show that every causal relation induces a pair of maps which are Galois adjoint (see Stubbe 2002). For an explicit operationally motivated definition of the causal relation underlying the Sasaki adjunction we refer to Coecke & Smets (2004) and Smets (2001b). The dynamic ingredient in case of $c \overset{\varphi_a}{\leadsto} b$ is the specific quantum measurement φ_a, in Amira *et al.* (1998) a more general framework is offered using the notion of "induction" to capture this dynamic ingredient in terms of "an externally imposed change of a particular physical system". As such an act of induction can for instance impose a state/property-transition in the case of a measurement, may assure free evolution or in case one would consider several physical (sub)systems, as in Coecke (2000), the interaction of those subsystems can be seen as an induction of one subsystem on another. This points to the ingredients which lead us into dynamic operational quantum logic. Taking up the perspective of (H, R_H) given above, we introduce first the Sasaki adjunction on a higher level by means of the

---

[5]The concept "first kind measurement" refers to a non-demolition measurement of which the repetition yields the same result. The concept "ideal measurement" refers to the abstraction over imperfections of the measurement apparatus and the fact that such a measurement does not change properties compatible with the one it measures. Ideal measurements of the first kind do not change classical systems, they have an effect on quantum systems (due to the existence of incompatible properties) and in particular allow us in the latter case to calculate the perturbation suffered by the system (see Piron 1976; Smets 2001b).

[6]We abstract here over the cases in which a' is an actual property when we want to perform φ_a, i.e. $a' \overset{\varphi_a}{\leadsto} 0$. In such a case, one can for instance interpret φ_a as an action which destroys the system Smets (2001b).

following diagram:

$$\begin{array}{ccccccc}
\mathcal{L}_1 & \stackrel{\varphi_a^*}{\to} & \mathcal{L}_2 : & b & \mapsto & \varphi_a^*(b) \\
| & & | & | & & | \\
\xi & & \xi & \xi & & \xi \\
\downarrow & & \downarrow & \downarrow & & \downarrow \\
H_1 & \stackrel{\hat{\varphi}_a^*}{\to} & H_2 : & \downarrow b & \mapsto & \downarrow \hat{\varphi}_a^*(b)
\end{array}$$

Where alternatively we can characterize $\hat{\varphi}_a^*$ as:

$$H_1 \to H_2 : B \mapsto \bigvee_H \{\downarrow \varphi_a^*(b) \mid b \in B\}.$$

Note that the existence of the join-preserving map $\varphi_a^* \colon \mathcal{L}_1 \to \mathcal{L}_2$ forces $\hat{\varphi}_a^*$ to satisfy a join continuity condition, (see Coecke *et al.* 2004, 2001; Coecke & Stubbe 1999a):

$$\bigvee_{\mathcal{L}} A = \bigvee_{\mathcal{L}} B \Rightarrow \bigvee_{\mathcal{L}} \hat{\varphi}_a^*(A) = \bigvee_{\mathcal{L}} \hat{\varphi}_a^*(B).$$

The introduction of this continuity condition is important since not all maps $\hat{\varphi}_a^* \colon H_1 \to H_2$ one could consider independently from φ_a^*, are physically meaningful. One has to keep track of underlying superposition properties. In terms of R_H the condition becomes:

$$R_H(A) = R_H(B) \Rightarrow R_H(\hat{\varphi}_a^*(A)) = R_H(\hat{\varphi}_a^*(B)).$$

With respect to the Sasaki hook, given $\varphi_a^* \dashv \varphi_{a,*} \colon \mathcal{L}_1 \to \mathcal{L}_2$ and $\hat{\varphi}_a^* \dashv \hat{\varphi}_{a,*} \colon H_1 \to H_2$, then $\downarrow \varphi_a^*(-) = \hat{\varphi}_a^*(\downarrow -)$ on $\mathcal{L}_1$ implies $\downarrow \varphi_{a,*}(-) = \hat{\varphi}_{a,*}(\downarrow -)$ on $\mathcal{L}_2$ (Coecke & Smets 2003). This allows us to denote the Sasaki hook on H as follows:

$$\hat{\varphi}_{a,*} : H_2 \to H_1 : B \mapsto \hat{\varphi}_{a,*}(B).$$

The causal relation can now be introduced on the level of H, i.e. $\stackrel{\varphi_a}{\leadsto} \subseteq H \times H$:

$$\forall a \in \mathcal{L}, B, C \in H : \hat{\varphi}_a^*(B) \subseteq C \Leftrightarrow B \stackrel{\varphi_a}{\leadsto} C \Leftrightarrow B \subseteq \hat{\varphi}_{a,*}(C).$$

Returning to implications, in the static case the implication on H is intuitionistic and as such satisfies the strengthened law of entailment:

$$B \Rightarrow_H C = \downarrow 1 \quad \Leftrightarrow \quad B \subseteq C$$

This static implication itself can alternatively be denoted as (see Coecke *et al.* 2004 and Coecke & Smets 2003):

$$\begin{aligned}
B \Rightarrow_H C \;&:=\; \bigvee_H \{A \in H \mid \forall d \in A : (d \in B \Rightarrow d \in C)\} \\
&=\; \{a \in \mathcal{L} \mid \forall d \leq a : (d \in B \Rightarrow d \in C)\}.
\end{aligned}$$

In the dynamic case, using the causal relation instead of the inclusion, i.e., $\stackrel{\varphi_a}{\leadsto}$, we look for an implication such that:

$$B \Rightarrow_{\varphi_a} C = \downarrow 1 \;\Leftrightarrow\; B \overset{\varphi_a}{\rightsquigarrow} C\,.$$

The candidate which naturally arises for this labeled dynamic implication is (see also Coecke 2003; Coecke *et al.* 2004; Coecke & Smets 2003; Smets 2001a):

$$\begin{aligned} B \Rightarrow_{\varphi_a} C &:= \bigvee_H \{A \in H \mid \forall d \in A : (d \in B \Rightarrow \hat{\varphi}_a^*(\downarrow d) \subseteq C)\} \\ &= \{a \in \mathcal{L} \mid \forall d \leq a : (d \in B \Rightarrow d \in \hat{\varphi}_{a,*}(C))\} \\ &= B \Rightarrow_H \hat{\varphi}_{a,*}(C). \end{aligned}$$

The Sasaki hook can as such be recovered as: $\downarrow 1 \Rightarrow_{\varphi_a} - = \hat{\varphi}_{a,*}(-)$. We know that the conjunction is adjoint to the static implication, this is exactly what condition (3) entails:

$$(A \bigwedge_H -) \dashv (A \Rightarrow_H -).$$

Similarly we can introduce a labelled dynamic conjunction as adjoint for each labeled dynamic implication (see Coecke 2003; Coecke *et al.* 2004; Coecke & Smets 2003; Smets 2001a):

(7) $$(A \otimes_{\varphi_a} -) \dashv (A \Rightarrow_{\varphi_a} -)$$

where $(A \otimes_{\varphi_a} -)$ encodes the Sasaki projection when $A = \downarrow 1$. Let us remark that in the specific case when $a = 1$, the implication $- \Rightarrow_{\varphi_a} -$ reduces to a static implication $- \Rightarrow_H -$ since in that case no actions need to be taken in consideration.

We extended (H, R_H) with dynamic conjunctions and dynamic implications, i.e. $(\forall a \in \mathcal{L} : \Rightarrow_{\varphi_a}, \otimes_{\varphi_a})$. Moreover, each adjoint pair (7) encodes a form of the deduction theorem. It is this logical feature of "deduction" together with the fact that static quantum logic arises in our view as a kind of "modal logic", in particular as an intuitionistic logic with a modal quantum operator, which allows us to give a positive answer to the question whether quantum logic is really a logic. The overview we presented indicates a line of research in the domain of quantum logic in which an explicit distinction between the corresponding closed subspaces and projectors on a Hilbert space $\mathcal{H}$, although they correspond, can be made — see Coecke & Smets (2004). Indeed, for every Sasaki projection we can say that φ_a^* is induced by the projector P_a, which orthogonally projects the vectors of $\mathcal{H}$ onto the subspace a (see Piron 1976). So, by means of labeled logical operators we encoded the actions of orthogonal projectors on subspaces in $\mathcal{H}$.

Free University of Brussels
Center for Logic and Philosophy of Science
e-mail:sonsmets@vub.ac.be

REFERENCES

Amira, H., B. Coecke and I. Stubbe (1998). How Quantales Emerge by Introducing Induction within the Operational Approach. *Helvetica Physica Acta* **71**, 554–572.

Birkhoff, G. and J. von Neumann (1936). The Logic of Quantum Mechanics. *Annals of Mathematics* **37**, 823–843. Reprinted in: C. A. Hooker (ed.), *The Logico-algebraic Approach to Quantum Mechanics*, vol. 1., pp. 1–26 (1975). Dordrecht: D. Reidel Publishing Company.

Coecke, B. (2000). Structural Characterization of Compoundness. *International Journal of Theoretical Physics* **39**, 581–590.

Coecke, B. (2002). Quantum Logic in Intuitionistic Perspective. *Studia Logica* **70**, 411–440.

Coecke, B. (2003). Do we have to Retain Cartesianity in Topos-Approaches to Quantum Theory and Quantum Gravity?, workingpaper.

Coecke, B., D. J Moore and I. Stubbe (2001). Quantaloids Describing Causation and Propagation for Physical Properties. *Foundations of Physics Letters* **14**, 123–145.

Coecke, B., D. J. Moore and S. Smets (2003a). From Operationality to Logicality: Philosophical and Formal Preliminaries, submitted for publication.

Coecke, B., D. J. Moore and S. Smets (2003b). From Operationality to Logicality: Syntax and Semantics, submitted for publication.

Coecke, B., D. J. Moore and S. Smets (2004). Logic of Dynamics & Dynamics of Logic: Some Paradigm Examples. In: S. Rahman, J. Symons, D. Gabbay and J. P. Van Bendegem (eds.), *Logic, Epistemology and the Unity of Science*, pp. 527–555. Kluwer Academic Publishers.

Coecke, B. and S. Smets (2002). Interpreting Orthomodularity Dynamically. Poster presented at the sixth biannual IQSA meeting.

Coecke, B. and S. Smets (2004). The Sasaki Hook is not a [Static] Implicative Connective but Induces a Backward [in Time] Dynamic One that Assigns Causes. *International Journal of Theoretical Physics* **43**, nr 7–8, 1705–1736.

Coecke, B. and I. Stubbe (1999a). Operational Resolutions and State Transitions in a Categorical Setting. *Foundations of Physics Letters* **12**, 29–49.

Coecke, B. and I. Stubbe (1999b). On a Duality of Quantales Emerging from an Operational Resolution. *International Journal of Theoretical Physics* **38**, 3269–3281.

Curry, H. B. ([1963] (1977)). *Foundations of Mathematical Logic*. New York: Dover Publications Inc.

Dalla Chiara, M. L. (1986). Quantum Logic. In: D. Gabbay and F. Guenthner (eds.), *Handbook of Philosophical Logic*, vol. III, pp. 427–469. Dordrecht: D. Reidel Publishing Company.

Dalla Chiara, M. L. and R. Giuntini (2003). Quantum Logics. In: *Handbook of Philosophical Logic*. 2nd edition.

Dunn, J. M. and G. H. Hardegree (2001). *Algebraic Methods in Philosophical Logic*. New York: Oxford University Press Inc.

Einstein, A., B. Podolsky and N. Rosen (1935). Can Quantum-Mechanical Description of Physical Reality be Considered Complete? *Physical Review* **47**, 777–780. Reprinted in: J. A. Wheeler and W. H. Zurek (eds.), *Quantum Theory and Measurement*, pp. 138-141 (1983). New Jersey: Princeton University Press.

Foulis, D. J. and C. H. Randall (1984). A Note on Misunderstandings of Piron's Axioms for Quantum Mechanics. *Foundations of Physics* **14**, 65–81.

Hardegree, G. M. (1975). Stalnaker Conditionals and Quantum Logic. *Journal of Philosophical Logic* **4**, 399-421.

Hardegree, G. M. (1979). The Conditional in Abstract and Concrete Quantum Logic. In: C. A. Hooker (ed.), *The Logico-Algebraic Approach to Quantum Mechanics*, vol. 2. Dordrecht: D. Reidel Publishing Company.

Hardegree, G. M. (1981). Material Implication in Orthomodular (and Boolean) Lattices. *Notre Dame Journal of Formal Logic* **22**, 163–182.

Herman, L., E. L. Marsden and R. Piziak (1975). Implication Connectives in Orthomodular Lattices. *Notre Dame Journal of Formal Logic* **XVI**, 305–328.

Kalmbach, G. (1983). *Orthomodular Lattices*. London/New York: Academic Press.

Malinowski, J. (1990). The Deduction Theorem for Quantum Logic: Some Negative Results. *The Journal of Symbolic Logic* **55**, 615–625.

Moore, D. J. (1999). On State Spaces and Property Lattices. *Studies in History and Philosophy of Modern Physics* **30**, 61–83.

Piron, C. (1976). *Foundations of Quantum Physics*. Massachusetts: W. A. Benjamin Inc.

Piron, C. ([1990] 1998) *Mécanique quantique: Bases et applications*. Presses polytechniques et universitaires romandes. Lausanne.

Smets, S. (2001a). *The Logic of Physical Properties in Static and Dynamic Perspective*. PhD-thesis. Free University of Brussels.

Smets, S. (2001b). On Causation and a Counterfactual in Quantum Logic: The Sasaki Hook. *Logique et Analyse* **173–175**, 307–325. Appeared 2003.

Smets, S. (2003). In Defense of Operational Quantum Logic. *Logic and Logical Philosophy* **11**, 191–202.

Stubbe, I. (1999). *Physics and Categories*. Mémoire de DEA, Faculté des sciences. Université Catholique de Louvain.

Stubbe, I. (2002). Causal Duality: What it is, and what it is Good For. Poster presented at the sixth biannual IQSA meeting.

von Neumann, J. (1932). *Grundlagen der Quantenmechanik*. Berlin: Springer Verlag. English translation: *Mathematical Foundations of Quantum Mechanics* (1996). New Jersey: Princeton University Press.

Erik Weber

ARE THERE ONTOLOGICAL EXPLANATIONS?*

ABSTRACT. There is a huge philosophical literature on scientific explanation, and no one seriously denies that the sciences explain in one way or another. But what about ontology? I will argue that ontological laws and ontological theories can explain. And I will point at the differences between ontological explanations and their scientific counterparts.

Introduction

My research field is the philosophy of science, more precisely causation and scientific explanation. Because Jerzy Perzanowski has done a lot of research in ontology, I thought it a good idea to write a paper for this volume on a question that has popped up in my head several times during the last couple of years: are there ontological explanations, and what to they look like?

There is a huge philosophical literature on scientific explanation, and no one seriously denies that the sciences explain in one way or another. In order to understand my basic question properly, we should distinguish between two kinds of ontology. On the one hand, scientific laws and theories have an underlying ontology, i.e. a set of assumptions about which kind of entities exist in their domain and which kind of properties these entities have. The scientific law or theory based on the ontology describes relations between these properties and objects. For instance, classical mechanics assumes that there are material bodies and that these bodies can have a position, a velocity, an acceleration, etc., and describes the relations between these properties. Ontology in this first sense does not explain: it is the base on which explanatory theories and laws are built. The assumptions themselves do not have explanatory power.

Ontologies of the second kind are laws and theories which describe how certain fundamental physical properties (such as: length, width, height, weight, ...) behave when subjected to certain operations. My basic claim in this paper is that ontological laws and theories of the second kind can explain. In order to argue for this claim I will first discuss ontological explanations that use only ontological laws (ontological covering law explanations, see Section 1). In Section 2, I discuss explanations based on ontological theories (theoretical ontological explanations).

*I thank Maarten Van Dyck for his comments on the first draft of this paper.

In: J. Malinowski and A. Pietruszczak (eds.), *Essays in Logic and Ontology* (*Poznań Studies in the Philosophy of the Sciences and the Humanities*, vol. 91), 277–283. Amsterdam/New York, NY: Rodopi, 2006.

1. Ontological covering law explanations

I start with an example (1.1) which I use to clarify the general characteristics of ontological covering law explanations (1.2). In 1.3 I compare them with scientific covering law explanations.

1.1. *An example*

Let a, b, c and d be names for building blocks. e and f are names for towers built with these blocks. We observe that e is higher than f (formally: $e > f$). This can be explained as follows.

C_1: e is the result of putting b on top of a (formally: $e = a \oplus b$).
C_2: f is the result of putting d on top of c (formally: $f = c \oplus d$).
C_3: a has the same height as c (formally: $a \sim c$).
C_4: $b > d$.
L_1: For all blocks x, y and z: if $x > y$, then $x \oplus z > y \oplus z$.
L_2: For all blocks x, y and z: if $x \sim y$, then $x \oplus z \sim y \oplus z$.
L_3: For all blocks x and y: $x \oplus y \sim y \oplus x$.
L_4: For all blocks x, y and z: if $x \sim y$ and $y \sim z$, then $x \sim z$.
L_5: For all blocks x, y and z: if $x \sim y$ and $y > z$, then $x > z$.
L_6: For all blocks x, y and z: if $x > y$ and $y \sim z$, then $x > z$.
T_1: $b \oplus a > d \oplus a$ (from C_4 & L_1)
T_2: $a \oplus d \sim c \oplus d$ (from C_3 & L_2)
T_3: $d \oplus a \sim a \oplus d$ (from L_3)
T_4: $a \oplus b \sim b \oplus a$ (from L_3)
T_5: $d \oplus a \sim c \oplus d$ (from L_4, T_3 & T_2)
T_6: $a \oplus b > d \oplus a$ (from L_5, T_4 & T_1)
T_7: $a \oplus b > c \oplus d$ (from L_6, T_6 & T_5)

E: $e > f$ (from T_7, C_1 & C_2)

Covering law explanations (both scientific and ontological ones) subsume their explanandum under one or more general laws: they show that it can be derived from the law(s) together with a number of initial conditions. In the example C_1–C_4 are the initial conditions, L_1–L_6 the laws. Together they constitute the explanans. T_1–T_7 are intermediate steps which clarify how the explanandum E follows from the explanans.

1.2. *Characteristics of ontological covering law explanations*

In our example, the explanandum is a relation between two objects, rather than a property of one single object. This is a first characteristic of ontological covering law explanations. Furthermore, it is claimed that the explanandum is determined by the structure of the objects and by the properties of their components. This is a second typical feature of ontological covering law explanations.

1.3. *Comparison with structure-interaction explanations*

To clarify the above-mentioned general characteristics, I will contrast ontological covering law explanations with structure-interaction explanations. These are the type of scientific explanation that are most similar to them.

Let us first look at an example, inspired by an example of Hempel (1965, p. 246). A mercury thermometer is rapidly immersed in hot water. We observe a temporary drop of the mercury column, followed by a swift rise. The core of the explanation of this fact is:

C_1: This thermometer has been rapidly immersed in hot water.

C_1: This thermometer consists of a glass tube which is partly filled with mercury.

L: In all thermometers consisting of a glass tube which is partly filled with mercury, the mercury column drops and then rises if immersed in hot water.

E: The mercury column in this thermometer drops and then rises.

This explanation is an instance of the following general format:

C_1: Object x causally interacted with object y which had property P.

C_2: Object x has structural properties $S_1, \ldots, S_n$.

L: All objects with structural properties $S_1, \ldots, S_n$ display property Q after a causal interaction with an object with property P.

E: Object x displays property Q.

I call explanations which fit this format structure-interaction explanations (S-I explanations). The most important characteristics of S-I explanations are:

(a) The explanandum fact is a property of one single object x.

(b) It is claimed that the explanandum fact is *triggered by an interaction* between x and another object y. It is also claimed that the way the object x reacts to this triggering interaction is determined by its structure. As a consequence, the covering law describes a causal relation between a triggering cause and a structuring cause on the one hand, and the effect to be explained on the other hand.

If we compare this with ontological covering law explanations, the differences are obvious. We explain individual properties instead of relations. Furthermore, the information about structure is used in a different way: it is combined with a information about interactions, rather than with information about properties of the components.

2. Theoretical ontological explanations

In 2.1 I present a theoretical explanation of the law used in the example in 1.3. In 2.2 I argue that, while ontological theories can also explain laws, they do it in a radically different way.

2.1. *An explanation of the thermometer law*

L can be derived from three more general laws: the law of thermal expansion of fluids, the law of thermal expansion of solids and the law of heat conduction. The law of thermal expansion of fluids is:

$$v = v_o + v_o\beta T$$

(v: final volume; v_o: initial volume, β: coefficient of expansion of the fluid considered; T: rise (+) or fall (-) of the temperature). The law of thermal expansion of solids is:

$$l = l_o + l_o\alpha T$$

(l: final length; l_o: initial length; α: coefficient of linear expansion of the solid considered). Finally, the law of heat conduction says that the quantity of heat dQ (joules) which, in a time period dt, flows between any two plane surfaces normal to the direction of heat flow, is given by the equation

$$dQ = -KA(\delta\theta/\delta n)dt$$

(K: thermal conductivity coefficient (joule/°C.sec.m); A: area the surfaces (m^2); $\delta\theta/\delta n$: temperature gradient (°C/m)). The law "In glass tubes which are partly filled with mercury and are rapidly immersed in hot water, the mercury level first drops and then rises" can be derived from these three laws. The derivation is as follows. The law of thermal expansion of fluids gives us the following equation for mercury:

$$v = v_o + v_o \times 0.00018 \times T_m$$

(v: final volume of the mercury; v_o: initial volume of the mercury; 0.00018: coefficient of expansion of mercury; T_m: rise of the temperature of the mercury).

On the other hand, the tube of the thermometer will expand in accordance to the following equation, which is an implementation of the law of thermal expansion of solids:

$$r = r_o + r_o \times 0.000008 \times T_g$$

(r: final radius of the tube; r_o: initial radius of the tube; 0.000008: coefficient of linear expansion of glass; T_g: rise of the temperature of the tube). Glass is a bad thermal conductor: its thermal conductivity (and thus also dQ) is small. As a consequence, in a first phase, only the temperature of the tube will rise: $T_m = 0$ and thus $v = v_o$. The relation between the level of the mercury and its volume is given by the geometrical formula

$$h = v/\pi r^2$$

(h: height of the mercury column; v: volume of the mercury; r: radius of the tube). From this formula and $v = v_o$ we derive that

$$h = v_o/\pi r^2.$$

The initial mercury level is given by

$$h_o = v_o/\pi r_o^2.$$

Because the temperature of the tube rises ($T_g > 0$), r is larger than r_o. The two last equations and $r > r_o$ imply that

$$h < h_o.$$

In other words: the mercury level drops in a first phase. But as a result of heat conduction, the temperature of the mercury will start to rise after a few instants; the mercury expands (law of thermal expansion of fluids). Because the coefficient of expansion of mercury is large, the expansion will be greater that the extension of the tube. Consequently, the mercury level will rise.

2.2. *How do ontological theories explain?*

In the explanation presented in Section 2.1, the behaviour of the thermometer is reduced to laws about the behaviour of its parts: the expansion and thermal conductivity of the glass, and the expansion of the mercury. Ontological theories explain in a different way: instead of reducing the law to properties of entities at a lower level (micro-reduction), ontological theories unify the laws that can be used in ontological covering law explanations.

As an example, consider the theory consisting of the following axioms and definitions (based on Forge 1993, p. 91–93; Forge has taken his axioms and definitions from Balzer, Moulines & Sneed 1987):

A_1: For all blocks x, y and z: $x \oplus (y \oplus z) \sim (x \oplus y) \oplus z$.
A_2: For all blocks x and y: $x \oplus y \sim y \oplus x$.
A_3: For all blocks x, y and z: $x \geq y$ if and only if $x \oplus z \geq y \oplus z$.
A_4: For all blocks x, y and z: if $x \geq y$ and $y \geq z$, then $x \geq z$.

The first three axioms describe the properties of the $\oplus$ operator, the fourth says that $\geq$ is a transitive relation. The definitions explicate how $\geq$ and $\sim$ relate to one another, and introduce a third relation:

D_1: For all blocks: $x > y$ if and only if $x \geq y$ and $y \not\geq x$.
D_2: For all blocks: $x \sim y$ if and only if $x \geq y$ and $y \geq x$.

All the laws we have used in Section 1.1 (and many other laws) can be derived from this theory. Let us see how. Law L_1 can be derived as follows (I suppress the expression "For all blocks ..."):

(1) If $x > y$ then $x \geq y$. (from D_1)
(2) If $x > y$ then $x \oplus z \geq y \oplus z$. (from (1) and A_3)
(3) If $x > y$ then $y \not\geq x$. (from D_1)
(4) If $x > y$ then $y \oplus z \not\geq x \oplus z$. (from (3) and A_3)
(5) If $x > y$ then $x \oplus z > y \oplus z$. (from (2), (4) and D_1)

Law L_2 can be derived as follows:

(1) If $x \sim y$ then $x \geq y$. (from D_2)
(2) If $x \sim y$ then $x \oplus z \geq y \oplus z$. (from (1) and A_3)
(3) If $x \sim y$ then $y \geq x$. (from D_2)
(4) If $x \sim y$ then $y \oplus z \geq x \oplus z$. (from (3) and A_3)
(5) If $x \sim y$ then $x \oplus z \sim y \oplus z$. (from (2), (4) and D_2)

L_3 is nothing but the second axiom.
L_4 can be proved by reductio as absurdum:

(1) $x \sim y$ (assumption)
(2) $y \sim z$ (assumption)
(3) $x \nsim z$ (assumption)
(4) $x \geq y$ (from (1) and D_2)
(5) $y \geq x$ (from (1) and D_2)
(6) $y \geq z$ (from (2) and D_2)
(7) $z \geq y$ (from (2) and D_2)
(8) $x \geq z$ (from (4), (6) and A_4)
(9) $z \geq x$ (from (7), (5) and A_4)
(10) $x \sim z$ (from (8), (9) and D_2)

Since we have obtained a contradiction between line (3) and line (10), the proof shows that the assumptions are not consistent. This implies that L_4 is true. Similar proofs can be given for L_5 and L_6.

In the proofs, the specific laws we have used in the explanation in Section 1.1 are not reduced to a more fundamental level. Instead they are subsumed under a small set of axioms and definitions. In other words: we have unification rather than micro-reduction.

3. Conclusions

In Section 1 I have shown that ontological laws can explain. Ontological covering law explanations differ from their scientific counterparts because they explain relations between objects rather than properties of individual objects, and because they use information about structure in a different way.

In Section 2 I have shown that ontological theories can unify ontological laws, and, in that sense, can explain them. Here too there is a difference with explanations of scientific laws by scientific theories: the latter can be either micro-reductions (as in our thermometer example) or unifications (similar to our ontological example). For ontological theories, there is only one option.

Universiteit Gent
Centre for Logic and Philosophy of Science
Blandijnberg 2
B-9000 Ghent
Belgium
e-mail: erik.weber@UGent.be

REFERENCES

Balzer, W., C. U. Moulines and J. Sneed (1987). *An Architectonic for Science*. Dordrecht: Reidel.

Forge, J. (1993). How Should We Explain Remote Correlations? *Philosophica* **51**, 83–103.

Hempel, C. G. (1965). *Aspects of Scientific Explanation and other Essays in the Philosophy of Science*. New York: Free Press.

Marek Rosiak

FORMAL AND EXISTENTIAL ANALYSIS OF SUBJECT AND PROPERTIES

ABSTRACT. The paper is a contribution to the object ontology. The general approach assumed in the investigation is that of Roman Ingarden's *The Controversy Over the Existence of the World* where an object is the subject-of-properties. The analysis of the form and the mode of existence of properties leads to the rejection of both negative and general properties. Each property is an individual qualitative moment of a particular object. Its form reveals existential heteronomy: the quality of the property is not immanent but refers to the object. The subject of properties has not its own qualitative content: its form is just the internal causality establishing the unity of an object. An object is not causally isolated from other objects, but external causation differs from internal either by being ramified in case of the composition and destruction of an object or reciprocal in case of interaction between coexisting objects.

Everyday we talk about properties, and moreover, are convinced that we can see, at least, some of them. Therefore it seems that their existence cannot be questioned. However, do we really know how they exist and what they are? There are philosophers who arrived at the conclusion that properties are only *entia rationis* and nothing corresponds to them in reality. Apparently there must be considerable problems with grasping their proper characteristic. In my attempt to analyze certain ontic aspects of properties I refer to the approach of Roman Ingarden who introduced analytical tools of rare subtlety and vast applicability in his *opus magnum* Ingarden (1964–1965).[1] For him formal analysis in ontology means the analysis of forms of entities such as, e.g. object, subject, property, etc. Such an analysis should always start with basic intuitions and in this sense, at least in the beginning, can be informal, i.e. non-formalized. I think that in my conclusions I sometimes differ from Ingarden. This is the result of the influence of the major work of contemporary metaphysics, perhaps the only one comparable with Ingarden's book, Whitehead's *Process and Reality*.

1. What is the most common context like in which we encounter properties, become aware of them, or at least believe that we are facing them? Once I have realized that there is something before me I start to regard it as an (individual) object. It can be said that existence is primarily the individual object's

[1] The complete text available in Ingarden (1987, 1964–1965). An extensive part of its existential-ontological part has been translated into English in Ingarden (1964). Referring to Ingarden I give the number of paragraph of the German edition (1964–1965) (the same in Polish 3rd edition 1987).

In: J. Malinowski and A. Pietruszczak (eds.), *Essays in Logic and Ontology* (*Poznań Studies in the Philosophy of the Sciences and the Humanities*, vol. 91), 285–299. Amsterdam/New York, NY: Rodopi, 2006.

business (hence convertibility of transcendental notions of being and unity). But after noticing that something exists we usually start to realize that it differs from something. This difference can be twofold: synchronic or diachronic. In the first case the object in question differs from another object existing simultaneously, in the other case we say that now the object is different than before, which means that it has changed. In my opinion we become aware of object's properties in these two kinds of circumstances first of all. If we agree on this we can say that properties' business is to make differences between objects. Wherever objects differ one from another and whenever an object changes they must be some properties belonging to these very objects. A classical (scholastic) statement that *attributum non est ens sed entis* is in perfect accordance with this principle. Properties themselves are not objects; they just make objects differ.

Of course, these two factors: existence and being different are interdepenent. Not only it is necessary to exist in order to be different but also an object must be different from everything else to establish existence of its own. Properties are as if limits[2] of existence: an existing being marks its presence with the help of its properties, properties are contrasts between the object and its surroundings. The standpoint of a numerical monism seems to deserve some attention here. If there is only one unchanging entity it cannot differ from anything and *eo ipso* has no properties. Its existence has no limits then. However, I would not dare say (as once Hegel did) that such an indifferent being is just pure nothingness. Its existence must be the existence of something; existence having no limits still has the source. Indeed, talking of the limits we should always remember that they necessarily refer to something limited; limits alone do not exhaust object's existence, they only localize it. Hence properties alone cannot establish existence of the object: it is not just a bundle of properties, there must be something prior to them in the object.

From this introductory account of properties some consequences seem to follow both with regard to their existential and formal characteristics. Let us take as an example of a property the weight of an object. As a property it must always be 1°: qualitative[3] and 2°: something's weight. Being qualitative implies that lack of a property is not itself a property. From this it follows that so called negative properties cannot be just lacks of certain properties. If we agree that when two objects differ they differ mutually (the relation of difference is symmetrical) we arrive at a conclusion that two objects differ with respect to two properties always and not just one. Otherwise x would differ from y because $P(x)$ and $\sim P(y)$ but y would have nothing different from x. Moreover, it seems that properties

[2]Properties determine their object and Latin "determinare" means "to delimit", "to mark", "to end". A shape of a spatial object is here of course the first example but let us try to consider the whole complex of an object's actual properties as its "shape" in a more general sense.

[3]In the broadest possible sense as Ingarden says, cf. Ingarden (1964–1965, § 34).

must group themselves in classes with regard of their making difference. You can say that x differs from y with regard to weight or color but you cannot say that they differ because one of them has a color and the other a weight.

2. Here a question can be raised: are not classes of properties (as, e.g., color in general) properties, as well? This opinion is quite widespread and some of its adherents regard it even self-evident. Nevertheless I think it cannot be upheld. For one thing, if an object has not only a particular instance of color but, in addition, color in itself it should have the most general "property", namely, the property "property", as well. For another, if red and color(ed) are two properties, they partly coincide because to have the property red is the same as having a red color. Red and colored stand in determinate/determinable relation. Because of this one can say not only that x differs from y with respect to color; x's and y's colors differ with the same respect as well. Doesn't it sound strange to say that two colors differ with respect to color?[4] Another argument against general properties is that such a property would be common to many objects. This does not fit our basic intuition that a property primary business is to make difference between things. With this respect a general property would be deficient: it would still need other properties to do the job. An additional argument comes from the fact that having the same property seems to be a sufficient condition of identity of an object. Properties make a difference between their objects and at the same time prove the inner identities of these objects. Having some general properties in common would imply unity of their objects and eventually a numerical monism. If an object had general properties they could sometimes disappear. Now, consider a case of the destruction of an object. If its general properties disappear altogether (but still exist in other objects), the properties in question are not general but individual. And if they continue to exist (being strictly the same in other still existing objects) we are forced to admit that a property can survive its subject which seems to violate the basic existential feature of a property.[5] If you try to say that the general properties in question continue to exist but only the number of their subjects changed, this time the result is the violation of the basic formal characteristic of the property, its dependence on the subject because we would have something like a partial independence of a (general) property of its subject(s).

Arguments like these are sometimes adopted by nominalists but they could serve them only on the assumption that universals could have been nothing but properties. Only then the conclusion follows that universals do not exist. This assumption, in turn, can be a consequence of some meta-categorial mistake plus

[4]Of course one can say that two colors differ with respect to their quality but the fact behind both expressions is the same because the quality mentioned is just color.

[5]They behave like objects of Wittgenstein's *Tractatus*, being existentially independent of any particular subject but nevertheless absolutely dependent — cf. 2.0122.

the standpoint of metaphysical moderate realism ("universalia in rebus"). The mistake consists in blurring the difference between categories (which are universals or predicates) and entities which fall under them (individual objects, property-instances, etc.). "A category" sometimes means a plurality of entities and sometimes their common character (e.g. being a substance) which is a universal. Now Platonic realism ("universalia ante rem") can be understood as the thesis that universals are peculiar individuals (substances). If we understand Aristotelian critique of this standpoint as shifting universals from the category of substances to that of non-substances we still assume that universals fall under some category whereas categories themselves (as well as lower genera and species) are universals which never fall under any category (although corresponding pluralities of entities are contained in relevant categories). Expression "in rebus" is somehow unfortunate but we must be aware that it just like "ante rem" and "post rem" cannot have literal meaning. I would interpret these three expressions rather as referring to existential dependence: "ante rem" means that "res" are dependent on universals but not the other way round (universals can exist without individuals), "post rem" — the opposite and "in rebus" means that the dependence is mutual: universals and individuals require each other to exist.[6]

Let us still take into consideration certain existential asymmetry between an object and its property. It is commonly recognized that properties need existential support from their subject. We can say that a property is existentially dependent on its subject and that this dependence is one-sided because the subject can still exist without this property. But we must acknowledge that instead of having this very property it has to gain some other relevant property. Now, if we admit general properties, then the dependence seems to lose its asymmetry because it is only true that a general property cannot exist without any subject whatsoever while it can still exist without any particular of its many subjects. I think that without existential dependence of a property on only one subject as in case of property-instances one cannot retain the asymmetry of existential dependence here.

As far as I know, the main argument in favor of general properties is the existence of different objects qualitatively identical with certain respect.[7] E.g. two different cars can have the very same color. The respective property-instances in both objects would differ only with respect to their subjects but not in themselves. Thus in themselves they are not many but only one general property. I

[6]Universals require individuals to propagate their exemplifications — according to Aristotle a new substance of a given natural kind can be produced only by another individual of the same kind; universals themselves have no effective power. Individuals need universals to be determined, which is the necessary condition of existence.

[7]I owe much to discussions on this subject with prof. Ingvar Johansson from Umea University during his visit in Poland in 2001.

want to compare this argument with the following one: let us imagine two qualitatively identical objects which differ only in respect of their localization. Would we say that this is one and the same general object or rather that there are still two objects? I suppose that the latter answer is more plausible; what prevents us from identifying the objects is their different localization. Is not the case of qualitatively identical properties exactly similar? They, too, should remain different because of localization in different objects. Having different localization is enough to be different for both objects and properties. Having a single subject is characteristic for a property as such: we can have a common friend, even a common girl-friend, but not a single common property. Nevertheless I agree that in case of properties as well as objects the relevant qualities are identical. We usually discern qualities from objects but it is equally important not to identify qualities with properties.[8] A criterion of identity for a property is to have equal qualitative matter and being possessed without interruption by the same subject.

3. Properties belong to their subjects not as an umbrella belongs to an English gentleman; their belonging consists in one-sided existential dependency on a unique subject. Quality of a property does not belong to it as a kind of a second order property (cf. Ingarden 1964–1965, p. 42) — it is its content or matter. The form of a property consists in qualifying its subject, i.e. in attributing the quality to the subject. Existential dependence of a property on its subject is rooted in its form or rather in the form of an object which is a subject-of-properties. The form of a property and the form of a subject are two aspects of a complete form of an object (cf. Ingarden 1964–1965, p. 39). The quality of a property doesn't qualify the property — being weighty is not itself weighty — it qualifies the subject of a property. There is a kind of exchange between a property and its subject: the property gives the subject a qualification and receives existence.[9] Although a property only transmits its quality to a subject and its existence has a dependent character, it cannot be dispensed with. Without properties an object would be monolitic, unable to change nor even to differ from other objects. As long as an object has the property the matter of the property is a qualification of the object and not of the property. But when the property disappears the relevant qualification of the object disappears as well.

If properties are so indispensable how can we make sense of the metaphysical standpoint claiming that there are only objects and (in particular) no properties whatsoever? This so called reism was put forward by Brentano and later

---

[8]The case of indiscernible qualities has of course interesting epistemological context which deserves closer inspection on another occasion. At the moment I can only mention the following: Indiscernible qualities do not fulfill the transitivity principle: A' can be indiscernible from A'', A'' — from A''' and yet A' may be well discernible from A'''. This cannot happen with one and the same universal, so it seems that sensible qualities cannot be universals.

[9]Has the Hegelian master-slave dialectic any correspondence to this?

restated by Tadeusz Kotarbiński. Taken literally it is simply unintelligible but subjected to some restrictions with respect to its negative thesis it seems to deserve attention. Let us agree that in every possible interpretation the class of existents ("res") has some generic characteristic which properties lack. Thus we can present possible interpretations of reism giving this requisite of existence as: being an individual (standard nominalism), being material (materialism), being sensible (metaphysical version of empiricism[10]), being able to exist separately[11] (existential monism in Ingarden's terms), or being existentially autonomous (cf. Ingarden 1964–1965, § 12). The first requirement would eliminate properties only on account that they all were universals; I have just rejected even a weaker standpoint. The second claim could be troublesome rather for some materialists than for the ontology of properties:[12] if you want everything to be material in the sense of "being made of matter" (no matter what the matter is) then surely you cannot approve of properties having no mass and being able to disappear without even a small puff of smoke. Being sensible can be incompatible with being a property if you reject possibility of sensory abstraction (the first kind of abstraction listed by Berkeley in the Introduction to his Treatise). But the objection that properties are not sense-data can be understood in a more sophisticated way: only sense-qualities are given to the senses and not their being properties: a categorial form is an intellectual component of knowledge. The next possible interpretation of reism uses the conceptual apparatus of Ingarden's existential ontology. To exist separately (as a separate whole[13]) is not to have to be a component of some broader unity. According to this interpretation what exists can exist in isolation. A probable premisses of this standpoint could be the following: to exist is to be able to act (see the etymology of German *Wirklichkeit*) and only a complete being can act. Existence is here conceived of as some effective power. But on the other hand it is hard to deny that an existing and acting object is complex. Consequently an existing being has nonexisting components. To avoid such nonintuitive conclusions one should perhaps agree that properties yet have some kind of half-existence (subsistence).[14] The requirement of being existentially autonomous can be taken weaker or stronger. The weaker ("non-fiction") condition is satisfied when a being is not purely intentional. This is

[10]Which is basically an epistemological standpoint.

[11]I.e. containing the moment of *Selbstandigkeit* in their mode of existence — cf. Ingarden 1964–1965, § 14 (translated as separateness in Ingarden 1987).

[12]So called "vulgar" in the orthodox marxist-leninist interpretation. Of course nothing can be troublesome for truly dialectical materialists.

[13]The notion of a whole Ingarden uses here is not his own technical notion of a so called summative whole but a broader notion corresponding to Husserlian notion of a whole from 3rd *Logical Investigation*; cf. Ingarden 1964–1965, § 43.

[14]Leibniz in *Monadology* seems to choose another solution: his monads have different properties (perceptions), yet are regarded simple.

perhaps the weakest possible interpretation of reism. It would eliminate properties only if they were *phænomena sine fundamento in rebus* — pure illusions. But the stronger interpretation is much more demanding: it takes Ingarden's condition of autonomy literally: a being is only internally determined, i.e. its essential determinates do not refer to anything else. A determinate of something can fail to fulfill this condition in two ways: First, the object can be determined by some external factor: e.g. being a bank note is determined by the decision of a national bank and lasts only until the withdrawal of this decision. A piece of paper can be a true bank note but its status must be externally supported by the authority of a bank.[15] Of course, a bank note is existentially heteronomous not in the sense of being a fiction, something unreal. Second, the determinates of X can determine some different Y; being determined as X implies to determine Y: e.g. being a man-eating tiger implies digesting some considerable piece of human body from time to time. And the very nature of a property is to determine (as red, hot, square, etc.) its subject. In case of a man-eating tiger we could perhaps say that to be a man-eater is not essential to a tiger (we would probably face general disagreement of tiger-eaten men), but we cannot say that to qualify a subject is not essential for a property. A property, just as a property, determines something different from itself. Being a property is not BEING QUALIFIED SO and so in itself: it is to QUALIFY something different.[16]

Can this feature be called a property-intentionality? Or maybe rather, a property transparency? To refer its content outside itself may look strange and can explain, at least psychologically, reistic critique of the ontological status of a property. Nevertheless, we ought to remember that this is to some extent the matter of perspective: we look on the form subject-of-properties from the side of properties. The same form seen from the other end does not make that amazing impression. It is easier to consider object (having properties) than properties (possessed by an object). Using Brentanian terms: we can refer to an object *modo recto*, but to its properties only *modo obliquo*. We can now better see the empiricist deformation of the content of knowledge: it is impossible to see properties directly, still less possible — to see pure qualities; what we see is always one object (in the foreground) having properties (in the background). The property-transparency can explain why properties cannot have (second order) properties: a would-be second order property cannot qualify a property of the first order because this latter refers all determinations directly to its subject.

4. According to Ingarden the form of a subject corresponds to still another qualitative matter than the matter of properties — the subject has as if its own

[15]On the other hand, when you e.g. are born, being born cannot be cancelled by any subsequent event.

[16]Perhaps this heterogenic character of properties inclined Aristotle to deny them the essence and consequently to reject the possibility of science of accidents.

immediate quality. It qualifies it prior to all other qualifications, establishes the "what" of an object, its so called constitutive nature (cf. Ingarden 1964–1965, p. 40). Without this primary qualification an object could not have been qualified by properties. Being qualified by properties alone would be like having borders without the interior. When there is nothing to limit, the limits make no sense. If we attribute existence to an object it must be something different from just a collection of properties which exist — otherwise we would have not one being but at best as many beings as we have properties. Plurality of properties is the plurality IN the object but not OF the object and this is so because properties are only aspects of some basic unity. But surely we cannot see this primary qualification behind the properties as if a bottom in a bottle of whisky.[17] The Cartesian idea of matter as *res extensa* supplemented with the distinction attribute(extension)/modi(other qualities) developed by Locke into an empiricist distinction between primary and secondary qualities proved to be involved in a regress: whatever quality you discover behind given secondary qualities, must be regarded — being sensory — secondary, as well. And even without the empiricist presupposition, the very hypothesis that the factor of unity is some quality seems to imply the Bradleyan paradox of relations: unity would need another unity to unify and so on.

This paradox ruined Twardowski's Part-Whole Theory and it was Husserl who tried to overcome it in his 3rd *Logical Investigation*. He distinguished the relation of foundation holding among parts which cannot exist in separation, the standard example is color and spatial extension. According to him the unity of all parts of an object is grounded in this relation but the relation itself is not a part of the object because it has not a qualitative character and constitutes the form of the object, not its matter. Its being formal consists in the pure existential dependency of the relata. This dependency is always subordinated to some general rule: particular properties m and n are not dependent as particulars but only as representatives of respective species. E.g. this particular shade of blue is founded on the piece of surface it actually covers not as the shade of blue but as a color and the surface is its fundament only as surface in general. Each shade of every possible color needs to cover some particular piece of surface but not this or that piece — just anyone. Thus properties of an object make one whole because of their subordination to the general law of foundational connection. The universals constituting the principle of this connection are not themselves connected with the subordinated properties. The universal principle of connection is not a metaphysical glue; founded qualities are connected with founding qualities directly and the principle of connection is a constituent, a reason of the connection, but not its component, a factor. The FACT of their connection does not contain the RULE of the connection.

[17] Polish traditional drink cannot serve as an example here — it is transparent.

If everything is a whole composed of parts it should be possible to analyze the form subject-of-properties as its special case. The form of a property, its subject-dependence cannot be just being founded. Every property has to be a property of a subject, and not e.g. of another property while a whole chain of subsequently founded parts is possible. Moreover, being a founded part does not imply to refer the founded quality to the fundament as its determination. From the formal point of view Husserl distinguishes two kinds of parts: founded — moments, and not founded — pieces. A piece can exist independently which assumes its full determination while the subject of a property cannot exist by itself, instead of this property it must have another one. Examples of foundational connections given by Husserl belong to at least two quite different groups. On the one hand, there are qualities which constitute the phenomenological matter of real objects (like color and extension) and on the other hand, in a fragment presenting theses about the so called intermediate parts there are already formed objects of different level, e.g. each piece, capable of independent existence, must be an object. In short, examples start too low on the subobjectual level (matter) and then jump too high on the superobjectual level (objects of higher order) and what we need is the analysis of an object of the first order (originally individual in Ingardenian terms).

5. Probably this gap in Husserl's analyses is not accidental. In my opinion the whole-part analysis can be effective only when the form of an entity can be analyzed in abstraction from its mode of existence. The form of a real object preserving identity in changes involves dynamic dependencies of properties on their subject while the form of a whole provides us only with the simple static coexistence of moments and their fundaments. The property transfers its qualitative matter to the subject but nevertheless does not dissolve in it. This is so because an object having a property can in principle change: lose it and acquire another one while preserving self-identity. Thus self-identity of an object must be immune to changes on the one hand but confined to the limits of the object on the other.[18] Ingarden presents a fourfold classification of properties (cf. Ingarden 1964–1965, § 57) which can help us to find this basic characteristic of an individual. In my opinion the classification is the result of different divisions. First, properties can be divided into effective and non-effective, the latter have purely comparative character like e.g. "bigger than" and do not assume any interaction between compared object.[19] We may say that the reason of holding of the relation of comparison is having relevant properties by some objects. But, of course, neither this reason is an effective cause nor the relation is an effect. This class, interesting in itself, is not relevant at the moment. Effective properties

[18] Without this last requirement we would encounter some of the problems of general properties analyzed before.

[19] Nor do they assume any mental operation of the knowing subject.

assume the relation of effective causation holding inside the subject or between the subject and another object. In the first case we have an absolutely proper property.[20] An external factor causing a property in a given subject can have the character of an event or can be a number of parallel processes resulting in a state of equilibrium needed for preserving the same property. The property produced by an event is called acquired because another event is needed to make it disappear. A processual interaction produces an externally conditioned property – it lasts as long as the state of equilibrium of processes (see the following diagram).

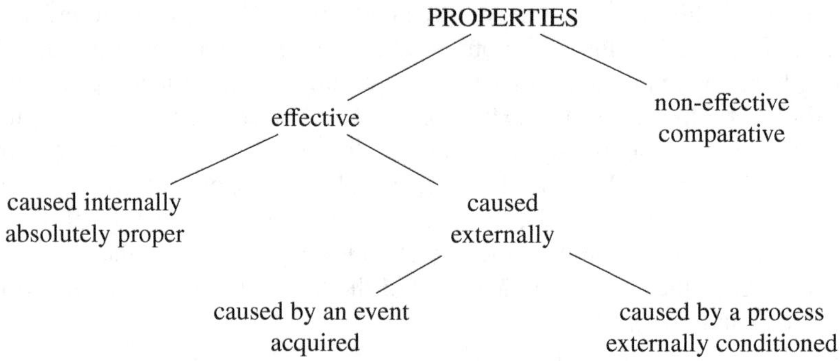

Let us consider absolutely proper properties. It is possible that in certain circumstances a real object can have a property as long as it exists. But is it possible that a real object necessarily has a particular property as long as it exists? In my opinion there cannot be necessary properties in real objects. Such a property would have been an isolated aspect of an object otherwise involved in constant causal interactions both within itself and with the environment. It would be doubtful if such an isolated part can be integrated with the rest of the object. An attempt at reducing identity of an object to certain set of its properties cannot be successful. This was realised already by Locke who distinguished real and nominal essence of a thing, acclaiming arbitrariness of the latter (and uncognizability of the first).[21] A property (in the sense of a property-instance I have discussed earlier) is always causally conditioned, although it can be initial, i.e. produced together with its subject. From the very beginning of its existence an object must have complete property characteristic but in the course of events it may change completely. "To change completely" does not mean here "to become unrecognizable"; we must remember that even slightest differences in quality result in different property-instances.

[20] Ingarden regards the existence of the subject alone as a sufficient condition of the existence of absolutely proper properties. Consequently the subject cannot acquire or lose them. I disagree with him on this point for reasons discussed below.

[21] Cf. his (1979), vol. II, part III, ch. 3, §§ 13–19.

6. Changing properties can nevertheless constantly exemplify the same universals: whatever changes of color a thing undergoes, it is still colored, etc. But color in general cannot be something peculiar to one thing only. Moreover, when a property is localized in an object it must have been produced in this object by some causal action and the object because of having this property can exert causal action upon something. But a universal cannot act, only individuals can.[22] Behind changing properties of an individual there must be some factor ruling the changes. Leibniz referred to it in his *Monadology* by the name of tendency and Locke pointed to it as powers and faculties of a substance (Locke 1979, vol. I, part II, ch. 23, §§ 7–11). It is responsible for producing subsequent new properties as a consequence of given ones. Is this factor individual or general? Leibniz describes it as an "internal principle [...] of change" adding that it must be accompanied by "something particular which changes (un detail de ce qui change)" (Leibniz, §§11–12). An argument for its individuality is its internal character (acclaimed both by Locke and Leibniz). But as a principle ruling subsequent changes (not just identical with its sequence) it cannot be individual in the same sense as a single property belonging to the sequence. (Locke is well aware of this and has problems how to classify powers of a substance; cf. Locke 1979, vol. I, part II, ch. 23, § 7.) It is not as a particular property because it transforms properties: makes them appear in given circumstances. Nor should we regard this principle as some mysterious hidden quality (scholastic *qualitas occulta*). It is rather the factor responsible for the coordination of the properties of an object. Each new external event or process in which the object is involved disturbs the state of equilibrium among its properties. E.g. the process of heating the gas in a container of a stated capacity makes the parameters of the gas (temperature, pressure) change until a new state of equilibrium is established. The event of having slipped on a banana skin makes me move chaotically until regaining balance. In all cases like these the appearance of some externally caused property causes events or processes ending in acquiring new properties which last till the next disturbance.

Sometimes an impulse can result in destroying the object: having slipped on a banana skin I can fell and kick the bucket. Equilibrium is reached, but by another object, my body instead of me. What is the difference between the two cases? Of course, dead body lacks certain properties of its deceased owner, but there is no particular property(-instance!) decisive for being alive. Each given property can be replaced by another property of the same object without loss of its identity. And even more: All properties together can be replaced by a complete set of other properties while the object remains the same. Instead of particular properties rather certain sets or ranges of properties can determine

---

[22] An universal acting in one of its exemplifications only and not in the other would be either inconsistent or divided.

possible limits of an object identity. But determining respective ranges of all relevant properties would not be enough: a man can be 2.30 m. long and a man can weight 20 kgs but not both.[23] So not ranges of properties independently determined but rather a subset of the Cartesian product of sets of homogenic properties is needed.

Let $\mathfrak{R} = \{S, T, U, \ldots\}$ be the complete (possibly infinite) set of kinds of properties needed[24] for the full qualitative determination of a real object where $s_i \in S$ $(i \in I)$ is a particular qualitative determination of a property of the kind S (e.g. particular shade of red belonging to the kind color). Let $\prod \mathfrak{R}$ be the Cartesian product of the family $\mathfrak{R}$. The complete qualitative determination of x is some $\mathbf{Q}(x) \subset \prod \mathfrak{R}$, $\mathbf{Q}(x) \neq \prod \mathfrak{R}$ where $d \in \mathbf{Q}(x)$ is, e.g., the present qualitative characteristic of x. For each x, $\mathbf{Q}(x)$ is uniquely determined and if $x \neq y$ then $\mathbf{Q}(x) \cap \mathbf{Q}(y) = \emptyset$. A 1-1 relation of the direct internal causality $\mathrm{C}^{\mathrm{d}}_x$ holds in every $\mathbf{Q}(x)$.[25] $\mathrm{C}^{\mathrm{d}}_x$ must be a partial function for every x because it can have dead ends: determinations leading directly to the object's destruction. The determination of new object(s) after the destruction is still the effect of the preceding determination but it is no longer the internal causation. The external causation (dotted arrows in the diagram below) holds where a determination of one object produces directly determinations of many objects (decomposition of x into u and v) or the other way round (composition of v out of x and y):[26] I assume here that non-ramified generation or corruption is unfeasible (the opinion held, e.g. by Leibniz in *Monadology*, § 6). Not every generation, however, is a composition: e.g. u is generated through the decomposition of x: u emerges from x. Similarly, not every corruption must be a decomposition: e.g. y's destruction takes part in the generation of v: y merges into v.[27] Such generation and corruption happens if there are originally individual objects (not composed of parts) which are not existentially original (cf. Ingarden 1964–1965, §§ 13, 38, 43). Of course, composable and decomposable objects can be generated and destroyed in both ways. In case of causal interactions between coexisting objects, like x and y, there is a mutual causation: C_{j+1} of x determines D_{k+1} of y, and vice versa. This type of external causation (mutual modification of objects) is distinguished from one-sided external causation (production or destruction of objects)

[23] A once popular example of female wisdom and beauty is incorrect — at least politically.

[24] They should be understood like Wolffian *essentialia* and *modi*; *attributa* are superfluous as already determined by essentialia. Cf. Wolff (1962), §§ 120–123.

[25] The causation between different individuals is the external causation. These two types of causation were often distinguished, however not always clearly. E.g. Leibniz in his *Monadology* explicitly denies external causation but accepts internal (it is the tendency — *appetition* — of a monad).

[26] At the moment this is a very crude idea which needs further elaboration. You cannot e.g. treat the conception of a child (or its birth) as the destruction of its parents.

[27] Cf. the oriental metaphor of a soul losing its individuality in the Absolute: a drop is returning to the ocean.

in Kantian table of categories: in IIIrd group (categories of relation) we have the category of cause/effect and the category of mutual interaction. In Ingarden's existential ontology we find the difference between derivative and dependent modes of being which also corresponds to the two types of external causation (cf. Ingarden 1964–1965, §§ 13, 14). The scheme of the one-sided external causation is also consistent with the classical thesis *corruptio unius est generatio alterius*: generation has always its substrate(s) and corruption its product(s).

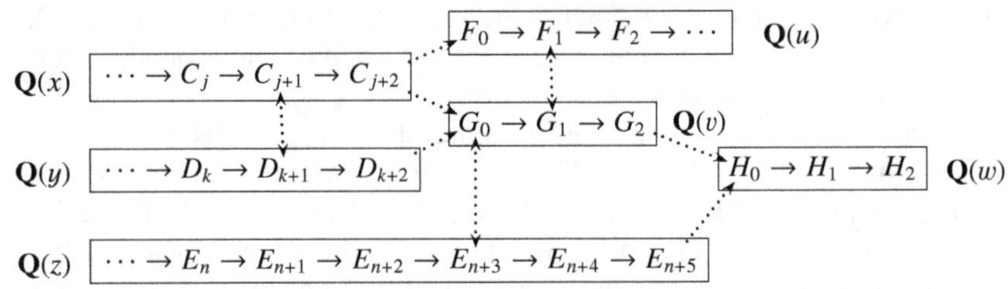

The unity of the set $\mathbf{Q}(x)$ of all possible determinations of x is established by the internal causation (continuous arrows): the causation which neither produces, destroys, nor involves objects in mutual interaction. Even when there is no single general principle of all determinations of x, its unity can be structurally determined as a maximal set of determinations causally but not externally connected with the present determination of x. The external determination can be defined as being mutual (p is a direct antecendent of b iff it is its direct consequent, as well) or such that a consequent either has more than one direct antecedent or is not the only direct consequent of its antecedent.

Direct causal relation $\mathrm{C^d}$ is a relation in $\prod \mathfrak{R}$. It has two species: external ($\mathrm{C_e}$) and internal direct ($\mathrm{C^{id}}$) causation, $\mathrm{C_e} \cup \mathrm{C^{id}} = \mathrm{C^d}$, $\mathrm{C_e} \cap \mathrm{C^{id}} = \emptyset$. External causation consists of two subspecies: $\langle c, d\rangle \in \mathrm{C_e}$ iff $\langle c, d\rangle \in \mathrm{C_m}$ or $\langle c, d\rangle \in \mathrm{C_o}$, $\mathrm{C_m} \cap \mathrm{C_o} = \emptyset$, where mutual external causation $\mathrm{C_m}$:

$$\langle c, d\rangle \in \mathrm{C_m} \quad \text{iff} \quad \langle c, d\rangle \in \mathrm{C^d} \text{ and } \langle d, c\rangle \in \mathrm{C^d},$$

and one-sided external causation $\mathrm{C_o}$:

$$\langle c, d\rangle \in \mathrm{C_o} \quad \text{iff} \quad \text{either } \langle c, d\rangle \in \mathrm{C^d} \text{ and } \exists_{e \in \prod \mathfrak{R}, e \neq c} \langle e, d\rangle \in \mathrm{C^d}$$
$$\text{or } \langle c, d\rangle \in \mathrm{C^d} \text{ and } \exists_{e \in \prod \mathfrak{R}, e \neq d} \langle c, e\rangle \in \mathrm{C^d}.$$

Transitional closure $\mathrm{C_i}$ of the internal direct causation $\mathrm{C^{id}}$ allows for the definition of the complete qualitative characteristic $\mathbf{Q}(x)$ of an object x:

if D is some qualitative characteristic of x then

$$d \in \mathbf{Q}(x) \quad \text{iff} \quad d = D \text{ or } \langle d, D\rangle \in \mathrm{C_i} \text{ or } \langle D, d\rangle \in \mathrm{C_i}.$$

7. The interplay of the internal and external causality is the driving factor of the universe: the internal activity consists in constant attempts of reaching the balance in determinations of objects while the external causation, on the other hand, disturbs internal balance of objects because of their mutual interactions, destructions and generations. None of these two factors would work in isolation. This constant swing of the internal balance and external action can be compared to the ball bouncing off a floor.[28] Its velocity in the highest position is null and reaches maximum at the floor. The first state can be interpreted as its internal balance (no deformation), the other being the state of interaction with the floor resulting in deformation. Its bouncing off is the result of the tendency to regain the internal balance (removing deformation).

Ingarden says that the constitutive nature of an object is its "what" and he assumes that it is something qualitative, although not a property. A subject qualified by its nature would be something like a hub of a spoked wheel or a dummy wearing different clothes: one immovable thing (permanent qualitative characteristic) which supports movable things (changing properties). As far as properties were existentially autonomous (having their qualifications in themselves) the constitutive nature would have to be "subject's own" qualification. The relation of properties to their subject becomes similar to a relation between objects, where properties are something like objects of higher order. It is not the case when heterogenic properties transfer their qualities directly to the subject — in this case it is qualified immediately by them and does not need any "deeper" qualification, like Lockean primary qualities. A subject is not a "sub-object" ("bare object"[29]), it is only a unifying factor of an object. Nor does it have to be container-like (I think that the term "substratum" suggests this). Properties of an object are not unified by the underlying subject like the components of a meal put on one plate: their unity has dynamic character provided by causal dependencies. The constitutive nature of x is the internal causation connecting elements of $\mathbf{Q}(x)$. As such it has no qualitative character, it is formal in the sense of being the connection and not what is connected. But it is an individual form, connecting only the elements of $\mathbf{Q}(x)$. It is not qualitative but it expresses itself qualitatively: produces new determinations out of given ones. Each determination $d \in \mathbf{Q}(x)$ is the determination of the momentary limits of x in this sense that it is the result of the constant interplay of internal and external factors. The difference between the qualitative matter of a property and the form of a property is that between an element of $\mathbf{Q}(x)$ taken in abstraction from its causal connections and causal connections in abstraction from what they connect. Heteronomy of a property consisting in referring the quality of the property to its subject is just

[28] The idea of the constant swing of the universe between internal and external is Whiteheadian. Cf. Whitehead (1979, p. 151).

[29] Like Gustav Bergmann's "bare particular".

referring the qualities in themselves to laws of nature which govern the transition of given properties into new ones. Thus the nature of an individual is subordinate to the Nature as a whole.

University of Łódź
Department of Logic
ul. Kopcińskiego 16/18
90-232 Łódź
Poland
e-mail: rosiak@filozof.uni.lodz.pl

REFERENCES

Ingarden, R. (1964–1965). *Der Streit um die Existenz der Welt.* 2nd edition. Tübingen: Niemeyer.

Ingarden, R. (1964). *Time and Modes of Being*. Translated by H. Michejda. Springfield Ill: Charles C. Thomas Publ. Fragments of only existential ontology from Ingarden (1964–1965).

Ingarden, R. (1987). *Spór o istnienie świata*, vol. 1–2. 3rd edition. Warszawa: PWN.

Locke, J. (1979 [1689]). *An Essay Concerning Human Understanding*. Oxford: Oxford Univ. Press.

Leibniz, G. W. (1714). *Monadologie*. Vienna.

Whitehead, A. N. (1979). *Process and Reality*. 2nd corrected edition. NY, London: The Free Press.

Wolff, Ch. (1962). *Philosophia prima sive Ontologia*. Hildesheim: Olms.

Tadeusz Szubka

THE METAPHYSICAL REALISM DEBATE: WHAT IS AT STAKE?*

ABSTRACT. The realism debate concerns the relationship of our beliefs, thoughts and language to the world or universe, and hence involves a number of fundamental questions ranging from metaphysics through epistemology to semantics and philosophy of language. While a few philosophers take it as an inevitable feature of the debate and try to advance it by coping simultaneously with all those questions, a number of others insists that the approach of this kind leads merely to confusions and misunderstandings. They usually suggest that the metaphysical or ontological aspects of it should be kept separate from such epistemological and semantic issues as the possibility of absolute knowledge, the correspondence theory of truth, or the truth-conditional theory of meaning. In other words, there is such a thing as pure or simple metaphysical realism that may be endorsed and defended, or undermined and rejected. The aim of the paper is to raise some doubts about that metaphilosophical strategy, and to argue that the comprehensive approach to the realism debate, in which the metaphysical issues are combined with — at least — some epistemological matters, is not so much caused by confusions and misunderstandings, but forced, as it were, by its subject matter and the philosophical nature of the debate.

In spite of various recent efforts to diminish its importance, the realism debate seems to remain at the center of philosophical discussions. To some extent that is not surprising since the debate concerns the relationship of our beliefs, thoughts and language to the world or universe, and hence involves a number of fundamental questions ranging from metaphysics through epistemology to semantics and philosophy of language. While a few philosophers (e.g. Hilary Putnam) take it as an inevitable feature of the debate and try to advance it by coping simultaneously with all those questions, a number of others insists that the approach of this kind leads merely to confusions and misunderstandings, and — for the most part — is responsible for the apparent lack of progress in the debate. They usually suggest that the metaphysical or ontological aspects of it should be kept separate from such epistemological and semantic issues as the possibility of absolute knowledge, the correspondence theory of truth, or the truth-conditional theory of meaning. For instance, John R. Searle boldly claims that it is a major confusion "to suppose that realism is identical with or at least implies the correspondence theory of truth," and a serious misconception "to suppose that there is something epistemic about realism", as well as an outright mistake "to sup-

*The paper is dedicated to Professor Jerzy Perzanowski, on the occasion of his 60th birthday, with appreciation and admiration for his contribution to analytical philosophy.

In: J. Malinowski and A. Pietruszczak (eds.), *Essays in Logic and Ontology* (*Poznań Studies in the Philosophy of the Sciences and the Humanities*, vol. 91), 301–316. Amsterdam/New York, NY: Rodopi, 2006.

pose that realism is committed to the theory that there is one best vocabulary for describing reality, that reality itself must determine how it should be described" (Searle 1995, p. 154–155).

Pronouncements of that kind provide support for the attempts to formulate the relevant aspects of the realism debate in exclusively metaphysical terms and encourage the idea that only by keeping them as far as possible from epistemological and semantic questions, a genuine progress can be made in it. In other words, there is such a thing as pure or simple metaphysical realism that may be endorsed and defended, or undermine and rejected.[1]

The aim of my paper is to raise some doubts about that metaphilosophical strategy. In order to do that I shall first consider various accounts of metaphysical realism found in recent literature and try to identify its crucial concepts. One of them is the notion of objectivity. Therefore, subsequently, I shall distinguish a few senses of that notion and eventually conclude that it is of not much use in the account of what the metaphysical realism debate is about. Then I turn to an attempt to analyze apparently the most significant concept for metaphysical realism, namely the concept of mind-independence, and shall argue that it is far more complicated than it is usually assumed. To make it philosophically relevant one has to understand that notion transcendentally, and that inevitably introduces some epistemological issues into the metaphysical realism debate. Hence the comprehensive approach to the realism debate, in which the metaphysical issues are combined with — at least — some epistemological matters, is not so much caused by confusions and misunderstandings, but forced, as it were, by its subject matter and the philosophical nature of the debate.

1. The Accounts of Metaphysical Realism

In its most basic sense, metaphysical realism is a doctrine that certain entities or items in a given domain are real, that is to say, genuinely or really exist. The domain in question can be all-inclusive, nothing less than the whole world or universe, and if so the resulting realism will be global. But although such a comprehensive doctrine may be held[2], usually one encounters more restricted and partial metaphysical realisms: concerning common sense or ordinary objects, concerning the entities postulated by scientific theories, concerning mental states, concerning moral properties, and so on. As a result of this, we have

[1] Except Searle, another vigorous advocate of that idea is Michael Devitt. See his (1997).

[2] It is certainly an exaggeration to maintain that "except in the vulgar sense, one is not a realist *tout court*: one is a realist with respect to some or other type of subject matter" (McGinn 1979, p. 247). As Alston notices, one is tempted to react to that claim by invoking the memorable words of Thomas Reid: "In this division, to my great humiliation, I find myself classed with the vulgar" (Alston 1997, p. 60).

common-sense realism, scientific realism, mental realism, moral or ethical realism, and so on.[3]

But regardless of its scope, the gist of metaphysical realism consists in the way of spelling out the claim that the entities in a given domain are "real, that is to say, genuinely or really exist". Let us consider a few examples of definitions of metaphysical realism that somehow unfold this enigmatic and potentially ambiguous phrase.

Perhaps the most often encountered accounts of realism and real existence are in terms of some kind of independence. Thus for John Haldane metaphysical or ontological realism is "the thesis that there is a world, independent of thought and language" (Haldane 1993, p. 15). The notational variant of the same account alludes John Heil in his survey of recent work in realism and anti-realism: "realists hold, while anti-realists deny, that the world is what it is independent of minds, languages, or conceptual schemes" (Heil 1989, p. 66). Another way of putting the same idea is to say that one central image essential to realism is of "a world that is there *anyway*, independently of us" (Railton & Rosen 1995, p. 433).[4] And more recently John Searle describes metaphysical realism as the view "that there exists a real world that is totally and absolutely independent of all of our representations, all of our thoughts, feelings, opinions, language, discourse, texts, and so on" (1998, p. 14). Realism of that sort, he claims, is not committed to any particular view about the nature of reality (whether it is e.g. physical or mental), but merely to the view "*that there is a way that things are that is logically independent of all human representations*" (Searle 1995, p. 155).

The more elaborated and detailed accounts of realism tend to introduce at least one additional idea: the idea of objectivity or the idea of objective existence.[5] Michael Devitt, who focuses his attention on metaphysical realism about the physical world defines it as a view according to which "tokens of most common-sense, and scientific, physical types objectively exist independently of

---

[3] Searle introduces here still another term, namely "external realism" that refers to realism about the physical world broadly conceived, as opposed to realism about mathematical objects or about ethical facts (Searle 1998, p. 13).

[4] The idea of a world being there anyway does not introduce here, as it may appear at first glance, a conceptual component different from the idea of an independent world. It is just another, more figurative manner of saying that the world in question is a certain way, "logically" independent from its being believed or known by us to be so. For clarificatory remarks about that expression see Stroud (2000), pp. 23–24.

[5] One may encounter also accounts that explicitly invoke the idea of externality, of something that exists out there, that is, outside mind or language. But that idea carries some spatial connotations that are not suitable for descriptions of certain forms of realism. John Searle, who makes heavy use of that idea, appears sometimes to be aware of that danger, and then suggests that he gives only an account of metaphysical realism in respect to the physical world (see footnote 3) But there are passages in which he seems to apply the idea across the board. For instance, he writes: "I use the metaphor of 'external' to mark the fact that the view in question holds that reality exists outside of, or external to, our system of representation" (1995, p. 154).

the mental" (Devitt 1999, p. 90). In emphasizing the objectivity of the world, metaphysical realists want to convey the idea that the world

> is not *constituted by* our knowledge, by our epistemic values, by our capacity to refer to it, by the synthesizing power of the mind, nor by our imposition of concepts, theories, or languages; it is not limited by what we can believe or discover (Devitt 1999, p. 91).

The idea of objectivity plays also a crucial role in more comprehensive characterizations of realism proposed, respectively, by Crispin Wright and Simon Blackburn. Wright gives the following account of it:

> Realism is a mixture of modesty and presumption. It modestly allows that humankind confronts an objective world, something almost entirely not of our making, possessing a host of occasional features which may pass altogether unnoticed by human consciousness and whose innermost nomological secrets may remain forever hidden from us. However, it presumes that we are, by and large and in favourable circumstances, capable of acquiring knowledge of the world and of understanding it (Wright 1987, p. 1).

For Blackburn realism may be described as a doctrine that maintains:

> Facts are facts. The world is largely independent of us and the way we are; it has its things, facts, properties, events, and states of affairs that have their own objective nature; our responses are explained by the world and are intended to match the world; our judgments are supposed to reflect, or depict, or mirror, or correspond to the way the world is; success in judgment is a matter of getting the way of the world right; or successfully describing how things stand (Blackburn 1999, p. 48).

It is rather obvious that those two comprehensive accounts combine the idea of metaphysical realism with some features of distinctively epistemological realism, that is a doctrine about the nature and scope of our knowledge of the world. For the sake of argument I shall put aside right now the latter, although eventually I will have to conclude that some epistemological ingredients are inseparable from metaphysical realism.

The above examples taken from the vast recent literature on realism suggest that the elucidation of metaphysical realism is essentially based on two notions: independence and objectivity. On the face of it they seem to be quite distinct ideas, each of them distinctively contributing to the task of clarification of what the thesis of metaphysical realism amounts to. But I think that in this case the appearances are deceiving. Objectivity does not seem to do much work here.

2. Objectivity

There are several distinctive notions of objectivity and, unfortunately, they are quite often conflated. To begin with, one should clearly distinguish and put aside the *ordinary* or *plain* notion of objectivity. According to that notion our inquiries and their results are objective when they are guided by undistorted evidence, when our judgments are unbiased and shaped not by our subjective predilections but mostly by impartial reasons, etc. It is implausible that objectivity in this sense is one of the key concepts defining metaphysical realism. It is pretty non-committal category that governs our discourse already on pre-philosophical level, and it is available both for advocates and opponents of metaphysical realism, as well as for all those who do not enter that philosophical debate.

Perhaps more pertinent to the question of metaphysical realism is not plain objectivity in itself, but rather its philosophical explication and justification. Such an elaboration yields a distinctive sense of objectivity, that may be called *epistemic or epistemological* objectivity. Here is one characteristic example of that elaboration, proposed by Joseph Raz (1999).

Raz starts his account with idea that we distinguish between various types or domains of thoughts and propositions, some of which we consider as objective and some as subjective. He does not attempt to define that objectivity ("domain-objectivity" is his term) or to give necessary and sufficient conditions for a domain to be objective. He elucidates that notion by formulating a number of conditions that partially characterize it. Let me mention four of them. The first condition states the possibility of knowledge, namely, a domain is objective only if it can express knowledge, or to be about something of which one can have knowledge. The second condition concerns the possibility of error: "a domain is objective only if thoughts which belong to it obey the distinction between seeming and being, that is, only if the fact that I think that things are so and so does not constitute their being so and so" (Raz 1999, p. 123). The third condition ensures the existence of proper reasons, namely it requires from a domain of objective thoughts that there should be facts or other considerations that are reasons for endorsing those thoughts as true or correct. The fourth condition is about independence: "thoughts are objective only if they are about a reality which exists independently of them, that is, whose truth is independent of the fact that those thoughts are entertained" (Raz 1999, p. 25).[6]

The fourth condition may be found by many to be a controversial one. Raz makes an attempt to weaken it slightly by insisting that he takes the expression "about a reality" in a very broad but also non-committal sense. For instance, mathematical propositions concerning numbers and their relations are to the same extent about reality as statements concerning physical objects, yet without entailing any particular view about the nature of mathematical entities. Keeping that in mind, one should presumably agree, Raz maintains, that there are only marginal and degenerate exceptions to that condition. But whether one

[6]Why one does not cut that long story short by providing the following brief and simple account of epistemological objectivity: "A subject matter is objective or relates to an objective reality if and only if there are questions about it that admit of answers that are substantially true" (Wiggins 1990, p. 66)? The answer can also be brief and simple: because it is rather difficult to specify what "substantially true" exactly amounts to, and because that account presupposes more ontological, and perhaps more basic, idea of objectivity, indicated by the expression "an objective reality".

would agree or not is beyond the point in the present context. What matters here is that one of the underlying ideas of domain-objectivity is the notion of independent existence of a reality, or truth independence, that boils down to the idea that propositions or beliefs are true in virtue of a reality independent of a thinker's activity of entertaining those beliefs or propositions. Thus the notion of objectivity so explicated clearly overlaps with the idea of mind-independence, or perhaps is even parasitic upon it.[7] Therefore it cannot have any distinctive contribution to the clarification of metaphysical realism. If one expects that the idea of objectivity is able to make such a distinctive contribution, one should look for it by considering other senses of this idea.

There are some senses of objectivity considered to be mainly ontological, and *prima facie* they seem to be more pertinent for the task of defining metaphysical realism. For instance, the distinction is often made between subjective states and occurrences and objective ones. The former are taken to be parts of mental life of a given subject whose mode of existence makes them available in a special privileged way for that subject, and unavailable in that way for other subjects. Those states and occurrences include feelings, beliefs, and desires of a given subject. The latter are all other states and facts; they are in principle intersubjectively accessible, provided they are not unknowable. The paradigmatic examples of them are various physical facts and, more generally, whatever happens in the common public space, somewhere outside the subject in the external world.

How closely is this distinction between subjective and objective related to the meaning of the words "objective" and "objectively" that one encounters in the definitions of metaphysical realism? Certainly more closely than objectivity either in the plain sense, or in the epistemological sense. But still not close enough to support the claim that those words apply only to the essentially non-subjective or intersubjective phenomena. There is no reason why only they are able to be items, to repeat Devitt's words, "not *constituted by* our knowledge, by our epistemic values, by our capacity to refer to it, by the synthesizing power of the mind, nor by our imposition of concepts, theories, or languages". That requirement is also met by such subjective states as my pain or fear. For instance, in no sense is my really painful, but surely subjective, headache constituted by my knowledge. For Wright an objective world of the realist is the world "almost entirely not of our making". Presumably various subjective states and occurrences, as pains and fears, belong to it, because they are not made by us. There is also no obstacle why they cannot have, to use Blackburn's phrase, "their own objective nature". Apparently my being in pain due to a toothache is different

[7] To establish a parasitic claim one would have to show that other conditions Raz mentions also invokes, or alludes to, the idea of an independently existing reality. I think that is a feasible task, although I am unable to undertake it here.

from my being in pain due to a headache. And the difference consists not merely in distinct causes of those two mental states but also in their qualitative character. To put it briefly, those two occurrences of pain have different objective natures.

Thus the above-mentioned philosophers must have in mind a rather different ontological sense of objectivity. Unfortunately, they do not make it fully explicit. However, an attempt to do that contains another recent account of realism proposed by Phillip Pettit (1991). He sees realism as a doctrine consisting of three main components, one of which is objectivism. What does objectivism require — Pettit asks — from a given area of discourse whose participants posit the existence of distinctive entities, as well as assert various things about them? It requires that

> the objects posited exist and have their character fixed independently of the dispositions of participants in the discourse to assert and believe things about them. Thus the epistemic states of the participants have no causal influence on the existence or character of those objects, nor are the objects non-causally dependent — say, dependent in a supervenient way — on such epistemic states. In short, the entities posited in the discourse enjoy a substantial kind of objectivity (Pettit 1991, p. 590).

The gist of this elucidation of objectivism and objectivity boils down to the claim asserting lack of both causal and non-causal dependence of the entities posited in a given discourse from some mental states (that is, epistemic states) of participants in this discourse. To simplify the matter a bit, we can say that those entities are mind-independent. If this is indeed so, then one is entitled to conclude that the idea of ontological objectivity, conceived in such a way, does not contribute any distinctive content at all to the account of metaphysical realism. It looks like then that everything hangs on the idea of mind-independence, or of independence from mental states of some sort.[8]

3. Mind-independence

We are usually assured that the independence in question is not the ordinary causal independence. Even the staunch realist, as anyone else for that matter, is prepared to admit that the world depends in various ways upon the mental activities of human beings. The evidence of that dependence abounds: we throw stones, break bones, make cars, and so on. What sort of independence then is at

[8]Given that, it is rather surprising how little attention this notion has received in recent work on metaphysical realism, although it is quite often emphasized that the ontological independence of things is crucial or essential component of realism (see e.g. Rescher 2000, pp. 96 and 104). One exception is John Wright (1997) who tries to built a comprehensive account of the realism debate around the theory of dependence and independence. However, for rather unconvincing reasons he rejects a counterfactual version of that theory (endorsed in this paper), and argues for its explanatory version (where dependence and independence is conceived in terms of explanatory priority), that seems to me more vulnerable to objections of various sorts.

stake here? It is a special kind of independence, that may be called, for want of a better term, existential or ontological independence.[9] It may be provisionary defined as follows:

(1) entities in a given domain are mind-independent if and only if they would have existed if there were no minds.[10]

For various reasons this is not a satisfactory formulation. The main source of trouble here is a very unspecific way in which mind is invoked in the definition. Of course, what is relevant here is not some kind of independence from *all* possible aspects of the mental. If it were so, then all philosophical accounts of mind (excluding eliminativism perhaps) would have to be antirealist. All philosophical theories of cultural and social phenomena would be antirealist as well. It seems to me that the best and simple way to block those absurd consequences is to say that what really matters here is not global mind-independence or mind-independence in general but independence merely from some mental states.[11] This is a rather obvious point but it is not always adequately acknowledged in the literature that discusses the issue of realism in general (in the above considered definitions, it is taken into account in Searle's description of realism and Pettit's account of objectivity). This leads us to the following amendment:

(2) entities in a given domain are independent of certain mental states, that is, epistemic states,[12] if and only if they would have existed if there were no such mental states.

Arguably, this is much better account but still far from satisfactory. For the realism debate about entities in a given domain is focused not on the independence from any epistemic states whatsoever but on the independence of those entities from the states whose aim is to form justified beliefs about them, or whose content amounts — at least in part — to representation of those entities. If one calls such states epistemic states *relevant* for the entities of a given domain, one arrives at the following formulation:

---

[9]Both those adjectives are not particularly illuminating. Nevertheless they are at least more appropriate than the adjective "logical" sometimes used in that context (for instance by Searle 1995, p. 155, and Stroud 2000, p. 22). The idea of logical independence makes sense as applied to the relationships holding between propositions or statements, but not as applied to ones holding between our minds and the world.

[10]Obviously this definition and its further modifications are suitable for a local realism. But they can easily be reformulated in the way appropriate for global realism.

[11]Another way of introducing that constraint is to say that what is at stake here is a special kind of mind-independence that may be called, say, cognitive or representational mind-independence. That strategy seems to be suggested by Schmitt (1995, p. 12).

[12]The category of epistemic states has to be conceived here as broadly as possible. It includes not only justified true beliefs, but also beliefs that do have those two features, as well as states that are perhaps not belief states in fully accomplished sense (for instance, basic perceptual experiences). Epistemic states are simply states with representational or intentional content.

(3) entities in a given domain are independent of epistemic mental states relevant for those entities, if and only if they would have existed if there were no such mental states.

However, one can point out that this account is still unsatisfactory for another reason. What seems to be at stake in the metaphysical realism debate is not the bare existence of entities, but usually the question whether those entities would have existed unchanged if there were no relevant epistemic states, that is whether they were exactly the same entities if there were no relevant epistemic states. Thus still further modification of the formulation has to be made in order to accommodate this point. But this modification has to be made very carefully. Otherwise the anti-realist may easily win in each case by the following simple argument: it is impossible for the entities in a given domain to be exactly the same in absence of the relevant epistemic states. They could not be the same, since they would lose some of their relations, namely they would no longer be related to those epistemic states. This quick attempt of making metaphysical realism an impossible position may be undermined by including an appropriate proviso in the definition. Thus one ends up with the following formulation:

(4) entities in a given domain are independent of epistemic mental states relevant for those entities, if and only if they would have existed unchanged if there were no such mental states, except the minor change involved in no longer being related to those states.

Although the proviso incorporated in that account appears to be a bit *ad hoc* and the idea of entities existing unchanged is rather unclear, one cannot perhaps improve upon the formulation without invoking a heavy ontological machinery. So maybe as a general and relatively non-committal definition of existential mind-independence it should be satisfactory. However, there is a danger that without further elaboration of the idea of relevant epistemic states, the proposed definition may yield a rather distorted picture what the metaphysical realism debate is really about.

4. Naturalism *versus* Transcendentalism

The proposed account of mind-independence is encapsulated in a counterfactual conditional stating what would happen or obtain under the unfulfilled assumption that there were no relevant mental states.[13] Thus the question arises: does this conditional really hold good and if so, how does one settle the matter?

[13]The counterfactual is sometimes presented as a "consequence" of describing realism in terms of mind-independence. For instance, John Searle writes: "I have defined realism as the view that the world exists independently of our representations of it. This has the consequence that if we had never existed, if there had never been any representations — any statements, beliefs, perceptions, thoughts, etc. — most of the world would have remained unaffected" (1995, p. 153).

There is no need to worry much about that, one might reply. Counterfactuals are almost an indispensable tool in clarification and definition of various philosophical categories, and there are many ways of grasping and assessing their content. Moreover, there are ways to moderate the contrary-to-fact character of the conditional in question by pointing out that indeed its antecedent is satisfied in plenty of local cases when a given cognitive mental state has not yet come into existence or no longer exists, but presumably the object or objects onto which it is directed remain almost completely unchanged. Even in the global case it is not too difficult to imagine a period of time well back in the past when there were no minds and the world in many of its underlying features was more or less as it is today. It is also quite easy to imagine some future unfortunate course of events causing disappearance of all mental life and leaving the rest of our environment mostly intact.

The reply of that kind tends inevitably to suggest that realism is simply an ordinary doctrine on a par with various other scientific or historical doctrines, claiming, for example, that given atoms are really distinct ones, that a certain biological species is independent and distinct from another, that two battles that took place in a certain war were indeed separate battles, and so on. The guiding picture beyond all those examples and perhaps even beyond that approach to the issue of mind-independence is an intuitive image of two spatially or temporally distinct entities. One may even say that the whole metaphysical realism debate begins to look like an extremely general scientific debate "how one physical part of the world is related to other physical parts of the world" (Stroud 2000, p. 6). In that spirit various naturalistically-oriented philosophers defend metaphysical realism: there are, they say, some complicated biological organisms having mental states, on the one hand, and, on the other hand, out there in their environment exist various objects that those states represent. It is obvious that the objects in question are autonomous; they are not part of the mental states representing them and they are not constituted by them. That line seems to be endorsed by Michael Devitt in his naturalistic defense of realism. The metaphysical realism he defends amounts to nothing more than simple acceptance of objective and independent existence of entities postulated by common sense and science. That means, he explains, that

> Science and common sense are not, for the most part, to be "reinterpreted". It is not just that our experiences are *as if* there are cats, there are cats. It is not just that the observable world is *as if* there are atoms, there are atoms (Devitt 1999, p. 91).

But does such a simple realism, being nothing more than general endorsement of common sense and science, need any defense? After all, any denial of metaphysical realism conceived in that way, has unavoidably to be, as Blackburn puts it, "not a subtle philosophical position, but a simple ignorance of empirical and scientific fact" (1999, p. 51). It is extremely implausible that the advocates of various versions of metaphysical antirealism reject to admit that there are cats, atoms and other mundane objects, and that their existence is independent from what we think or believe about them.

Blackburn suggests that what is at stake in the realism debate is not mind-independence in that ordinary or scientific sense ("first-order" mind-independence). This is because the realism debate is a reflective philosophical debate conducted on a different level than ordinary and scientific discourse. The positions taken in that debate are in a certain sense "metatheoretical doctrines" that involve "such concepts as description, reflection, truth, and meaning" (Blackburn 1999, p. 51). And although we describe the debate in familiar ordinary terms, by saying, for example, that it concerns the relationship between ourselves, or our minds, and the world, we do not expect to resolve it by finding some common characteristics of our cognitive interactions with the world, both in ordinary life and in science. We rather try to step outside those particular interactions and from such a vantage point of view establish what is the relationship of mind to the world, that is to say, decide in what sense the world is from that perspective mind-independent or mind-dependent. Thus the metaphysical realism debate is the debate about that kind of "second-order" mind-independence.

The particularity and uniqueness of the realism debate is also emphasized by Stroud. He claims that what we try to achieve in it is

> a certain kind of understanding of ourselves and of reality that will make intelligible to us in general terms the relation in which we stand to that reality — or perhaps the relation in which we really stand to it (Stroud 2000, p. x).

Of course, in the realism debate one is not interested in the relation of all aspects of our human nature to the world but rather in the relation in which our own perceptions, thoughts, and beliefs, stand to the way the world is. One cannot settle it just by describing in the most general terms the way the world is, and by providing some general account of our cognitive responses. What is really needed here is an attempt to answer the question about the relation of those responses to the world from the point of view that is not immersed in them, that is as it were located outside of them. We are required to stand back from our perceptions, thoughts, and beliefs, while at the same time reflect upon the way they are related to the world. And this is where the special character of the kind of understanding relevant for the realism debate lies, and not in its other features (for instance, in its abstractness and generality). However, it is extremely hard to pin down and describe the nature of that kind of understanding. One cannot simply say that this just a detached understanding, since it is a detached understanding of a very special sort. Indeed, one of the greatest difficulties of the realism debate is "the ease with which we shift back and forth between philosophical and more mundane questions about reality without noticing the differences between them" (Stroud 2000, p. 14).

What is suggested by Blackburn and Stroud about the status of the realism debate, including its metaphysical version, sounds extremely plausible.[14] However, their imagery may be misleading and it requires some elaboration and illustration. The source of misunderstandings may be in particular Blackburn's distinction between the first-order notions and questions and the second-order ones, as well as between ordinary or scientific theories of the world and their philosophical metatheory. The distinction strongly suggests the following picture of the realism debate. We have, on the one hand, various particular discourses that deal directly with the world: they try to describe and explain its different regions. On the other hand, we have philosophy; however, while doing it we are not engaged in one of those first-order discourses, since there is no such thing as a specific area of the world that only philosophy is able to investigate. The subject matter of philosophy, including the realism debate, is not the world in itself, or some aspect of it, but rather our cognitive enterprises and their relation to the world. Thus philosophy does not issue in theories about the world, but in general theories about those theories, i.e., in metatheories. If so, then the realism debate, even the metaphysical one, is not about the way the world is, but rather about the way our thoughts, judgments and theories are related to the world. That picture seems to have two shortcomings. First, it does not put enough emphasis on the fact that in order to describe the relation of our cognitive enterprises to the world, one has to commit oneself to some general idea of the way the world is, and, in order to avoid the charge of being uncritical, make a good case for it. Hence, to some extent at least, the metaphysical realism debate is a first-order debate, however peculiar and unspecific. Second, it encourages one to conceive philosophy on a par with other more familiar metatheories, as metalogic or metamathematics, and thus to distort its real nature. After all, the metaphysical realism debate is quite different in character, and not only in subject matter, from the debate about the most effective strategy of proving consistency of a given formal system, or from any other specific metatheoretical debate.

The imagery invoked by Stroud in his account of uniquely philosophical character of the realism debate is perhaps less prone to cause misunderstandings, but it suffers from being extremely indirect. That is, Stroud is concerned to tell us what the philosophical realism debate is not, but avoids to answer the direct question, what it really is. Hence the reader may be left with the impression

[14]It is not surprising that Devitt, criticized by Blackburn for missing the point of the genuinely philosophical realism debate, thinks otherwise. He maintains that Blackburn's insistence upon uniqueness of the realism debate is simply an indirect way of expressing his emotional uneasiness that a straightforward first-order approach to the realism debate, within the framework of naturalism, will take "the philosophical fun out of the realism debate", will remove "a lot of the debate's *excitement*, a lot of its *aura of significance*" (Devitt 1999, p. 105). I take it as obvious that Blackburn is for something more important here: the issue is not about emotional excitement or feeling of significance, but about the possibility of a certain theoretical approach.

that when Stroud talks about a special way in which the detachment necessary for the realism debate should be understood, he is driven by the idea of some trans-empirical or transcendental subject located outside the world but somehow grasping it. Since it is widely agreed among contemporary philosophers that such an idea of a separate knowing subject is just a myth, and any attempt to put oneself in its position, is will-o'-the-wisp, the urgent question arises: what else may be at issue in Stroud's insistence upon the uniqueness of the philosophical realism debate?

I think the best way to answer that question is to begin with dissociating the attempts to achieve a transcendental point of view from the dubious and incredible metaphysics of the trans-empirical subject. In other words, one should not assume that the idea of some detached point of view — described metaphorically as God's eye view, the view from nowhere, or the perspective of the "cosmic exile" — necessarily presupposes the idea of a mysterious subject or mind being outside the world and enjoying a unique epistemic access to it. The transcendental point of view may be achieved by gradually imposing various constraints upon our ordinary epistemic processes. One may first try to identify and exclude "anthropocentric" modes of knowing, that is, those inextricably tied up with peculiarities of our human perspective upon the world. For instance, it is quite likely that one will come to the conclusion that our perception of things as coloured does not reflect the features of the world in itself, but rather the way in which it impinges upon our sense of sight. And that would lead to antirealism about colour, which is a version of metaphysical local antirealism.[15] Moreover, one may also try to achieve the transcendental point of view by drastically raising the epistemic standards. Perhaps the best known requirement in that respect is expressed in the claims that to take the transcendental point of view means to confine one's attention to the epistemic states and procedures that enjoy indubitability, infallibility or incorrigibility. No matter how incredible that requirement may sound today, it was emphasized in the twentieth century by phenomenology and by those members of that movement who made various attempts to resolve the metaphysical realism debate. Phenomenologists were inclined to think that those privileged epistemic states are components of "pure" or transcendental consciousness. The metaphysical problem of realism appears in that light as follows:

> it pertains to the existence or the nonexistence of the real world, but it results from the difference between the fundamental dubitability of the real world and the indubitability of our consciousness, or from the transcendence of the real world and its components in relation to pure conscious experience. If it could be demonstrated that *no real world exists*

[15] It is extremely difficult to identify and specify the anthropocentric elements in our cognition. We have to rely basically upon what we are able to imagine and conceive, and those abilities themselves may be contaminated and limited by our local human perspective. For some remarks about that predicament see Stroud 2000, pp. 32–41.

> in fact, although the actual course of all the conscious experiences in which that world and its components are given remained unchanged, there would be only *one* domain of being — that of conscious experience (Ingarden 1964, pp. 14–15).

It is well know that the founder of phenomenology, Edmund Husserl, came eventually to the conclusion that from the transcendental perspective it can be demonstrated that the real world does not exist, and more precisely, that it is constituted by the acts of pure consciousness. Thus his position was clearly a version of fairly global metaphysical antirealism, according to which the world is mind-dependent. Of course, it did not amount to the plainly absurd view that rocks, cats, atoms, genes, and other ordinary and scientific objects do not really exist or are simply made by us. It was a sophisticated, though perhaps ultimately indefensible, philosophical position that by taking the transcendental point of view one is committed to metaphysical antirealism.

5. Conclusion

What I have suggested above amounts essentially to distinguishing two senses of mind-independence — natural or ordinary, and transcendental or philosophical — and to proposing a bifurcation of the realism debate. Of course, that is reminiscent of the well-known Kantian distinction between empirical realism and empirical idealism on the one side, and transcendental realism and transcendental idealism on the other side. However, it needs to be stressed that at the empirical or natural level we are unable to generate the realism debate in the proper philosophical sense. We just get a strange opposition of two views, one of which is a trivial generalization over ordinary and scientific opinions, and the other its patently absurd and false denial. No doubt, one may attempt to provide a philosophical account of why any other approach to the realism debate is impossible (some naturalists try to provide that). But while giving such an account one should make clear that one takes minimalist, or quietist, or skeptical attitude towards the realism debate as it has been traditionally understood by philosophers.[16] However engaging in the strictly philosophical realism debate means that even in its metaphysical variety, one is absolutely unable to keep it metaphysically pure and simple. It will always be entangled with epistemology, because what is at stake in it is the question of independence of the world from certain epistemic states envisaged from a special epistemological point of view.[17]

---

[16]This is just what Devitt does not make clear. As Blackburn notices, he confusingly presents a straightforward minimalism "as a kind of vindication of realism" (1999, p. 51).

[17]Various versions of that paper were presented as seminar talks at the Institute for Advanced Studies in the Humanities, University of Edinburgh (27 April 2000), Center for Philosophy of Science, University of Pittsburgh (24 October 2000), Loyola Marymount University (16 March 2001),

Catholic University of Lublin
Faculty of Philosophy
Al. Racławickie 14
PL 20-950 Lublin
Poland
e-mail: szubka@kul.lublin.pl

Szczecin University
Institute of Philosophy
Krakowska 61/69
PL 71-017 Szczecin
Poland

REFERENCES

Alston, W. P. (1997). Realism and the Tasks of Epistemology. In: Kulp (1997), pp. 53–94.

Blackburn, S. (1999). Introduction to the Realism Debates. In: Hales (1999), pp. 47–51.

Devitt, M. (1997). *Realism and Truth*. 2nd edition with a new afterword. Princeton, NJ: Princeton University Press.

Devitt, M. (1999). A Naturalistic Defense of Realism. In: Hales (1999), pp. 90–105 (includes: Postscript to "A Naturalistic Defense of Realism", pp. 104–105).

Haldane, J. (1993). Mind-World Identity Theory and the Anti-Realist Challenge. In: J. Haldane and C. Wright (eds.), *Reality, Representation, and Projection*, pp. 15–37. New York: Oxford University Press.

Hales, S. D., ed. (1999). *Metaphysics: Contemporary Readings*. Belmont, CA: Wadsworth.

Heil, J. (1989). Recent Work in Realism and Anti-Realism. *Philosophical Books* **30**, 65–73.

Ingarden, R. (1964). *Time and Modes of Being*. Springfield: Thomas.

Kulp, C. B., ed. (1997). *Realism/Antirealism and Epistemology*. Lanham: Rowman & Littlefield.

McGinn, C. ([1979] 1999). An A Priori Argument for Realism. In: C. McGinn, *Knowledge and Reality: Selected Essays*, pp. 247–266. Oxford: Clarendon Press.

Pettit, P. (1991). Realism and Response-Dependence. *Mind* **100**, 587–626.

Putnam, H. (1990). *Realism with a Human Face*. Cambridge, MA: Harvard University Press.

Railton, P. & G. Rosen (1995). Realism. In: J. Kim and E. Sosa (eds.), *A Companion to Metaphysics*, pp. 433–437. Oxford: Blackwell.

---

and the University of Memphis (30 March 2001). On all those occasions valuable and stimulating comments were made. For further advice and encouragement I am grateful to Kit Fine, Terry Horgan, Leemon McHenry, and Mark Timmons. The paper was completed during my visit at the University of Pittsburgh, with much appreciated assistance of Fulbright Senior Research Grant.

Raz, J. (1999). Notes on Value and Objectivity. In: J. Raz, *Engaging Reason: On the Theory of Value and Action*, pp. 118–160. Oxford: Oxford University Press.

Rescher, N. (2000). *Process Philosophy: A Survey of Basic Issues*. Pittsburgh: University of Pittsburgh Press.

Schmitt, F. F. (1995). *Truth: A Primer*. Boulder: Westview Press.

Searle, J. R. (1995). *The Construction of Social Reality*. New York: Free Press.

Searle, J. R. (1998). *Mind, Language and Society: Philosophy in the Real World*. New York: Basic Books.

Stroud, B. (2000). *The Quest for Reality: Subjectivism and the Metaphysics of Colour*. New York: Oxford University Press.

Wiggins, D. (1990). Moral Cognitivism, Moral Relativism and Motivating Moral Beliefs. *Proceedings of the Aristotelian Society* **91**, 61–85.

Wright, C. (1987). *Realism, Meaning and Truth*. Oxford: Blackwell.

Wright, J. (1997). *Realism and Explanatory Priority*. Dordrecht: Kluwer.

Tomasz Jarmużek
Maciej Nowicki
Andrzej Pietruszczak

AN OUTLINE OF THE ANSELMIAN THEORY OF GOD

ABSTRACT. The article presents a formalization of Anselm's so-called Ontological Arguments from *Proslogion*. The main idea of our research is to stay to the original text as close as is possible. We show, against some common opinions, that

(i) the logic necessary for the formalization must be neither a purely sentential modal calculus, nor just non-modal first-order logic, but a modal first-order theory;
(ii) such logic cannot contain logical axiom $\ulcorner A \rightarrow \Diamond A \urcorner$;
(iii) none of Anselm's reasonings requires the assumptions that God is a consistent object or that existence of God is possible (in symbols "$\Diamond$Eg");
(iv) no such thing as the so-called Anselm's Principle (in symbols "$\Box$(Eg $\rightarrow$ $\Box$Eg)") is involved in any of the proofs;[1]
(v) Anselm's claims (that God exists in reality and that God necessarily exists in reality) can be obtained independently, hence there is no need for presenting them in an opposite order than Anselm did.[2]

Moreover we show a single line of reasoning underlying the whole *Proslogion* and allowing Anselm to deduce many theorems concerning God's nature. Last but not least we study the possibility of proving the uniqueness of God within the outlined theory.

1. Introduction

Let us begin with the reminder of the crucial fragments of *Proslogion*. The key part of the second chapter runs as follows:

> Indeed, we believe You to be something than which nothing greater can be thought. Or is there, then, no such nature [as You], for the Fool has said in his heart that God does not exist? But surely when this very same Fool hears my words "something than which nothing greater can be thought," he understands what he hears. And what he understands is in his understanding, even if he does not understand [i.e., judge] it to exist. For that a thing is in the understanding is distinct from understanding that [this] thing exists. For example, when a painter envisions what he is about to paint: he indeed has in his understanding that which he has not yet made, but he does not yet understand that it exists. But after he has painted [it]: he has in his understanding that which he has made, and he understands that it exists. So even the Fool is convinced that something than which nothing greater can be

[1] The assumption that existence of God is possible and "Anselm's Principle" are in (Hartshorne 1961; 1962) and in others; see, for example, (Perzanowski 1995).

[2] A different situation may be found in the reconstructions in Hartshorne style: first, the thesis "$\Box$Eg" is obtained and then "Eg" is detached by the logical axiom "$\Box$Eg $\rightarrow$ Eg".

In: J. Malinowski and A. Pietruszczak (eds.), *Essays in Logic and Ontology* (*Poznań Studies in the Philosophy of the Sciences and the Humanities*, vol. 91), 317–330. Amsterdam/New York, NY: Rodopi, 2006.

> thought is at least in his understanding; for when he hears of this [being], he understands [what he hears], and whatever is understood is in the understanding. But surely that than which a greater cannot be thought cannot be only in the understanding. For if it were only in the understanding, it could be thought to exist also in reality — something which is greater [than existing only in the understanding]. Therefore, if that than which a greater cannot be thought were only in the understanding, then that than which a greater cannot be thought would be that than which a greater can be thought! But surely this [conclusion] is impossible. Hence, without doubt, something than which a greater cannot be thought exists both in the understanding and in reality (Anselm 2000, pp. 93–94).

Moving forward to chapter III one can read that:

> Assuredly, this [being] exists so truly [i.e., really] that it cannot even be thought not to exist. For there can be thought to exist something which cannot be thought not to exist; and this thing is greater than that which can be thought not to exist. Therefore, if that than which a greater cannot be thought could be thought not to exist, then that than which a greater cannot be thought would not be that than which a greater cannot be thought — [a consequence] which is contradictory. Hence, something than which a greater cannot be thought exists so truly that it cannot even be thought not to exist (Anselm 2000, p. 94).

It is clear enough that Anselm wants to prove there two facts:

(a) that God exists in reality and

(b) that God exists so truly that He cannot even be thought not to exist.

Against some common opinions we claim that in *Proslogion* one can find not just one but two different and independent arguments concerning existence of God. In what follows we present a formalization of both of them. It should be stated however that it is not only a mere interpretation but an exact translation of the arguments into the realm of logic.[3] Of course many authors have claimed the same, but most of their efforts seem to be mere misinterpretations.

2. The Theory $\mathcal{A}$

2.1. *The Language of* $\mathcal{A}$

In order to sketch the Anselmian theory of God $\mathcal{A}$ we shall work with a system of modal first-order logic. Hence the alphabet of the language of $\mathcal{A}$, $L_{\mathcal{A}}$, contains individual variables: "x", "y", "z", ..., as well as logical symbols: "$\neg$", "$\wedge$", "$\vee$", "$\rightarrow$", "$\leftrightarrow$", "$\forall$" and "$\exists$". These symbols are to be understood as in classical logic.[4] Moreover, as $\mathcal{A}$ is supposed to be a modal theory, a kind of modality has

[3]We may say so even though we shall omit some steps Anselm made in his original proofs.

[4]Note that one does not necessarily have to read "$\rightarrow$" as the material implication, since it would be quite unreasonable to believe that this concept can be found in *Proslogion*; we rather suggest reading "$A \rightarrow B$" as "B follows from A". Therefore "$A \leftrightarrow B$" can be read as "A follows from B and B follows from A".

to be introduced. Thus let us "□" and "◇" represent epistemic (or *a priori*) necessity and possibility concerning what can be thought.

We shall also need some specific symbols. They are two predicates — one unary "E" and one binary ">". They should be interpreted as "exists in reality" and "is greater than", respectively. Finally, we take one individual constant "g"; it shall stand for "God".

2.2. *The Logic of the Theory $\mathcal{A}$*

Our modal theory $\mathcal{A}$ is built upon a certain modal lower predicate logic which is a development of the ordinary non-modal lower (first-order) predicate calculus, LPC. Thus, its logical theses and rules contain all theses and rules of LPC.[5] We also use a certain axiom and a certain rule characterizing epistemic necessity and possibility. The modal logic obtained this way is an LPC-version of the pure sentential modal logic M (cf. Chellas 1980, p. 234), therefore let us denote it by LPC-M_w.

The only modal axiom of LPC-M_w is

(df◇) $\quad \Diamond A \leftrightarrow \neg\Box\neg A\,.$

Moreover, we require that the set of theses of LPC-M_w is closed under the following LPC-version of *weak* monotonicity rule[6]

(RM_w) $\quad$ if $\ulcorner A \rightarrow B\urcorner \in$ LPC then $\ulcorner \Box A \rightarrow \Box B\urcorner \in$ LPC-M_w.

Hence LPC-M_w is LPC-monotonic in the sense that the monotonicity rule can be applied only to theses of LPC.

The following formulæ are among theses of LPC-M_w:

(df□) $\quad \Box A \leftrightarrow \neg\Diamond\neg A,$

(M) $\quad \Box(A \wedge B) \rightarrow (\Box A \wedge \Box B),$

(M◇) $\quad \Diamond A \vee \Diamond B \rightarrow \Diamond(A \vee B)\,.$

Moreover, the set LPC-M_w is closed under the following *weak* monotonicity rule:

($RM\Diamond_w$) if $\ulcorner A \rightarrow B\urcorner \in$ LPC then $\ulcorner \Diamond A \rightarrow \Diamond B\urcorner \in$ LPC-M_w.

Remark 1. Observe that the logic LPC-M_w is epistemic in the sense that none of its theses begins with "□" (cf. Lemmon's logics "E" in his (1957)).

---

[5] Of course, the set of theses of LPC includes the set of theses of propositional calculus, PC, and is closed under the rule of *modus ponens* (MP): $A \rightarrow B, A \vdash B$.

[6] The logic M is closed under the *strong* monotonicity rule

(RM) $\quad$ if $\ulcorner A \rightarrow B\urcorner \in$ M then $\ulcorner \Box A \rightarrow \Box B\urcorner \in$ M.

Remark 2. As "$\Diamond$" and "$\Box$" are epistemic modalities, they deal with what can be thought and with what must be thought, respectively. However, from a certain point of view, everything can be thought. Hence we rather suggest reading "$\Diamond A$" and "$\Box A$" as "it can be (consistently) thought that A" and "it is not the case that $\ulcorner \neg A \urcorner$ can be (consistently) thought", respectively.

Remark 3. Almost every modal system taken in order to formalize Anselm's Ontological Proof contains the following equivalent formulæ:

(T) $\Box A \to A$,

(T$\Diamond$) $A \to \Diamond A$,

thus being a stronger system than LPC-M_w. It should be clear however, that Anselm would rather not agree with them, since he would never claim that whatever is the case, can also be (consistently) thought. There are many things in the world that are far beyond human's conceptual capabilities. Thus neither (T$\Diamond$) nor (T) are suitable for the formalization of the notion of Anselmian conceivability.

2.3. *Deduction in* LPC-M_w. *Thesis of* $\mathcal{A}$

A formula A is *deducible* from a set of formulæ $\mathcal{S}$ in the logic LPC-M_w (written $\mathcal{S} \vdash_{\text{LPC-M}_w} A$) iff there is a finite sequence $A_1, \ldots, A_n$ such that $A_n = A$ and for each i, $1 \leqslant i \leqslant n$, either $A_i \in \mathcal{S}$, or A is a thesis of LPC-M_w, or there are $j, k < i$ such that $A_j = \ulcorner A_k \to A_i \urcorner$. In other words, A is *deducible* from $\mathcal{S}$ in LPC-M_w iff A can be obtained from $\mathcal{S}$ and theses of LPC-M_w by *modus ponens* (MP).

We obtain the following fact (see Lemmon 1977, p. 16): $\mathcal{S} \vdash_{\text{LPC-M}_w} A$ iff for some $\{B_1, \ldots, B_m\} \subseteq \mathcal{S}$ ($m \geqslant 0$) we have $\ulcorner (B_1 \wedge \cdots \wedge B_m) \to A \urcorner \in$ LPC-M_w. Hence we use the standard notion of *being deducible* in a modal logic; see e.g. (Chellas 1980; Lemmon 1977).

A formula of $L_{\mathcal{A}}$ is a *thesis of* $\mathcal{A}$ iff it is deducible from the set specific axiom of $\mathcal{A}$ (see below the axioms (A1) and (A2)) in the logic LPC-M_w, i.e., it can be obtained from {(A1), (A2)} and theses of the logic LPC-M_w by (MP)

2.4. *Specific Axioms of* $\mathcal{A}$

Besides the above theses of LPC-M_w the theory $\mathcal{A}$ has the following specific axiom:

(A1) $\neg \Diamond \exists_x\, x \succ \mathsf{g}\,.$

This sentence states the famous Anselmian phrase *aliquid quo nihil maius cogitari possit*; it shall be read explicitly as "it is not the case that it can be thought that there is an object greater than God".

Now, (A1) allows Anselm to deduce many theorems concerning the nature of God. Firstly, we may obtain the following lemma.

LEMMA 1. Let A be any formula of the language of $\mathcal{A}$. Then the formula

$$(\star) \qquad \neg\Diamond\exists_x(A \wedge x > \mathsf{g})$$

is thesis of $\mathcal{A}$, i.e., (A1) $\vdash_{\text{LPC-M}_\text{W}} (\star)$

PROOF. we obtain the following derivation:

1.	$\neg\Diamond\exists_x\, x > \mathsf{g}$	(A1)
2.	$\neg\Diamond\exists_x\, x > \mathsf{g} \rightarrow \neg\Diamond\exists_x(A \wedge x > \mathsf{g})$	LPC-M_W[7]
3.	$\neg\Diamond\exists_x(A \wedge x > \mathsf{g})$	1, 2 , and (MP) $\Box$

The strategy Anselm takes is now as follows. In order to prove that God possesses a certain property π assume that

$$\text{(A2)} \qquad \neg\pi\mathsf{g} \rightarrow \Diamond\exists_x(\pi x \wedge x > \mathsf{g}).$$

This statement is very intuitive. It says that if God does not have a certain property, then it can be thought an object having this property and, given that the property in question is a great-making one, being greater than God (see previously cited fragments of *Proslogion*). Of course, (A2) does not assume possibility or consistency of God. If God is simply inconsistent and $\pi\mathsf{g}$ does not hold then it is quite natural to think of an object which (i) would be consistent, (ii) would possess π, and hence (iii) would be greater than God (since God is inconsistent and does not possess π). If God is consistent then it is still natural to think of Him as having the property π; then this object of thought (or using Anselm's phrase: existing in mind) would be simply greater. We only require that the property under consideration must be a greatness-making one.

Hence we add the following axioms to $\mathcal{A}$. Let π be any perfection. Then every formula of the form (A2) is a specific axiom of $\mathcal{A}$.

The weak point in this formulation of the axioms (A2) is the term "perfection". We understand any perfection to be a greatness-making property, i.e., a property which is, according to Anselm, non-relational and positive (cf. Anselm 2000, pp. 25–27). For example, among perfections there are such items as existence in reality ("Ex"), necessary existence ("$\Box$Ex"), existence without beginning, omnipotence, benevolence and omniscience.[8] However a controversy arose in the literature on perfections in Anselm.

Remark 4. It is also problematic whether perfections are great-making properties only for God or for any object.[9] If one wishes to pick the second option, then the formula in the axiom (A2) can be strengthened to the following sentence:

$$\forall_y(\neg\pi y \rightarrow \Diamond\exists_x(\pi x \wedge x > y)).$$

---

[7]Indeed, $\ulcorner\exists_x(A \wedge x > \mathsf{g}) \rightarrow \exists_x\, x > \mathsf{g}\urcorner \in$ LPC. Thus $\ulcorner\Diamond\exists_x(A \wedge x > \mathsf{g}) \rightarrow \Diamond\exists_x\, x > \mathsf{g}\urcorner \in$ LPC-M_W, by (RM$\Diamond_\text{W}$), and we use PC.

[8]See, for example, *Monologion* 15, 16, 31, 36, and *Proslogion* 5, 18 for more details.

[9]See, for example, (Hopkins 1976, pp. 114–115).

Hence it is obvious that some kind of a detailed theory of perfections is required here. However, for the sake of this paper we assume that the notion of perfection is intuitive.

Most of the claims about God including the famous arguments concerning His existence hold due to the above assumptions. Together they allow to unfold a single line of reasoning, becoming *unum argumentum* Anselm speaks of in the preface to *Proslogion*.

3. God Truly Exists

The apparatus sketched above suffices for proving

THEOREM 1. Eg (God exists in reality).

It is a well-known fact that Anselm is committed to the presupposition that existence is a greatness-making property, i.e. perfection.[10] Hence the instantiation of (A2) playing crucial role here is "$\neg \mathsf{Eg} \rightarrow \Diamond \exists_x (\mathrm{E}x \wedge x \succ \mathsf{g})$". We have: (A1), (A2) $\vdash_{\mathrm{LPC\text{-}M_W}}$ Eg.

PROOF. We obtain the following derivation:[11]

1. $\neg \mathsf{Eg} \rightarrow \Diamond \exists_x (\mathrm{E}x \wedge x \succ \mathsf{g})$ — (A2)
2. $\neg \Diamond \exists_x (\mathrm{E}x \wedge x \succ \mathsf{g})$ — ($\star$), the sequence in Lemma 1 for A = "Ex"
3. Eg — 1, 2, and PC □

4. God Cannot Be Thought Not to Exist

We may proceed to prove the main claim of the third chapter now.

THEOREM 2. $\neg \Diamond \neg \mathsf{Eg}$, i.e., $\Box \mathsf{Eg}$.

Another presupposition to be found here is that the so-called necessary existence is a perfection (*vide* step 1 in the following proof).

PROOF. We obtain the following derivation:

1. $\neg \Box \mathsf{Eg} \rightarrow \Diamond \exists_x (\Box \mathrm{E}x \wedge x \succ \mathsf{g})$ — (A2) for πx = "□Ex"
2. $\neg \Diamond \exists_x (\Box \mathrm{E}x \wedge x \succ \mathsf{g})$ — ($\star$), the sequence in Lemma 1 for A = "□Ex"
3. $\Box \mathsf{Eg}$ — 1, 2, and PC
4. $\neg \Diamond \neg \mathsf{Eg}$ — 3, (df□), and PC □

---

[10] In fact this is the premise that has been under many attacks from the critics of the argument.

[11] Both Anselm's original proofs of Theorems 1 and 2 are made by *reductio ad absurdum*. Our proofs can be easily transformed to such a form if *modus ponens* is used instead of the following rule: $\neg A \rightarrow B, \neg B \vdash A$.

Remark 5. Of course, as it was earlier said (see Remark 3), (T) cannot be assumed, hence "$\Box$Eg" does not entail "Eg" in $\mathcal{A}$. However, it is obvious that there is no need for proving Theorem 1 from Theorem 2 since, as we have shown, the former can be obtained independently.

5. Is God Unique?

5.1. *Extension of the Logic* LPC-M_w

In order to consider the problem of uniqueness of God we have to modify the theory $\mathcal{A}$. Firstly, we have to strengthen the logic underlying $\mathcal{A}$.

The new logic is an LPC-version of the pure sentential modal logic R (cf. Chellas 1980, p. 234), therefore let us denote it by LPC-R_w. The only modal axiom of LPC-R_w is (df$\Diamond$). Moreover, we require that the set of theses of LPC-R_w is closed under the following LPC-version of *weak* regularity rule[12]

(RR$_W$) if $\ulcorner(A \wedge B) \rightarrow C\urcorner \in$ LPC, then $\ulcorner(\Box A \wedge \Box B) \rightarrow \Box C\urcorner \in$ LPC-R_w.

Of course, the logic LPC-R_w is also closed under the rule (RM_w). Moreover, it contains the following formulæ:

(K) $\Box(A \rightarrow B) \rightarrow (\Box A \rightarrow \Box B)$,

(R) $\Box(A \wedge B) \leftrightarrow (\Box A \wedge \Box B)$,

(R$\Diamond$) $\Diamond(A \vee B) \leftrightarrow (\Diamond A \vee \Diamond B)$,

(K$\Diamond$) $(\neg\Diamond A \wedge \Diamond B) \rightarrow \Diamond(\neg A \wedge B)$,

(1) $(\Box A \wedge \Diamond B) \rightarrow \Diamond(A \wedge B)$,

(2) $\Box(A \rightarrow B) \rightarrow (\Diamond A \rightarrow \Diamond B)$.

Remark 6. LPC-R_w is also the smallest logic which includes LPC and the formulæ (df$\Diamond$) and (K), and is closed under the rule (RM_w). Thus, LPC-R_w is an LPC-version of the pure sentential modal Routley's logic C1.[13] Therefore LPC-R_w can be also denoted by LPC-C1.

---

[12]The logic R is closed under the *strong* regularity rule

(RR) if $\ulcorner(A \wedge B) \rightarrow C\urcorner \in$ R, then $\ulcorner(\Box A \wedge \Box B) \rightarrow \Box C\urcorner \in$ R.

[13]By C1 Routley (1968) denoted the smallest pure sentential modal logic which includes PC, (df$\Diamond$) and (K), and is closed under (MP) and (RM_w).

By E1 and E2 Lemmon (1957) denoted the smallest pure sentential modal logics which includes PC, (df$\Diamond$), (K), and (T), and is closed under (MP) and moreover under (RM_w) or (RM), respectively. Thus, C1 is the logic E1 without (T). In (Lemmon 1966), by C2 is denoted the logic E2 without (T). So C2 is R.

Secondly, we have to employ a predicate standing for identity, "≡". This can be simply obtained when one takes (non-modal) lower (first-order) predicate calculus with identity, $LPC_{\equiv}$, instead of LPC. Hence LPC-R_W is extended to $LPC_{\equiv}$-R_W.[14]

Remark 7. The logics LPC-R_W and $LPC_{\equiv}$-R_W are epistemic in the sense that none of their theses begins with "□".

5.2. *Deduction in* LPC-R_W *and* $LPC_{\equiv}$-R_W. *The Theory* $\mathcal{A}'$

The notion of *being deducible* in the logic LPC-R_W (resp. $LPC_{\equiv}$-R_W) is defined in the same way as for the logic LPC-M_W. Thus, a formula A is deducible from a set $\mathcal{S}$ in LPC-R_W (resp. $LPC_{\equiv}$-R_W) iff A can be obtained from $\mathcal{S}$ and theses of LPC-R_W (resp. $LPC_{\equiv}$-R_W) by (MP).

Now the theory $\mathcal{A}'$ is defined in the old manner, simply by adding (A1) and (A2) to the logic $LPC_{\equiv}$-R_W.

A formula of $L_{\mathcal{A}}$ is a *thesis of* $\mathcal{A}'$ iff it is deducible from {(A1), (A2)} in the logic $LPC_{\equiv}$-R_W, i.e., it can be obtained from {(A1), (A2)} and theses of $LPC_{\equiv}$-R_W by (MP). Of course, all theses of $\mathcal{A}$ are theses of $\mathcal{A}'$.

We shall also extend the language $L_{\mathcal{A}}$ with a unary predicate "G" standing for "is god". The predicate is intended to be applied only to objects satisfying (A1), therefore we define it as

(df G) $\qquad \forall_x(\mathsf{G}x \leftrightarrow \neg\Diamond\exists_y\, y > x).$

Given (A1), we have of course that Gg.

5.3. *Some Preliminary Remarks*

There is probably no way one can prove that there is at most one God within $\mathcal{A}'$. Nevertheless there are several ways of proving that

THEOREM 3. It must be thought that there is at most one God, i.e.

(†) $\qquad \forall_{x,y}(\mathsf{G}x \wedge \mathsf{G}y \ \rightarrow\ \Box\, x \equiv y)$

or, equivalently given that Gg,

(‡) $\qquad \forall_x(\mathsf{G}x \ \rightarrow\ \Box\, x \equiv \mathsf{g}).$

FACT 1. 1. (†), $\mathsf{Gg} \vdash_{\mathrm{LPC}}$ (‡)

2. (‡) $\vdash_{LPC_{\equiv}\text{-}R_W}$ (†)

PROOF. 1. Obvious. For 2.

1. $\mathsf{G}x \wedge \mathsf{G}y$ additional hypothesis

[14] Of course, now we allow to apply the rules of $LPC_{\equiv}$-R_W to theses of $LPC_{\equiv}$ as well.

2.	(‡)	assumption
3.	(‡) $\rightarrow$ $(\mathsf{G}x \rightarrow \Box\, x \equiv \mathsf{g})$	LPC
4.	(‡) $\rightarrow$ $(\mathsf{G}y \rightarrow \Box\, y \equiv \mathsf{g})$	LPC
5.	$\Box\, x \equiv \mathsf{g} \wedge \Box\, y \equiv \mathsf{g}$	1–4 and PC
6.	$\Box\, x \equiv \mathsf{g} \wedge \Box\, y \equiv \mathsf{g} \;\rightarrow\; \Box\, x \equiv y$	$\mathrm{LPC}_{\equiv}\text{-}\mathrm{R_W}$
7.	$\Box\, x \equiv y$	5, 6, and (MP)

Moreover, we use the Deduction Theorem and the rule of generalization. □

One needs only to make certain assumptions concerning the nature of the $\succ$ relation. Of course all of them must, in a sense, correspond to Anselm's views.

Remark 8. We shall put "□" at the beginning of each claim characterizing $\succ$ relation as we believe that any statement concerning the nature of $\succ$ is not merely true but necessary, i.e., holds in every epistemically possible world; cf. (Hughes & Cresswell 1996, pp. 349–350).

A careful reader of Anselm will easily find out that he explicitly states that

God is not greater than God (cf. Anselm 2000, p. 503).

An almost similar claim can be immediately derived from (A1) by means of LPC and ($\mathrm{RM}\Diamond_{\mathrm{W}}$):

Fact 2. $\neg\Diamond\; \mathsf{g} \succ \mathsf{g}$.

Proof. The sentence "$\Diamond\; \mathsf{g} \succ \mathsf{g} \rightarrow \Diamond\exists_x\; x \succ \mathsf{g}$" is a thesis of $\mathrm{LPC}\text{-}\mathrm{M_W}$, by LPC and ($\mathrm{RM}\Diamond_{\mathrm{W}}$). Since $\neg\Diamond\; \mathsf{g} \succ \mathsf{g}$, by (A1) and PC. □

Hence whatever the properties of $\succ$ are, this relation is assuredly not reflexive. It is the relation $\succcurlyeq$ (where $x \succcurlyeq y$ iff $x \equiv y$ or $x \succ y$) that bears this property.

5.4. *The First Way*

The most natural way from the point of Anselm's works is to assume that the relation $\succ$ is connected, i.e.

($\Box\mathrm{con}_{\succ}$) $\Box\forall_{x,y}(x \succ y \vee y \succ x \vee x \equiv y)$.

First Proof of Theorem 3. ($\Box\mathrm{con}_{\succ}$), (df G) $\vdash_{\mathrm{LPC}\text{-}\mathrm{R_W}}$ (†):

1.	$\mathsf{G}x \wedge \mathsf{G}y$	additional hypothesis
2.	$\neg\Diamond\exists_z\; z \succ x \wedge \neg\Diamond\exists_z\; z \succ y$	1, (df G), and PC
3.	$\neg\Diamond\exists_z\; z \succ x \;\rightarrow\; \neg\Diamond\; y \succ x$	$\mathrm{LPC}\text{-}\mathrm{M_W}$[15]
4.	$\neg\Diamond\exists_z\; z \succ y \;\rightarrow\; \neg\Diamond\; x \succ y$	$\mathrm{LPC}\text{-}\mathrm{M_W}$
5.	$\neg\Diamond\; y \succ x \wedge \neg\Diamond\; x \succ y$	2, 3, 4, and PC
6.	($\Box\mathrm{con}_{\succ}$)	assumption

[15] Indeed, since "$y \succ x \rightarrow \exists_y\; y \succ x$" $\in$ LPC, so "$\Diamond\; y \succ x \rightarrow \Diamond\exists_y\; y \succ x$" $\in \mathrm{LPC}\text{-}\mathrm{M_W}$, by ($\mathrm{RM}\Diamond_{\mathrm{W}}$).

7. $(\Box\mathrm{con}_{\succ}) \;\to\; \Box(x \succ y \lor y \succ x \lor x \equiv y)$ — LPC-$\mathrm{M_w}$
8. $\Box(x \succ y \lor y \succ x \lor x \equiv y)$ — 6, 7, and (MP)
9. $\Box(A \lor B \lor C) \land \neg\Diamond A \land \neg\Diamond B \;\to\; \Box C$ — LPC-$\mathrm{R_w}$ where $A/x \succ y$, $B/y \succ x$, $C/x \equiv y$
10. $\Box\, x \equiv y$ — 5, 8, 9, and PC

Moreover, we use the Deduction Theorem and next the rule of generalization.

Now we prove that (A1), $(\Box\mathrm{con}_{\succ})$, (df G) $\vdash_{\mathrm{LPC\text{-}R_w}}$ (‡):

1. $\mathsf{G}x$ — additional hypothesis
2. $\neg\Diamond\exists_y\, y \succ x$ — 1, (df G), and PC
3. $\neg\Diamond\exists_y\, y \succ x \;\to\; \neg\Diamond\; \mathsf{g} \succ x$ — LPC-$\mathrm{M_w}$
4. $\neg\Diamond\; \mathsf{g} \succ x$ — 2, 3, and (MP)
5. (A1) — assumption
6. $\neg\Diamond\exists_x\, x \succ \mathsf{g} \;\to\; \neg\Diamond\; x \succ \mathsf{g}$ — LPC-$\mathrm{M_w}$
7. $\neg\Diamond\; x \succ \mathsf{g}$ — 5, 6, and (MP)
8. $(\Box\mathrm{con}_{\succ})$ — assumption
9. $(\Box\mathrm{con}_{\succ}) \;\to\; \Box(x \succ \mathsf{g} \lor \mathsf{g} \succ x \lor x \equiv \mathsf{g})$ — LPC-$\mathrm{M_w}$
10. $\Box(x \succ \mathsf{g} \lor \mathsf{g} \succ x \lor x \equiv \mathsf{g})$ — 8, 9, and (MP)
11. $\Box(A \lor B \lor C) \land \neg\Diamond A \land \neg\Diamond B \;\to\; \Box C$ — LPC-$\mathrm{R_w}$ where $A/x \succ \mathsf{g}$, $B/\mathsf{g} \succ x$, $C/x \equiv \mathsf{g}$
12. $\Box\, x \equiv \mathsf{g}$ — 4, 7, 11, and PC

Moreover, we use the Deduction Theorem and the rule of generalization. □

5.5. *The Second Way*

However, the claim stated by $(\Box\mathrm{con}_{\succ})$ may be objected as it is quite strong indeed. Hence we can try to weaken it by introducing the notion of conditional connectivity.

First, let us again extend the language of $\mathcal{A}'$ with a binary predicate "$\approx$". The expression "$x \approx y$" shall be read as "x has the same characteristics as y" (cf. Jarmużek & Nowicki 2004). Hence we define conditional connectivity in such way that it shall hold for any two objects with the same characteristics, i.e.

$(\Box\mathrm{ccon}_{\succ})$ $\Box\forall_{x,y}(x \approx y \to (x \succ y \lor y \succ x \lor x \equiv y))$.

Since "$\forall_{x,y}(x \succ y \lor y \succ x \lor x \equiv y)$" entails "$\forall_{x,y}(x \approx y \to (x \succ y \lor y \succ x \lor x \equiv y))$" in LPC, so, by ($\mathrm{RM_w}$), we have:

Fact 3. $(\Box\mathrm{con}_{\succ})$ entails $(\Box\mathrm{ccon}_{\succ})$ in the logic LPC-$\mathrm{M_w}$.

Moreover, since Gg, it would be also quite natural to assume that any object which is god, must be thought to be similar to g with respect to all properties

(A3) $\forall_x(\mathsf{G}x \to \Box\, x \approx \mathsf{g})$.

SECOND PROOF OF THEOREM 3. (A1), (A3), ($\Box$ccon$_{>}$), (df G) $\vdash_{\text{LPC-R}_W}$ (‡):

1.	$\mathsf{G}x$	additional hypothesis
2.	(A3)	assumption
3.	(A3) $\to$ ($\mathsf{G}x \to \Box\, x \approx \mathsf{g}$)	1, (A3), and LPC
4.	$\Box\, x \approx \mathsf{g}$	1–3 and PC
5.	($\Box$ccon$_{>}$)	assumption
6.	($\Box$ccon$_{>}$) $\to$ $\Box(x \approx \mathsf{g} \to (x > \mathsf{g} \vee \mathsf{g} > x \vee x \equiv \mathsf{g}))$	LPC
7.	$\Box(x \approx \mathsf{g} \to (x > \mathsf{g} \vee \mathsf{g} > x \vee x \equiv \mathsf{g}))$	5, 6, and (MP)
8.	7 $\to$ (4 $\to \Box(x > \mathsf{g} \vee \mathsf{g} > x \vee x \equiv \mathsf{g})$)	(K)
9.	$\Box(x > \mathsf{g} \vee \mathsf{g} > x \vee x \equiv \mathsf{g})$	4, 7, and PC
10.	Steps 2-7, 11 and 12 in the first proof of (A1), (df G), ($\Box$con$_{>}$) $\vdash_{\text{LPC-R}_W}$ (‡)	

Moreover, we use the Deduction Theorem and the rule of generalization. □

5.6. *The Third Way*

One can even think of dropping the assumption that connectivity is somehow connected with $>$. Now the supposition that for any two elements of the ontological space there is an upper bound seems to be more neutral

($\Box$dir$_{\geqslant}$) $\Box\forall_{x,y}\exists_z(z \geqslant x \wedge z \geqslant y)$.

The condition ($\Box$dir$_{\geqslant}$) is weaker than the condition ($\Box$con$_{>}$). Firstly,

FACT 4. 1. (con$_{>}$) $\vdash_{\text{LPC}_{\equiv}}$ (dir$_{\geqslant}$)

2. ($\Box$con$_{>}$) $\vdash_{\text{LPC}_{\equiv}\text{-R}_W}$ ($\Box$dir$_{\geqslant}$)

PROOF. 1. Let x and y be any members of the ontological space. From (con$_{>}$) we have: $y \geqslant x$ or $x \geqslant y$. Hence, ether $y \geqslant x$ and $y \geqslant y$ or $x \geqslant x$ and $x \geqslant y$. In both cases there is a z such that $z \geqslant x$ and $z \geqslant y$.

2. By the Deduction Theorem and the rule (RM$_W$). □

THIRD PROOF OF THEOREM 3. ($\Box$dir$_{\geqslant}$), (df G), (df $\geqslant$) $\vdash_{\text{LPC}_{\equiv}\text{-R}_W}$ (†):

1.	$\mathsf{G}x \wedge \mathsf{G}y$	additional hypothesis
2.	$\neg\Diamond\exists_z\, z > x \wedge \neg\Diamond\exists_z\, z > y$	1, (df G), and PC
3.	($\Box$dir$_{\geqslant}$)	assumption
4.	($\Box$dir$_{\geqslant}$) $\to$ $\Box\exists_z(z \geqslant x \wedge z \geqslant y)$	LPC-R$_W$
5.	$\Box\exists_z(z \geqslant x \wedge z \geqslant y)$	3, 4, and (MP)
6.	5 $\to$ $\Box\exists_z[(z > x \vee z \equiv x) \wedge (z > y \vee z \equiv y)]$	LPC-R$_W$, by (df $\geqslant$)
7.	$\Box\exists_z[(z > x \vee z \equiv x) \wedge (z > y \vee z \equiv y)]$	5, 6, and (MP)
8.	$\Box\exists_z((A \vee B) \wedge (C \vee D)) \wedge \neg\Diamond\exists_z\, A \wedge \neg\Diamond\exists_z\, C \to \Box\exists_z(B \wedge C)$ where $A/z > x$, $B/z \equiv x$, $C/z > y$, $D/z \equiv y$	LPC-R$_W$
9.	$\Box\exists_z(z \equiv x \wedge z \equiv y)$	2, 7, 8, and PC

10.	$\Box\exists_z(z \equiv x \wedge z \equiv y) \rightarrow \Box\, x \equiv y$	$\mathrm{LPC}_{\equiv}$-$\mathrm{R_w}$
11.	$\Box\, x \equiv y$	9, 10 and (MP)

Moreover, we use the Deduction Theorem and the rule of generalization. □

5.7. *The Fourth Way*

Finally we shall look at the problem from another point of view than Anselm's: by taking

$$(A1') \qquad \neg\Diamond\exists_x \neg\, \mathsf{g} \succcurlyeq x$$

instead of (A1). Thus Anselmian "definition" of God is changed. Now, God is not only maximal but greatest as well. As before, the predicate "is god" can be applied only to objects satisfying $(A1')$. Therefore it is defined as

$$(\text{df}'\, \mathsf{G}) \qquad \forall_x(\mathsf{G}x \leftrightarrow \neg\Diamond\exists_y \neg\, x \succcurlyeq y).$$

Of course, by $(A1')$, Gg.

From $(A1')$, (df◇), ($\mathrm{RM_w}$), and LPC we obtain

$$(3) \qquad \Box\,\forall_x\, \mathsf{g} \succcurlyeq x\,.$$

We also assume that $\succ$ is asymmetrical, i.e.

$$(\Box\mathrm{as}_{\succ}) \qquad \Box\forall_{x,y}(x \succ y \;\rightarrow\; \neg\, y \succ x).$$

Thus the relation $\succ$ is irreflexive and the relation $\succcurlyeq$ is antisymmetrical, i.e.

$$(\Box\mathrm{irr}_{\succ}) \qquad \Box\forall_x \neg\, x \succ x\,,$$
$$(\Box\mathrm{antis}_{\succcurlyeq}) \quad \Box\forall_{x,y}(x \succcurlyeq y \wedge y \succcurlyeq x \;\rightarrow\; x \equiv y).$$

Now, on the assumption that $(A1')$ and $(\Box\mathrm{as}_{\succ})$, we can prove that there is at most one God in the sense of the new definition (df′ G).

FOURTH PROOF OF THEOREM 3. $(A1')$, $(\Box\mathrm{antis}_{\succcurlyeq})$, $(\text{df}'\,\mathsf{G}) \vdash_{\mathrm{LPC\text{-}R_w}} (\ddagger)$:

1.	$\mathsf{G}x$	additional hypothesis
2.	$\neg\Diamond\exists_y \neg\, x \succcurlyeq y$	1, (df′ G)
3.	$\neg\Diamond\exists_y \neg\, x \succcurlyeq y \;\rightarrow\; \neg\Diamond\neg\, x \succcurlyeq \mathsf{g}$	$\mathrm{LPC\text{-}M_w}$
4.	$\neg\Diamond\neg\, x \succcurlyeq \mathsf{g}$	2, 3, and (MP)
5.	$(A1')$	assumption
6.	$\neg\Diamond\exists_x \neg\, \mathsf{g} \succcurlyeq x \;\rightarrow\; \neg\Diamond\neg\, \mathsf{g} \succcurlyeq x$	$\mathrm{LPC\text{-}M_w}$
7.	$\neg\Diamond\neg\, \mathsf{g} \succcurlyeq x$	5, 6, and (MP)
8.	$(\Box\mathrm{antis}_{\succcurlyeq})$	assumption

9. $(\Box\text{antis}_{\geqslant}) \rightarrow \Box(x \geqslant \mathsf{g} \wedge \mathsf{g} \geqslant x \rightarrow x \equiv \mathsf{g})$ LPC-$\mathrm{M_w}$
10. $\Box(x \geqslant \mathsf{g} \wedge \mathsf{g} \geqslant x \rightarrow x \equiv \mathsf{g})$ 8, 9, and (MP)
11. $\Box(A \wedge B \rightarrow C) \wedge \neg\Diamond\neg A \wedge \neg\Diamond\neg B \rightarrow \Box C$ LPC-$\mathrm{R_w}$ where $A/x \geqslant \mathsf{g}$, $B/\mathsf{g} \geqslant x$, $C/x \equiv \mathsf{g}$
12. $\Box\, x \equiv \mathsf{g}$ 4, 7, 10, 11, and (MP)

Moreover, we use the Deduction Theorem and the rule of generalization. □

We shall also show that $(A1')$ and $(\Box\text{as}_{>})$ entail (A1). It means that the new axioms $(A1')$ and $(\Box\text{as}_{>})$ allow to derive Theorems 1, 2 and 3 (in the logic LPC-$\mathrm{R_w}$).

Fact 5. $(A1')$ and $(\Box\text{as}_{>})$ entail (A1) in the logic LPC-$\mathrm{R_w}$.

Proof. $(\Box\text{as}_{>}), (\text{df} \geqslant) \vdash_{\text{LPC-M}_\text{w}} (A1') \rightarrow (\text{A1})$:

1. $(\Box\text{as}_{>})$ assumption
2. $\Box\, \forall_{x,y}[x > y \rightarrow (\neg\, y > x \wedge x \not\equiv y)]$ 1 and LPC-$\mathrm{M_w}$
3. $\Box\, \forall_{x,y}(x > y \rightarrow \neg\, y \geqslant x)$ 2, (df $\geqslant$), and LPC-$\mathrm{M_w}$
4. $\Box\forall_x(x > \mathsf{g} \rightarrow \neg\, \mathsf{g} \geqslant x)$ 3 and LPC-$\mathrm{M_w}$
5. $\Box(\exists_x\, x > \mathsf{g} \rightarrow \exists_x \neg\, \mathsf{g} \geqslant x)$ 4 and LPC-$\mathrm{M_w}$
6. $\Box(A \rightarrow B) \rightarrow (\Diamond A \rightarrow \Diamond B)$ (2), where $A/\exists_x\, x > \mathsf{g}$, $B/\exists_x \neg\, \mathsf{g} \geqslant x$
7. $\Diamond\exists_x\, x > \mathsf{g} \rightarrow \Diamond\exists_x\neg \mathsf{g} \geqslant x$ 5, 6, and (MP)
8. $\neg\Diamond\exists_x\neg\, \mathsf{g} \geqslant x \rightarrow \neg\Diamond\exists_x\, x > \mathsf{g}$ 7 and PC □

5.8. *Summary*

We have shown four ways of extending or modifying $\mathcal{A}'$ in order to prove Theorem 3. It is independently entailed by: $(\Box\text{con}_{>})$, $(\Box\text{ccon}_{>})$ with (A3), $(\Box\text{dir}_{\geqslant})$, as well as by $(A1')$ with $(\Box\text{as}_{>})$.

Hence let us end with putting the following problem: *Find more (and the most) natural way for proving the uniqueness of God.*

Nicolaus Copernicus University
Department of Logic
ul. Asnyka 2
87-100 Toruń
Poland
e-mail: {jarmuzek,mnowicki,pietrusz}@uni.torun.pl

REFERENCES

Anselm of Canterbury (2000). *Complete Philosophical and Theological Treatises of Anselm of Canterbury*, J. Hopkins, H. Richardson (transl.), Minneapolis, Minn.: The Arthur J. Banning Press.

Chellas, B. (1980). *Modal Logic. An Introduction.* Cambridge: Cambridge University Press.

Hartshorne, Ch. (1961). The Logic of the Ontological Argument. *Journal of Philosophy* **58**, 471–473.

Hartshorne, Ch. (1962). *The Logic of Perfection.* La Salle: Open Court.

Hopkins, J. (1976). *Anselm of Canterbury: Volume Four: Hermeneutical and Textual Problems in the Complete Treatises of St. Anselm.* New York: Mellen Press.

Hughes, G E., and M. J. Cresswell (1996). *A New Introduction to Modal Logic.* London: Routledge.

Jarmużek T., and M. Nowicki (2004). *Ratio Anselmi.* In: J. Malinowski and A. Pietruszczak (eds.), *Wokół filozofii logicznej* [Around Logical Philosophy], pp. 165–178. Toruń: Nicolaus Copernicus University Press.

Lemmon, E. J. (1957). New Foundations for Lewis Modal Systems. *The Journal of Symbolic Logic* **22**, 176–186.

Lemmon, E. J. (1966). Algebraic Semantics for Modal Logics I. *The Journal of Symbolic Logic* **31**, 46–56.

Lemmon, E. J., in collaboration with D. Scott (1977). *"Lemmon Notes": An Introduction to Modal Logic.* Edited by K. Segerberg, no. 11 in the American Philosophical Quarterly Monograph Series, Oxford: Basil Blackwell.

Perzanowski, J. (1995). O wskazanych przez Ch. Hartshorne'a modalnych krokach w dowodzie ontologicznym św. Anzelma [Hartshorne's Modal Steps in Anselm's Ontological Argument]. In: J. Perzanowski, A. Pietruszczak, C. Gorzka (eds.), *Filozofia/Logika: Filozofia Logiczna 1994* [Philosophy/Logic: Logical Philosophy 1994], pp. 77–96. Toruń: Nicolaus Copernicus University Press.

Routley, R. (1968). Decision Procedure and Semantics for C1, E1 and S0.5°. *Logique et Analyse* **44**, 468–469.

Roman Murawski

PHILOSOPHY OF MATHEMATICS IN THE 20TH CENTURY: MAIN TRENDS AND DOCTRINES*

ABSTRACT. The aim of the paper is to present the main trends and tendencies in the philosophy of mathematics in the 20th century. To make the analysis more clear we distinguish three periods in the development of the philosophy of mathematics in this century: (1) the first thirty years when three classical doctrines: logicism, intuitionism and formalism were formulated, (2) the period from 1931 till the end of the fifties — period of stagnation, and (3) from the beginning of the sixties till today when new tendencies putting stress on the knowing subject and the research practice of mathematicians arose.

1. The First 30 Years — the Rise of Classical Conceptions

Philosophy of mathematics and the foundations of mathematics entered the 20th century in the atmosphere of a crisis.[1] It was connected mainly with the problem of the status of abstract objects. In set theory created by Georg Cantor in the last quarter of the 19th century antinomies were discovered. Some of them were already known to Cantor — e.g., the antinomy of the set of all sets or the antinomy of the set of all ordinals (the latter was independently discovered by C. Burali-Forti). He was able to eliminate them by distinguishing between classes and sets. New difficulties appeared when B. Russell discovered the antonomy of irreflexive classes (called today Russell's antinomy) in the system of logic of G. Frege (to which the latter wanted to reduce the entire mathematics). Besides those antinomies one discovered also so called semantical ones (G. D. Berry, K. Grelling) connected with semantical notions such as reference, sense and truth.

In this situation there was a need to find a new solid and safe foundations for mathematics. Attempts to provide such a foundations and to overcome the crisis led, among others, to the rise of new directions and conceptions in the philosophy of mathematics, namely of logicism, intuitionism and formalism. One used here of course achievements of earlier mathematicians, especially of those living in the 19th century. In particular important were the very rise of set theory (G. Cantor), the idea of the arithmetization of analysis (A. Cauchy, K. Weierstrass, R. Dedekind), axiomatization of the arithmetic

*The paper was written in the framework of Polish-Flemish research project BIL 01/80.

[1] It is usually called the second crisis and distinguished from the first one which took place in the antiquity and was connected with the discovery of incommensurables by Pythagoreans.

In: J. Malinowski and A. Pietruszczak (eds.), *Essays in Logic and Ontology* (*Poznań Studies in the Philosophy of the Sciences and the Humanities*, vol. 91), 331–347. Amsterdam/New York, NY: Rodopi, 2006.

of natural numbers (G. Peano), non-Euclidean geometries (N. I. Lobachevsky, J. Bolayi, C. F. Gausss) and the complete axiomatization of systems of geometry (M. Pasch, D. Hilbert) and, last but not least, the rise and the development of mathematical logic (G. Boole, A. De Morgan, G. Frege, B. Russell).

New doctrines in the philosophy of mathematics looking for solid, sure and safe foundations of mathematics saw them first of all in mathematical logic and set theory. The most important proposals are called logicism, intuitionism and formalism.

1.1. *Logicism*

Logicism was founded by Gottlob Frege (1848–1925) and developed by Bertrand Russell (1872–1970) and Alfred North Whitehead (1861–1947). It grew out from the tendency to arithmetize the analysis — this tendency was characteristic for foundational studies in the 19th century. The aim of it was to show that the theory of real numbers underlying the analysis can be developed on the basis of the arithmetic of natural numbers. The realization of this task led to a new one, namely to find foundations for the arithmetic of natural numbers, i.e., to found it on a simpler, more elementary theory. This task was undertaken just by G. Frege who tried to reduce arithmetic to logic (cf. Frege, 1884, 1893/1903). Unfortunately the system of logic used by him turned out to be inconsistent what was discovered by B. Russell in 1901 (Russell discovered that one can construct in this system the antinomy of irreflexive classes called today Russell's antinomy). In this situation Russell together with A. N. Whitehead undertook anew the task of reducing arithmetic to a simpler theory. They introduced so called ramified theory of types and showed that the whole mathematics can be reduced to it. This was the beginning of a mature form of logicism. Its main thesis says that (the whole) mathematics is reducible to logic, i.e., mathematics is only a part of logic and logic is an epistemic ground of all mathematics. This thesis can be formulated as the conjunction of the following three theses: (1) all mathematical concepts (in particular all primitive notions of mathematical theories) can be explicitly defined by purely logical notions, (2) all mathematical theorems can be deduced (by logical deduction) from logical axioms and definitions, (3) this deduction is based on a logic common for all mathematical theories, i.e., the justification of theorems in all mathematical theories refers to the same basic principles which form one logic common for the whole of mathematics (here is also included a thesis that any argumentation in mathematics should be formalized). Hence theorems of mathematics have uniquely determined contents and it is a logical contents. Moreover, mathematical theorems are analytic (similarly as logical theorems).

Russell and Whitehead showed in their monumental work *Principia Mathematica* (vol. I – 1910, vol. II – 1912, vol. III – 1913) how to reduce the whole of mathematics to the system of logic given by the ramified theory of types. Some difficulties by the attempts to prove certain properties of natural numbers forced the necessity of introducing principles of a non-logical character, in particular the axiom of infinity, reducibility axiom and the axiom of choice (they have in fact a set-theoretical character). This led to the modern version of logicism which states that the whole of mathematics is reducible to logic and set theory.

1.2. *Intuitionism*

Second doctrine of the philosophy of mathematics formulated at the beginning of the 20th century was intuitionism founded by Dutch mathematician Luitzen Egbertus Jan Brouwer (1881–1966). It used some ideas which appeared by Leopold Kronecker (1823–1891), Henri Poincaré (1854–1912) and by a group of French mathematicians called the Paris group of intuitionism or French semi-intuitionists (R. L. Baire, E. Borel, H. L. Lebesgue and the Russian mathematician N. N. Lusin).

Brouwer presented his philosophical views for the first time in his doctoral dissertation *Over de Grondslagen de Wiskunde* (1907) and developed them in the papers (Brouwer, 1912 and 1949). The main aim of him (and of the intuitionism) was to avoid inconsistencies in mathematics. Brouwer proposed means to do that which turned out to be very radical and led in the effect to the deep reconstruction of the whole of mathematics.

The main fundamental thesis of Brouwer's intuitionism is the rejection of platonism in the philosophy of mathematics, i.e., of the thesis about the existence of mathematical objects which is independent of time, space and human mind. The proper ontological thesis is the conceptualism. According to Brouwer mathematics is a function of the human intelect and a free activity of the human mind, it is a creation of the mind and not a theory or a system of rules and theorems. Mathematical objects are mental constructions of an (idealized) mathematician. Mathematics is based on the fundamental intuition of the a priori time (one can see here the connection with the philosophy of I. Kant). This intuition is a basis for the mental construction of natural numbers. Moreover this construction is the basic of mathematical activity from which all other mathematical activity springs. One of the consequences of this is the thesis that arithmetical statements are synthetic a priori judgements.

One should reject the axiomatic-deductive method as a method of developing and founding mathematics. It is not sufficient to postulate only the existence of mathematical objects (as it is done in the axiomatic method) but one must first construct them. One must also reject the actual infinity. An infinite set can be understood only as a law or a rule of forming more and more of its elements, but they will never exist as forming an actual totality. Hence there are no uncountable sets and no cardinal numbers other than $\aleph_0$.

The conceptualistic thesis of intuitionism implies also the rejection of any nonconstructive proofs of the existential theses, i.e., of proofs giving no constructions of the postulated objects. In fact in intuitionism "to be" equals "to be constructed". Brouwer claimed that just nonconstructive proofs were the source of all troubles and mistakes in mathematics. Since such proofs are usually based on the law of the excluded middle ($p \vee \neg p$) as well as on the law of double negation ($p \equiv \neg\neg p$), intuitionists could not any longer use the classical logic. Moreover they claimed that logic is neither a basis for mathematics nor a starting point of it. Brouwer said that logic is based on mathematics, that it is secondary and dependent on mathematics and not vice versa, i.e., logic is a basis for mathematics as the logicists assert.

A mathematical theorem is a declaration that a certain mental construction has been completed. All mathematical constructions are independent of any language (the only role of a language is to enable the communication of mathematical constructions to others). Hence there is no language (formal or informal) which would be safe for mathematics and would protect it from inconsistencies. It is a mistake to analyse the language of mathematics instead of analysing the mathematical thinking.

The ideas of Brouwer were popularized by his student Arend Heyting (1898–1980). He attempted to explain them in a language usually used in the reflection on mathematics — note that Brouwer expressed his ideas in a language far from the standards accepted by mathematicians and logicians and therefore not always understandable. It seems that without those attempts of Heyting the ideas of Brouwer would soon have disappeared. Heyting constructed also the first formalized system of the intuitionistic propositional calculus, i.e., of the propositional calculus satisfying principles of the intuitionism. This system has never been accepted by Brouwer.

1.3. *Formalism*

The third classical doctrine in the philosophy of mathematics of the 20th century is formalism created by David Hilbert (1862–1943). Hilbert was of the opinion that the attempts to justify and found mathematics undertaken hitherto, especially by the intuitionism, were unsatisfactory because they led to the restriction of mathematics and to the rejection of various parts of it, in particular those considering infinity (cf. Hilbert, 1926). He undertook an attempt to justify the actual infinity by finitistic methods referring only to concrete, finite objects being the starting point of mathematics. Those objects are natural numbers meant as numerals (systems of signs). They are given immediately and clearly. Actual infinity on the other hand is only an idea of reason in Kant's sense, hence a notion for which there is no real basis because it transcends every experience. Hilbert proposed to justify the mathematics using the notion of an actual infinity (in Hilbert's terminology — infinitistic mathematics) by finitistic methods. He wanted to do it via proof theory (*Beweistheorie*). To do this one should formalize the whole mathematics, i.e., to represent it as a formalized system (or a set of formalized systems) and then to study such systems as systems of symbols only (hence as systems of concrete and visible objects clearly and immediately given) governed by certain rules referring only to the form and not to the sense or meaning of formulas. This can be done by finitistic and consequently safe methods. In particular one should show in this way the consistency and conservativeness of infinitistic mathematics. This was the task of proof theory.

One should note here that formalization was for Hilbert only an instrument used to prove the correctness of (infinitistic) mathematics. Hilbert did not treat mathematical theories as games on symbols or collections of formulas without any contents. Formalization was only a methodological tool in the process of studying the properties of the preexisting mathematical theories.

2. Between 1931 and the End of the Fifties

In 1931 a paper by a young Austrian mathematician Kurt Gödel (1906–1978) was published (cf. Gödel 1931). It turned out to be revolutionary and played a crucial role in the foundations and philosophy of mathematics. Gödel proved in it that arithmetic of natural numbers and all formal systems containing it are essentially incomplete provided they are consistent (and based on a recursive, i.e., effectively recognizable set of axioms). He announced also a theorem stating that (roughly speaking) no such theory can prove its own consistency, i.e., to prove the consistency of a given theory T containing arithmetic one needs methods and assumptions stronger than those of the theory T. Hence in particular one cannot prove the consistency of an infinitistic theory by finitistic methods. In this way Gödel's incompleteness theorems indicated certain epistemological limitations of axiomatic method. In particular they showed that one cannot completely axiomatize neither the arithmetic of natural numbers nor other richer theories (what was required by Hilbert's programme) and that there are no absolute inconsistency proofs for mathematical theories.[2]

Since one of the main aims of doctrines in the philosophy of mathematics formulated at the beginning of the century was to show that mathematics is consistent and that it is free of antinomies, one sees that Gödel's results were in fact a shock for them.[3] Indeed after 1931 one can observe certain stagnation in the philosophy of mathematics which lasted till the end of the fifties. There were formulated new conceptions but they were not so significant as logicism, intuition-

[2]For Gödel's theorems and the discussion of their philosophical and methodological consequences see, e.g., Murawski, 1999.

[3]One can find opinions that Gödel's theorems have definitely deprived mathematics of certainty, absolute truth and necessity (cf. Kline, 1980).

ism and formalism. One should mention here first of all works by Willard Van Orman Quine (1908–2000), Ludwig Wittgenstein (1889–1951) and Kurt Gödel.

2.1. *Quine's Philosophy of Mathematics*

Quine's philosophy of mathematics can be characterized as holistic. According to it mathematics should be considered not in separation from other sciences but as an element of the collection of theories explaining the reality (cf. Quine 1951a, 1951b, 1953). Mathematics is indispensible there, in particular in physical theories, hence its objects do exist.[4] In this way Quine attacked the anti-realist and anti-empiricist approaches to the philosophy of mathematics. This cleared the way for empiricist approaches (see below Section 3). Quine accepted only one way of existence. One does not find by him physical, mathematical, intensional, conceptual etc. existence but simply existence. He rejected also the possibility of dividing a scientific theory into an analytical (hence purely conventional) part and a synthetic one (dealing with the reality).

2.2. *Wittgenstein's Philosophy of Mathematics*

Wittgenstein's ideas concerning mathematics can be reconstructed from his remarks made at various periods — hence they are not uniform, moreover they are even inconsistent (cf. Wittgenstein 1953 and 1956). They grew out from his philosophy of language as a game. He was against logicism, and especially against Russell's attempts to reduce mathematics to logic. He claimed that by such reductions the creative character of a mathematical proof disappears. A mathematical proof cannot be reduced to axioms and rules of inference, because it is in fact a rule of constructing a new concept. Logic does not play so fundamental role in mathematics as logicism claims — its role is rather auxiliary. Mathematical knowledge is independent and specific in comparison with logic. Mathematical truths are a priori, synthetic and constructive. Mathematicians are not discovering mathematical objects and their properties but creating them. Hence mathematical knowledge is of a necessary character. One can easily see here the connections of Wittgenstein's philosophy of mathematics and Kant's ideas as well as the ideas of intuitionists.

2.3. *Gödel's Philosophy of Mathematics*

Gödel formulated his philosophical ideas concerning mathematics especially in connection with some problems of set theory (cf. Gödel 1944 and 1947). His philosophy of mathematics can be characterized as platonism. He claimed that mathematical objects exist in the reality independently of time, space and the

[4]This argumentation is known in the literature as Quine's-Putnam indispensability argument.

knowing subject. He stressed the analogy between logic and mathematics on the one hand and natural sciences on the other. Mathematical objects are transcendental with respect to their representation in mathematical theories. The basic source of mathematical knowledge is intuition[5] though it should not be understood as giving us the immediate knowledge. It suffices to explain and justify simple basic concepts and axioms. Mathematical knowledge is not the result of a passive contemplation of data given by intuition but a result of the activity of the mind which has a dynamic and cumulative character. Data provided by the intuition can be developed by a deeper study of mathematical objects and this can lead to the adoption of new statements as axioms. Those new axioms can be justified from outside, i.e., by their consequences (whether they enable us to solve problems which are unsolvable so far, whether they make possible to simplify the proofs or to obtain new interesting corrolaries). Gödel admitts here consequences both in mathematics itself as well as in physics. This is, according to him, the second (besides intuition) criterion of the truth of mathematical statements.

As stated above, Gödel presented and explained his philosophical ideas mainly in connection with set theory, in particular with the problem of continuum. As a realist he was convinced that the continuum hypothesis has a determined logical value, i.e., it is true or false (though we are not able now to decide which one). So Gödel assumed that there exists an absolute universe of sets which we try to describe by axioms of set theory. The fact that we can neither prove nor disprove the continuum hypothesis indicates that the axioms does not describe this universe in a complete way. Hence the necessity of adopting new axioms in mathematics, in particular in set theory. Gödel suggested that new infinity axioms postulating the existence of new large cardinals should be considered. He claimed that such axioms can imply not only some corollaries concerning the continuum hypothesis but also some new results in arithmetic of natural numbers.

2.4. *The Influence of the Philosophy of Mathematics on the Foundations of Mathematics*

Above we indicated the significance and the influence of mathematical logic on the rise and development of doctrines in the philosophy of mathematics, in particular in the case of the classical conceptions, i.e., logicism, intuitionism and formalism. But also another dependence should be noted, namely the influence of philosophical doctrines and conceptions on foundational studies. This was the case especially after 1931.

---

[5]Gödel never explained what he meant by intuition. Hence various interpretations are possible. Since 1959 he was interested in Husserl's phenonomenology. This probably influenced his theory of mathematical intuition.

Under the influence of the logicism and platonism Alfred Tarski (1901–1983) founded and developed set-theoretical semantics. It gave rise to the very important part of the foundations of mathematics, namely to model theory. Hilbert's formalism stimulated researches in the proof theory, hence in the finitistic metamathematics. Though Gödel's results indicated that Hilbert's programme of justification of mathematics cannot be realised in the original form but soon appeared new results and ideas which showed that this can be partially done. Proof theory became an independent discipline in the foundations of mathematics. Under the influence of intuitionism various systems of constructive mathematics as well as various constructivistic trends were developed.

Constructivism is a common name for various doctrines the main thesis of which is the demand to restrict mathematics to the consideration of constructive objects and to constructive methods only. Hence constructivism is a normative attitude the aim of which is not to build appropriate foundations for and to justify the existing mathematics but rather to reconstruct the latter according to the accepted principles and to reject all the methods and results which do not fulfil them. The constructivistic tendencies appeared in the last quarter of the nineteenth century as a reaction against the intensive development of highly abstract methods and concepts in mathematics inspired by Cantor's set theory.

There are various constructivist programmes and schools. They differ by their interpretation and understanding of the concept of constructivity. One of the most developed schools is intuitionism discussed above (see Section 1.2). Others are finitism, ultraintuitionism (called also ultrafinitism or actualism), predicativism, classical and constructive recursive mathematics, Bishop's constructivism.

One should also note that intensive studies on intuitionistic mathematics and logic took place. Recently those researches became strong impulse from the computer science since it turned out that intuitionistic logic provides a very useful tool in theoretical informatics.

3. Philosophy of Mathematics After 1960

In the early sixties certain renaissance of interests in the philosophy of mathematics can be observed. The old doctrines: logicism, intuitionism and formalism, still prevail but there appear some new conceptions based on another presuppositions and having another aims.

Formalism and intuitionism did not abandon in principle the ideas of their founders though the development of mathematical logic and the foundations of mathematics influenced them and forced some modifications of the original theses.

As indicated above, Gödel's incompleteness results struck Hilbert's formalistic programme and showed that it cannot be fully realized. In this situation two ideas appeared: One suggested to extend the scope of the admissible methods and allow general constructive ones instead of only finitistic methods (the idea was explicitely formulated for the first time by Paul Bernays) — this led to the so called generalized Hilbert's programme. The second idea was based on the following reasoning: If the entire infinitistic mathematics cannot be reduced to and justified by finitistic mathematics then one can ask for which part of it is that possible. In other words: how much of infinitistic mathematics can be developed within formal systems which are conservative over finitistic mathematics with respect to real sentences? This question constitutes the relativized version of the programme of Hilbert. Recently results of the so called reverse mathematics developed mainly by H. Friedman and S. G. Simpson contributed very much to this programme.[6] In fact they showed that several interesting and significant parts of classical mathematics are finitistically reducible. This means that Hilbert's programme can be partially realized.

Under the influence of intuitionism several constructivistic doctrines were developed — we wrote about them above.

Logicism appears as so called pluralistic logicism and is represented by H. Mehlberg and H. Putnam. According to it the main task in mathematics is the construction of proofs in axiomatic systems. By the deduction theorem if a sentence φ is a theorem of an axiomatic theory T then there exist axioms $\varphi_1, \varphi_2, \ldots, \varphi_n$ of this theory such that the formula $\varphi_1 \wedge \varphi_2 \wedge \cdots \wedge \varphi_n \longrightarrow \varphi$ is a theorem of logic. Hence mathematical theories are nothing more than a source of logical laws. It is abolutely inessential what axioms one adopts, there do not exist better and worse theories — important are only the logical interconnections between axioms and theorems. Hence for example there is (from the point of view of the pure mathematics) no difference between Euclidean or non-Euclidean geometry.[7]

Researches in mathematical logic and the foundations of mathematics indicated the crucial role played in mathematics by set theory. Therefore so much attention was devoted to it not only in foundational studies but also in the philosophical context. This attitude was strengthened by results of Gödel (1939) and Paul Cohen (1963) who showed that two most discussed principles of set theory: the axiom of choice and the continuum hypothesis are consistent and indepen-

[6] Detailed description of the results of the reverse mathematics and of their meaning for the philosophy of mathematics is given in (Murawski 1994) (one can also find there an extensive bibliography).

[7] One can see certain similarity of this idea and Aristotle's thesis that the necessity cannot be found in any single statement about mathematical objects but in hypothetical statements saying that if a certain proposition is true then a certain other proposition is also true. Hence (using today's terminology) we can say that for Aristotle the necessity of mathematics was that of logically necessary hypothetical propositions. Similar ideas one finds also by Russell before 1900, later he has changed his mind.

dent of other axioms of Zermelo-Fraenkel set theory. Hence various mutually inconsistent axiomatic set theories are possible and consequently there is no unique set-theoretical base for mathematics. The situation is even more complicated: Zermelo-Fraenkel axiomatization of set theory is not the only possible, there are also other axiomatic theories based on another principles — e.g., Gödel-Bernays set theory (with sets and classes), Quine's system of *New Foundations*, theory of semi-sets of P. Vopěnka and his alternative theory of sets (based on the ideas of Bolzano instead of those of Cantor). All this led anew to the discussion between formalism and realism in set theory (and consequently in mathematics). By the formalistic approach there is no necessity of justifying the accepted axioms (one can in fact adopt any consistent axioms) — cf. Cohen (1971). The realistic one pretends that axioms provide the description of the real world of sets. The latter was represented, e.g., by Gödel (see above). In the eighties a naturalized version of Gödel's ideas has been developed by Penelope Maddy (cf. Maddy 1980, 1990a, 1990b, 1997). Gödel thought we can intuit abstract sets, Maddy claims that we can see sets of concrete objects whose members are before our eyes. We perceive sets of concrete physical objects by perceiving their elements (physical objects). Sets are located in the space-time real world. In this way we can know "simple" sets, i.e., hereditarily finite sets. More complicated sets are treated by Maddy as theoretical objects in physics — they and their properties can be known by metatheoretical considerations. One sees that in this conception Gödel's mathematical intuition has been replaced by the usual sensual perception. The advantage of Maddy's approach is that it unifies the advantages of Gödel's platonism enabling us to explain the evidence of certain mathematical facts with Quine's realism taking into account the role that mathematics plays in scientific theories.

The most characteristic feature of the development of the philosophy of mathematics in the second half of the 20th century is the anti-foundational tendency and the empiricist approach. They were the reaction to the limitations and one-sidedness of the classical views which are giving one-dimensional static picture of mathematics as a science and are trying to provide indubitable and infallible foundations for mathematics. They treat mathematics as a science in which one automatically and continuously collects true proved propositions. Hence they provide only one-sided reconstructions of the real mathematics in which neither the development of mathematics as a science nor the development of mathematical knowledge of a particular mathematician would be taken into account. New conceptions challenge the dogma of foundations and try to reexamine the actual research practices of mathematicians and those using mathematics and to avoid the reduction of mathematics to one dimension or aspect only. They want to consider mathematics not only in the context of justification but to take into account also the context of discovery.

One of the first attempts in this direction was the conception of Imre Lakatos (1922–1974). He attempted to apply some of Popper's ideas about the methods of natural science to episodes from the history of mathematics.[8]

Lakatos claims that mathematics is not an indubitable and infallible science — on the contrary, it is fallible. It is being developed by criticising and correcting former theories which are never free of vagueness and ambiguity. One tries to solve a problem by looking simultaneously for a proof and for a counterexample. New proofs explain old counterexamples, new counterexamples undermine old proofs. By proofs Lakatos means here usual non-formalized proofs of actual mathematics. In such a proof one uses explanations, justifications, elaborations which make the conjecture more plausible, more convincing. Lakatos does not analyse the idealized formal mathematics but the informal one actually developed by "normal" mathematicians, hence mathematics in process of growth and discovery. His main work *Proofs and Refutations* (1963–1964) is in fact a critical examination of dogmatic theories in the philosophy of mathematics, in particular of logicism and formalism. Main objection raised by Lakatos is that they are not applicable to actual mathematics. Lakatos claims that mathematics is a science in Popper's sense, that it is developed by successive criticism and improvement of theories and by establishing new and rival theories. The role of "basic statements" and "potential falsifiers" is played in the case of formalized mathematical theories by informal theories (cf. Lakatos 1967).

Another attempt to overcome the limitations of the classical theories of philosophy of mathematics is the conception of Raymond L. Wilder (1896–1982). He presented his ideas in many papers and lectures — complete version of them can be found in two of his books: Wilder 1968 and 1981. Wilder's main thesis says that mathematics is a cultural system. Mathematics can be seen as a subculture, mathematical knowledge belongs to the cultural tradition of a society, mathematical research practice has a social character. Thanks to such an approach the development of mathematics can be better understood and the general laws of changes in a given culture can be applied in historical and philosophical investigations of mathematics. It also enables us to see the interrelations and influences of various elements of the culture and to study their influence on the evolution of mathematics. It makes also possible to discover the mechanisms of the development and evolution of mathematics. Wilder's conception is therefore sometimes called evolutionary epistemology. He has proposed to study mathematics not only from the point of view of logic but also using methods of anthropology, sociology and history. Wilder maintains that mathematical concepts should be located in the Popper's "third world". Mathematics investigates

[8]The reference to the history of mathematics is one of the characterisic features of new trends in the philosophy of mathematics.

no timeless and spaceless entities. It cannot be understood properly without regarding the culture in the framework of which it is being developed. In this sense mathematics shares many common features with ideology, religion and art. A difference between them is that mathematics is science in which one justifies theorems by providing logical proofs and not on the basis of, say, general acceptance.

The thesis that mathematical knowledge is a priori, absolutely certain and indubitable as well as infallible was criticized also in the quasi-empiricism of Hilary Putnam. He claimed (cf. Putnam 1975) that mathematical knowledge is not a priori, absolute and certain, that it is rather quasi-empirical, fallible and probable, much like natural sciences. He argues that quasi-empirical mathematics is logically possible and that ordinary mathematics has been quasi-empirical all along. In (1967) Putnam proposed a modal picture of mathematics according to which mathematics does not study any particular objects themselves but rather possibilities involving any objects whatsoever. Hence mathematics studies the consequences of axioms and asserts also the possibility of its axioms having models. The introduction of modalities opened the door to new epistemologies of mathematics.

The new anti-foundational trends in the philosophy of mathematics described above should not be treated as competitive with respect to old theories. They should be rather seen as complements of logicism, intuitionism and formalism. One is looking here not for indubitable, unquestionable and irrefutable foundations of mathematics, one tries not to demonstrate that the actual mathematics can be reconstructed as an infallible and consistent system but one attempts to describe the actual process of building and constructing mathematics (both in individual and historical aspect).

The quasi-empirical or even empirical approach to mathematics became recently a new dimension through the broader and broader application of computers. This phenomenon puts new challenges to the philosophy of mathematics.

Computers are applied in mathematics in at least six different ways: (1) to do numerical calculations, (2) to solve (usually approximatively) algebraical or differential equations and systems of equations, to calculate integrals, etc. (3) in automated theorem proving, (4) in checking mathematical proofs, (5) in proving theorems (one speaks in such situations about proofs with the help of computers), (6) to do experiments with mathematical objects. Special interests and controversies awake applications of the type (5). One of the main problems considered in this context is the problem of the epistemological status of theorems obtained in this way and consequently — in the case of accepting such theorems as legitimated mathematical theorems — the problem of the character of mathematical knowledge.

This problem is usually discussed in the context of the four-color theorem (cf. for example Tymoczko 1979). This theorem — proved in 1976 — is the first mathematical theorem whose proof is essentially based on the application of a computer program. We have no proof of this theorem not using computer. Moreover, some of the deciding ideas used in the proof were obtained and improved with the help of computer experiments, by a "dialogue" with the computer. Hence the four-color theorem is the first theorem of the new type, i.e., it has (at least so far) no traditional proof and its only proof we know is based on a(n) (computer) experiment (more exactly: it uses computer computations). So one should ask whether the mathematical knowledge can be still classified as an a priori knowledge and one should choose between the following two options: (1) to extend the scope of admissible proof methods and admit also the usage of computers (hence a certain experiment) in proving theorems or (2) to decide that the four-color theorem has not been proved so far and does not belong to the domain of mathematical knowledge. Observe that the first option implies the thesis that mathematical knowledge is not an a apriori knowledge any more but has in fact a quasi-empirical character. The thesis of the quasi-empirical character of mathematical knowledge can be found, e.g., by R. Hersh, Ph. Kitcher and E. R. Swart (cf. Swart 1980). They argue that mathematics always admitted empirical elements and had in fact an empirical character. On the other hand one attempts also to defend the a priori character of mathematics dispite the usage of computers in it by arguing that proofs using computers can be tranformed into traditional proofs by adding new axioms or that a computer is in fact a mathematician and it knows the result proved deductively or that procedures similar to applying computer programs have been used in mathematics for a long time, hence the applications used in the proof of the four-color theorem are in fact nothing essentially new (cf. for example Levin 1981, Krakowski 1980 or Detlefsen and Luker 1980).

It is worth noting here that not only the usage of computers is the source of new challenges in the philosophy of mathematics but also the theoretical computer science, in particular the computation theory, raises new problems and simultaneously provides new tools to analyse known results and to give new interpretations of them. As an example can serve here the information theory and its applications to the epistemology of mathematics (cf. Chaitin 1982 and 1998). In particular the analysis of Gödel's incompleteness theorems from the point of view of the information theory leads to the conclusion that mathematical knowledge has a quasi-empirical character and is in fact similar to sciences, in particular in the case of the criteria of accepting new axioms and methods.

Considering new conceptions in the philosophy of mathematics one must also mention structuralism.[9] It can be characterized as a doctrine claiming

[9]The presentation and analysis of structuralistic doctrines can be found in the doctoral dissertation by I. Bonecka-Krzykowska — cf. Bondecka-Krzykowska (2002).

that mathematics studies structures and that mathematical objects are featureless positions in these structures. As forerunners of such views one can see R. Dedekind, D. Hilbert, P. Bernays and N. Bourbaki. The latter is in fact a pseudonym of a group of (mainly French) mathematicians who undertook in the thirties the task of a systematization of the whole of mathematics (the result of their investigations was the series of books under the common title *Éléments de mathématique*). Their work refered to Russell's idea of reconstructing mathematics as one system developed on a firm (logical) basis. For bourbakists the mathematical world is the world of structures. The very notion of a structure was explained by them in terms of set theory. They distinguished three principal types of mathematical structures: algebraic, order and topological structures.

The idea of treating mathematics as a science about structures is being developed nowadays by Michael Resnik, Stewart Shapiro and Geoffrey Hellman. Resnik claims (cf. Resnik 1981, 1982, see also Resnik 1997) that mathematics can be viewed as a science of patterns with its objects being positions in patterns. The identity of mathematical objects is determined by their relationships to other positions in the given structure to which they belong. He does not postulate a special mental faculty used to acquire knowledge of patterns (they are not seen through a mind's eyes). We go through a series of stages during which we conceptualize our experience in successively more abstract terms. This process do not necessarily yield necessary truths or a priori knowledge. Important is here our tendency to perceive things as structured. The transition from experience to abstract structures depends upon the culture in which it takes place. Add that the transition from simple patterns to more complicated ones and the development of pure theories of patterns rely upon deductive evidence.

S. Shapiro claims (cf. Shapiro 1989, 1991) that there is a strict connection between objects of mathematics and objects of natural sciences. An explanation of it can be provided in his opinion just by structuralism according to which mathematics studies not objects *per se* but structures of objects. Hence objects of mathematics are only "places in a structure". The advantage of such an approach is that it enables us to explain the phenomenon of applicability of abstract mathematical theories in natural sciences as well as the interrelations between various domains of mathematics itself. It enables also a holistic approach to mathematics and science.

G. Hellman argues (cf. Hellman 1989) that one can interpret mathematics (in particular arithmetic and analysis) as nominalistic theories concerned with certain logically possible ways of structuring concrete objects. He uses by such interpretations second-order logic and modal operators (hence his approach is sometimes called modal-structural).

4. Conclusions

The overview of the development of the philosophy of mathematics in the 20th century given above shows on the one hand the strong influence of and interconnections between mathematical logic and the foundations of mathematics on the one hand and the philosophy of mathematics on the other. It indicates also the dominating (especially in the second half of the century) tendency to provide not only a reconstruction of the existing mathematics and to show in this way that it is safe and consistent (what was the aim of the classical doctrines developed at the beginning of the century) but to study the real research practice of mathematicians. The latter leads to the development of quasi-empirical trends and doctrines in the philosophy of mathematics. Those tendencies — reconstruction of mathematics on a safe base on the one hand and the study of the mathematical practice on the other — should not be treated as competitive but as complementary. They attempt to show and to investigate new aspects of the phenomenon of mathematics not covered by the others.

Adam Mickiewicz University
Faculty of Mathematics and Comp. Sci.
ul. Umultowska 87
61-614 Poznań, Poland
e-mail: rmur@amu.edu.pl

REFERENCES

Bondecka-Krzykowska, I. (2002). *Koncepcje strukturalistyczne we współczesnej filozofii matematyki: Analiza krytyczno-porównawcza* [Structuralistic Conceptions in the Modern Philosophy of Mathematics: A Critical-Comparative Analysis], PhD Thesis, Poznań.

Brouwer, L. E. J. (1907). *Over de Grondslagen der Wiskunde.* Amsterdam: Mass en van Suchtelen.

Brouwer, L. E. J. (1912). *Intuitionisme en formalisme.* Groningen: Noordhoff. Also in: *Wiskundig Tijdschrift* **9** (1912), 180–211. English translation: Intuitionism and Formalism, *Bulletin of the American Mathematical Society* **20** (1913), 81–96.

Brouwer, L. E. J. (1948). Consciousness, Philosophy and Mathematics. In: E. W. Beth, H. J. Pos, H. J. A. Hollak (eds.), *Library of the Tenth International Congress in Philosophy*, vol. 1, pp. 1235–1249. Amsterdam: North-Holland Publ. Comp.

Chaitin, G. (1982). Gödel's Theorem and Information. *International Journal of Theoretical Physics* **21** (12), 941–954.

Chaitin, G. (1998). *The Limits of Mathematics: A Course on Information Theory and the Limits of Formal Reasoning.* Singapore: Springer Verlag Singapore Pte. Ltd.

Cohen, P. J. (1971). Comments on the Foundations of Set Theory. In: D. Scott (ed.), *Axiomatic Set Theory*, Proceedings of Symposia in Pure Mathematics, vol. XIII, Part 1, pp. 9–16. Rhode Island: American Mathematical Society, Providence.

Detlefsen, M. and M. Luker (1980). The Four-Color Theorem and Mathematical Proof. *Journal of Philosophy* **77**, 803–820.

Frege, G. (1884). *Die Grundlagen der Arithmetik: Eine logisch mathematische Untersuchung über den Begriff der Zahl.* Breslau: Wilhelm Koeber.

Frege, G. (1893). *Grundgesetze der Arithmetik: Begriffsschriftlich abgeleitet.* Bd. 1. Jena: Hermann Pohle.

Frege, G. (1903). *Grundgesetze der Arithmetik: Begriffsschriftlich abgeleitet.* Bd. 2. Jena: Hermann Pohle.

Gödel, K. (1931). Über formal unentscheidbare Sätze der 'Principia Mathematica' und verwandter Systeme, I. *Monatshefte für Mathematik und Physik* **38**, 173–198. Reprinted with English translation: On Formally Undecidable Propositions of Principia Mathematica and Related Systems. In: K. Gödel, *Collected Works*, vol. I (edited by S. Feferman *et al.*), pp. 144–195 (1986). Oxford–New York: Oxford University Press and Clarendon Press.

Gödel, K. (1944). Russell's Mathematical Logic. In: P. A. Schilpp, (ed.), *The Philosophy of Bertrand Russell*, pp. 123–153. Evanston: Northwestern University. Reprinted in: K. Gödel, *Collected Works*, vol. II (edited by S. Feferman *et al.*), pp. 119–141 (1990). New York and Oxford: Oxford University Press.

Gödel, K. (1947). What is Cantor's Continuum Problem? *The American Mathematical Monthly* **54**, 515–525. Second revised version in: P. Benacerraf and H. Putnam, *Philosophy of Mathematics: Selected Readings*, pp. 258–273 (1964). New Jersey: Prentice-Hall, Inc., Englewood Cliffs. Reprinted in: K. Gödel, *Collected Works*, vol. II (edited by S. Feferman *et al.*), pp. 176–187 (1990). New York and Oxford: Oxford University Press.

Hellman, G. (1989). *Mathematics Without Numbers: Towards a Modal-Structural Interpretation.* Oxford: Clarendon Press.

Hilbert, D. (1926). Über das Unendliche. *Mathematische Annalen* **95**, 161–190. English translation: On the Infinite. In: J. van Heijenoort (ed.), *From Frege to Gödel: A Source Book in Mathematical Logic, 1879–1931*, pp. 367–392 (1967). Cambridge, Mass.: Harvard University Press.

Kline, M. (1980), *Mathematics: The Loss of Certainty.* New York: Oxford University Press.

Krakowski, I. (1980), The Four-Color Problem Reconsidered. *Philosophical Studies* **38**, 91–96.

Lakatos, I. (1963–1964), Proofs and Refutations: The Logic of Mathematical Discovery. *British Journal for the Philosophy of Science* **14**. As a book: Cambridge 1976.

Lakatos, I. (1967). A Renaissance of the Empiricism in the Recent Philosophy of Mathematics? In: I. Lakatos (ed.), *Problems in the Philosophy of Mathematics*, pp. 199–202. Amsterdam: North-Holland Publ. Comp. Extended version in: I. Lakatos, *Philosophical Papers, vol. 2: Mathematics, Science and Epistemology* (edited by J. Worall and G. Currie), pp. 24–42 (1978). Cambridge–London–New York–Melbourne: Cambridge University Press.

Levin, M. (1981). On Tymoczko's Arguments for Mathematical Empiricism. *Philosophical Studies* **39**.

Maddy, P. (1980). Perception and Mathematical Intuition. *Philosophical Review* **89**, 163–196.

Maddy, P. (1990a). *Realism in Mathematics.* Oxford: Clarendon Press.

Maddy, P. (1990b). Physicalistic Platonism. In: A. D. Irvive (ed.), *Physicalism in Mathematics*, pp. 259–289. Dordrecht: Kluwer Academic Publishers.

Maddy, P. (1997). *Naturalism in Mathematics*. Oxford: Oxford University Press.

Murawski, R. (1994). Hilbert's Program: Incompleteness Theorems vs. Partial Realizations. In: J. Woleński (ed.), *Philosophical Logic in Poland*, pp. 103–127. Dordrecht–Boston–London: Kluwer Academic Publishers.

Murawski, R. (1999). *Recursive Functions and Metamathematics*. Dordrecht–Boston–London: Kluwer Academic Publishers.

Putnam H. (1967). Mathematics Without Foundations. *Journal of Philosophy* **64**, 5–22. Revised version in: H. Putnam, *Mathematics, Matter and Method: Philosophical Papers*, vol. I, pp. 43–59. Cambridge-London-New York–Melbourne: Cambridge University Press.

Putnam H. (1975), What Is Mathematical Truth? In: H. Putnam, *Mathematics, Matter and Method: Philosophical Papers*, vol. I, pp. 60–78. Cambridge–London–New York–Melbourne: Cambridge University Press.

Quine, W. V. O. (1951a). Two Dogmas of Empiricism. *Philosophical Review* **60** (1), 20–43. Also in: W. V. O. Quine, *From a Logical Point of View*, pp. 20–46. Cambridge, Mass.: Harvard University Press.

Quine, W. V. O. (1951b). On Carnap's Views on Ontology. *Philosophical Studies* **2**.

Quine, W. V. O. (1953). On What There Is. In: *From a Logical Point of View*, pp. 1–19. Cambridge, Mass.: Harvard University Press.

Resnik, M. D. (1981). Mathematics as a Science of Patterns: Ontology and Reference. *Noûs* **15**, 529–550.

Resnik, M. D. (1982). Mathematics as a Science of Patterns: Epistemology. *Noûs* **16**, 95–105.

Resnik, M. D. (1997). *Mathematics as a Science of Patterns*. Oxford: Clarendon Press.

Shapiro, S. (1989). Structure and Ontology. *Philosophical Topics* **17**, 145–171.

Shapiro, S. (1991). *Foundations Without Foundationalism*. Oxford: Clarendon Press.

Swart, E. R. (1980). The Philosophical Implications of the Four-Color Problem. *American Mathematical Monthly* **87**, 697–707.

Tymoczko, Th. (1979). The Four-Color Problem and its Philosophical Significance. *The Journal of Philosophy* **76**, 57–83.

Whitehead, A. N. and B. Russell (1910–1913). *Principia Mathematica*. 3 volumes. Cambridge: Cambridge University Press.

Wilder R. L. (1968). *Evolution of Mathematical Concepts: An Elementary Study*. New York: John Wiley & Sons.

Wilder R. L. (1981). *Mathematics as a Cultural System*. Oxford: Pergamon Press.

Wittgenstein, L. (1953). *Philosophical Investigations*. Oxford: Basil Blackwell.

Wittgenstein, L. (1956). *Remarks on the Foundations of Mathematics*. Oxford: Basil Blackwell.

Krzysztof Wójtowicz

INDEPENDENCE AND JUSTIFICATION IN MATHEMATICS

ABSTRACT. In the article the problem of independence in mathematics is discussed. The status of the continuum hypothesis, large cardinal axioms and the axiom of constructablility is presented in some detail. The problem whether incompleteness is really relevant for ordinary mathematics and for empirical science is investigated. Another aim of the article is to give some arguments for the thesis that the problem of reliability and justification of new axioms is well-posed and worthy of attention. In my opinion, investigations concerning the status of independent sentences give insight into our understanding of mathematical concepts, of mathematical knowledge and of the role of mathematics in empirical science.

1. Gödel's Theorems

The aim of this article is to discuss (or — at least — indicate) some problems concerning incompleteness in mathematics.

According to Gödel's theorems, virtually every nontrivial mathematical theory is incomplete. In particular, PA (Peano artithmetic — the first-order formalization of number theory), and set theory ZFC are incomplete theories. As both these theories are in a certain sense fundamental for mathematics (in fact, virtually all mathematical concepts can be formally reconstructed within ZFC), Gödel's theorems seem to imply, that incompleteness and independence are phenomena, which are essential to the whole of mathematics.

2. The Incompleteness of PA *versus* the Incompleteness of ZFC

The most well-known examples of incomplete theories, which are subject to independence phenomena are Peano arithmetic (PA) and Zermelo-Fraenkel set theory (ZFC). Many number-theoretic or combinatorial sentences, which are independent form PA can be proved using set-theoretic notions.[1] One could say, that ZFC is the "oracle" for sentences independent from PA, and makes possible to settle many (if not most) of the problems unsolvable in PA.[2] In particular,

[1]E.g. the Paris-Harrington or Kirby-Goodstein, which are independent of PA, nevertheless true about ω (which is provable in ZFC, and many weaker systems as well).

[2]Not all sentences $\alpha \in L_{PA}$, independent from PA are provable in ZFC. E.g. the sentence Con_{ZFC} (i.e. ZFC is consistent), which can be expressed in L_{PA} after a suitable arithmetization of syntax, is independent from ZFC.

In: J. Malinowski and A. Pietruszczak (eds.), *Essays in Logic and Ontology* (*Poznań Studies in the Philosophy of the Sciences and the Humanities*, vol. 91), 349–373. Amsterdam/New York, NY: Rodopi, 2006.

the notion of “standard model for PA” (i.e. the natural numbers structure) is definable within ZFC, so the intuitive notion of arithmetical truth can be given a formal definition within ZFC (as “truth in the standard model”). The fact, that a certain sentence σ is independent from PA is a fact, which shows, that no complete description of our notion of “natural number” can be given using finitary, number-theoretic notions only. But we can always use some stronger methods to solve (at least some of) these problems.

The situation of ZFC is quite different. One cannot think of the standard model for ZFC, and there is no “supertheory”, which enables one to decide questions undecidable within ZFC.[3] This is particularly clear when one considers open set-theoretic problems — if we want to really “solve” them, not only to settle their metamathematical status, we have to abandon ZFC and look for new, stronger axioms. As ZFC is the commonly accepted formalization of mathematics, accepting new axioms amounts to accepting new standards of mathematical reasoning. It is not obvious, how this should be done, and there are several opinions concerning the question, whether these problems are well-posed at all.

3. CH: The “Paradigm of an Independent Sentence”

3.1. *Is the Continuum Problem Still an Open Problem?*

A paradigmatic example of an “open question” in set theory is the continuum hypothesis (CH). It looks elementary and quite innocent, nevertheless it proved to be an extremely hard problem to solve. The metamathematical status of CH has been established 80 years after its formulation by Cantor, and it has been a constant source of inspiration for generations of set theorists.[4]

ZFC is the commonly accepted formal framework (or: formal idealization) for mathematics. That means, that ZFC embodies all legitimate methods of mathematical reasoning. But CH cannot be decided in ZFC. Is it therefore reasonable to maintain, that the question “What is the true value of the continuum?” is well-posed? Should one not rather claim, that the continuum problem has been solved by works of Gödel and Cohen — and that this is the only solution one can expect? Or are we allowed to consider it to be still an open problem (in a suitable sense of this word)? These questions are important, as CH is — in a sense — a paradigmatic example of an independent sentence (i.e. of an “open problem”).

---

[3]One could say, that the standard model for ZFC is the whole set-theoretic universe V. But nobody knows, how it looks like.

[4]“The whole transfinite landscape can be viewed as the result of Cantor’s attempt to articulate and solve the Continuum Problem” (Kanamori 1996, p. 1).

Of course, the opinions here differ quite strongly, and depend on one's philosophical standpoint. Here I present some examples of the arguments and opinions, which can be found in the literature concerning the current discussion. Most of these opinions concern CH, but they also have a more general character.

(1) Gödel was quite aware of the fact, that CH is independent from ZFC.[5] But he often stressed the possibility (and necessity) of finding new axioms which could settle CH. In particular, he thought that large cardinal axioms could provide a solution. Unfortunately, soon after Cohen's discovery of forcing, Levy and Solovay proved, that large cardinal axioms do not allow one to solve this problem (Levy & Solovay 1967).[6] Moreover, Easton proved, that the value of the continuum can be almost arbitrary.[7] It became clear, that large cardinal axioms do not make the solution of the continuum problem possible.

(2) Cohen, who is a formalist considers the existing solution quite satisfactory. In his opinion, the question about the "true value of the continuum" lacks sense (cf. Albers *et al.* 1990, p. 54; Cohen 1971).

(3) Dales and Woodin express a quite contrary opinion: they consider the program of looking for natural, plausible axioms, which would enable one to solve the continuum problem as one of important goals of set theory (Dales & Woodin 1987, pp. 90–91].

(4) Mathias expresses the view about the universe, according to which

> within one enormous universe there are many inner models, and the various "independence arguments" may be reworked to give positive information about the way the various inner models relate to each other. Far from undermining the unity of the set-theoretic view, the various techniques available for building models actually promote that unity (Mathias 1992, p. 116).

(5) Martin admits, that his attitude towards CH is naive and uncritical — he considers it a genuine problem concerning the set-theoretic universe. However, he admits that there are strong arguments in favor of the opposite standpoint.

> Those who argue that the concept of set in not sufficiently clear to fix the truth-value of CH have a position which is at present difficult to assail. As long as no new axiom is found which decides CH, their case will continue to grow stronger, and our assertion that the meaning of CH is clear will sound more and more empty (Martin 1976, pp. 90–91).

---

[5]He proved it to be consistent with ZFC, nevertheless considered it to be false, therefore independent from ZFC.

[6]Levy and Solovay showed, that assuming the existence of a measurable, compact, Ramsey cardinals is both consistent with CH and ¬CH (cf. e.g. Felgner 1979, p. 186–187; Levy & Solovay 1967).

[7]Easton proved, that for any function F, such, that:
i) F is a non-decreasing function from the class of regular cardinal numbers into cardinal numbers;
ii) for any κ: $\kappa < \mathrm{cf}(F(\kappa))$;
it is possible to construct a model for ZFC such that $2^{\kappa} = F(\kappa)$, for any regular cardinal κ (Easton 1970).

(6) Mac Lane expresses a quite sceptical view about the reasonableness of the CH problem.

> I suspect that it is futile to wonder now, whe he [i.e. Gödel] imagined to be the "real" cardinal of the continuum. Those earnest specialist who still search for that cardinal may call to mind that infamous image of the philosopher — a blind man in a dark cellar looking for a black cat that is not there (Mac Lane 1992, p. 121).

(7) A similar opinion is expressed by Mostowski:

> Of course, if there are a multitude of set-theories then none of them can claim the central place in mathematics. Only their common part could claim such a position; but it is debatable whether this common part will contain all the axioms needed for a reduction of mathematics to set theory (Mostowski 1979, pp. 288–289).

(8) Kanamori observes, that metamathematical investigations made the problem of truth and consistency into

> matters of manipulation as in algebra. [...] From this point of view, that the ZFC axioms do not determine the cardinality of the reals seems an entirely satisfactory state of affairs. [...] the situation today with CH is actually quite like the situation with the parallel axiom and non-Euclidean geometries (Kanamori 1996, p. 46).

(9) A similar opinion is expressed by Dehorney, who claims, that — as ZFC is incomplete "the point is not longer to actually prove the properties, but rather to CALIBRATE them in a scale of increasingly strong logical axioms" (Dehorney 1996, p. 379)

(10) Scott claims, that

> the models are all just models of the first-order axioms, and first-order logic is weak. I still feel that it ought to be possible to have strong axioms which would generate these types of models as submodels of the universe, but where the universe can be thought of as something absolute (Scott 1977, p. xiv).

In particular, the questions about truth and falsity of independent sentences (such as CH) are legitimate.

Even this short presentation shows, that the opinions about the status of independent sentences differ quite strongly. Nevertheless, arguments in favor of the plausibility of new axioms extending ZFC are discussed. Before I shall discuss the problem of independence in the general setting, I would like to give some examples — just to give the "feeling" of the arguments, which are formulated in these discussions.[8]

[8]More arguments of this kind can be found in Maddy (1988a, 1988b), where a systematic analysis of the heuristic principles ("rules of thumb") governing the acceptance of axioms is given.

3.2. *What is the True Value of the Continuum?*

(1) Cohen as a formalist rejects the question of the truth of falsity of CH, but nevertheless makes some speculations on the value of the *continuum*. He notes, that the set $\aleph_1$ is simply the set of countable ordinals, but

> this is merely a special and the simplest way of generating a higher cardinal. The set C is, in contrast, generated by a totally new and powerful principle, namely the Power Set Axiom. [...] Thus C is greater than $\aleph_n$, $\aleph_\omega$, $\aleph_\alpha$, where $\alpha = \aleph_\omega$, etc. This point of view regards C as an incredibly rich set given to us by a bold new axiom, which can never be approached by any piecemeal process of construction (Cohen 1966, p. 151).

(2) Gödel rejected CH. In his opinion, CH has some implausible consequences, which make CH implausible as well. CH implies in particular, that there are subsets of **R** of the cardinality of the continuum, which are "small" in a suitable sense of this word (see Gödel [1947] 1964, pp. 266–267).

> "[A]gainst the many plausible propositions which imply the negation of the continuum hypothesis, not one plausible proposition is known which would imply the continuum hypothesis (Gödel [1947] 1964, p. 268).

Gödel tried to find and justify some new axioms, which could make the solution of the continuum problem possible (Gödel [1970a] 1995, [1970b] 1995), but his efforts were not quite successful (cf. Ellentuck 1975; Solovay 1995).

(3) According to Wang

> Some set theorist state that if $\aleph_1 = 2^{\aleph_0}$, than there must be a surprisingly delicate balance between the reals and the countable ordinals (Wang [1974] 1983, pp. 549–550).

This argument does not seem very compelling, as — independent of the cardinality of the continuum — there is always a balance between the reals and ordinals of a certain type.

(4) Friedman notes, that CH can be interpreted in two ways: as the statement, that $\mathcal{P}(\mathbf{N})$ is a small set (the smallest possible); on the other hand, as the statement, that there are many countable ordinal numbers (Friedman 1971). A similar observation is made by Drake: $\aleph_1$ is the number of countable well orders, and the continuum is the number of countable linear orders. But "being a well-ordered set" is a much stronger notion than "being a linear order," it seems, that the continuum should not be as small as $\aleph_1$. These facts are considered to be arguments against CH.

(5) In Freiling (1986) an interesting thought experiment is described: darts are thrown at random at the unit interval $[0, 1]$. Of course, it is quite unlikely to hit a number from a previously fixed countable subset of $[0, 1]$. E.g., if we throw two darts, it is virtually impossible, that the second dart hits a rational multiple of the first. This fact depends only on the countability of this particular set, not on other properties. Let us therefore assume, that we assign a countable

subset $f(x)$ of the interval [0, 1] to each number $x \in [0, 1]$ (i.e. we have a function $f\colon [0,1] \to \mathcal{P}_{\aleph}([0,1])$, where $\mathcal{P}_{\aleph}([0,1])$ is the family of countable subsets of the interval). It is quite impossible, that the second dart x_2 hits $f(x_1)$. By symmetry (as the line cannot tell the order of the throws), it is also impossible, that the first dart x_1 hits $f(x_2)$.That leads us to the following axiom:

$$(\mathrm{A}_{\aleph}) \qquad \forall f\colon [0,1] \to \mathcal{P}_{\aleph}([0,1])\ \exists x_1, x_2\, (x_1 \notin f(x_2) \wedge x_2 \notin f(x_1))$$

which seems quite convincing. But this axiom implies in a very straightforward way the negation of CH.[9] If we accept the intuitions about random darts, we should consequently reject CH.

A similar argument (which avoids some of weaknesses of Freiling's argumentation) is considered by Simms in (1989, 1991). Simms formulates an intuitive argument (concerning random infinite 0-1 sequences), which shows, that CH cannot be true (as it leads either to a contradiction or to a rejection of quite intuitive principles). His considerations concern a thought experiment, "conducted" in the possible world, where supertasks (such as ω-tosses) are possible. As we obviously have some intuitions concerning infinite random processes, we gain some intuitions concerning the world of set theory.

I do not think, that the arguments of Freiling and Simms are very convincing, but I will not discuss their weaknesses here. Nevertheless, these arguments can serve as an inspiration for the discussion of the problem of interplay between "ordinary" and "set theoretical" intuitions. Freiling's and Simms' thought experiments motivate us to ask the question about the legitimacy of thought experiments in mathematics and the plausibility of arguments of this kind.

4. V = L *versus* MC

4.1. V = L

Another example of a discussion of this kind is the discussion concerning the axiom of constructibility (V = L) and large cardinal axioms. V = L settles the continuum problem (ZF + V = L proves CH, and even GCH (Gödel 1940), and many other open problems in set theory, so it might appear to be a quite attractive axiom candidate. Nevertheless, V = L is not commonly accepted — on the contrary, there seems to be a kind of consensus in the mathematical (set-theoretical) community, that it is not a very plausible axiom. What kinds of arguments for and against V = L are usually formulated in this discussion?

---

[9] Assume CH, and let $<$ be a well-ordering on [0, 1] of the length $\aleph_1$. Define $f(x) := \{y : y < x\}$, which is a function from [0, 1] into $\mathcal{P}_{\aleph}([0,1])$. If $(\mathrm{A}_{\aleph})$ is true, then there are two numbers x_1 and x_2 such that $x_1 < x_2$ and $x_2 < x_1$, a contradiction.

(1) V = L is usually considered to restrictive. In particular, it restricts the notion of the subset of the set of natural numbers (as only constructible sets are allowed to exist). Such an opinion is expressed by Moschovakis (1980, p. 610). This restriction, which forbids the existence of all sets, which appear on a certain stage of the construction of the universe is not considered plausible (Drake 1974, p. 131). A similar opinion can be found in Foreman (1988, p. 13). In short: V = L does not account for all of the possible behaviour of sets or other mathematical objects.

(2) Maddy claims, that V = L expresses a very special way of thinking about mathematics, which she calls *Definabilism*. According to this view, only explicitly definable objects are legitimate in mathematics. Maddy distinguishes two main methodological positions concerning mathematics, which she calls *Definiabilism* and *Combinatorialism*. According to *Combinatioralism*, the possibility of giving an explicit definition of the investigated object is not essential. This is typical for modern mathematics. Maddy rejects *Definiabilism*, which — in her opinion — is a "discredited methodological maxim" (Maddy 1993, p. 41). But V = L is really an expression of the doctrine of *Definiabilism*. Therefore it also should be rejected.[10]

(3) According to Steel, a new axiom should be an

> assumption to be adopted by all, as part of a broadest point of view. The "broadest point of view" privis is meant to exclude from attention the temporary adoption of restrictive assumptions as a convenient device for avoiding irrelevant structure. V = L is often assumed temporarily for such reasons by set theorists who do not believe it (Steel 2000, p. 422).

Accepting V = L is a sterile move, as it solves mathematical problems by giving them a new interpretation: a sentence φ is interpreted as φ^{L}. But this reinterpretation can be done just in any theory. Accepting V = L is really equivalent to restricting ourselves to a certain part of the mathematical universe — it can be considered not to be a statement concerning the nature of the mathematical world, but as a decision, that we will investigate only a fragment of this world. But "it simply prevents us from asking many questions, since we are then forbidden to ask about the world outside L" (Steel 2000, p. 423). But this is restrictive, since there are many interesting mathematical structures outside L. Therefore V = L is not attractive, and should be rejected.

(4) Devlin is a proponent of V = L. In his opinion, it is a "natural axiom, closely bound up with what we mean by 'set'. [...] It is known not to contradict the more basic assumptions about sets" (Devlin 1977, p. iv). Accepting it enables us to solve many problems (in algebra, measure theory or topology), which are not solvable in ZFC. In Devlin's opinion, the situation is similar to the case of AC, which was also accepted for pragmatic reasons.

---

[10]In Maddy (1997) Maddy gives some formal criteria which lead to the rejections of V = L (cf. paragraph 5.7).

(5) Quine claims, that V = L is an elegant axiom. He notes, that for the purposes of ordinary (in particular: applied) mathematics it is enough to use ZF + V = L. V = L maximally restricts the size of the universe — there are only definable objects, there are not too many large cardinals.[11] V = L should be accepted for reasons of ontological economy, as it "inactivates the more gratuitous flights of higher set theory" (Quine 1992, p. 95).

Friedman expresses the opinion, that the point of view of a "normal mathematician" is quite similar. The set theorist is looking for possibly deep set theoretic phenomena, "so V = L is anathema, since it restricts the set theoretic universe so drastically that all sorts of phenomena are demonstrably not present" (Friedman 2000, p. 436). But for most mathematicians, set theory is not interesting in itself, and set theoretic phenomena and results are not really relevant for mathematics proper. "For the mathematician, mathematics is emphatically not a branch of set theory" (Friedman 2000, p. 435). For the "normal mathematician", set theory is only a tool, it provides methods of formalizing mathematics, and to establish mathematical rigour. That means, that the less set theory, the better — so the "normal mathematician" would rather accept V = L (if he was interested in problems of this kind at all, which he normally is not.)

4.2. *MC*

V = L is inconsistent with the axiom of measurable cardinals (MC) (see Scott 1961). V = L is usually rejected. Does it mean, that large cardinal axioms should be accepted as a plausible alternative? What are the arguments in favor of large cardinals?

The view of the set-theoretic universe as open-ended leads in a natural way to large cardinal axioms. Above every level of the set-theoretic hierarchy there is a new level, which is not accessible by the means we have at our disposal. This point of view was expressed e.g. by Zermelo in (1930). Now, e.g. Kanamori and Magidor divide the arguments in favor of the existence of large cardinals into there groups:

1. Four abstract principles, motivating the formulation of large cardinal axioms:

1.1. Generalization. From the point of view of hereditary finite sets, ω is a huge object. So if we believe, that there are objects, which — "from the point of view" of smaller sets — look like ω from the point of view of finite sets, we are naturally motivated to formulate large cardinal axioms.

1.2. Reflection. The set theoretic universe V has the reflection property — if a sentence φ is true in V, it is also true in a certain initial segment V_α. This motivates the investigations of cardinal numbers κ with a similar property: if

[11] Inaccessible cardinals are compatible with V = L, but measurable cardinals are not.

something is true about V_κ, it is also true about V_α, for a certain $\alpha < \kappa$. Of course, κ has to be big enough, as V_κ has to resemble the whole universe V.[12]

1.3. RESEMBLANCE. If we think, that the universe is so big, that there are similar levels V_α and V_β, we are naturally led to elementary embeddings and indiscernibility arguments, which are connected with large cardinal axioms.

1.4. RESTRICTION. Many set-theoretic sentences (in particular combinatorial principles concerning "small" cardinals or the real numbers) are equivalent to certain large cardinal numbers. These axioms are the weakest axioms, which must be accepted in order to prove a certain combinatorial sentence. That means, that large cardinal axioms provide us with a measure of consistency strength of the extensions of ZFC.

2. The efficacy and esthetics of concepts. Large cardinals can be used as natural measure of consistency strength of ZFC + φ — for various statements φ of set theory. The known large cardinals

> fit into a linear order via consistency strength, and so provide an abiding esthetics in neatly categorizing statements about sets (Kanamori & Magidor 1978, p. 105).

3. The third kind of arguments is the fruitfulness of the methods, leading to new theorems.

> Psychologically, work in the theory of large cardinals might result in an intuitive picture of V as rich and populated by a multitude of distinctive entities, and in such a fertile landscape, glimmerings of new interconnections are bound to emerge, indicating possible new applications of known methods and ultimately leading to new theorems (Kanamori & Magidor 1978, p. 105).

Similar opinion is expressed by Hauser. He claims, that

> there is essentially only one way of extending the theory provided in ZFC which remains faithful to our primary intuitions of the concept of set, namely by postulating the existence of large cardinals (Hauser 2002, p. 269).

His two main arguments to support this thesis are:

(i) The logical strength of independent sentences can be calibrated by large cardinal axioms.

(ii) Set theories with large cardinal axioms are mutually interpretable with other set theories (which is an outcome of the inner models program).

Both these facts show, that large cardinal axioms can be considered quite plausible, and are a natural superstructure for ZFC. In particular, the concept of set is not an inherently vague notion, as set theory develops in one direction only.

A similar argument is also given by Foreman. He stresses the importance of the idea of reflection as expressesing the idea, that the set formation process

[12] The interplay between reflection principles and large cardinal axioms was first studies in Levy (1960). Shapiro in (1987) investigates reflection principles for second order set theory and shows their equivalence with large cardinal axioms. Reflection was also considered by Gödel to be one of important heuristic principles governing the formation of the axioms of set theory.

has no natural stopping point. We can always take the collection (union) of all already formed sets and form a new set. "Hence any property that holds in the mathematical universe should hold of many set-approximations of the mathematical universe" (Foreman 1988, p. 14). Dehorney claims, that self-similarity is a strong form of infinity.

> It is [...] natural [...] to iterate the process of going from finite to infinite and consider higher infinities that behave with respect to usual infinities the way the latter behave with respect to finite objects (Dehorney 1996, p. 380).

5. Philosophy and New Axioms

Whether one considers the problem of justifying new axioms legitimate or not, depends certainly on one's philosophical standpoint. Here I want to discuss shortly the problem of truth (and justification) of new axioms from the point of view of some typical philosophical positions.

IF-THENISM. From this point of view, the question about truth or falsity of independent statements does not make any sense. The only relevant problem is the problem of their metamathematical status (in particular of their relative consistency). The question, whether φ is provable using a new axiom ψ is treated as a purely formal problem, and we are not in the least concerned with truth (or even plausibility[13]) of ψ. In particular, the question, what the mathematical universe is really like, does not even arise, as mathematics is treated as a purely formal game.

FICTIONALISM. For the fictionalist, mathematical sentences are just useful fictions (Field 1980, 1989). We use them in science, because they make our inferences shorter, but they are — at least in principle — dispensable.[14] We are free to use any mathematical theory we consider useful. As all these theories are fictions, questions concerning truth do not arise at all. The justification of new axioms is — if this problem arises at all — a pracical problem, solved by pragmatic considerations. If we are to choose between $T + \varphi$ and $T + \neg\varphi$, our choice depends only on the fact, which of these theories is more useful.

CLASSICAL PLATONISM. The most typical and prominent example of a platonist is Gödel. Gödel believed in the objective existence of a set-theoretic universe.

---

[13] Of course, the if-thenist will not deny the fact, that mathematicians consider some axioms plausible, and some not. But he considers it to be a purely psychological phenomenon, which has nothing to do with truth.

[14] In Field (1980) author tries to justify the thesis, that mathematics is conservative with respect to nominalistic theories.

Its existence is independent from the activities of the mathematician. Mathematical concepts "form an objective reality of their own, which we cannot create or change, but only perceive and describe" (Gödel [1951] 1995, p. 320). That means in particular, that mathematical axioms have a truth value: Gödel claimed, that mathematics proper is not a hypothetico-deductive system, but a "body of mathematical propositions which hold in an absolute sense, without any further hypothesis" (Gödel [1951] 1995, p. 305). In particular, we should try to determine the truth of all possible mathematical propositions. Gödel was of course aware of the fact, that due to his theorems, no mathematical theory can encompass all mathematical truths, but he was convinced, that new, plausible axioms can be found to settle open problems in mathematics. He stressed the inexhaustibly of mathematics — which in particular implies, that there can never be an end to the procedure of forming new axioms, because "the very formulation of the axioms up to a certain stage gives rise to the next axiom" (Gödel [1951] 1995, p. 307). In particular, the problem of finding an justifying new axioms is a well-posed problem, and in fact it is one of the most important tasks of set theorists. Gödel was quite convinced, that set-theoretical techniques could be used e.g. in number theory — in particular techniques involving new large cardinal axioms. The two main methods of finding and criteria of justifying new axioms are: (i) intuitive insight;[15] (ii) fruitfulness in solving problems.[16]

FULL-BLOODED PLATONISM. The formalist claims, that no mathematical theory has an interpretation. Balaguer's full-blooded platonist claims, that every consistent mathematical theory (in particular every variant and every extension of set theory) has an interpretation (Balaguer 1998). The mathematical universe is not one huge object, in which several models live. It is rather a "bunch" of different mathematical (in particular: set-theoretic) universes, in which all possible mathematical concepts are realized. But in this case it does not make sense to ask, whether a theory T truly describes the mathematical universe, as every consistent theory T describes a certain universe. That means, that a really full-blooded platonist would not bother about the problem of plausibility of new axioms, as his ontology is rich enough to offer an interpretation for every possible (consistent) theory.

---

[15] "[D]espite their remoteness from sense experience, we do have something like a perception also of the objects of set theory, as is seen from the fact that the axioms force themselves upon us as being true" (Gödel [1947] 1964, p. 271).

[16] "Even disregarding the intrinsic necessity of some new axiom, and even in case it has no intrinsic necessity at all, a probable decision about its truth is possible also in another way, namely, inductively by studying its 'success'. Success here means fruitfulness in consequences. [...] There might exist axioms so abundant in their verifiable consequences, shedding so much light upon a whole field, and yielding such powerful methods for solving problems [...] that, no matter whether of not they are intrinsically necessary, they would have to be accepted at least in the same sense as any well-established physical theory" (Gödel [1947] 1964, p. 265).

So from the point of view of looking for new axioms, full-blooded platonism does not offer any motivation. In this respect, it does not really differ from formalism, but it strongly differs from Gödel's platonism.

QUINE'S QUASI-EMPIRICISM. From Quine's point of view, the truth conditions for mathematical statements are recognizable *via* empirical sciences. Truth has nothing to do with intuition, but it has much to do with applications of mathematics. In particular, new axiom candidates should be evaluated in this context. Quine's views will be discussed in more detail in paragraph 7.

SET-THEORETIC NATURALISM. Maddy (1997) claims, that our choice between mathematical theories should be based only on internal mathematical criteria — we should not take into account any argument of philosophical or scientific nature.

Maddy admits, that is enough to use only a part of ZFC to reconstruct applicable mathematics. But — in her opinion — this fact is not very essential for the development of mathematics, and it would be artificial to restrict one's attention to ZFC only. The question, whether it is possible to find and justify new axioms should be answered with a certain methodological maxim in mind: the development of mathematics should be free of any restrictions. This general statement applies in particular to set theory, which does not need any external justification — in particular its internal problems and goals are essential. One's philosophical standpoint is not important at all, as the only important factor is the mathematical practice. In particular one need not to be a realist to consider the problems of justifying new axioms well-posed. A platonist wants to justify the thesis, that mathematical theories are true, but the set-theoretic naturalist "doesn't care about these philosophical niceties; what matters are the intra-mathematical goals ant the effectiveness of various means of achieving them" (Maddy 2000, p. 417).

The aim of Maddy's program is to formulate some methodological criteria, which would allow us to choose between competing theories (like e.g. ZFC+MC and ZFC + V = L). These criteria should be based on intramathematical goals only, not on philosophical considerations, or considerations concerning the role of mathematics in empirical sciences. In particular, these criteria should be independent from one's philosophical standpoint.

Two main methodological principles are: MAXIMIZE and UNIFY. Due to the first principle, the chosen theory should provide as many isomorphism types as possible (loosely speaking, it means, that a rich class of mathematical object can be defined). The second principle states, that it should be possible to chose one particular theory as the fundamental set theory. Maddy then gives several technical definitions of notions such as "a fair interpretation of T_1 in T_2", "proper maximalization", "strong maximalization", "restrictive theory", and finally she formulates a technical criterion (using some technical results concerning relative consistency results in set theory). In particular she claims, that ZFC + V = L is restrictive, and therefore cannot be considered an attractive theory candidate.

6. Are the Independence Phenomena Really Relevant?

From Gödel's theorems we know, that — loosely speaking — every non-trivial theory is incomplete. This fact is often considered to be an argument to support the thesis, that our knowledge of the world is inherently incomplete, and that we will never get a complete description. This point of view seems to be justified, but — on the other hand — it is not quite clear, whether the independent sentences in question are genuine scientific problems. It might be claimed, that some sentences just happen to be independent, but that they have no real bearing on our growth of knowledge. I would like to discuss this problem in two respects: (i) its importance for mathematics; (ii) its importance for natural science.

6.1. *Are the Independence Phenomena Really Relevant for Ordinary Mathematics?*

It is clear, that independence phenomena are central for set theory and metamathematics. Questions like "Is ZFC + φ consistent?", "Is there an inner model of ZFC + ψ?", "Does some particular combinatorial principle involving e.g. $\aleph_\omega$ follow from a certain large cardinal axiom?", questions concerning descriptive set theory etc. are the stuff, of which set theory is made. If there are open questions in set theory, they should be solved, or at least their metamathematical status should be established — e.g. by calibrating the independent sentences by axioms of increasing strength (which is also often considered to be a solution of the problem).

But are the independence phenomena relevant for the rest of mathematics? Remember, that the common formal idealization of mathematics is ZFC, and that is suffices for all purposes of usual mathematical practice.[17] From the point of view of the usual mathematical practice, the fact that some set-theoretic sentences are independent of ZFC is not really relevant. These sentences have no bearing on the work a differential geometer or a functional analyst, even if they are of primary importance for the set theorist. Set theory serves as a kind of a

---

[17] What does "usual mathematical practice" mean? Of course, it is a matter of taste. Simpson gives the following characterization, which — I think — would be accepted by the majority of the mathematical community:

"We identify as ordinary or non-set-theoretic that body of mathematics which is prior to or independent of the introduction of abstract set-theoretic concepts. We have in mind such branches as geometry, number theory, calculus, differential equations, real and complex analysis, countable algebra, the topology of complete separable metric spaces, mathematical logic, and computability theory" (Simpson 1999, p. 1).

formal idealization, allows to give a formal reconstruction of all the notions of ordinary mathematics. Nevertheless, its internal conceptual and technical problems have no bearing on the development of normal mathematics, as set theory is not really a part of "genuine" mathematics.

To a certain extent, it is the question of one's taste and his notion of naturalness and epistemological relevance of mathematical problems. There is no consensus about these matters. It looks like, that there can be no consensus — as in many other matters, which are connected with one's philosophical attitude. Nevertheless, even if a conclusive answer is not possible to attain, I think that the problem is worth paying attention. Here I present two different answers to the question 6.1, given by two prominent logicians: Friedman and Feferman.

6.1.1. *Answer 1: Yes, they are! (Friedman)*

Many examples of independent sentences, which are claimed to have bearing on the usual mathematical practice have been given by Friedman. In Friedman (1981,1986) he has presented sentences, which are independent not only from ZFC, but even from ZFC + V = L. One has to make assumptions concerning the existence of Mahlo cardinals in order to prove them.[18] The sentences from Friedman (1981) concern Borel functions from the Hilbert cube into the unit interval.[19] In Friedman's opinion

> these propositions provide examples of interesting theorems whose proofs necessarily involve the outer limits of what is commonly accepted as valid principles of mathematical reasoning (Friedman 1981, p. 209).

In (Friedman 1986) several examples of sentences concerning combinatorial properties of some finite objects are presented. These sentences are provable only under the assumption of the existence of Mahlo cardinals. In Friedman's opinion, this fact is a quite convincing argument, that methods of strong abstract set theory will prove to play an essential role in more standard finite mathematical contexts. Friedman stresses the fact, that — unlike in the case of CH — one cannot make the problem disappear by restricting the notion of set to constructible sets, as these sentences are independent not only from ZFC, but from

---

[18] It is necessary to assume, that $\forall n \exists \kappa(\kappa$ is n-Mahlo). It is not even enough to assume the existence of a n-Mahlo cardinal for a fixed n.

[19] Consider the Hilbert cube Q (i.e. $[0,1]^{\omega}$ equipped with the product topology). Consider the group H consisting of such permutations of ω, which are constant for almost all $n \in \omega$ (i.e. for $g \in$ H, $g(n) = n$ for all $n \in \omega$, except for a finite number of n's.). The group H acts on Q^n : $g \bullet (x_1, \ldots, x_n) = (g \bullet x_1, \ldots, g \bullet x_n)$. Let $\approx$ be the following equivalence relation on $Q^n : x \approx y$ iff x, y are in the same orbit of this action. The independent sentence is:

Let $F\colon Q^{n+1} \to [0,1]$ be a Borel function such, that for $x \in Q$, $y, z \in Q^n$, $y = z$, implies $F(x,z) = F(x,y)$. Then there exists a sequence $\{x_k\}_k$ of elements of Q, of finite or countable length α, such that for all $s < t_1 < \cdots < t_n \leqslant \alpha$, $f(x_s, x_{t_1}, \ldots, x_{t_n})$ equals the first coordinate of x_{s+1}.

ZFC + V = L as well. If the necessity of the use of abstract set theoretic assumptions (like large cardinals assumptions) in ordinary mathematics could be proved, this

> would open up a foundational crisis of nearly unprecedented magnitude since we seem to have no way of convincing ourselves of the correctness or consistency of such set theoretic principles short of faith in our very uneasy intuition about them (Friedman 1986, p. 93).

The problem is also discussed in Friedman (2000), where the author mentions some recent independence results in the theory of Boolean relations. In Friedman's opinion, these results show, that abstract set theory proves to be necessary in ordinary mathematical contexts. The problem of finding and justifying new axioms ceases to be an internal problem of set theory, and becomes an essential problem for the whole of mathematics. Friedman (2000, pp. 439–440) formulates certain general conditions, which should be fulfilled by these new axioms, so that they could be accepted:

> (a) They should be entirely mathematically natural. [...]
> (b) They should be concrete. At least within infinitary discrete mathematics. [...]
> (c) They should be thematic. [...] For example, the same theme may already be inherent in well known basic, familiar, and useful facts.
> (d) They should have points of contact with a great variety of mathematics.
> (e) They should be open-ended. I.e. the pain will never end until the adoption of large cardinals.
> (f) They should be elementary. [...] That way, even scientists and engineers can relate to it, so it is harder for the math community to simply bury them.
> (g) Their derivations should be accessible, with identifiable general techniques. That way, the math community can readily immerse itself in hands on crystal clear uses of large cardinals that beg to be removed — but cannot.
> (h) They should be used in normal mathematics as pursued before such thematic results.

There are two important features of Friedman's standpoint:

(i) Friedman admits, that purely set-theoretical problems are not important for the whole of mathematics. Independence phenomena can be essential only when they occur in natural, "ordinary" mathematical contexts.

(ii) Friedman claims, that in fact there are such natural problems, which cannot be solved using ZFC only. To solve them, we have to "go outside" of the common standards of mathematical justification. We cannot restrict our mathematical practice to ZFC alone — we have to abandon ZFC and use its extensions. Friedman's conclusion is quite strong:

> We believe that the results reported on here are the early stages of an evolutionary process in which new axioms for mathematics will be commonly used in an essential way in the more concrete parts of mathematics (Friedman 1998, p. 804).

6.1.2. *Answer 2: No, they are not! (e.g. Feferman)*

Feferman's views about the relevance of independence phenomena for usual mathematics are quite different. He is quite skeptical about the program of looking for new axioms. Feferman notes, that the research in large cardinals have made possible to obtain many new results within descriptive set theory, but they have not given the solution of the continuum's problem. In Feferman's opinion, this has to do with the fact, that CH is not a definite problem — it is "inherently vague", as it concerns the power set of natural numbers, which is not a definite mathematical object (see Feferman 2000, p. 405).

What about he independence phenomena in general? Are they really relevant to the mathematical practice? It is clear, that using the arithmetization of syntax one can find some undecidable sentences, which state the existence of solutions of certain polynomials. To be exact, there are true, but unprovable expressions of the form:

$$\forall x, z_1, \ldots z_n\, [p(x, z_1, \ldots, z_n) \neq q(x, z_1, \ldots, z_n)].$$

Such number-theoretic problems seem to be as natural as they can be. But, in Feferman's opinion, these problems are "still very distant form the bread-and-butter problems of everyday number theory ('everyday' over the last 300 years)" (Feferman 1987, p. 196). These problems are remote form classic unsettled problems of number theory. The diophantic propositions in question have been "cooked-up" by metamathematical considerations — just to demonstrate the incompleteness of some formal systems. But they have no interesting mathematical content in themselves (like e.g. Goldbach's conjecture). It is important to note, that these number-theoretic independent propositions are insensitive to the formal system in question. The diophantine equation, which appears in the sentence equivalent to Con(PA) is not really different from the equations appearing in sentences expressing Con(ZFC) or Con(ZFC + MC). So the mathematical content of the diophantine problem has nothing to do with the internal mathematical content of the respective formal theories (Feferman 1987, p. 196).

Feferman does not agree with Friedman, who claims, that his independent sentences are entirely natural. Friedman's sentence from Friedman (1986) even if it "does make use only of concepts from ordinary finitary mathematics, it is extremely remote from any ordinary mathematical reading" (Feferman 1987, p. 202). Friedman's claim, that the use of strong set-theoretic methods will prove essential in standard finite mathematics is not justified, as Friedman's sentence is not natural, and it is fabricated to justify his thesis. It surely cannot be called "standard". Feferman's conclusion is even stronger: "the case remains to be established that any use of the Cantorian transfinite beyond $\aleph_0$ is necessary for the mathematics of the finite in the everyday sense of the word" (Feferman 1987, p. 202).

A similar opinion is expressed in Feferman (2000) (where Friedman's new results connected with the theory of Boolean relations are discussed). In Feferman's opinion, the conception of the cumulative hierarchy (where the power set operation is iterated in the transfinite) is inherently vague and that in general it is not clear, what is a fact of the matter.[20] Only the crudest features of the conception of cumulative set-theoretic hierarchy are really needed to build up a coherent body of results. In particular,

> there is not a shred of evidence so far that we will need anything beyond ZFC — or even much weaker systems — to settle outstanding combinatorial problems of interest to the working mathematician [...] nor is there any evidence that the kind of metamathematical work we've seen from Paris-Harrington to Friedman will bear any relevance to the solutions of these problems (Feferman 2000, p. 407).

Feferman also discusses some more general problems concerning incompleteness. In particular, he discusses "Gödel's doctrine", by which he means Gödel's view about the incompleteness phenomena (according to Gödel, the true reason of the existence of these phenomena is the fact, that "the formation of higher types can be continued into the transfinite" (Gödel 1931, p. 180, footnote 48a)).[21]

If we add suitable second-order axioms to PA, we are able to define the notion of "truth" for PA and prove Con(PA), which of course is not provable in PA alone. In this way, adding higher types allows one to prove new sentences. This procedure can also be applied in the case of many other theories: adding new axioms concerning "higher types" to the theory S in question allows one to produce a new theory S^*, which proves Con(S) (usually not provable in S alone).

In the case of set theory this procedure has a slightly different character: we add the axiom stating the existence of an inaccessible cardinal. $\mathrm{ZFC}+\exists\kappa\, \mathtt{Inacc}(\kappa)$ allows one to prove the existence of a model for ZFC (it is V_κ). In the general case, more and more large cardinal axioms can be added.

These facts seem to confirm Gödel's doctrine: in fact, adding new types yields new theorems and solves many problem. But Feferman claims, that it is not REALLY NECESSARY to add higher types to obtain these new results. He examines here the example of PA. If we are intuitively convinced, that the Peano axioms are true in the intended structure, we can simply introduce a truth predicate $\mathrm{Tr}_1(x)$. But it is not necessary to define this predicate in the second-order language, as it can be considered a primitive notion. If we add new axioms concerning Tr_1, it will be possible to prove Con(PA). This extended theory Tr_1(PA) allows one to prove several sentences unprovable in PA. A similar procedure can be applied in the case of Tr_1(PA): it is incomplete, but we can simply add the truth predicate Tr_2 for Tr_1(PA) together with appropriate axioms. This allows to prove Con(Tr_1(PA)) (and many other sentences). Of course, this procedure can be continued (cf. Feferman 1987, pp. 190–194).

---

[20] Feferman even claims, that it is not a fact of the matter, whether all projective sets are Lebesgue measurable (Feferman 2000, p. 405).

[21] Gödel stressed the importance of introducing new axioms in order to solve open mathematical problems. In particular, he believed, that new set-theoretic axioms will make it possible to solve many open number-theoretic and set-theoretic problems, like CH.

This kind of reasoning leads to the formulation of a certain alternative doctrine, so that Gödel's doctrine can be rejected. On could also maintain that — contrary to Gödel's views — the "real reason" of independence phenomena is the fact, that the definition of truth for a formal language is not expressible in that language, and that adding the notion of truth together with suitable axioms allows to prove new sentences. In fact, it is even enough to add suitable reflection axioms of the form $\mathrm{Prov}_S(\ulcorner\varphi\urcorner) \Rightarrow \varphi$, for any sentence φ. Informally, these axioms express the faith in the correctness of S — without any new truth predicates at all. But they allow to prove new sentences — in particular Con(S). This leads to another variant of Gödel's doctrine:

> the "true reason" for the incompleteness phenomena is that though a formal system S may be informally recognized to be correct, we must adjoin formal expression of that recognition by means of a reflection principle in order to decide Gödel's undecidable statements (Feferman 1987, p. 194).

Both these alternatives are — according to Feferman — equally plausible. Of course, a platonist would consider the strengthening by adding higher types (e.g. second-order or large cardinal axioms) more plausible, than adding new primitive predicates. But Feferman claims, that "unless one is not ALREADY a set-theoretical platonist, Gödel's doctrine does not by itself provide compelling reasons to embrace the Cantorain transfinite" (Feferman 1987, p. 194).

I do not think, that Feferman's arguments are convincing. Of course, it is possible to produce axioms not involving higher types, which allow to prove new lower-level statements (the simplest example is Con(S)). But the question arises, what are the deep reasons to consider sentences like Con(S) true. I think, that — on the informal level — we believe in the consistency of S (e.g. PA or ZFC) due to the fact, that it is possible to "point to" (i.e. define) a structure, in which S is true. We are intuitively convinced, that PA is true, because we can think of an intended interpretation for PA — not just because we happen to believe, that these axioms are true. I think, that adding a new truth predicate Tr is really motivated by the belief, that PA is true in the intended structure (not even "true in ANY structure"). But to "point to" (i.e. define) this structure, one has to assume some stronger theory, e.g. ZFC.[22] In this case one should admit, that — if we want to formally "grasp" these intuitions — we have to formulate some higher-type axioms, which allow to define these models. Otherwise, the informal considerations about truth in the intended structure have to be treated in a purely instrumentalists fashion — we really think about truth in a structure, and justify our beliefs considering the properties of this structure (e.g. that the PA axioms are true in it, and that all its objects are standard, etc.). But then we

[22]Feferman admits, that the reason, which lead us to the acceptance of sentences with the truth predicate Tr_1 are our informal beliefs in the truth of PA in a certain structure.

try to forget about these structures, formulate axioms involving new predicates, and pretend, that these predicates are primitive.[23]

6.1.3. *Summary*

The opinions presented above are two extreme examples. Friedman takes for sure, that abstract set theory will prove essential in ordinary mathematical practice, and that the program of looking for new axioms is no longer an internal business of set theory — the whole mathematical community should be concerned (in this form or another) with this program. Set-theoretic problems are essential for the whole of mathematics.

On the other hand, Feferman claims, that independent sentences presented by Friedman are artificial and have nothing to do with mathematical practice. They are just metamathematical statements in disguise. These remarks apply to undecidable diophantine propositions as well — they have no natural mathematical content. Feferman even claims, that set-theoretical notions are superfluous in mathematics — ordinary mathematics can be reconstructed within much weaker systems, where very little set theory is needed.[24]

It is not possible to settle the argument. It is not quite probable, that the mathematical community will be completely converted to one or the other point of view — it is difficult to expect, that a set-theorist will consider his problems artificial, or that an expert in financial mathematics will consider relative consistency results involving large cardinals relevant for "real mathematics". The differences in opinions are deeply rooted in the differences in mathematical taste.

7. Are the Independence Phenomena Really Relevant for Natural Science?

In which sense are the independence phenomena relevant for natural science? To put it in an extreme form: are there any facts about our empirical world, which cannot be known because of the independence phenomena in mathematics — e.g. we cannot know, whether an elementary particle exists, because a certain sentence σ is independent from PA or ZFC?

I do not want to go into issues, what it exactly means, that a sentence has empirical content. But there are examples of sentences, which are claimed to have

[23]Notice, that a similar argumentation could be applied to ZFC. The existence of an inaccessible number κ allows one to prove the existence of model for ZFC. That means, that our informal consideration involving the notion of "truth of ZFC within V_κ"-allow to justify Con(ZFC). But the claim, that we can simply accept Con(ZFC) as a plausible new axiom, completely forgetting our considerations involving the existence of κ is not quite convincing. It does not seem to be honest to forget about κ, as our "feeling" that Con(ZFC) is true would really be based on our "feeling", that κ exists.

[24]A large part of "ordinary mathematics" can be reconstructed within Z_2, cf. e.g. (Simpson 1999).

physical meaning, and which are independent from ZFC.[25] Regardless of the fact, whether these sentences really have physical, empirical content, or whether one regards them as artificial (in the manner Feferman thinks Friedman's sentences are artificial) one can easily imagine a situation, where a certain physical sentence (e.g. a prediction of the behaviour of a physical system) depends on mathematical assumption — and in particular, that it might happen, that this sentence is independent from ZFC.[26]

I think, that it is quite natural to discuss problems of this kind in the context of Quine's indispensability argument and his theory of ontological commitments. Quine's indispensability argument in favor of mathematical realism rests on the fact, that mathematics is useful (and — in fact — indispensable) in natural science. In a sense, empirical theories assure us of the truth of mathematical statements. Therefore it seems natural to consider independence phenomena as (potentially) relevant to the empirical science, if they can be found in that part of mathematics, which has applications in physics.

On Quine's holistic view, our theories are tested as a whole. No division into analytic and synthetic statements can be made. This has important consequences for our decisions about the ontology we assume. The mathematical part of our theories cannot be "explained away" in a instrumentalist (or conventionalist) manner, and our ontology must include not only observable entities, but also the theoretical and mathematical objects as well.[27] In particular, the class of epistemologically relevant mathematical problems depends on the fact, which mathematical questions are relevant for natural science. Consider a math-

---

[25]The following result has been proved by da Costa and Doria. Within ZFC: (1). There is an expression for a motion $m(t)$ on $\mathbf{R}^2$ such that the assertion "$m(t)$ is ergodic in $\mathbf{R}^2$", can neither be proved nor disproved in ZFC. (2) There is an expression for a differentiable dynamical system v such that the statement "v has a Smale horseshoe" can neither be proved nor disproved in ZFC. (3) There is a flow X so that the statement "X is a Bernoullian flow" can neither be proved nor disproved in ZFC. (4) If φ is a ZFC predicate, that gives a nontrivial characterization for chaos in a dynamiacal system, then there is an expression Z for a dynamical system such that neither $\varphi(Z)$, nor $\neg\varphi(Z)$ can be proved in the formal theory (da Costa & Doria 1996, p. 112). In a series of articles both authors present many results of similar kind, concerning classical mechanics, general relativity or even game theory. In their opinion "Gödel incompleteness is no outlandish phenomenon; it is an essential part of the way we conceive mathematics" (da Costa & Doria 1994, p. 188).

[26]It is not unproblematic to stipulate, which of the hypotheses are empirical. The example of Freiling'a and Simms' thought experiments (cf. 3.2) show, that even in an ω-world, the problem of CH cannot be considered an empirical question (or at least: it is not obvious, that it has empirical meaning).

[27]"Total science is like a field of force whose boundary conditions are experience. A conflict with experience at the periphery occasions readjustments in the interior of the field. Truth values have to be redistributed over some of our statements [...]. Having reevaluated one statement we must reevaluate some others, which may be statements logically connected with the first or may be statements of logical connections themselves. [...] No particular experiences are linked with any particular statements in the interior of the field, except indirectly through considerations of equilibrium affecting the field as a whole" (Quine [1951] 1953, pp. 42–43).

ematical theory T, formulated within a certain language L (e.g. in the language of set theory or analysis or arithmetic, etc.), which is used in empirical science (e.g. in physics). Let $\varphi \in L$ be a sentence independent from T. The question, whether we should believe φ or rather $\neg\varphi$ depends only on the fact, which of the theories $T + \varphi$, $T + \neg\varphi$ proves to be more successful in explaining the empirical phenomena. If $S_{T+\varphi}$ is the physical theory based on $T + \varphi$, and $S_{T+\neg\varphi}$ is the physical theory based on $T + \neg\varphi$, then the criterion of our choice between φ and $\neg\varphi$ is merely the fact, whether $S_{T+\varphi}$ or $S_{T+\neg\varphi}$ is more empirically adequate. If φ is an open (independent) mathematical sentence, which has no bearing on the description of the empirical phenomena (or: there is no way to tell, which of the theories $S_{T+\varphi}$, $S_{T+\neg\varphi}$ is really empirically adequate) we have no way to make a reasonable choice between these theories. In particular, if it happened, that φ is equivalent to a certain empirical hypothesis H_e (stating e.g. the existence of an object), which (e.g. for technical reasons) cannot be decided experimentally, than it would not be reasonable to decide, that H_e is true, just because of our feeling, that φ is (as a mathematical sentence) more plausible, than $\neg\varphi$.

Of course, I do not claim, that mathematics is an empirical science, and that we solve mathematical problems like “Is φ true?” experimentally. The statement above only means, that our choice of mathematical tools used in science is based on the empirical adequacy of respective theories. Whether we treat them instrumentally or not, is a different problem.

If there is no difference between $S_{T+\varphi}$ and $S_{T+\neg\varphi}$, than the question “Is φ true?” lacks content. The question about the truth and/or plausibility of independent sentences makes sense only in the context of empirical theories. In particular, if the formalization within ZFC (or even a weaker system, e.g. second-order arithmetic Z_2)[28] proves to be sufficient for the purposes of physics, then the questions about the truth of CH or V = L or MC will not be considered really relevant. To use Quine’s words, they will be considered interesting mathematical recreation, without ontological (and epistemological) rights. On the other hand, if one of the theories $T+\varphi$, $T+\neg\varphi$ (say, $T+\varphi$) will prove successful, then the question about the truth of φ will be given a natural answer: as we assume the existence of mathematical object only if they are a part of the ontology of our physical theory, we should assume the truth of φ. Mathematical statements (as well as logical principles) are a part of our web of beliefs and their truth is established together with the truth of physical hypotheses.

From this point of view, it does not make any sense to ask, whether φ is mathematically natural, plausible, whether it fits our intuitions, etc. What matters is the fact, that $T+\varphi$ proves to be empirically adequate. The internal mathematical motivations and considerations would not play any important role here.

[28] Cf. Simpson (1999), where many results concerning the equivalence of “ordinary” mathematical statements with subsystems of Z_2 can be found.

8. Concluding Remarks

It was not my purpose to answer the question, whether measurable cardinals exist, whether $\mathfrak{c} = \aleph_2$ or not. My aim was rather to convince the reader, that they really are problems, which deserve attention. Even if the discussion concerning particular questions ("Is MC true?", "Is CH true?") will not be conclusive (and it probably will not be — at least not for all sentences of this kind), I think, that such discussions can deepen our understanding of mathematical concept, give insight into the sources of mathematical intuitions and mathematical knowledge and the interplay between mathematics and the empirical science.

Uniwersytet Warszawski
Instytut Filozofii
Krakowskie Przedmieście 3
00-047 Warszawa
Poland
e-mail: wojtow@uw.edu.pl

REFERENCES

Albers, D. J., G. L. Alexanderson and C. Reid (1990). *More Mathematical People*. Boston: Harcourt Brace Jovanovich.

Balaguer M. (1998). *Platonism and Anti-Platonism in Mathematics*. New York, Oxford: Oxford University Press.

Benacerraf, P. and H. Putnam eds. (1964). *Philosophy of Mathematics*. New Jersey: Prentice-Hall, Englewood Cliffs.

Benacerraf, P. and H. Putnam, eds. (1983). *Philosophy of Mathematics*. 2nd edition. Cambridge: Cambridge University Press.

Cohen, P. J. (1966). *Set Theory and the Continuum Hypothesis*. New York – Amsterdam: W. A. Benjamin, Inc.

Cohen, P. J. (1971). Comments on the Foundations of Set Theory. In: D. Scott (ed.), *Axiomatic Set Theory*. Proceedings in Symposia in Pure Mathematics, vol. 13, part 1. Rhode Island: AMS, Providence.

Da Costa, N. C. A. and F. A. Doria (1994). Suppes Predicates an the Construction of Unsolvable Problems in the Axiomatized Sciences. In: P. Humpreys (ed.), *Patric Suppes: Scientific Philosopher*, pp. 151–193. Dordrecht: Kluwer Academic Publishers.

Da Costa, N. C. A. and F. A. Doria (1996). Structures, Suppes Predicates, and Boolean-Valued Models in Physics. in: P. I. Bystrov and V. N. Sadovsky (eds.), *Philosophical Logic and Logical Philosophy*, pp. 91–118. Dordrecht: Kluwer Academic Publishers.

Dales, H. G. and W. H. Woodin (1987). *An Introduction to Independence for Analyst.* Cambridge: Cambridge University Press.

Dehorney, P. (1996). Another Use of Set Theory. *Bulletin of Symbolic Logic* **2**, 379–391.

Devlin, K. (1977). *The Axiom of Constructability.* Lecture Notes in Mathematics, vol. 617. Springer-Verlag.

Drake, F. R. (1974). *Set Theory: An Introduction to Large Cardinals.* North Holland.

Easton, W. B. (1970). Powers of Regular Cardinals. *Annals of Mathematical Logic* **1**, 139–178.

Ellentuck, E. (1975). Gödel's Square Axioms for the Continuum. *Mathematische Annalen* **216**, 29–33.

Feferman, S. (1987). Infinity in Mathematics: Is Cantor Necessary? In: G. T. Di Francia (ed.), *L'infinito nella scienza.* Roma: Instituto della Enciclopedia Italiana.

Feferman, S. (2000). Why the Programs for New Axioms Need to be Questioned. *The Bulletin of Symbolic Logic* **6**, 401–413.

Felgner, U. (1979). Bericht über die Cantorsche Kontinuums-Hypothese. In: U. Felgner (ed.), *Mengenlehre*, pp. 166–205. Darmstadt: Wissenschaftliche Buchgesellschaft.

Field, H. (1980). *Science Without Numbers.* Oxford: Basil Blackwell.

Field, H. (1989). *Realism, Mathematics and Modality.* Oxford, Cambridge: Blackwell.

Foreman, M. (1988). Generic Large Cardinals: New Axioms for Mathematics? In: *Documenta Mathematica: Jahrbuch der Deutschen Mathematiker-Vereinigung*, pp. 11–21. Proceedings of the International Congress of Mathematicians, Berlin 1988, vol. II. Rosenheim.

Freiling, C. (1986). Axioms of Symmetry: Throwing Darts at the Real Number Line. *Journal of Symbolic Logic* **51**, 190–200.

Friedman, H. (1971). The Generalized Continuum Hypothesis is Equivalent to the Generalized Maximization Principle. *Journal of Symbolic Logic* **36**, 39–54.

Friedman, H. (1981). On the Necessary use of Abstract Set Theory. *Advances in Mathematics* **41**, 209–280.

Friedman, H. (1986). Necessary Uses of Abstract Set Theory in Finite Mathematics. *Advances in Mathematics* **60**, 92–122.

Friedman, H. (1998). Finite Functions and the Necessary Use of Large Cardinals. *Annals of Mathematics* **148**, 803–893.

Friedman, H. (2000). Normal Mathematics Will Need New Axioms. *The Bulletin of Symbolic Logic* **6**, 434–446.

Gödel, K. (1931). Über formal unentscheidbare Sätze der *Principia Mathematica* und verwandter Systeme I. *Monatshefte für Mathematik und Physik* **38**, 173–198. English translation in: Gödel (1986), pp. 144–195.

Gödel, K. (1940). *The Consistency of the Axiom of Choice and of the Generalized Continuum Hypothesis with the Axioms of Set Theory.* Annals of Mathematical Studies, vol. 3. Princeton: Princeton University Press. Reprinted in: Gödel (1990), pp. 33–101.

Gödel, K. ([1947] 1964). What is Cantor's Continuum Problem? *American Mathematical Monthly* **54**, 515–525. Reprinted in: Benacerraf & Putnam (1964), pp. 258–273.

Gödel, K. ([1951] 1995). Some Basic Theorems on the Foundations of Mathematics and their Implications. In: Gödel (1995), pp. 304–323.

Gödel, K. ([1970a] 1995). Some Considerations Leading to the Probable Conclusion, That the True Power of the Continuum is $\aleph_2$. In: Gödel (1995), pp. 420–421.

Gödel, K. ([1970b] 1995). A Proof of Cantor's Continuum Hypothesis from a Highly Plausible Axiom About Orders of Growth. In: Gödel (1995), pp. 422–423.

Gödel, K. (1986). *Collected Works*, vol. 1. Edited by S. Feferman *et al.* New York, Oxford: Oxford University Press, Clarendon Press.

Gödel, K. (1990). *Collected Works*, vol. 2. Edited by S. Feferman *et al.* Oxford University Press.

Gödel, K. (1995). *Collected Works*, vol. 3. Edited by S. Feferman *et al.* Oxford University Press.

Hauser, K. (2002). Is Cantor's Problem Inherently Vague? *Philosophia Mathematica* (3), vol. 10, 257–285.

Kanamori, A. (1996). The Mathematical Development of Set Theory from Cantor to Cohen, *Bulletin of Symbolic Logic* **2**, 1–71.

Kanamori, K. and M. Magidor (1978). The Evolution of Large Cardinal Axioms in Set Theory. In: G. H. Müller and D. S. Scott (eds.), *Higher Set Theory*, pp. 99–275. Lecture Notes in Mathematics, vol. 669. Berlin: Springer-Verlag.

Levy, A. (1960). Principles of Reflection in Axiomatic Set Theory. *Fundamenta Mathematicæ* **49**, 1–10.

Levy, A. and R. M. Solovay (1967). Measurable Cardinals and the Continuum Hypothesis. *Israel Journal of Mathematics* **5**, 234–248.

Mac Lane, S. (1992). Is Mathias an Ontologist? In: H. Judah, W. Just and H. Woodin (eds.), *Set Theory of the Continuum*. Springer-Verlag.

Maddy, P. (1988a). Believing the Axioms I. *Journal of Symbolic Logic* **53**, 481–511.

Maddy, P. (1988b). Believing the Axioms II. *Journal of Symbolic Logic* **53**, 736–764.

Maddy, P. (1993). Does V equal L? *Journal of Symbolic Logic* **58**, 15–41.

Maddy, P. (1997). *Naturalism in Mathematics*. Oxford: Clarendon Press.

Maddy, P. (2000). Does Mathematics Need New Axioms? *The Bulletin of Symbolic Logic* **6**, 413–422.

Martin, D. A. (1976). Hilbert's First Problem: the Continuum Hypothesis. in: F. E. Browder (ed.), *Mathematical Developments Arising from Hilbert's Problems*, pp. 81–92. Proceedings of Symposia in Pure Mathematics, vol. 28. Rhode Island: AMS, Providence.

Mathias, A. R. D. (1992). What is Mac Lane Missing? In: H. Judah, W. Just and H. Woodin (eds.), *Set Theory of the Continuum*, pp. 113–118. Springer-Verlag.

Moschovakis, Y. (1980). *Descriptive Set Theory*. North-Holland.

Mostowski, A. (1979). Recent Results in Set Theory. In: U. Felgner (ed.), *Mengenlehre*, pp. 276–302. Darmstadt: Wissenschaftliche Buchgesellschaft.

Quine, W. V. O. ([1951] 1953). Two Dogmas of Empiricism. Reprinted in *From a Logical Point of View*, pp. 20–46. Cambridge, MA: MIT Press.

Quine, W. V. O. (1992). *Pursuit of Truth*. Cambridge, MA: Harvard University Press.

Scott, D. (1961). Measurable Cardinals and Constructible Sets. *Bull. Acad. Polon. Sci.* **9**, 521–524.

Scott, D. (1977). Introduction to: J. L. Bell, *Boolean-Valued Models and Incependence Proofs in Set Theory*. Oxford: Clarendon Press.

Shapiro, S. (1987). Principles of Reflection and Second-Order Logic. *Journal of Philosophical Logic* **16**, 309–333.

Simms, J. C. (1989). Traditional Cavalieri Principles Applied to the Modern Notion of Area. *Journal of Philosophical Logic* **18**, 275–314.

Simms, J. C. (1991). Why the Continuum Hypothesis Is False. *Jahrbuch 1990 der Kurt-Gödel-Gesellschaft*. Wien.

Simpson, S. G. (1999). *Subsystems of Second Order Arithmetic*. Springer.

Solovay, R. M. (1995). Introductory Note to Gödel 1970a, 1970b, 1970c. In: Gödel (1995), pp. 405–420.

Steel, J. R. (2000). Mathematics Needs New Axioms. *The Bulletin of Symbolic Logic* **6**, 422–433.

Wang, H. ([1974] 1983). The Concept of Set. Reprinted in: Benacerraf & Putnam (1983), pp. 530–570.

Zermelo, E. (1930). Über Grenzzahlen und Mengenbereiche: Neue Untersuchungen über die Grundlagen der Mengenlehre, *Fundamenta Mathematicæ* **16**, 29–47.

Tomasz Bigaj

DO QUANTUM-MECHANICAL SYSTEMS ALWAYS POSSESS DEFINITE PROPERTIES DICTATED BY THEIR STATES?

ABSTRACT. In the article the possibility of breaking the eigenvalue-eigenstate link in quantum mechanics is considered. An argument is presented to the effect that there are some non-maximal observables for which the implication from eigenstates to eigenvalues is not valid, i.e. such that although the probability of revealing certain value upon measurement is one, they don't possess this value before the measurement. It is shown that the existence of such observables leads to contextuality, i.e. the thesis that one Hermitean operator can represent more than one physical observable. Finally, contextuality brought about by these considerations is compared with contextuality suggested by the Kochen-Specker paradox.

1. Introduction

It wouldn't be far from the truth if we said that the issue of interpreting quantum mechanics comes down almost entirely to the question of how to interpret probabilities introduced by this theory. More specifically, what is at stake here is the relationship between descriptions in terms of probabilities of measurement results on the one hand, and descriptions in terms of objective properties possessed by a given system on the other. This is what some refer to as the "state-attributions" vs. "value-attributions" problem, or the eigenstate-eigenvalue link.[1] For example, according to the Copenhagen school interpretation of QM, the link between state-attributions and value-attributions can be presented in the form of the following equivalence: we can ascribe a measurable property to a system if and only if the probability of obtaining this property as a result of measurement on this system equals 1. As a consequence, we must recognize that there are situations in which the system possesses no definite value

[1]The first terminology was introduced by B. van Fraassen, whereas the second is used for example by R. Clifton. I admit that the latter is probably unfortunate, and that it can cause some misunderstandings. Namely, one can think that we will be here entertaining the possibility of weakening the relationship between eigenvalues and eigenstates *as they are defined in quantum-mechanical formalism*. It is well known that in quantum formalism an eigenvector of a particular operator **A** is defined with the help of the equation $\mathbf{A}v = av$, where a is a complex number, called an eigenvalue for this particular eigenvector. Obviously, for each eigenvector there is an appropriate eigenvalue, and *vice versa* — each eigenvalue corresponds to some (often more than one) eigenvectors. But in present context by the eigenstate-eigenvalue link we understand the relationship between being in a particular quantum state and having an objective property, which is of course not a matter of a terminology, or better a mathematical formalism, but of (hopefully) empirical facts.

In: J. Malinowski and A. Pietruszczak (eds.), *Essays in Logic and Ontology* (*Poznań Studies in the Philosophy of the Sciences and the Humanities*, vol. 91), 375–394. Amsterdam/New York, NY: Rodopi, 2006.

for a given observable. However, there are interpretations which differ radically in this matter from the Copenhagen interpretation. For instance, all realist, or hidden-variables interpretations of QM (including Bohm's famous theory) have it that the "only if" clause in the above equivalence should be rejected, and that in every situation all parameters which can properly describe the system take definite values.

A more modest way of breaking down the equivalence between eigenstates and eigenvalues is proposed in the so-called modal interpretation of QM. Van Fraassen — one of its proponents — claims that the equivalence holds only for pure states, but if a system is in a mixed state, then it possesses some of the properties whose probabilities of occurrence defined by the state are nevertheless less than 1 (van Fraassen 1973, 1991). In other words, the actual value state of a system is more informative than its dynamic state. Assuming this, van Fraassen tries to solve the notorious problem of measurement by claiming that after measurement the system under the measurement is in a mixed dynamic state, which is a reduct of a composite state dictated by Schrödinger's equation, but its value attributions are as if it were in one of the pure dynamic states allowed by the mixture. It is worth noting that neither the Copenhagen, nor the modal interpretations question the validity of the implication in the opposite direction: from state attributions to value attributions.[2] However, R. Clifton (1996) has suggested that if the modal interpretations of QM were to satisfy certain plausible metaphysical requirements, then they should also reject the universal validity of this implication, otherwise the Kochen-Specker paradox would threaten. Clifton also argued, on more general grounds, that from the fact that some event will necessarily occur in the future it doesn't follow that right now the system possesses a definite property "responsible" for this occurrence. In my paper I would like to strengthen this claim and to propose an elementary argument in favor of the thesis that being in an eigenstate of an observable, a system can nevertheless lack an appropriate eigenvalue of the observable in question. More specifically, I am going to formulate a simple "measurement paradox", which in my opinion can be solved only in two ways: either by abandoning the rules of classical logic (and for example accepting one of the versions of quantum logic), or by rejecting the above-mentioned link from eigenstate to eigenvalue. Analyzing further consequences of this last move, I would like to show that it leads to a mild version of contextuality, i.e. the claim that some Hermitean operators can represent more than one physical quantity. Finally, I shall compare contextu-

[2]In Bohm's theory the situation is a little different, because for him some measurement results are not "genuine" in the sense that they do not correspond to any real observable. However, if we restrict the above-mentioned implication to the case of genuine observables, then it will remain true that, if the probability of revealing some particular value of observable equals 1, this value must have corresponded to the observable before the measurement. See Albert (1992, pp. 134–179); Clifton (1996, p. 389).

ality brought about by our considerations with contextuality of hidden-variable theories required by the Kochen-Specker theorem.

2. A "Measurement Paradox"

Let us start with a simple example of an observable A having a finite number of values $\{a_1, \dots, a_n\}$. Quantum formalism represents this observable with the help of a Hermitean operator $\mathbf{A}$, which has n mutually orthogonal eigenvectors $\{v_1, \dots, v_n\}$. Measuring A consists of a certain experimental procedure, the result of which is one of the numbers $a_1, \dots, a_n$. In classical situations we would expect that if the measurement was carried out properly, the revealed value a_i of A should characterize the physical system under consideration not only after the measurement, but also at least just before it took place. In other words, the results of classical measurements can be projected back to the immediate past of the physical system. As it is well known, this is not the case in quantum mechanics. Repeated measurements on a number of particles prepared (apparently) in the same initial state don't in general yield the same results. Therefore the probability distribution over the set of possible results is invoked in the quantum-mechanical description. In particular, when the initial state of a system is represented by a vector v being a linear combination of eigenvectors $v_1, \dots, v_n$, namely $v = \sum_{i=1}^{n} c_i v_i$, where at least two different coefficients $c_1, \dots, c_n$ are non-zero, the value of A before the measurement is considered as not determined by the rules of quantum mechanics. It can only be said that for every value of A there is certain probability that the measurement will yield it, that probability being given by the square of the norm of an appropriate coefficient $|c_i|^2$.

Consider now a somewhat different experimental procedure. Assume that one prepares a measuring device exactly as in the case of a measurement of the observable A, but instead of recording a particular outcome, he/she merely notes whether this outcome is of *any* particular value. If this is the case, we can conventionally assume that the overall result of this procedure is 1 (or "yes"), but if it were otherwise, the result would be 0 ("no"). In that way we have operationally introduced a new observable B with two possible values (this sort of observables is sometimes referred to as "propositions" or "questions"). But now the question is: what is the meaning of this new observable B? A fairly natural answer is that B is an observable, which measures whether A possesses any definite value at all. This is not an unusual move in quantum mechanics. Consider for example projector operators, which are commonly interpreted as representing questions of the type "Does the value of an observable A lie within certain set Δ?" In our example we would have essentially the question of the same type, where the set Δ would comprise all the spectrum of A.

So we can be tempted to accept the following equivalence as a characteristic of the observable B:

> B possesses value 1 (“yes”) iff
> A has any definite value from its spectrum.

But now things begin to go in a wrong direction. Note that because the measurement of A must yield a definite value, the probability of obtaining the result “yes” for the observable B is 1. This means not only that after a particular measurement B has the value “yes”, but that B will *necessarily* have the value “yes” after any measurement properly carried out. But in quantum formalism it implies that before the measurement the system was in the eigenstate of the observable B corresponding exactly to the value “yes”, and this fact in turn proves that B possessed the value “yes” before the measurement. And returning to the interpretation of the observable B we see that this means that before the measurement of B (and of A, because both consist essentially of the same experimental procedure) the observable A must have had a *definite value*. But from this it follows that there is a value a_i which is actually taken by A, and that, contrary to our previous assumption, the initial state of the system is not υ, but υ_i.

Let us present this reasoning in a detailed way, to see where the source of the difficulty is. The main interpretative assumption of standard quantum mechanics is the following equivalence (the eigenvalue-eigenstate link):

(1) An observable A for a system S possesses a determinate value iff S is in an eigenstate for that observable A,

where an eigenstate of A is a state in which there is a value a_i of A such that the probability of obtaining a_i equals 1. The next assumption is a consequence of (1):

(2) If a system is in the state $\upsilon = \sum_{i=1}^{n} c_i \upsilon_i$, where at least two coefficients are not zero, then no sentence of the form “$A = a_i$” is true (the property A does not have any definite value), assuming that there is no degeneracy.

As the third step, we invoke the equivalence, which is the suggested interpretation of the observable B. We assume that for every observable A there is an observable B with two values “yes” (1) or “no” (0) and such that

(3) B has value “yes” iff A has some definite value a_i.

The next assumption follows exactly from the operational definition of B.

(4) The probability of obtaining value “yes” for B is 1, so the system is always in the eigenstate of B for this value.

From (4) and (1) it follows that:

(5) *B* always takes the definite value "yes",

which together with (3) gives that

(6) The sentence "*A* has some definite value a_i" is always true.

Assuming conventional logic, we must derive from (6) that

(7) At every moment there is a definite value a_i such that the sentence "*A* has a_i" is true.

Now, when the system is in state v, (7) plainly contradicts (2). So the case is proven and now we can turn to the solutions.

3. The immediate solution: what does *B* really measure?

An immediate response to the paradox of this kind can be (and in fact very often is) to point at the equivalence (3), supposedly defining the observable *B*, as the source of trouble. One can namely question this equivalence on the basis of the observation that it goes beyond what is actually measured in the experimental procedure described earlier. The way we measure *B* entitles us only to the following, unquestionable equivalence:

(3*) A measurement of *B* reveals the value 1 iff this measurement reveals any value of *A*.

Now, of course this equivalence is insufficient to determine the value of *B* when *B* is not measured. But even according to the standards of the orthodox interpretation of quantum mechanics, limiting ourselves to this kind of characteristics only has definitely too strong positivistic flavor. After all, from time to time we can assign with certainty a particular value to a physical quantity, even when no actual measurement is done. So the question is: how to extend the equivalence (3*) for other cases? I think that two possibilities can be considered. One possibility would be to accept the limitations of the positivistic approach and to contend that *B* has the value 1 if and *only* if it is currently measured. Before the measurement the value of *B* is simply not defined. Hence *B* cannot tell us anything about whether *A* had any value before the measurement, and therefore the problem disappears.

This solution has an immediate consequence of violating the eigenstate-egenvalue link, or the equivalence (1). For in that case a physical system would always be in an eigenstate for *B* (the probability of obtaining the value "yes" in an experiment is unquestionably 1), yet before the measurement *B* would lack any definite value. All in all, that is what I will be ultimately arguing for, but for now it would be premature to pronounce the victory. So let us turn to another possible interpretation of *B*. One can just simply say that the equivalence (3*) is the primary way of defining operationally the observable *B*, but we should add that all cases in which *B* is not subject to a measurement are covered by the standard interpretive rules of quantum mechanics, i.e. by the equivalence (1). Because the probability of obtaining the answer "yes" is inevitably 1, we should ascribe the value "yes" to the question *B* also before the measurement. However, from this it doesn't follow that before the experiment the observable *A* had any definite value, for the equivalence (3) no longer holds. So in consequence *B* cannot be interpreted as the question "Does *A* possess any definite value?". In fact, it is now difficult to connect *B* with any intuitively clear question (one such possibility may be the question "Will a measurement of *A* reveal any definite value?"[3]). From this perspective the previous question given by (3) does not constitute a proper quantum observable. In that way the paradox is finally solved (some would even say that the paradox has never occurred) and the peace of mind is finally restored.

But is this solution fully satisfactory? I would say that in this way we haven't really settled the issue, but rather swept it under the rug, so that it no longer bothers us. On the level of observables we have achieved an agreement between what is true now and what will be certainly true after an appropriate measurement. But on the level of propositions, the gap still exists. Consider, again, the proposition "*A* has some definite value". After measuring *A*, this proposition must necessarily turn out to be true, yet in spite of this quantum mechanics tells us that before the measurement one should expect it not to be true (which doesn't necessarily mean that it is false). Now, it could be claimed that there is a property, let's call it S, which is connected with the above proposition in such a way that the system possesses this property iff the proposition in question is true. Obviously, after a measurement of *A* the system will have this property S, whereas it should lack it before the measurement. Of course, this fact by itself does not violate the eigenstate-eigenvalue link in the form of the equivalence (1), for we assume that there is no proper quantum observable which corresponds to the property S. But this seems to be a somehow arbitrary way to defend (1). Note that we can find more propositions and properties of this kind. One of them would be just the equivalence (3). It is straightforward that because (3*) is true, (3) must turn out true under the joint measurement of *A* and *B*. Yet according to the current solution, it doesn't hold when there is no measurement, for then *B* has the value 1, whereas *A* lacks any definite value. So here again we should admit that there

[3]This interpretation of the observable *B* is susceptible to the following objection. When we have measured the observable *A* and recorded its value, the fact that *B* has the value "yes" in fact tells us more than only that *the next* measurement of *A* will reveal a particular definite value. Normally we would agree that under this particular condition the positive answer to *B* implies that *right now A* has a definite value.

is no proper quantum question, such that the answer to this question is "yes" iff the proposition (3) is true.

I would add to what was said that the consequence that there are no proper observables of such a kind as the one fulfilling the equivalence (3) could be seen as an unnecessary restriction on what is expressible in standard formalism of quantum mechanics. On many occasions observables like our *B* are introduced in foundational analyses of QM. Let me mention one particular example, drawn from the eminent author, David Albert. In his book (1992, p. 116*ff*) he is considering different versions of no-collapse measurement theories. In particular, on the way to his famous "many-minds" theory he is touching upon what he calls "the bare theory". Let us consider a measuring device of a two-valued observable (e.g. spin) and an observer, noting the results of experiment. According to a dynamic evolution, when there is no collapse, then after the measurement the entire complex system will be in a superposition of two components — each corresponding to one possible result of experiment. Hence we must admit that the observer cannot arrive at any definite belief concerning the value of the measured observable. However — and here is the catch — Albert introduces a new question, applicable to the observer, namely "Does the observer have any particular definite belief about the value of the measured observable?". Appealing to the fact that when an observable has the same value in two different states, it will have this very same value in the superposition of these states, Albert argues that the answer to this question must be inevitably "yes", in spite of our previous assumption that the observer is unable to say exactly what is the result of the measurement. In the next steps he is trying to show that this simple observation, together with some other facts, can in principle account for our fundamental intuitions concerning measurement — that every measurement has a definite result, that the relative frequencies of particular results are given by the initial state of the system, etc. — without introducing any collapse at all.

Further details of this reasoning are not important to us, but the crucial thing is that in order to make his point, Albert must assume that there is a proper observable *C* such that *C* has value "yes" iff the observer has a particular belief concerning the result of the experiment. However, if we endorse the view analyzed above, this move is unjustified. If the observer *h* is in one of two eigenstates corresponding to his particular belief concerning spin of the particle, then the observable *C* having the value "yes" truly expresses the fact that *h* has a particular belief. But when *h* is in a superposition, *C* of course has still the value "yes", which nevertheless no longer tells us that *h* has *any* belief at all — it merely tells us that if you could do a measurement (with a collapse!) then at the end it would turn out that *h* has a definite belief concerning the property of the particle. But this is definitely too little for Albert's purpose of defending the view that even without collapse we can pretend *as if* measurements had definite results and observers had definite beliefs.

I am not saying of course that an argument of this kind can definitely settle the issue. Maybe there are some arguments showing that Albert's approach to the measurement problem is not tenable. However, I think that I said just enough to be convinced that it can be worthwhile to look for some other ways of solving the initial problem.

4. Two Non-Standard Solutions

Here I will consider two other proposals to avoid the initial measurement paradox. The first makes an appeal to non-classical logic, and thereby blocks the path from (6) to (7), whereas the second requires altering the standard interpretation of observables and experimental outcomes in quantum mechanics, which leads to a modification of the assumption (1). To the end of the article we will be mostly concerned with the second solution, but in the beginning a few words about the first.

The "quantum-logical" solution is the following. First, let us reformulate the somehow ambiguous proposition that the observable A has a definite value. It means that the following disjunction must be true even before the experiment: $A = a_1 \vee A = a_2 \vee \cdots \vee A = a_n$. From this we can infer in classical logic that one of the disjuncts must be true, i.e. that there is a value a_i such that the proposition $A = a_i$ is true. And this is the consequence which is not coherent with the standard quantum interpretation. But we can block this consequence by rejecting the classical metalogical rule that the truth of a disjunction implies the truth of at least one of its disjuncts. In fact, in quantum logic, where logical connectives are defined with the help of appropriate operations on subspaces of Hilbert space, this rule is not valid.[4]

The most famous feature of this particular type of quantum logic is the failure of the law of distributivity: $p \wedge (q \vee r) \equiv (p \wedge q) \vee (p \wedge r)$. Much attention has been given to this particular fact by philosophical commentators.[5] However, it is easy to verify that quantum logic displays also the required characteristic: namely it doesn't allow us to conclude from the truth of the disjunction $p \vee q$ that one of its disjuncts must be true. Thus for example, when the state vector lies in the span $L_p \oplus L_q$, but neither in L_p nor in L_q, then $p \vee q$ is true but neither p nor

[4]The quantum-logical approach is presented for example in Greechie & Gudder (1973), Hughes (1989), van Fraassen (1981).

[5]See e.g. Putnam (1969); Jammer (1974, pp. 346–361). Putnam tried to employ quantum logic for the purpose of defending realistic interpretation of QM, but this attempt was clearly misguided. He argued that two incompatible observables can have simultaneously definite values, but the sentence of the form of the conjunction "observable A has a definite value a and observable B has a definite value b" is nevertheless always false. Putnam's mistake comes from the fact that he overlooked that in quantum logic the derivation from «'$p \vee q$' is true» to «p is true or q is true» is not valid. And besides, in quantum logic the truth of both conjuncts guarantees the truth of the entire conjunction!

q is. And this is exactly the case in our previously analyzed situation. When we take an observable A, then all the subspaces corresponding to propositions $A = a_i$ together span the entire Hilbert space, which means that the disjunction $A = a_1 \vee \cdots \vee A = a_n$ must be true in every quantum state. However, when the state vector is a superposition of eigenvectors of A, no single proposition of the form $A = a_i$ can be true. Therefore the problem is solved: a system can have definite value for the observable B, and simultaneously no value for A.

Let me mention incidentally that for our purpose it is not necessary to accept full non-Boolean quantum logic. It can be shown that with a little modification of semantic rules even classical logic (i.e. logic with all classical tautologies) may accommodate breaking of the implication from the truth of the disjunction to the truth of one of its disjuncts. I analyzed such a logic elsewhere (Bigaj 2001), where I compared it with the standard quantum logic and showed that certain counterintuitive features of the latter can be avoided in the former. In particular, I have shown that standard quantum logic can be easily accused of an unjustified change of meaning for classical connectives (which leads for example to such counterintuitive consequences as to the denial of the rule that when a conjunction of two sentences is false and one of the conjuncts is true, the other must be false). On the contrary, the classical logic modified semantically in a certain way, which allows introducing the third logical value, can be shown to preserve classical meanings of logical connectives almost intact. But let us return to the main story.

Assume that one is not willing to sacrifice even a tiny portion of the old, classical logic (as most physicists probably wouldn't want to). Then immediately, the only solution left is to change the rule equating being in an eigenstate with having an appropriate eigenvalue. We must reject the principle (1), and, as announced earlier, this will involve rejection of one of its conditionals: if the probability of obtaining a given value of an observable equals 1, then this observable must have possessed this value before the measurement. Returning to our paradoxical reasoning, this would mean that the assumption (5) could be false: in spite of being in an eigenstate, B nevertheless can lack any definite value before the measurement. In that way we can reconcile the definition (3) of the observable B with the fact that A has no value before the experiment.[6]

[6]Actually, we are going here beyond what is covered by the equivalence (3). Namely, from (3) we can only infer that when A lacks any definite value, B has *not* the value "1", which leaves us with two choices: either B has the value "0", or has no value at all. At first sight, it seems intuitive that when A has no definite value, the answer to B should be nevertheless definitive, namely "no". After all, we insisted to read B as a question "Does A possess any definite value?". However, this move would lead to the most extreme case of violation of the implication from eigenvalue to eigenstate, when it is accepted that although an observable possesses a definite value before the measurement, the probability of obtaining this value is 0. For that reason I decided to choose the second option. Under my interpretation B can be defined fully as follows: B has the value 1 iff A has a definite value from its spectrum, B has the value 0 iff A has a definite value from the outside of its spectrum (which

A fairly obvious consequence of this move is contextuality. Namely, one and the same Hermitean operator can represent physically different observables. The observable B from our example is clearly represented by the identity operator **I**. It means that every vector in a Hilbert space of states is an eigenvector of **I** with the value 1 (meaning "yes"). This in turn follows from the fact that it is necessary that the result of a measurement of B is 1 ("yes"). But from our argument we see that in fact sometimes B has no definite value at all; namely when and only when the initial state is not an eigenstate for A. So if the initial state is one of v_i, then B has the value 1. But if a system is in a superposition of v_i, B has no value: neither 1 ("yes"), nor 0 ("no"). Every vector in a Hilbert space H is an eigenvector for B, but only for some of them B has a definite value.

Now consider an observable A', incompatible with A (such that there are some eigenvectors of A which are not eigenvectors of A'. Eigenvectors for A' will be $\{v'_1, \ldots, v'_n\}$ and eigenvalues $\{a'_1, \ldots, a'_n\}$. Let us introduce an observable B' in the same way as B, namely; B' has value "yes" iff A' has a definite value. A simple argument will convince us that the operator representing B' is the same as for B — it is the identity operator **I**. Consider the situation in which a system is in an eigenstate v_i for A, which is not an eigenstate for A'. It means that in fact observable B has value 1 (because A has some definite value), whereas B' has no value. Therefore B and B' are different observables, represented by the same operator **I**. We can write it down more formally. Let $[A]v$ denote the actual value of an observable A in the state v. Then we have: $[B]v_i = 1$, but $[B']v_i =$ not determined, and simultaneously $[B]v'_i =$ not determined, $[B']v'_i = 1$, under condition that v_i is not an eigenvector for A' and v'_i is not an eigenvector for A'. In fact we can write our assumption as follows: $[B]v = 1$ iff $v = v_i$, and otherwise $[B]v$ is not determined, and analogously $[B']v = 1$ iff $v = v'_i$, and otherwise $[B']v$ is not determined. So the contextuality is inescapable: for a given Hermitean operator (in this case the identity operator **I**) there exist two distinct observables represented by it, which have different values at the same times, although the probabilities of obtaining similar results for both of them are the same.

5. Illustrations

The most obvious illustration of what was said before is of course the case of spin ½. So let the observable A be spin S_z and the observable A' — spin S_x. The question B would be: "Does S_z have either the value ½ or −½?", with an-

is impossible) and B has no determinate value iff A has no determinate value. In other words, and more intuitively: we are now interpreting the question B not simply as "Does A have any determinate value?", but rather as "Does A have any determinate value from its spectrum, or does it have any determinate value not included in its spectrum?" (I am grateful to Lawrence Sklar for drawing my attention to this important problem).

swers "yes" or "no", and the question B': "Does S_x have either the value ½ or −½?". Every vector in two-dimensional complex Hilbert space can be presented either as the combination $c_1|S_z = ½\rangle + c_2|S_z = -½\rangle$ or in another basis as: $\frac{1}{\sqrt{2}}(c_1 + c_2)|S_x = ½\rangle + \frac{1}{\sqrt{2}}(c_1 - c_2)|S_x = -½\rangle$. In order to answer the question B, measurement of A must be performed, in which one of two results will be revealed: $S_z = ½$ or $S_z = -½$, with respective probabilities $|c_1|^2$ and $|c_2|^2$. In both cases the answer for the question B would be positive, so we conclude that the observable B will have the value "yes" with probability 1. The same goes with the question B' with the exception that the measurement carried out in this case must refer to S_x. And again there are only two possibilities: that the result will be ½ (with probability $\frac{1}{2}|c_1 + c_2|^2$) or −½ (with probability $\frac{1}{2}|c_1 - c_2|^2$), so the answer for B' is inevitably "yes". Now from the fact that the probability of obtaining a value for a given observable is 1 we usually derive that this value was present before the measurement. Hence before the measurement it was already true that S_z had one of the values ½ or −½, and it was simultaneously true that S_x had one of the values ½ or −½. At this point only non-classical logic can prevent us from accepting the consequence that before the measurement both (incompatible) components of spin had definite values. But if we, as it was suggested, reject the move from the probability 1 to the definiteness of value before the experiment, then we don't have to resort to such drastic measures as rejecting classical logic. We will suggest that B has value "yes" only if the particle is in one of the eigenstates $|S_z = ½\rangle$ or $|S_z = -½\rangle$, but in the superposition of these two B has no value at all. Similarly with B', which leads to the conclusion that B and B' can never have definite value simultaneously.

Let us consider closely the relation between B and B'. We have the experimentally confirmed fact that for every state of the particle, the probability of obtaining any result for the observable B is exactly the same as the probability of obtaining the same result for B' (1 for "yes", 0 for "no"). It is striking that in quantum mechanical formalism such observables are identified. But note that in principle we can't even measure them simultaneously, because it is impossible to perform two measurements for different spin components at once! So for example from the operational point of view B and B' constitute two different measurable properties. And this observation is confirmed by our proposal. There are situations in which B has value "yes" but B' has no value at all, and *vice versa*. The representation of B and B' by the same identity operator $\mathbf{I}$ tells us only about the results of appropriate measurements for them, but not that B and B' are the same observables.

We can find illustrations of our paradox also for the case of continuous observables. Consider the position X of a particle. Strictly speaking, the operator $\mathbf{X}$ has no eigenvalues at all, so we cannot introduce a new observable B asking if X has any definite value. But we can overcome this difficulty in the following way. Let ψ be a wave function for a given particle. Divide the space in which ψ is greater than zero into two separate areas L_1 and L_2 (which of course means that we can present the wave function as a sum: $\psi = \psi(L_1) + \psi(L_2)$). Now we can ask the following question (and this will constitute our observable B): "Is the particle either in L_1 or in L_2?". In order to answer this question, we must measure position of the particle, and of course it will appear in one of the two areas L_1 or L_2. Therefore the answer must be "yes", but again if we concluded from this that before the measurement the particle was in one of the two areas, it would mean that one of the components of the wave function $\psi(L_1)$ or $\psi(L_2)$ was zero, which contradicts our initial assumption. Moreover, introducing not two but as many as we like areas $L_1, \ldots, L_n$ we could "prove" that the particle was localized before the measurement within an arbitrarily small region of space.

6. Generalization – the Case of Non-Maximal Observables

Somebody might still not be satisfied with our solution to the measurement paradox, arguing that this solution is *ad hoc*. Why the link leading from the eigenstate to the eigenvalue breaks only for the observables like B? And moreover, what does this phrase "like B" really mean? Does it apply to all observables, or only to observables represented by the identity operator? The answer lies in between. The key is the notion of maximality of operators. An operator A is maximal iff it has distinct eigenvalues for every two orthogonal eigenvectors (if v_i, v_j – eigenvectors of A and $\langle v_i|v_j\rangle \neq 0$, then $a_i \neq a_j$). Maximality is sometimes called as "no degeneracy case". First note that our observables B and B' are not maximal. In fact we have here the most extreme example of degeneracy, with all eigenvalues being equal. And it is exactly non-maximality which is responsible for breaking the usual equivalence. We can easily reproduce our paradox for the case of observables C and C', which are not represented by the identity operator, but nevertheless by degenerate operators. Consider the following definition. Let us take two observables A and A' represented by maximal operators, and such that there is a proper subset of eigenvectors of A and a proper subset of eigenvectors of A', which span the same subspace of H (denote these eigenvectors respectively by $v_{i_1}, \ldots, v_{i_k}$ and $v'_{i_1}, \ldots, v'_{i_k}$, and their appropriate eigenvalues by $a_{i_1}, \ldots, a_{i_k}$ and $a'_{i_1}, \ldots, a'_{i_k}$, where $k < n$). Let's then define an observable C in such a way that it takes value 1 (or "yes") when A has one of the values a_{i_1} $\ldots, a_{i_k}$, and value 0 ("no") otherwise. And similarly define an observable C': $C' = 1$ iff $A' = a'_{i_1}, \ldots, a'_{i_k}$, and $C' = 0$ otherwise. We see that C acts as a discriminator of two states: in the first A has one of the values $a_{i_1}, \ldots, a_{i_k}$ and in the second it has some other value. Now consider the case in which a system is in an initial state which is a superposition of states with A having values from $a_{i_1}, \ldots, a_{i_k}$ (i.e. when the state vector equals $v = \sum_{l=1}^{k} c_{i_l} v_{i_l}$). As before, in this state the probability of obtaining the result "yes" for the observable is 1, but we cannot infer from this that before the measurement the observable A had any particular value within the range $a_{i_1}, \ldots, a_{i_k}$. Moreover, if the initial vector v happens to be one of the eigenvectors $v'_{i_1}, \ldots, v'_{i_k}$ of A', then not only after appropriate measurement will the observable C' have the value "yes", but we can assume that this value was assigned to C' even before the measurement. Hence, we have the situation of two observables C and C' represented by the same Hermitean operator (in this case this operator is a projection operator P on the subspace spanned by $v_{i_1}, \ldots, v_{i_k}$ or $v'_{i_1}, \ldots, v'_{i_k}$, which is equivalent to the fact that the probability distribution of obtaining the same results from measuring both C and C' is the same. Nevertheless these two observables should be considered different, for we have $[C]v \neq [C']v$ for some vectors v.

For examples we can look at the case of spin 1. Let us consider the x-component of spin 1 – S_x, with three eigenvalues: -1, 0, 1. We can define an observable C as follows: C has value 0 iff $S_x = 0$, and C has value 1 iff either $S_x = 1$ or $S_x = -1$. In other words, the observable B is a question: does the x-component of spin have value 0 or does it have one of the values 1, -1? Note also that we could have presented C in a simple form as $C = S_x^2$. Now assume that the state of a system is a superposition: $v = c_1|S_x = -1\rangle + c_2|S_x = 1\rangle$. In that case the likelihood that the result 1 obtains for C is certainty, but we cannot infer from this that before measurement S_x had one of the values 1, -1. Therefore, according to our solution, we must assume that when the state of the particle is v, C does not have a determinate value, although in that case it is certain that the result of a measurement will be 1.

Now let us consider another observable, given by the equation: $A' = aS_x^2 + bS_y^2 + cS_z^2$. The observable A' has three eigenvalues (which can be arranged as distinct numbers): $a + b$, $a + c$, and $b + c$, corresponding respectively to the following orthogonal vectors: $|S_z = 0\rangle$, $|S_y = 0\rangle$ and $|S_x = 0\rangle$. We can see that observables S_x and A' share one common eigenvector, namely $|S_x = 0\rangle$, whereas for the other two the following equations hold: $|S_z = 0\rangle = \frac{1}{\sqrt{2}}(|S_x = 1\rangle + |S_x = -1\rangle)$, and $|S_y = 0\rangle = \frac{1}{\sqrt{2}}(|S_x = 1\rangle - |S_x = -1\rangle)$. So when the value of S_x equals either 1 or -1, the value of A' is not determined. The observable C' will be defined as follows: $C' = 0$ iff $A' = a + b$ and $C' = 1$ iff $A' = a + c$ or $A' = b + c$. Now we can easily confirm that both observables C and C' are represented by the same Hermitean operator **C**. This is the operator which has eigenvalue 0 for the vector $|S_x = 0\rangle$ and eigenvalue 1 for all vectors orthogonal to $|S_x = 0\rangle$. However, as we have argued, when for example the state of the particle is $|S_x = -1\rangle$, the observable C has definite value 1, whereas C' not, for A' has no value determined except that it is not equal $a + b$.

7. Functional Dependence of Observables

Another way of formulating cases in which the link (1) breaks down is in terms of functional dependence between observables. This formulation can be shown to cover all cases discussed before, and as I suppose it can also give us some additional arguments for our solution. Let $A_1, \ldots, A_k$ be mutually compatible maximal observables for a system S. Consider an observable $B = f(A_1, \ldots, A_k)$ for which the function f is not one-to-one, i.e. there are at least two different sequences of eigenvalues $a^1_{i_1}, \ldots, a^1_{i_k}$ and $a^2_{i_1}, \ldots, a^2_{i_k}$, such that $f(a^1_{i_1}, \ldots, a^1_{i_k}) = f(a^2_{i_1}, \ldots, a^2_{i_k})$. B is obviously non-maximal, for there are two orthogonal vectors v_1 and v_2 such that for each of them B takes the same value (v_1 is the eigenvector for eigenvalues $a^1_{i_1}, \ldots, a^1_{i_k}$ and v_2 for $a^2_{i_1}, \ldots, a^2_{i_k}$; the value of B is of course $f(a^1_{i_1}, \ldots, a^1_{i_k}) = f(a^2_{i_1}, \ldots, a^2_{i_k})$). Now what happens, if the system is in a superposition $c_1v_1 + c_2v_2$? Obviously the probability of obtaining the result $f(a^1_{i_1}, \ldots, a^1_{i_k})$ when measuring B would in this case be 1. But it can be argued that because the observable B obtains its values only in virtue of its functional dependence on observables $A_1, \ldots, A_k$, B shouldn't have a definite value when $A_1, \ldots, A_k$ don't have any. The result of applying a function f to undetermined values can be no other than undetermined!

We can easily show that all cases previously analyzed can be presented as the cases of functional dependence of this sort. First we had an observable B which was equivalent to the question whether some maximal observable A has any definite value at all. This is the simplest functional dependence given by a constant function: $B = f_{\texttt{const}}(A)$, where $f_{\texttt{const}}(a_i) = 1$ for all i. So in every case when A has any value, the value of B equals 1, but this doesn't justify concluding that when the value of A is not determined, B still takes 1 as its value. The second example was an observable C which had the value 1 when A had one of the values $a_{i_1}, \ldots, a_{i_l}$ for some $l < n$, and 0 otherwise. In that case the function defining C is $f(a) = 1$, when $a \in \{a_{i_1}, \ldots, a_{i_l}\}$, and $f(a) = 0$ otherwise.

So in general the rule with regards to possessing values by an observable defined as a function of some other observables, which we adopt here is the following. Let $B = f(A_1, \ldots, A_k)$.

> B possesses any definite value in the state v iff all observables $A_1, \ldots, A_k$ possess values in that state v.

From this it follows that when a system is not in an eigenstate for observables $A_1, \ldots, A_k$, the observable B cannot take any definite value, even though the actual state of the system may be one of its eigenstates. Therefore assuming this rule leads to violation of the equivalence (1), but only in the case where B is not maximal. For maximal operators there is no need to question the eigenstate-eigenvalue equivalence, at least on the basis of our present considerations.

8. Contextuality and the Kochen-Specker Contradiction

There are obvious similarities between the approach to non-maximal operators proposed here and the debate over the famous Kochen-Specker no-go result. Kochen and Specker have proved that under certain assumptions the hidden-variable hypothesis leads to a mathematical contradiction (Kochen & Specker 1967). A purely mathematical part of their reasoning can be presented in the form of a proof that for spaces of dimensionality greater than 2 it is impossible to assign numbers 0 or 1 to each one-dimensional ray, in such a way that for every complete orthogonal set of rays exactly one receives the value 1. The physical meaning of this fact should be obvious. Each complete orthogonal set of rays represents a Hermitean operator, and hence a physical observable, whereas an assignment of 0 and 1 to these rays stands for a definite valuation on this particular observable. The crucial assumption leading from the hidden-variable hypothesis to the Kochen-Specker contradiction is that every one-dimensional ray receives the same value independently of what particular complete orthogonal set of rays it is considered a member. This assumption can be formulated in a different form as the thesis that every observable represented by a projection operator onto one-dimensional space receives its particular value independently of the context in which it is measured. So it is natural that some people saw the chance of blocking the derivation of the paradox by rejecting this postulate, and hence adopting some kind of contextuality (for a discussion see e.g. Redhead 1987, pp. 119–138).

Let us formulate it more precisely. Let P_i be a Hermitean operator projecting onto one-dimensional ray, represented by a normalized vector v_i. In a Hilbert space, whose dimensionality n is greater than 2, P_i is necessarily a non-maximal operator, because every vector orthogonal to v_i is an eigenvector for P_i, with corresponding eigenvalue 0. In the proposed way of blocking the Kochen-Specker contradiction it is assumed that there is no single physical observable corresponding to the projection operator P_i, but in fact infinitely (even uncountably) many distinct observables. Each particular physical interpretation of P_i is connected with a choice of an orthonormal basis for the eigenspace orthogonal to v_i (this eigenspace corresponds to the eigenvalue 0). So the hidden-variable theory cannot assign any definite value to the operator P_i; rather it assigns a value to P_i in the context of a particular choice of a basis for the eigenspace. In quantum formalism this can be conveniently presented with the help of the notion of functional dependence between operators. There are infinitely many different (and mutually incompatible) maximal Hermitean operators, on which P_i is functionally dependent in a certain specific way. For every such operator A the vector v_i is an eigenvector with corresponding eigenvalue a_i. The function f will therefore have the following simple form: $f(a_i) = 1$ and $f(x) = 0$ for all x different than a_i. Now let's consider two such maximal operators A and B, incompatible with each other. They will of course share one common eigenvector — namely v_i — but other eigenvectors in general will be different. The proposed way to block the Kochen-Specker argument and to rescue the hidden-variable theory from the logical inconsistency, requires that the projection operator P_i can receive different value when considered as a function $f(A)$, and different when presented as $g(B)$, with g defined as follows: $g(b_i) = 1$, $g(x) = 0$ otherwise (b_i is of course an eigenvalue corresponding to v_i). In other words, the fact that $f(A) = g(B) = P_i$ doesn't justify that $[f(A)] = [g(B)]$ (the actual values of $f(A)$ and $g(B)$ are the same). Consequently, we shouldn't even use the notation $[P_i]$ for symbolizing the actual value of P_i; rather, we ought to emphasize the dependence on the context by introducing indices: $[P_i]_A$ or $[P_i]_B$.

Comparing this solution with the present approach we can note striking similarities. We have earlier suggested that non-maximal operators can represent more than one physical observable, and that, depending on the context of measurement, these observables can lack a definite value, in spite of being in an eigenstate. Now let us apply this proposal to the currently analyzed case of projection operators. We begin with a maximal observable A with n different eigenvalues and n appropriate eigenstates. Suppose we are not interested in finding the exact value of A, but only in checking whether it is a particular value a_i, or some other value $a_1, \ldots, a_{i-1}, a_{i+1}, \ldots, a_n$. So we are asking the question "Does A possess a value a_i, or some other value?". The answer to this question is "yes" (or 1) if and only if measuring A reveals the value a_i, and "no" (0) iff the measurement reveals a value different than a_i. Naturally, this question is represented in quantum formalism as a projection operator P_i on the subspace v_i. Let us assume that the initial state v of the system is a non-trivial superposition of eigenvectors of the observable A, excluding the vector v_i. In such a situation it is certain that the answer to our question would be "No, the value of A is not a_i; it is some other value $a_1, \ldots, a_{i-1}, a_{i+1}, \ldots, a_n$". However, are we entitled to draw from this the conclusion that before measurement the observable represented by P_i had already the value "no"? It would mean that at that time the value of the observable A was one of the numbers $a_1, \ldots, a_{i-1}, a_{i+1}, \ldots, a_n$, and this contradicts the assumption that the system was initially in a superposition v of eigenstates (recall the necessity of an appeal to the metalogical rules of classical logic in this move). Hence we propose to admit that the observable represented by our question had no determinate value before measurement, notwithstanding the fact that this value was sure to be revealed upon measurement.

But what if we considered another maximal observable B, incompatible with A but sharing with it a common eigenvector v_i? Now it can happen that the initial state v of the system is an eigenstate of B. We can analogously define a question whether B has value b_i, which would be represented by exactly the same projection operator P_i. However, now it is perfectly reasonable to assume that before measurement this observable admits the definite value 0, because B

has already an eigenvalue corresponding to the eigenvector v. Hence it should be clear that we are talking in fact about two distinct observables, represented by the same projection operator P_i, which we can conveniently symbolize as $f(A)$ and $g(B)$. The difference between them lies in this that when the system is in an eigenstate for B but not for A, the actual value $[f(B)]$ equals 0, but $[g(A)]$ is not determined, and *vice versa*.

All this looks surprisingly similar to the contextuality suggested by the Kochen-Specker result. But there are important differences. The crucial difference is a result of different motivations for both solutions. Those who wish to block the Kochen-Specker contradiction are motivated mainly by the realism of values, or the hidden-variable hypothesis, whereas the motivation for the proposed form of contextuality comes from the entirely opposite direction, namely from the wish to preserve objective, ontological indeterminacy of values, as well as to preserve classical logic. Consider then two maximal observables A and B, as before. According to the contextuality motivated by the Kochen-Specker result, the difference between two observables $f(A)$ and $g(B)$ manifests itself only when the initial state vector v is a superposition of the common eigenvector v_i and other eigenvectors of A and of B. In such a situation, standard quantum mechanics doesn't assign any definite value to A and to B, therefore P_i will not receive any value, either. However, the contextual hidden-variable theory requires that the actual values for all operators exist, and that it is possible that $[A]v = a_i$, but $[B]v \neq b_i$, hence $[f(A)] = 1$, but $[g(B)] = 0$. But when the state vector v is either identical with v_i, or is orthogonal to v_i, both observables $f(A)$ and $g(B)$ receive the same value (either 1 or 0). On the other hand, the contextuality motivated by the Copenhagen interpretation gives rise to a difference between observables only when the state vector v is orthogonal to v_i, and it doesn't lead to the simultaneous assignment of different values to different observables $f(A)$ and $g(B)$, but only to the admission that one of the observables has a defitine value, whereas the other has no value at all. All this can be conveniently presented in Table 3.

9. Conclusions

As the reader might have expected, the answer to the title question defended in this article is negative. I have suggested that although the current state of a physical system tells us that the probability of revealing certain property upon measurement equals 1, it may still be possible that this property is not possessed by the system before the measurement. How radical is this view in terms of our common-sense ontological intuitions? Clifton argues in his article that we could find some general justifications for breaking the eigenstate-eigenvalue link, both in quantum and in classical world (pp. 388–389). In the quantum case, one can

Table 3.

Initial state v	Contextual hidden-variable theory	Contextual Copenhagen theory
$v = v_i$	$[f(A)] = [g(B)] = 1$	$[f(A)] = [g(B)] = 1$
$v \perp v_i$	$[f(A)] = [g(B)] = 0$	$[f(A)] = 0$, $[g(B)] =$ not determined, when v is an eigenvector of A; $[f(A)] =$ not determined, $[g(B)] = 0$, when v is an eigenvector of B; $[f(A)] = [g(B)] =$ not determined, when v is neither an eigenvector of A, nor of B
$v = av_i + bu$, where $a, b \neq 0$ and $u \perp v_i$	either $[f(A)] = 0$, $[g(B)] = 1$, or $[f(A)] = 1$, $[g(B)] = 0$	$[f(A)] = [g(B)] =$ not determined

generally expect that the measurement can disturb the system, possibly in such a way as to create the property previously absent in the system. But even in classical cases one may doubt — as Clifton points out — whether a disposition to react in a certain way under certain condition is necessarily connected with some "intrinsic" property possessed even in absence of that condition.

On the contrary, I would claim that breaking the implication from the eigenstate to the eigenvalue is something rather unexpected, and that it can affect our way of understanding measurement and its role. First of all, I don't think that the classical example appealing to a disposition to react in a certain way is an appropriate one. I might doubt if there is an intrinsic property responsible for (certain) shattering the glass plate when dropped on beds of nails, but in that case I definitely wouldn't call this event "a measurement". There is something unsettling in admitting that a properly carried *measurement*, which constantly reveals one and the same property, cannot justify the existence of this property beforehand. The disturbance conception of measurement may be good for explanation why we can't know simultaneously values of two incompatible observables, but I doubt that it could explain "*creatio ex nihilo*" of properties during experiments.

However, as I tried to argue in this article, the alternatives to the violation of the implication eigenstate → eigenvalue are no more attractive. The price we must pay for saving the eigenstate-eigenvalue link is either admitting that certain properties of a system cannot be represented by proper observables, or accepting the violation of some rules of classical logic.

As we saw, one of the consequences of our proposal is contextuality. Contextuality is often seen as a highly unwelcome feature, mainly because of its proliferation of observables, which some call "de-Ockamization" of quantum mechanics. I hope that my examples show that contextuality is not such a bizarre feature at all. There are some natural arguments that what we call "observable" is something more than can be grasped with the help of an appropriate Hermitean operator. If we appeal to intuitively clear operational procedures, then it should be expected that one and the same non-maximal Hermitean operator can represent more than one physical quantity. At least that is what I wanted to argue for.

Acknowledgements. The work on this article was supported by the grant KBN 1H01 A 029 19 received from Komitet Badań Naukowych (Committee of Scientific Research, Poland).

Uniwersytet Warszawski
Instytut Filozofii
Krakowskie Przedmieście 3
00-047 Warszawa
Poland

Rowan University
Department of Philosophy and Religion
201 Mullica Hill Road
Glassboro, NJ08012
USA
e-mail: bigaj@rowan.edu

REFERENCES

Albert, D. (1992). *Quantum Mechanics and Experience*. Cambridge: Harvard University Press.

Bigaj, T. (2001). Three-Valued Logic, Indeterminacy and Quantum Mechanics. *Journal of Philosophical Logic* **30**, 97–119.

Clifton, R. (1996). The Properties of Modal Interpretations of Quantum Mechanics. *British Journal for the Philosophy of Science* **47**, 371–397.

Greeche, R. J. and S. P. Gudder (1973). Quantum Logics. In: C. A. Hooker (ed.), *Contemporary Research in the Foundations and Philosophy of Quantum Theory*. Dordrecht: Reidel.

Hughes, R. I. G. (1989). *The Structure and Interpretation of Quantum Mechanics*. Cambridge – London: Harvard University Press.

Jammer, M. (1974). *The Philosophy of Quantum Mechanics: The Interpretation of Quantum Mechanics in Historical Perspective*. New York: John Wiley.

Kochen, S. and , E. P. Specker (1967). The Problem of Hidden Variables in Quantum Mechanics. *Journal of Mathematics and Mechanics* **17**, 59–87.

Putnam, H. (1969). Is Logic Empirical? In: R. S. Cohen and M. W. Wartofsky (eds.), *Boston Studies in the Philosophy of Science*, Vol. 5, pp. 199–215. Dordrecht: Reidel.

Redhead, M. (1987). *Incompleteness, Non-Locality and Realism.* Oxford: Oxford University Press.

van Fraassen, B. (1973). Semantic Analysis of Quantum Logic. In: C. A. Hooker (ed.), *Contemporary Research in the Foundations and Philosophy of Quantum Theory*. Dordrecht: Reidel.

van Fraassen, B. (1991). *Quantum Mechanics: An Empiricist View*. Oxford: Clarendon Press.

Renata Ziemińska

TWO NOTIONS OF THE INTERNAL AND GOLDMAN'S EPISTEMIC EXTERNALISM

ABSTRACT. Two concepts of the internal should be distinguished in the current epistemic internalism/externalism debate: (1) the internal in an introspective sense as what is accessible by introspection and (2) the internal in a biological sense as what is inside the organism's nervous system. When "internal" is meant in the introspective sense, Goldman's process reliabilism is externalist, but when "internal" is taken in the biological sense, Goldman's process reliabilism is internalist. Goldman as a naturalist prefers "internal" in the biological sense, but the concept is unsuitable for presenting the current epistemic internalism/externalism controversy. If one understands "internal" in the introspective sense, Goldman's reliabilism is strongly externalist.

Alvin I. Goldman rejects epistemic internalism (1999, p. 293) but he also claims that his "process reliabilism is not externalist" (1992a, p. 434). Although he is recognized as a prominent epistemic externalist, Goldman himself rejects the internal/external distinction as such.[1]

Kim (1993) tries to explain Goldman's epistemological position by saying that it is partly externalist and partly internalist. In this paper I shall try to show that Kim's account is incoherent and that Goldman's process reliabilism is strongly externalist given Kim's definition of the internal.

I think that the way to resolve the confusion about Goldman's epistemic externalism is the distinction between two notions of the internal which has often been blurred in the current debate on epistemic internalism and externalism. The two senses of "internal" are as follows:

(1) "internal" is that which is accessible by mere introspection or reflection (I shall call it internal in the introspective sense);

(2) "internal" is that which is within the organism's nervous system or, simply, within the organism (I shall call it internal in the biological sense).

The first, introspective, sense of "internal" is distinctive for the epistemic internalism/externalism controversy in the theory of knowledge and justification. The second, biological, sense of "internal" is distinctive for the semantic or content internalism/externalism controversy in the philosophy of mind and lan-

[1] Similarly, Pollock seems to be an epistemic externalist but he calls himself "naturalistic internalist".

In: J. Malinowski and A. Pietruszczak (eds.), *Essays in Logic and Ontology* (*Poznań Studies in the Philosophy of the Sciences and the Humanities*, vol. 91), 395–400. Amsterdam/New York, NY: Rodopi, 2006.

guage. In his theory of justification Goldman uses "internal" in the biological sense and this, in my opinion, has generated some misunderstandings.

1. Introspection

Externalism is complementary to internalism and the meaning of externalism depends on the meaning of internalism. The basic idea of epistemic internalism is that of cognitive access to justifiers (justification making properties of a belief). Alston says that internalists

> restrict justifiers to items that are *within* something, more specifically, within the subject. But, of course, not everything that is "within" a knowing subject will be admitted as a possible justifier by an internalist. Physiological processes within the subject, of which the subject knows nothing, will not be allowed (Alston 1989, p. 185).

Alston himself does not precisely determine the internal justifiers. He only says that they must be fairly directly accessible to their subject on reflection. Kim (1993, p. 305) says that epistemic justifiers accepted by most internalists are the justifying factors graspable by introspection (they are introspectible). He presupposes that knowledge achieved by reflection coincides with knowledge achieved by introspection, and I agree with him on this point. Even though reflection may include certain memories and reasoning, all results of reflection are graspable by introspection.

Introspection is a highly controversial concept. However, I think that it is the best one to grasp what the internalists mean by the subject's access to justifiers. Introspection is a kind of access to the subject's conscious states.

Internalists usually use the term "introspection" in a broad sense including memory (retrospection) and reason. Externalists are more strict in their use of the term "introspection": "the only facts that an agent can know by introspection are facts concerning what conscious states he is (or is not) currently in" (Goldman 1999, p. 278). I think that it is rather pre-introspective awareness that "shows" what conscious state I am *currently* in. Pre-introspective awareness[2] is not a kind of observation, because it is not at the center of our attention. Proper introspection as always occurs later than its subject is not observation either; it is rather retrospection (including some reasoning or even imagery) of the pre-introspective awareness. That is why introspection is not direct, not perception-like and not independent of our previous knowledge. And the results of introspection are not infallible or incorrigible. There is, however, a kind of knowledge achieved by introspection (so called first-person knowledge) or,

[2] When I think about water, the water is the only subject of my attention, but I am also pre-introspectively aware that I am thinking about water.

at least, some well-grounded beliefs concerning our own recent experience and thinking.[3]

Naturalistic epistemologists usually do not accept the traditional concept of introspection. The first reason is that the majority of cognitive processes are not available to introspection and, therefore, it is the data of cognitive science and not the results of introspection that are important for naturalists. Second, the borderline between what is accessible by introspection and the rest of the world is not important for the proponents of the theory that mental states are identical with physical states. And the borderline between the organism and the rest of the world is clear and important for them. This is why naturalists prefer the biological sense of "internal" and not the introspective one.

Goldman is a naturalist and he seems to avoid the concept of introspection when specifying his own position. However, he cannot avoid it when speaking about the proponents of epistemic internalism (1999, p. 277) since the point of the epistemic internalist/externalist controversy is the introspective access to justifiers. It is the introspective sense of "internal" not the biological one that refers to that access. The biological sense of "internal", although useful for the externalist, is unsuitable for presenting the standpoint of the traditional internalist and the epistemic internalism/externalism controversy.

2. Kim about Goldman

I accept Alston's general concept of justification that there are three conditions for a belief to be justified: (1) the belief must have grounds, (2) the grounds have to be adequate, and (3) there must be a basing relation between the belief and its grounds. As Kim accurately noticed, each of those three elements of justification provides scope for an internalism/externalism distinction depending on the introspective access criterion.[4] Therefore, we may say that epistemic externalism falls into three kinds: (1) ground externalism, (2) adequacy externalism, and (3) connection externalism. I shall focus here on ground externalism only, since it is the main concern of my argument; Kim claims that Goldman's process reliabilism is ground internalism and I shall try to show that it is ground externalism.

According to Kim, one is a ground internalist if one adheres to the idea that the only possible grounds for justification are psychological states accessible by introspection, whereas the externalist allows for some different justifying

[3]The epistemic internalists assumes that such introspective knowledge exists and that all justifiers should be available to the subject by introspection.

[4]A theory of justification is internalist if it requires the justification making properties to be accessible by introspection. And a theory of justification is externalist if it allows the justifiers to be not accessible by introspection.

grounds, e.g., a proper causal relation. He admits that to be an externalist it is enough to agree that "for some beliefs, their justifying grounds can be something external" (Kim 1993, p. 308). Furthermore, if that which is internal is accessible by introspection and that which is external is not (the concept of the internal and the external are, according to Kim, mutually exclusive and jointly exhaustive), such cognitive processes as unconscious happenings within the brain or retinal stimulation or neurophysiological processes are external. It is true that Goldman's process reliabilism allows such cognitive processes as justifying factors (Goldman 1992b, p. 156). Thus, the following argument can be formulated (see Ziemińska 1998, p. 58):

(1) in order to be a ground externalist it is enough to claim that some justifying grounds can be external (Kim 1993, p. 308);
(2) that which is external from an epistemic point of view is not accessible by introspection (Kim 1993, p. 305);
(3) unconscious cognitive processes are not accessible by introspection (assumption);
(4) Goldman's process reliabilism allows unconscious cognitive processes as justifying factors (Goldman 1992b, p. 156) and, therefore;
(5) Goldman's process reliabilism is ground externalism.

Kim does not follow this line of reasoning. He accepts the denial of my assumption (3). Although he admits that his standpoint involves certain problems, he finally disregards the difference between cognitive states accessible by introspection and those that are not. He says that all cognitive processes may be taken as accessible by introspection. This is an untenable oversimplification. The introspective access to all cognitive processes requires that any unconscious brain process is accessible by introspection. But this is not the case. Introspection is a kind of access to conscious states only. However, many cognitive processes, for instance brain processes, are not conscious.

Kim's main reason for recognizing Goldman's process reliabilism as ground internalism is the statement by the latter that only internal cognitive processes are relevant to his theory of justification. It is true that Goldman uses the term "internal" when discussing cognitive processes relevant to his theory of justification. However, Goldman's "internal" is not the same as Kim's "internal". Kim adopts the concept of the internal in the introspective sense (accessible by introspection), and Goldman uses the concept of the internal in the biological sense (within the organism's nervous system). Goldman's process reliabilism assumes that we are justified in believing something only if our belief is produced by a reliable cognitive process. Belief-forming processes are restricted to processes within the organism's nervous system and in this sense they are internal. Goldman does not claim that the only justifiers can be mental states accessible by introspection.

Kim does not distinguish between the two concepts of the internal. However, if "internal" is understood as "being inside the organism", then Goldman's process reliabilism is internalist, and if "internal" means "accessible by introspection", Goldman's process reliabilism is externalist. Interference of the two concepts has been in my opinion the source of Kim's confusion about Goldman's externalism; Goldman uses "internal" in the biological sense, while Kim in the introspective one.

3. Goldman's Externalism

Goldman states clearly that his "process reliabilism is not externalist, since it focuses on 'inner' processes" (1992a, p. 434). The term "externalist" in the quoted sentence refers to the concept of the internal in a biological sense (Goldman writes about processes that are "beneath the skin"). Goldman never declares himself to be an internalist in the introspective sense of "internal". On the contrary, he claims that epistemic internalism is an unattainable idea; he exposes and rejects the general structure of internalism: "I see no hope for internalism" (Goldman 1999, p. 293). For Goldman all internalist accessibility requirements are unclear and problematic (1999, p. 293), and he tries to avoid referring to the introspective notion of the internal. He himself seems to be unaware of the interference of the two concepts of the internal I discussed above. When speaking about epistemic internalists, Goldman uses "internal" in the introspective sense, but considering his own position he uses "internal" in the biological sense. Perhaps this is why Goldman rejects the internal/external distinction as such. The biological sense of the terms is unsuitable for presenting the epistemic internalism/externalism controversy. The introspective sense of "internal" denoting that what is accessible by introspection is in turn not suitable for naturalist epistemology. Perhaps this is a dilemma for Goldman to resolve.

Goldman introduces some internalist factors to his process reliabilism. The theory of normal worlds as the domain of reliability (to meet the Cartesian demon world counter-example) refers to our general beliefs about the actual world and is a kind of internalisation. Goldman's virtue reliabilism brings a further internalisation of reliability: beliefs are justified if we consider them to be produced by a reliable (virtuous) process. Perceptual beliefs of the victims of the Cartesian demon may be justified (although they are not produced by reliable processes) if the inhabitants of the Cartesian demon world *judge* perception to be a reliable process (Goldman 1992b, p. 158).

However, Goldman's epistemology is still externalist in the standard sense because there is no guarantee of introspective access to justifiers. Which processes are generally judged to be reliable in a certain world is something not accessible by introspection. If we continue to use "internal" in the introspective sense, Goldman's reliabilism is strongly externalist.

Uniwersytet Szczeciński
Instytut Filozofii
ul. Krakowska 61/69
71-017 Szczecin
Poland
e-mail: zieminre@univ.szczecin.pl

REFERENCES

Alston, W. P. (1989). *Epistemic Justification: Essays in the Theory of Knowledge*. Ithaca: Cornell University Press.

Goldman, A. I. (1992a). Reliabilism. In: J. Dancy and E. Sosa (eds.), *A Companion to Epistemology*, pp. 433–436. Oxford: Blackwell Publishers.

Goldman, A. I. (1992b). Epistemic Folkways and Scientific Epistemology. In: *Liaisons: Philosophy Meets the Cognitive and Social Sciences*, pp. 155–175. A Bradford Book. Cambridge MA: The MIT Press.

Goldman, A. I. (1999). Internalism Exposed. *The Journal of Philosophy* **96**, 271–293.

Kim, K. (1993). Internalism and Externalism in Epistemology. *American Philosophical Quarterly* **30**, 303–315.

Ziemińska R. (1998). What Is Epistemic Externalism? *Reports of Philosophy* **18**, 53–80.

POZNAŃ STUDIES IN THE PHILOSOPHY OF THE SCIENCES AND THE HUMANITIES

Contents of back issues

VOLUME 1 (1975)

Main topics:
The Method of Humanistic Interpretation; The Method of Idealization; The Reconstruction of Some Marxist Theories.
(sold out)

VOLUME 2 (1976)

Main topics:
Idealizational Concept of Science; Categorial Interpretation of Dialectics.
(sold out)

VOLUME 3 (1977)

Main topic:
Aspects of the Production of Scientific Knowledge.
(sold out)

VOLUME 4 (1978)

Main topic:
Aspects of the Growth of Science.
(sold out)

VOLUME 5 (1979)

Main topic:
Methodological Problems of Historical Research.
(sold out)

VOLUME 6 (1982)

SOCIAL CLASSES ACTION & HISTORICAL MATERIALISM

Main topics:
On Classes; On Action; The Adaptive Interpretation of Historical Materialism; Contributions to Historical Materialism.
(sold out)

VOLUME 7 (1982)

DIALECTICAL LOGICS FOR THE POLITICAL SCIENCE
(Edited by Hayward R. Alker, Jr.)

VOLUME 8 (1985)

CONSCIOUSNESS: METHODOLOGICAL AND PSYCHOLOGICAL APPROACHES
(Edited by Jerzy Brzeziński)

VOLUME 9 (1986)

THEORIES OF IDEOLOGY AND IDEOLOGY OF THEORIES
(Edited by Piotr Buczkowski and Andrzej Klawiter)

VOLUME 10 (1987)

WHAT IS CLOSER-TO-THE-TRUTH?
A PARADE OF APPROACHES TO TRUTHLIKENESS
(Edited by Theo A.F. Kuipers)

VOLUME 11 (1988)

NORMATIVE STRUCTURES OF THE SOCIAL WORLD
(Edited by Giuliano di Bernardo)

VOLUME 12 (1987)

POLISH CONTRIBUTIONS TO THE THEORY AND PHILOSOPHY OF LAW
(Edited by Zygmunt Ziembiński)

VOLUME 13 (1988)

DIMENSIONS OF THE HISTORICAL PROCESS
(Edited by Leszek Nowak)

VOLUME 14 (1990)

ACTION AND PERFORMANCE: MODELS AND TESTS.
CONTRIBUTIONS TO THE QUANTITATIVE PSYCHOLOGY
AND ITS METHODOLOGY
(Edited by Jerzy Brzeziński and Tadeusz Marek)

VOLUME 15 (1989)

VISIONS OF CULTURE AND THE MODELS OF CULTURAL SCIENCES
(Edited by Jerzy Kmita and Krystyna Zamiara)

VOLUME 16 (1990)

IDEALIZATION I: GENERAL PROBLEMS
(Edited by Jerzy Brzeziński, Francesco Coniglione, Theo A.F. Kuipers and Leszek Nowak)

VOLUME 17 (1990)

IDEALIZATION II: FORMS AND APPLICATIONS
(Edited by Jerzy Brzeziński, Francesco Coniglione, Theo A.F. Kuipers and Leszek Nowak)

VOLUME 18 (1990)

STUDIES ON MARIO BUNGE'S *TREATISE*
(Edited by Paul Weingartner and Georg J.W. Dorn)

VOLUME 19 (1990)

NARRATION AND EXPLANATION.
CONTRIBUTION TO THE METHODOLOGY OF THE HISTORICAL RESEARCH
(Edited by Jerzy Topolski)

VOLUME 20 (1990)

Jürgen Ritsert
MODELS AND CONCEPTS OF IDEOLOGY

VOLUME 21 (1991)

PROBABILITY AND RATIONALITY
STUDIES ON L. JONATHAN COHEN'S PHILOSOPHY OF SCIENCE
(Edited by Ellery Eells and Tomasz Maruszewski)

VOLUME 22 (1991)

THE SOCIAL HORIZON OF KNOWLEDGE
(Edited by Piotr Buczkowski)

VOLUME 23 (1991)

ETHICAL DIMENSIONS OF LEGAL THEORY
(Edited by Wojciech Sadurski)

VOLUME 24 (1991)

ADVANCES IN SCIENTIFIC PHILOSOPHY
ESSAYS IN HONOUR OF PAUL WEINGARTNER ON THE OCCASION OF THE 60TH ANNIVERSARY OF HIS BIRTHDAY
(Edited by Gerhard Schurz and Georg J.W. Dorn)

VOLUME 25 (1992)

IDEALIZATION III: APPROXIMATION AND TRUTH
(Edited by Jerzy Brzeziński and Leszek Nowak)

VOLUME 26 (1992)

IDEALIZATION IV: INTELLIGIBILITY IN SCIENCE
(Edited by Craig Dilworth)

VOLUME 27 (1992)

Ryszard Stachowski

THE MATHEMATICAL SOUL.
AN ANTIQUE PROTOTYPE OF THE MODERN MATEMATISATION OF PSYCHOLOGY

VOLUME 28 (1993)

POLISH SCIENTIFIC PHILOSOPHY:
THE LVOV-WARSAW SCHOOL
(Edited by Francesco Coniglione, Roberto Poli and Jan Woleński)

VOLUME 29 (1993)

Zdzisław Augustynek and Jacek J. Jadacki

POSSIBLE ONTOLOGIES

VOLUME 30 (1993)

GOVERNMENT: SERVANT OR MASTER?
(Edited by Gerard Radnitzky and Hardy Bouillon)

VOLUME 31 (1993)

CREATIVITY AND CONSCIOUSNESS.
PHILOSOPHICAL AND PSYCHOLOGICAL DIMENSIONS
(Edited by Jerzy Brzeziński, Santo Di Nuovo, Tadeusz Marek and Tomasz Maruszewski)

VOLUME 32 (1993)

FROM ONE-PARTY-SYSTEM TO DEMOCRACY
(Edited by Janina Frentzel-Zagórska)

VOLUME 33 (1993)

SOCIAL SYSTEM, RATIONALITY, AND REVOLUTION
(Edited by Marcin Paprzycki and Leszek Nowak)

VOLUME 34 (1994)

Izabella Nowakowa

IDEALIZATION V: THE DYNAMICS OF IDEALIZATIONS

VOLUME 35 (1993)

EMPIRICAL LOGIC AND PUBLIC DEBATE.
ESSAYS IN HONOUR OF ELSE M. BARTH
(Edited by E.C.W. Krabbe, R.J. Dalitz, and P.A. Smit)

VOLUME 36 (1994)

MARXISM AND COMMUNISM: POSTHUMOUS REFLECTIONS ON POLITICS, SOCIETY, AND LAW
(Edited by Martin Krygier)

VOLUME 37 (1994)

THE SOCIAL PHILOSOPHY OF AGNES HELLER
(Edited by John Burnheim)

VOLUME 38 (1994)

IDEALIZATION VI: IDEALIZATION IN ECONOMICS
(Edited by Bert Hamminga and Neil B. De Marchi)

VOLUME 39 (1994)

PROBABILITY IN THEORY-BUILDING.
EXPERIMENTAL AND NON-EXPERIMENTAL APPROACHES TO SCIENTIFIC RESEARCH IN PSYCHOLOGY
(Edited by Jerzy Brzeziński)

VOLUME 40 (1995)

THE HERITAGE OF KAZIMIERZ AJDUKIEWICZ
(Edited by Vito Sinisi and Jan Woleński)

VOLUME 41 (1994)

HISTORIOGRAPHY BETWEEN MODERNISM AND POSTMODERNISM. CONTRIBUTIONS TO THE METHODOLOGY OF THE HISTORICAL RESEARCH
(Edited by Jerzy Topolski)

VOLUME 42 (1995)

IDEALIZATION VII: IDEALIZATION, STRUCTURALISM, AND APPROXIMATION
(Edited by Martti Kuokkanen)

Idealization, Approximation and Counterfactuals in the Structuralist Framework – T.A.F. Kuipers, *The Refined Structure of Theories*; C.U. Moulines and R. Straub, *Approximation and Idealization from the Structuralist Point of View*; I.A. Kieseppä, *A Note on the Structuralist Account of Approximation*; C.U. Moulines and R. Straub, *A Reply to Kieseppä*; W. Balzer and G. Zoubek, *Structuralist Aspects of Idealization*; A. Ibarra and T. Mormann, *Counterfactual Deformation and Idealization in a Structuralist Framework*; I.A. Kieseppä, *Assessing the Structuralist Theory of Verisimilitude*. **Idealization, Approxima-tion and Theory Formation** – L. Nowak, *Remarks on the Nature of Galileo's Methodological Revolution*; I. Niiniluoto, *Approximation in Applied Science*; E. Heise, P. Gerjets and R. Westermann, *Idealized Action Phases. A Concise Rubicon Theory*; K.G. Troitzsch, *Modelling, Simulation, and Structuralism*; V. Rantala and T. Vadén, *Idealization in Cognitive Science. A Study in Counterfactual Correspondence*; M. Sintonen and M. Kiikeri, *Idealization in Evolutionary Biology*; T. Tuomivaara, *On Idealizations in Ecology*; M. Kuokkanen and M. Häyry, *Early Utilitarianism and Its Idealizations from a Systematic Point of View*. **Idealization, Approximation and Measurement** – R. Westermann, *Measurement-Theoretical Idealizations and Empirical Research Practice*; U. Konerding, *Probability as an Idealization of Relative Frequency. A Case Study by Means of the BTL-Model*; R. Suck and J. Wienöbst, *The Empirical Claim of Probality Statements, Idealized Bernoulli Experiments and their Approximate Version*; P.J. Lahti, *Idealizations in Quantum Theory of Measurement*.

VOLUME 43 (1995)

Witold Marciszewski and Roman Murawski
MECHANIZATION OF REASONING IN A HISTORICAL PERSPECTIVE

Chapter 1: *From the Mechanization of Reasoning to a Study of Human Intelligence*; Chapter 2: *The Formalization of Arguments in the Middle Ages*; Chapter 3: *Leibniz's Idea of Mechanical Reasoning at the Historical Background*; Chapter 4: *Between Leibniz and Boole: Towards the Algebraization of Logic*; Chapter 5: *The English Algebra of Logic in the 19th Century*; Chapter 6: *The 20th Century Way to Formalization and Mechanization*; Chapter 7: *Mechanized Deduction Systems*.

VOLUME 44 (1995)

THEORIES AND MODELS IN SCIENTIFIC PROCESSES
(Edited by William Herfel, Władysław Krajewski,
Ilkka Niiniluoto and Ryszard Wójcicki)

Introduction; **Part 1. Models in Scientific Processes** – J. Agassi, *Why there is no Theory of Models?*; M. Czarnocka, *Models and Symbolic Nature of Knowledge*; A. Grobler, *The Representational and the Non-Representational in Models of Scientific Theories*; S. Hartmann, *Models as a Tool for the Theory Construction: Some Strategies of Preliminary Physics*; W. Herfel, *Nonlinear Dynamical Models as Concrete Construction*; E. Kałuszyńska, *Styles of Thinking*; S. Psillos, *The Cognitive Interplay Between Theories and Models: the Case of 19th Century Optics*. **Part 2. Tools of Science** – N.D. Cartwright, T. Shomar, M. Suarez, *The Tool-Box of Science*; J. Echeverria, *The Four Contexts of Scienctific Activity*; K. Havas, *Continuity and Change; Kinds of Negation in Scientific Progress*; M. Kaiser, *The Independence of Scientific Phenomena*; W. Krajewski, *Scientific Meta-Philosophy*; I. Niiniluoto, *The Emergence of Scientific Specialties: Six Models*; L. Nowak, *Antirealism, (Supra-) Realism and Idealization*; R.M. Nugayev, *Classic, Modern and Postmodern Scientific Unification*; V. Rantala, *Translation and Scientific Change*; G. Schurz, *Theories and Their Applications – a Case of Nonmonotonic Reasoning*; W. Strawiński, *The Unity of Science Today*; V. Torosian, *Are the Ethics and Logic of Science Compatible?* **Part 3. Unsharp Approaches in Science** – E.W. Adams, *Problems and Prospects in a Theory of Inexact First-Order Theories*; W. Balzer, G. Zoubek, *On the Comparision of Approximative Empirical Claims*; G. Cattaneo, M. Luisa Dalla Chiara, R. Giuntini, *The Unsharp Approaches to Quantum Theory*; T.A.F. Kuipers, *Falsification Versus Effcient Truth Approximation*; B. Lauth, *Limiting Decidability and Probability*; J. Pykacz, *Many-Valued Logics in Foundations of Quantum Mechanics*; R.R. Zapatrin, *Logico-Algebraic Approach to Spacetime Quantization*.

VOLUME 45 (1995)

COGNITIVE PATTERNS IN SCIENCE AND COMMON SENSE.
GRONINGEN STUDIES IN PHILOSOPHY OF SCIENCE,
LOGIC, AND EPISTEMOLOGY
(Edited by Theo A.F. Kuipers and Anne Ruth Mackor)

L. Nowak, *Foreword*; **General Introduction** – T.A.F. Kuipers and A.R. Mackor, *Cognitive Studies of Science and Common Sense*; **Part I: Conceptual Analysis in Service of Various Research Programmes** – H. Zandvoort, *Concepts of Interdisciplinarity and Environmental Science*; R. Vos, *The Logic and Epistemology of the Concept of Drug and Disease Profile*; R.C. Looijen, *On the Distinction Between Habitat and Niché, and Some Implications for Species' Differentiation*; G.J. Stavenga, *Cognition, Irreversibility and the Direction of Time*; R. Dalitz, *Knowledge, Gender and Social Bias*; **Part II: The Logic of the Evaluation of Arguments, Hypotheses, Rules, and Interesting Theorems** – E.G.W. Krabbe, *Can We Ever Pin One Down to a Formal Fallacy?*; T.A.F. Kuipers, *Explicating the Falsificationist and Instrumentalist Methodology by Decomposing the Hypothetico-Deductive Method*; A. Keupink, *Causal Modelling and Misspecification: Theory and Econometric Historical Practice*; M.C.W. Janssen and Y.-H. Tan, *Default Reasoning and Some Applications in*

Economics; B. Hamminga, *Interesting Theorems in Economics*; **Part III: Three Challenges to the Truth Approximation Programme** – S.D. Zwart, *A Hidden Variable in the Discussion About 'Language Dependency' of Truthlikeness*; H. Hettema and T.A.F. Kuipers, *Sommerfeld's* Atombau: *A Case Study in Potential Truth Approximation*; R. Festa, *Verisimilitude, Disorder, and Optimum Prior Probabilities*; **Part IV: Explicating Psychological Intuitions** – A.R. Mackor, *Intentional Psychology is a Biological Discipline*; J. Peijnenburg, *Hempel's Rationality. On the Empty Nature of Being a Rational Agent*; L. Guichard, *The Causal Efficacy of Propositional Attitudes*; M. ter Hark, *Connectionism, Behaviourism and the Language of Thought.*

VOLUME 46 (1996)

POLITICAL DIALOGUE: THEORIES AND PRACTICE
(Edited by Stephen L. Esquith)

Introduction – S.L. Esquith, *Political Dialogue and Political Virtue*. **Part I: The Modern Clasics** – A.J. Damico, *Reason's Reach: Liberal Tolerance and Political Discourse*; T.R. Machan, *Individualism and Political Dialogue*; R. Kukla, *The Coupling of Human Souls: Rousseau and the Problem of Gender Relations*; D.F. Koch, *Dialogue: An Essay in the Instrumentalist Tradition*. **Part II: Toward a Democratic Synthesis** – E. Simpson, *Forms of Political Thinking and the Persistence of Practical Philosophy*; J.B. Sauer, *Discoursee, Consensus, and Value: Conversations about the Intelligible Relation Between the Private and Public Spheres*; M. Kingwell, *Phronesis and Political Dialogue*, R.T. Peterson, *Democracy and Intellectual Mediation – After Liberalism and Socialism*. **Part III: Dialogue in Practice** – S. Rohr Scaff, L.A. Scaff, *Political Dialogue in the New Germany: The Burdens of Culture and an Asymmetrical Past*; J.H. Read, *Participation, Power, and Democracy*; S.E. Bennett, B. Fisher, D. Resnick, *Speaking of Politics in the United States: Who Talks to Whom, Why, and Why Not*; J. Forester, *Beyond Dialogue to Transformative Learning: How Deliberative Rituals Encourage Political Judgement in Community Planning Processes*; A. Fatić, *Retribution in Democracy*.

VOLUME 47 (1996)

EPISTEMOLOGY AND HISTORY. HUMANITIES AS A PHILOSOPHICAL PROBLEM AND JERZY KMITA'S APPROACH TO IT
(Edited by Anna Zeidler-Janiszewska)

A. Zeidler-Janiszewska, *Preface*. **Humanistic Knowledge** – K.O. Apel, *The Hermeneutic Dimension of Social Science and its Normative Foundation*; M. Czerwiński, *Jerzy Kmita's Epistemology*; L. Witkowski, *The Frankfurt School and Structuralism in Jerzy Kmita's Analysis*; A. Szahaj, *Between Modernism and Postmodernism: Jerzy Kmita's Epistemology*; A. Grzegorczyk, *Non-Cartesian Coordinates in the Contemporary Humanities*; A. Pałubicka, *Pragmatist Holism as an Expression of Another Disenchantment of the World*; J. Sójka, *Who is Afraid of Scientism?*; P. Ozdowski, *The Broken Bonds with the World*; J. Such, *Types of Determination vs. the Development of Science in Historical Epistemology*; P. Zeidler, *Some Issues of Historical Epistemology in the Light of the Structuralist Philosophy of Science*; M. Buchowski, *Via Media: On the Consequences of Historical Epistemology for the Problem of Rationality*; B. Kotowa, *Humanistic Valuation and Some Social Functions of the Humanities*. **On Explanation and Humanistic Interpretation** – T.A.F. Kuipers, *Explanation by Intentional, Functional, and Causal Specification*; E. Świderski, *The Interpretational Paradigm in the Philosophy of the Human Sciences*; L. Nowak, *On the Limits of the Rationalistic Paradigm*; F. Coniglione, *Humanistic Interpretation between Hempel and Popper*; Z. Ziembiński, *Historical Interpretation vs. the Adaptive Interpretation of a Legal*

Text; W. Mejbaum, *Explaining Social Phenomena*; M. Ziółkowski, *The Functional Theory of Culture and Sociology*; K. Zamiara, *Jerzy Kmita's Social-Regulational Theory of Culture and the Category of Subject*; J. Brzeziński, *Theory and Social Practice. One or Two Psychologies?*; Z. Kwieciński, *Decahedron of Education (Components and aspects). The Need for a Comprehensive Approach*; J. Paśniczek, *The Relational vs. Directional Concept of Intentionality*. **The Historical Dimension of Culture and its Studies** – J. Margolis, *The Declension of Progressivism*; J. Topolski, *Historians Look at Historical Truth*; T. Jerzak-Gierszewska, *Three Types of the Theories of Religion and Magic*; H. Paetzold, *Mythos und Moderne in der Philosophie der symbolishen Formen Ernst Cassirers*; M. Siemek, *Sozialphilosophische Aspekte der Übersetzbarkeit*. **Problems of Artistic Practice and Its Interpretation** – S. Morawski, *Theses on the 20th Century Crisis of Art and Culture*; A. Erjavec, *The Perception of Science in Modernist and Postmodernist Artistic Practice*, G. Dziamski, *The Avant-garde and Contemporary Artistic Consciousness*; H. Orłowski, *Generationszugehörigkeit und Selbsterfahrung von (deutschen) Schriftstellern*; T. Kostyrko, *The "Transhistoricity" of the Structure of Work of Art and the Process of Value Transmission in Culture*; G. Banaszak, *Musical Culture as a Configuration of Subcultures*; A. Zeidler-Janiszewska, *The Problem of the Applicability of Humanistic Interpretation in the Light of Contemporary Artistic Practice*; J. Kmita, *Towards Cultural Relativism "with a Small 'r'"*; The Bibliography of Jerzy Kmita.

VOLUME 48 (1996)

THE SOCIAL PHILOSOPHY OF ERNEST GELLNER
(Edited by John A. Hall and Ian Jarvie)

J.A. Hall, I. Jarvie, *Preface*; J.A. Hall, I. Jarvie, *The Life and Times of Ernest Gellner*. **Part 1: Intelectual Background** – J. Musil, *The Prague Roots of Ernest Gellner's Thinking*; Ch. Hahn, *Gellner and Malinowski: Words and Things in Central Europe*; T. Dragadze, *Ernest Gellner in Soviet East*. **Part 2: Nations and Nationalism** – B. O'Leary, *On the Nature of Nationalism: An Appraisal of Ernest Gellner's Writings on Nationalism*; K. Minogue, *Ernest Gellner and the Dangers of Theorising Nationalism*; A.D. Smith, *History and Modernity: Reflection on the Theory of Nationalism*; M. Mann, *The Emergence of Modern European Nationalism*; N. Stagardt, *Gellner's Nationalism: The Spirit of Modernisation?* **Part 3: Patterns of Development** – P. Burke, *Reflections on the History of Encyclopaedias*; A. MacFarlane, *Ernest Gellner and the Escape to Modernity*; R. Dore, *Soverein Individuals*; S. Eisenstadt, *Japan: Non Axial Modernity*; M. Ferro, *L'Indépendance Telescopée: De la Décolonisation a L'Impérialisme Multinational*. **Part 4: Islam** – A. Hammoudi, *Segmentarity, Social stratification, Political Power and Sainthood: Reflections on Gellner's Theses*; H. Munson, Jr., *Rethinking Gellner's Segmentary Analysis of Morocco's Ait^c^Atta*; J. Baechler, *Sur le charisme*; Ch. Lindholm, *Despotism and Democracy: State and Society in PreModern Middle East*; H. Munson, *Jr. Muslism and Jew in Morocco: Reflections on the Distinction between Belief and Behaviour*; T. Asad, *The Idea of an Anthropology of Islam*. **Part 5: Science and Disenchantment** – P. Anderson, *Science, Politics, Enchantment*; R. Schroeder, *From the Big Divide to the Rubber Cage: Gellner's Conception of Science and Technology*; J. Davis, *Irrationality in Social Life*. **Part 6: Relativism and Universals** – J. Skorupski, *The Post-Modern Hume: Ernest Gellner's 'Enlightenment Fundamentalism'*; J. Wettersten, *Ernest Gellner: A Wittgensteinian Rationalist*; I. Jarvie, *Gellner's Positivism*; R. Boudon, *Relativising Relativism: When Sociology Refutes the Sociology of Science*; R. Aya, *The Empiricist Exorcist*. **Part 7: Philosophy of History** – W. McNeill, *A Swang Song for British Liberalism?*; A. Park, *Gellner and the Long Trends of History*; E. Leone, *Marx, Gellner, Power*; R. Langlois, *Coercion, Cognition and Production: Gellner's Challenge to Historical Materialism and Post-Modernism*; E. Gellner, *Reply to my Critics*; I. Jarvie, *Complete Bibliography of Gellner's Work*.

VOLUME 49 (1996)

THE SIGNIFICANCE OF POPPER'S THOUGHT
(Edited by Stefan Amsterdamski)

Karl Popper's Three Worlds – J. Watkins, *World 1, World 2 and the Theory of Evolution*; A. Grobler, *World 3 and the Cunning of Reason*. **The Scientific Method as Ethics** – J. Agassi, *Towards Honest Public Relations of Science*; S. Amsterdamski, *Between Relativism and Absolutism: the Popperian Ideal of Knowledge*. **The Open Society and its Prospects** – E. Gellner, *Karl Popper – The Thinker and the Man*; J. Woleński, *Popper on Prophecies and Predictions*.

VOLUME 50 (1996)

THE IDEA OF THE UNIVERSITY
(Edited by Jerzy Brzeziński and Leszek Nowak)

Introduction – K. Twardowski, *The Majesty of the University*. **I** – Z. Ziembiński, *What Can be Saved of the Idea of the University?*; L. Kołakowski, *What Are Universities for?*; L. Gumański, *The Ideal University and Reality*; Z. Bauman, *The Present Crisis of the Universities*. **II** – K. Ajdukiewicz, *On Freedom of Science*; H. Samsonowicz, *Universities and Democracy*; J. Topolski, *The Commonwealth of Scholars and New Conceptions of Truth*; K. Szaniawski, *Plus ratio quam vis*. **III** – L. Koj, *Science, Teaching and Values*; K. Szaniawski, *The Ethics of Scientific Criticism*; J. Brzeziński, *Ethical Problems of Research Work of Psychologists*. **IV** – J. Goćkowski, *Tradition in Science*; J. Kmita, *Is a "Creative Man of Knowledge" Needed in University Teaching?*; L. Nowak, *The Personality of Researchers and the Necessity of Schools in Science*. **Recapitulation** – J. Brzeziński, *Reflections on the University*.

VOLUME 51 (1997)

KNOWLEDGE AND INQUIRY: ESSAYS ON JAAKKO HINTIKKA'S EPISTEMOLOGY AND PHILOSOPHY OF SCIENCE
(Edited by Matti Sintonen)

M. Sintonen, *From the Science of Logic to the Logic of Science*. **I: Historical Perspectives** – Z. Bechler, *Hintikka on Plentitude in Aristotle*; M.-L. Kakkuri-Knuuttila, *What Can the Sciences of Man Learn from Aristotle?*; M. Kusch, *Theories of Questions in German-Speaking Philosophy Around the Turn of the Century*; N.-E. Sahlin, *'He is no Good for My Work': On the Philosophical Relations between Ramsey and Wittgenstein*. **II: Formal Tools: Induction, Observation and Identifiability** – T.A.F. Kuipers, *The Carnap-Hintikka Programme in Inductive Logic*; I. Levi, *Caution and Nonmonotic Inference*; I. Niiniluoto, *Inductive Logic, Atomism, and Observational Error*; A. Mutanen, *Theory of Identifiability*. **III: Questions in Inquiry: The Interrogative Model** – S. Bromberger, *Natural Kinds and Questions*; S.A. Kleiner, *The Structure of Inquiry in Developmental Biology*; A. Wiśniewski, *Some Foundational Concepts of Erotetic Semantics*; J. Woleński, *Science and Games*. **IV: Growth of Knowledge: Explanation and Discovery** – M. Sintonen, *Explanation: The Fifth Decade*; E. Weber, *Scientific Explanation and the Interrogative Model of Inquiry*; G. Gebhard, *Scientific Discovery, Induction, and the Multi-Level Character of Scientific Inquiry*; M. Kiikeri, *On the Logical Structure of Learning Models*. **V: Jaakko Hintikka: Replies**.

VOLUME 52 (1997)

Helena Eilstein
LIFE CONTEMPLATIVE, LIFE PRACTICAL.
AN ESSAY ON FATALISM

Preface. **Chapter One: Oldcomb and Newcomb** – 1. *In the King Comb's Chamber of Game*; 2. *The Newcombian Predicaments*. **Chapter Two: Ananke** – 1. *Fatalism: What It Is Not?*; 2. *Fatalism: What Is It?*; 3. *Fatalism and* a priori *Arguments*; 4. *Fatalism and 'Internal' Experience*; 5. *Determinism, Indeterminism and Fatalism*; 6. *Transientism, Eternism and Fatalism*; 7. *Fatalism: What It Does Not Imply?*. **Chapter Three: Fated Freedom** – 1. *More on Libertarianism*; 2. *On the Deterministic Concept of Freedom*; 3. *Moral Self and Responsibility in the Light of Probabilism*; 4. *Fated Freedom*. **Chapter Four: The Virus of Fatalism** – 1. *Fatalism and Problems of Cognition*; 2. *The Virus of Fatalism: Why Mostly Harmless?*

VOLUME 53 (1997)

Dimitri Ginev
A PASSAGE TO THE HERMENEUTIC PHILOSOPHY OF SCIENCE

Preface; **Introduction**: *Topics in the Hermeneutic Philosophy of Science*; **Chapter 0**: *On the Limits of the Rational Reconstruction of Scientific Knowledge*; **Chapter 1**: *On the Hermeneutic Nature of Group Rationality*; **Chapter 2**: *Towards a Hermeneutic Theory of Progressive Change in Scientific Development*; **Chapter 3**: *Beyond Naturalism and Traditionalism*; **Chapter 4**: *A Critical Note on Normative Naturalism*; **Chapter 5**: *Micro and Macrohermeneutics of Science*; *Concluding Remarks.*

VOLUME 54 (1997)

IN ITINERE. EUROPEAN CITIES AND THE BIRTH OF MODERN SCIENTIFIC PHILOSOPHY
(Edited by Roberto Poli)

Introduction – R. Poli, *In itinere. Pictures from Central-European Philosophy*. **Stages of the Tour** – K. Schuhmann, *Philosophy and Art in Munich around the Turn of the Century*; M. Libardi, *In itinere: Vienna 1870-1918*; W. Baumgartner, *Nineteenth-Century Würzburg: The Development of the Scientific Approach to Philosophy*; L. Dappiano, *Cambridge and the Austrian Connection*; J. Sebestik, *Prague Mosaic. Encounters with the Prague Philosophers*; J.J. Jadacki, *Warsaw: The Rise and Decline of Modern Scientific Philosophy in the Capital City of Poland*; J. Woleński, *Lvov*; L. Albertazzi, *Science and the Avant-Garde in Early Nineteenth-Century Florence*; F. Minazzi, *The Presence of Phenomenology in Milan between the Two World Wars. The Contribution of Antonio Banfi and Giulio Preti.*

VOLUME 55 (1997)

REALISM AND QUANTUM PHYSICS
(Edited by Evandro Agazzi)

Introduction – E. Agazzi. **Part One: Philosophical Considerations** – P. Horwich, *Realism and Truth*, E. Agazzi, *On the Criteria for Establishing the Ontological Status of Different Entities*; A. Baltas, *Constraints and Resistance*: *Stating a Case for Negative Realism*; M. Paty,

Predicate of Existence and Predicability for a Theoretical Object in Physics. **Part Two: Observability and Hidden Entities** – F. Bonsak, *Atoms: Lessons of a History*; A. Cordero, *Arguing for Hidden Realities*; B. d'Espagnat, *On the Difficulties that Attributing Existence to "Hidden" Quantities May Raise*; M. Pauri, *The Quantum, Space-Time and Observation.* **Part Three: Applications to Quantum Physics** – D. Albert, *On the Phenomenology of Quantum-Mechanical Superpositions*; G.C. Ghiraldi, *Realism and Quantum Mechanics*; M. Crozon, *Experimental Evidence of Quark Structure Inside Hadrons.*

VOLUME 56 (1997)

IDEALIZATION VIII: MODELLING IN PSYCHOLOGY
(Edited by Jerzy Brzeziński, Bodo Krause and Tomasz Maruszewski)

Part I: Philosophical and Methodological Problems of Cognition Process – J. Wane, *Idealizing the Cartesian-Newtonian Paradigm as Reality: The Impact of New-Paradigm Physics on Psychological Theory*; E. Hornowska, *Operationalization of Psychological Magnitudes. Assumptions-Structure-Consequences*; T. Bachmann, *Creating Analogies – On Aspects of the Mapping Process between Knowledge Domains*; H. Schaub, *Modelling Action Regulation.* **Part II: The Structure of Ideal Learning Process** – S. Ohlson, J.J. Jewett, *Ideal Adaptive Agents and the Learning Curve*; B. Krause, *Towards a Theory of Cognitive Learning*; B. Krause, U. Gauger, *Learning and Use of Invariances: Experiments and Network Simulation*; M. Friedrich, *"Reaction Time" in the Neural Network Module ART 1.* **Part III: Control Processes in Memory** – J. Tzelgov, V. Yehene, M. Naveh-Benjamin, *From Memory to Automaticity and Vice Versa: On the Relation between Memory and Automaticity*; H. Hagendorf, S. Fisher, B. Sá, *The Function of Working Memory in Coordination of Mental Transformations*; L. Nowak, *On Common-Sense and (Para-)Idealization*; I. Nowakowa, *On the Problem of Induction. Toward an Idealizational Paraphrase.*

VOLUME 57 (1997)

EUPHONY AND LOGOS
ESSAYS IN HONOUR OF MARIA STEFFEN-BATÓG AND TADEUSZ BATÓG
(Edited by Roman Murawski, Jerzy Pogonowski)

Preface. **Scientific Works of Maria Steffen-Batóg and Tadeusz Batóg** – *List of Publications of Maria Steffen-Batóg*; *List of Publications of Tadeusz Batóg*; J. Pogonowski, *On the Scientific Works of Maria Steffen-Batóg*; Jerzy Pogonowski, *On the Scientific Works of Tadeusz Batóg*; W. Lapis, *How Should Sounds Be Phonemicized*?; P. Nowakowski, *On Applications of Algorithms for Phonetic Transcription in Linguistic Research*; J. Pogonowski, *Tadeusz Batóg's Phonological Systems.* **Mathematical Logic** – W. Buszkowski, *Incomplete Information Systems and Kleene 3-valued Logic*; M. Kandulski, *Categorial Grammars with Structural Rules*; M. Kołowska-Gawiejnowicz, *Labelled Deductive Systems for the Lambek Calculus*; R. Murawski, *Satisfaction Classes – a Survey*; K. Świrydowicz, *A New Approach to Dyadic Deontic Logic and the Normative Consequence Relation*; W. Zielonka, *More about the Axiomatics of the Lambek Calculus.* **Theoretical Linguistics** – J.J. Jadacki, *Troubles with Categorial Interpretation of Natural Language*; M. Karpiński, *Conversational Devices in Human-Computer Communication Using WIMP UI*; W. Maciejewski, *Qualitative Orientation and Gramatical Categories*; Z. Vetulani, *A System of Computer Understanding of Text*; A. Wójcik, *The Formal Development of van Sandt's Presupposition Theory*; W. Zabrocki, *Psychologism in Noam Chomsky's Theory (Tentative Critical Remarks)*; R. Zuber, *Defining Presupposition without Negation.* **Philosophy of Language and Methodology of Sciences** – J. Kmita, *Philosophical Antifundamentalism*; A. Luchowska, *Peirce and Quine: Two Views on Meaning*; S. Wiertlewski, *Method According to Feyerabend*; J. Woleński, *Wittgenstein and*

Ordinary Language; K. Zamiara, *Context of Discovery – Context of Justification and the Problem of Psychologism*.

VOLUME 58 (1997)

THE POSTMODERNIST CRITIQUE OF THE PROJECT OF ENLIGHTENMENT
(Edited by Sven-Eric Liedman)

S.-E. Liedman, *Introduction*; R. Wokler, *The Enlightenment Project and its Critics*; M. Benedikt, *Die Gegenwartsbedeutung von Kants aufklärender Akzeptanz und Zurückweisung des Modells der Naturwissenschaft für zwischenmenschliche Verhältnisse: Verfehlte Beziehungen der Geisterwelt Swedenborgs*; S.-E. Liedman, *The Crucial Role of Ethics in Different Types of Enlightenment (Condorcet and Kant)*; S. Dahlstedt, *Forms of the Ineffable: From Kant to Lyotard*; P. Magnus Johansson, *On the Enlightenment in Psycho-Analysis*; E. Lundgren-Gothlin, *Ethics, Feminism and Postmodernism: Seyla Benhabib and Simone de Beauvoir*; E. Kiss, *Gibt es ein Projekt der Aufklärung und wenn ja, wie viele (Aufklärung vor dem Horizont der Postmoderne)*; E. Kennedy, *Enlightenment Anticipations of Postmodernist Epistemology*; L. Nowak, *On Postmodernist Philosophy: An Attempt to Identify its Historical Sense*; M. Castillo, *The Dilemmas of Postmodern Individualism*.

VOLUME 59 (1997)

BEYOND ORIENTALISM. THE WORK OF WILHELM HALBFASS AND ITS IMPACT ON INDIAN AND CROSS-CULTURAL STUDIES
(Edited by Eli Franco and Karin Preisendanz)

E. Franco, K. Preisendanz, *Introduction and Editorial Essay on Wilhelm Halbfass*; Publications by Wilhelm Halbfass; W. Halbfass, *Research and Reflections: Responses to my Respondents. Beyond Orientalism? Reflections on a Current Theme*. **Part I: Cross-Cultural Encounter and Dialogue** – F. X. Clooney, SJ, *Wilhelm Halbfass and the Openness of the Comparative Project*; F. Dallmayr, *Exit from Orientalism: Comments on Wilhelm Halbfass*; S. D. Serebriany, *Some Marginal Notes on India and Europe*; R. Sen (née Mookerjee), *Some Reflections on India and Europe: An Essay in Understanding*; K. Karttunen, *Greeks and Indian Wisdom*; D. Killingley, *Mlecchas, Yavanas and Heathens: Interacting Xenologies in Early Nineteenth-Century Calcutta*; W. Halbfass, *Research and Reflections: Responses to my Respondents. Cross-Cultural Encounter and Dialogue*. **Part II: Issues of Comparative Philosophy** – J. Nath Mohanty, *Between Indology and Indian Philosophy;* J. S. O'Leary, *Heidegger and Indian Philosophy*; S. Rao, *"Subordinate" or "Supreme"? The Nature of Reason in India and the West*; R. Ivekoviæ, *The Politics of Comparative Philosophy*; B.-A. Scharfstein, *The Three Philosophical Traditions*; W. Halbfass, *Research and Reflections: Responses to my Respondents. Issues of Comparative Philosophy*. **Part III: Topics in Classical Indian Philosophy** – J. E. M. Houben, *Bhartṛhari's Perspectivism: The Vṛtti and Bhartṛhari's Perspectivism in the First kāṇḍa of the Vākyapadīya*; J. Bronkhorst, *Philosophy and the Vedic Exegesis in the Mīmāṃsā*; J. Taber, *The Significance of Kumārila's Philosophy*; K. Harikai, *Kumārila's Acceptance and Modification of Categories of the Vaiœeṣika School*; V. Lysenko, *The Vaiœeṣika Notions of ākāsa and diœ from the Perspective of Indian Ideas of Space*; B. M. Perry, *Early Nyāya and Hindu Orthodoxy: ānvīkṣikī and adhikāra*; W. Halbfass, *Research and Reflections: Responses to my Respondents: Topics in Classical Indian Philosophy*. **Part IV: Developments and Attitudes in Neo-Hinduism** – A. O. Fort, *Jīvanmukti and Social Service in Advaita and Neo-Vedānta*; S. Elkman, *Religious Plurality and Swami Vivekananda*. **Part V: Indian Religion, Past and Present** – *M. Hara, A Note on dharmasya sūkṣmā gatiḥ*; A. Wezler, *The Story of Aṇī-Māṇḍavya as told in the Mahābhārata: Its significance for Indian*

Legal and Religious History; Y. Grinshpon, *Experience and Observation in Traditional and Modern Pātañjala Yoga*; F. J. Korom, *Language Belief and Experience in Bengali Folk Religion*; W. Halbfass, *Research and Reflection: Responses to my Respondents. Developments and Attitudes in Neo-Hinduism; Indian religion, Past and Present.*

VOLUME 60 (1998)

MARX'S THEORIES TODAY
(Edited by Ryszard Panasiuk and Leszek Nowak)

R. Panasiuk, *Introduction*; **Part I: On Dialectics and Ontology** – S.-E. Liedman, *Engels and the Laws of Dialectics*; R. Panasiuk, *On Dialectics in Marxism Again*; R. Albritton, *The Unique Ontology of Capital*; R. Washner, *It is not Singularity that Governs the Nature of Things. The Principle of Isolated Individual and Its Negation in Marx's Doctoral Thesis*; **Part II: On Historical Materialism and Social Theories** – Z. Cackowski, *The Continuing Validity of the Marxian Thought*; P. Casal, *From Unilineal to Universal Historical Materialism*; I. Hunt, *A Dialectical Interpretation and Resurrection of Historical Materialism*; W. Krajewski, *The Triumph of Historical Materialism*; L. Nowak, *The Adaptive Interpretation of Historical Materialism: A Survey. On a Contribution to Polish Analytical Marxism*; M. Kozłowski, *A New Look at Capitalism. Between the* Decommunisation *of Marx's and the* Defeudalisation *of Hegel's Visions of Capitalism*; F. Moseley, *An Empirical Appraisal of Marx's Economic Theory*; Ch. Bertram and A. Carling, *Stumbling into Revolution. Analytical Marxism, Rationality and Collective Action*; K. Graham, *Collectives, Classes and Revolutionary Potential in Marx*; U. Himmelstrand, *How to Become and Remain a Marxicising Sociologist. An Egocentric Report*; **Part III: On Axiology and the Socialist Project** – P. Kamolnick, *Visions of Social Justice in Marx: An Assessment of Recent Debates in Normative Philosophy*; W. Schmied-Kowarzik, *Karl Marx as a Philosopher of Human Emancipation*; H. J. Sandkühler, *Marx – Welche Rationalität? Epistemische Kontexte und Widersprüche der Transformation von Philosophie in Wissenschaft*; J. Kmita, *The Production of "Rational Reality" and the "Systemic Coercion"*; J. Bidet, *Metastructure and Socialism*; T. Andreani, *Vers une Issue Socialiste à la Crise du Capitalisme*; W. Becker, *The Bankruptsy of Marxism. About the Historical End of a World Philosophy*; D. Aleksandrowicz, *Myth, Eschatology and Social Reality in the Light of Marxist Philosophy*.

VOLUME 61 (1997)

REPRESENTATIONS OF SCIENTIFIC REALITY
CONTEMPORARY FORMAL PHILOSOPY OF SCIENCE IN SPAIN
(Edited by Andoni Ibarra and Thomas Mormann)

Introduction – A. Ibarra, T. Mormann, *The Long and Winding Road to Philosophy of Science in Spain*. **Part 1: Representation and Measurement** – A. Ibarra, T. Mormann, *Theories as Representations*; J. Garrido Garrido, T*he Justification of Measurement*; O. Fernandez Prat, D. Quesada, *Spatial Representations and Their Physical Content*; J.A. Díez Calzada, *The Theory-Net of Interval Measurement Theory*. **Part 2: Truth, Rationality, and Method** – J.C. García-Bermejo Ochoa, *Realism and Truth Approximation in Economic Theory*; W.J.Gonzáles, *Rationality in Economics and Scientific Predictions*; J.P. Zamora Bonilla, *An Invitation to Methodonomics*. **Part 3: Logics, Semantics and Theoretical Structures** – J.L.Falguera, *A Basis for a Formal Semantics of Linguistic Formulations of Science*; A. Sobrino, E. Trillas, *Can Fuzzy Logic Help to Pose Some Problems in the Philosophy of Science?*; J. de Lorenzo, *Demonstrative Ways in Mathematical Doing*; M. Casanueva, *Genetics and Fertilization: A Good Marriage*; C.U. Moulines, *The Concept of Universe from a Metatheoretical Point of View*.

VOLUME 62 (1998)

IN THE WORLD OF SIGNS
(Edited by Jacek Juliusz Jadacki and Witold Strawiński)

Introduction. *How to Move in the World of Signs*. **Part I: Theoretical Semiotics** – A. Bogusławski, *Conditionals and Egocentric Mental Predicates*; W. Buszkowski, *On Families of Languages Generated by Categorial Grammar*; K.G. Havas, *Changing the World – Changing the Meaning. On the Meanings of the "Principle of Non-Contradiction"*; H. Hiż, *On Translation*; S. Marcus, *Imprecision, Between Variety and Uniformity: The Conjugate Pairs*; J. Peregrin, P. Sgall, *Meaning and "Propositional Attitudes"*; O.A. Wojtasiewicz, *Some Applications of Metric Space in Theoretical Linguistic*. **Part II: Methodology** – E. Agazzi, *Rationality and Certitude*; I. Bellert, *Human Reasoning and Artifical Intelligence. When Are Computers Dumb in Simulating Human Reasoning?*; T. Bigaj, *Analyticity and Existence in Mathematics*; G.B. Keene, *Taking up the Logical Slack in Natural Languages*; A. Kertész, *Interdisciplinarity and the Myth of Exactness*; J. Srzednicki, *Norm as the Basis of Form*; J.S. Stepanov, *"Cause" in the Light of Semiotics*; J.A. Wojciechowski, *The Development of Knowledge as a Moral Problem*. **Part III: History of Semiotics** – E. Albrecht, *Philosophy of Language, Logic and Semiotics*; G. Deledalle, *A Philosopher's Reply to Questions Concerning Peirce's Theory of Signs*; J. Deledalle-Rhodes, *The Transposition of Linguistic Sign in Peirce's Contributions to "The Nation"*; R. E. Innis, *From Feeling to Mind: A Note on Langer's Notion of Symbolic Projection*; R. Kevelson, *Peirce's Semiotics as Complex Inquiry: Conflicting Methods*; J. Kopania, *The Cartesian Alternative of Philosophical Thinking*; Xiankun Li, *Why Gonsung Long (Kungsun Lung) Said "White Horse Is Not Horse"*; L. Melazzo, *A Report on an Ancient Discussion*; Ding-fu Ni, *Semantic Thoughts of J. Stuart Mill and Chinese Characters*; I. Portis-Winner, *Lotman's Semiosphere: Some Comments;* J. Réthoré, *Another Close Look at the Interpretant*; E. Stankiewicz, *The Semiotic Turn of Breal's "Semantique"*; **Part IV: Linguistic** – K. Heger, *Passive and Other Voices Seen from an Onomasiological Point of View*; L.I. Komlószi, *The Semiotic System of Events, Intrinsic Temporal and Deictic Tense Relations in Natural Language. On the Conceptualization of Temporal Schemata*; W. M. Osadnik, E. Horodecka, *Polysystem Theory, Translation Theory and Semiotics*; A. Wierzbicka, *THINK – a Universal Human Concept and a Conceptual Primitive*. **Part V: Cultural Semiotics** – G. Bettetini, *Communication as a Videogame*; W. Krysiński, *Joyce, Models, and Semiotics of Passions*; H. Książek-Konicka, *"Visual Thinking" in the Poetry of Julian Przyboś and Miron Białoszewski*; U. Niklas, *The Space of Metaphor*; M. C. Ruta, *Captivity as Event and Metaphor in Some of Cervantes' Writings*; E. Tarasti, *From Aestetics to Ethics: Semiotics Observations on the Moral Aspects of Art, Especially Music*; L. Tondl, *Is It Justified to Consider the Semiotics of Technological Artefacts?*; V. Voigt, *Poland, Finland and Hungary (A Tuatara's View)*; T.G. Winner, *Czech Poetism: A New View of Poetics Language*; J. Wrede, *Metaphorical Imagery – Ambiguity, Explicitness and Life*. **Part VI: Psycho-Socio-Semiotics** – E.M. Barth, *A Case Study in Empirical Logic and Semiotics. Fundamental Modes of Thought of Nazi Politician Vidkun Quisling, Based on Unpublished Drafts and Notebooks*; P. Bouissac, *Why Do Memmes Die?* W. Kalaga, *Threshold of Signification*, A. Podgórecki, *Do Social Sciences Evaporate?*

VOLUME 63 (1998)

IDEALIZATION IX: IDEALIZATION IN CONTEMPORARY PHYSICS
(Edited by Niall Shanks)

N. Shanks, *Introduction*; M. Bishop, *An Epistemological Role for Thought Experiments*; I. Nowak & L. Nowak, *"Models" and "Experiments" as Homogeneous Families of Notions*; S. French & J. Ladyman, *A Semantic Perspective on Idealization in Quantum Mechanics*; Ch. Liu, *Decoherence and Idealization in Quantum Measurement*; S. Hartmann, *Idealization in Quantum Field Theory*; R. F. Hendry, *Models and Approximations in Quantum Chemistry*; D. Howard, *Astride the Divided Line: Platonism, Empiricism, and Einstein's Epistemological Opportunism*; G. Gale, *Idealization in Cosmology: A Case Study*; A. Maidens, *Idealization, Heuristics and the Principle of Equivalence*; A. Rueger & D. Sharp, *Idealization and Stability:A Perspective from Nonlinear Dynamics*; D. L. Holt & R. G. Holt, *Towards a Very Old Account of Rationality in Experiment: Occult Practices in Chaotic Sonoluminescence*.

VOLUME 64 (1998)

PRAGMATIC IDEALISM. CRITICAL ESSAYS ON NICHOLAS RESCHER'S SYSTEM OF PRAGMATIC IDEALISM
(Edited by Axel Wüstehube and Michael Quante)

Introduction: A. Wüstehube, *Is Systematic Philosophy still Possible?*; T. Airaksinen, *Moral Facts and Objective Values*; L. Rodríguez Duplá, *Values and Reasons*; G. Gale, *Rescher on Evolution and the Intelligibility of Nature*; J Kekes, *The Nature of Philosophy*; P. Machamer, *Individual and Other-Person Morality: A Plea for an Emotional Response to Ethical Problems*; D. Marconi, *Opus Incertum*; M. Marsonet, *Scientific Realism and Pragmatic Idealism*; R. Martin, *Was Spinoza a Person?*; H. Pape, *Brute Facts, Real Minds and the Postulation of Reality: Resher on Idealism and the Ontological Neutrality of Experience*; J. C. Pitt, *Doing Philosophy: Rescher's Normative Methodology*; L. B. Puntel, *Is Truth "Ideal Coherence"?*; M. Quante, *Understanding Conceptual Schemes: Rescher's Quarrel with Davidson*; A. Siitonen, *The Ontology of Facts and Values*; M. Willaschek, *Skeptical Challenge and the Burden of Proof: On Rescher's Critique of Skepticism*; N. Rescher, *Responses*.

VOLUME 65 (1999)

THE TOTALITARIAN PARADIGM AFTER THE END OF COMMUNISM. TOWARDS A THEORETICAL REASSESSMENT.
(Edited by Achim Siegel)

A. Siegel, **Introduction**: *The Changing Fortunes of the Totalitarian Paradigm in Communist Studies*. **On Recent Controversies Over The Concept Of Totalitarianism** – K. von Beyme, *The Concept of Totalitarianism – A Reassessment after the End of Communist Rule*; K. Mueller, *East European Studies, Neo-Totalitarianism and Social Science Theory*; L. Nowak, *A Conception that is Supposed to Correspond to the Totalitarian Approach to Realsocialism*; E. Nolte, *The Three Versions of the Theory of Totalitarianism and the Significance of the Historical-Genetic Version*; E. Jesse, *The Two Major Instances of Totalitarianism: Observations on the Interconnection between Soviet Communism and National Socialism*. **Classic Concept Of Totalitarianism: Reassessment And Reinterpretation** – J.P. Arnason, *Totalitarianism and Modernity: Franz Borkenau's "Totalitarian Enemy" as a Source of Sociological Theorizing on Totalitarianism*; A. Sölner, *Sigmund Neumann's "Permanent Revolution": A Forgotten Classic of Comparative Research into Modern Dictatorship*; F. Pohlmann, *The "Seeds of Destruction" in Totalitarian Systems. An Interpretation of the Unity in Hannah Arendt's Political Philosophy*; W.J. Patzelt, *Reality Construction under Totalitarianism: An Ethnomethodological Elaboration of Martin Draht's Concept of Totalitarianism*; A. Siegel, *Carl Joachim Friedrich's Concept of Totalitarianism: A Reinterpretation*; M.R. Thompson, *Neither Totalitarian nor Authoritarian: Post-Totalitarianism in Eastern Europe*.

VOLUME 66 (1999)

Leon Gumański

TO BE OR NOT TO BE? IS THAT THE QUESTION?
AND OTHER STUDIES IN ONTOLOGY, EPISTEMOLOGY AND LOGIC

Preface; *The Elements of a Judgment and Existence*; *Traditional Logic and Existential Presuppositions*; *To Be Or Not To Be? Is That The Question?*; *Some Remarks On Definitions; Logische und semantische Antinomien*; *A New Approach to Realistic Epistemology*; *Ausgewählte Probleme der deontischen Logik*; *An Attempt at the Definition of the Biological Concept of Homology*; *Similarity*.

VOLUME 67 (1999)

Kazimierz Twardowski

ON ACTIONS, PRODUCTS AND OTHER TOPICS IN PHILOSOPHY
(Edited by Johannes Brandl and Jan Woleński)

Introduction; Translator's Note; Self-Portrait (1926/91); Biographical Notes. **I. On Mind, Psychology, and Language:** *Psychology vs. Physiology and Philosophy (1897)*; *On the Classification of Mental Phenomena (1898)*; *The Essence of Concepts (1903/24)*; *On Idio- and Allogenetic Theories of Judgment (1907)*; *Actions and Products (1912)*; *The Humanities and Psychology (1912/76)*; *On the Logic of Adjectives (1923/27)*. **II. On Truth and Knowledge:** *On So-Called Relative Truths (1900)*; A priori, *or Rational (Deductive) Sciences and* a posteriori, *or Empirical (Inductive) Sciences (1923)*; *Theory of Knowledge. A Lecture Course (1925/75)*. **III. On Philosophy:** *Franz Brentano and the History of Philosophy (1895)*; *The Historical Conception of Philosophy (1912)*; *On Clear and Unclear Philosophical Style (1920)*; *Symbolomania and Pragmatophobia (1921)*; *Address at the 25th Anniversary Session of the Polish Philosophical Society (1929/31)*; *On the Dignity of the University (1933)*. **Bibliography**.

VOLUME 68 (2000)

Tadeusz Czeżowski

KNOWLEDGE, SCIENCE AND VALUES. A PROGRAM FOR SCIENTIFIC PHILOSOPHY
(Edited by Leon Gumański)

L. Gumański, *Introduction.* **Part 1: Logic, Methodology and Theory of Science** – *Some Ancient Problems in Modern Form*; *On the Humanities*; *On the Method of Analytical Description*; *On the Problem of Induction*; *On Discussion and Discussing*; *On Logical Culture*; *On Hypotheses*; *On the Classification of Sentences and Propositional Functions*; *Proof*; *On Traditional Distinctions between Definitions*; *Deictic Definitions*; *Induction and Reasoning by Analogy*; *The Classification of Reasonings and its Consequences in the Theory of Science*; *On the so-called Direct Justification and Self-evidence*; *On the Unity of Science*; *Scientific Description*. **Part 2: The World of Human Values and Norms** – *On Happiness*; *How to Understand "the Meaning of Life"?*; *How to Construct the Logic of Goods?*; *The Meaning and the Value of Life*; *Conflicts in Ethics*; *What are Values?*; *Ethics, Psychology and Logic.* **Part 3: Reality–Knowledge–World** – *Three Attitudes towards the World*; *On Two Views of the World*; *A Few Remarks on Rationalism and Empiricism*; *Identity and the Individual in Its Persistence*; *Sensory Cognition and Reality*; *Philosophy at the Crossroads*; *On Individuals and Existence*. J.J. Jadacki, *Trouble with Ontic Categories or Some Remarks on Tadeusz Czeżowski's Philosophical Views*; W. Mincer, *The Bibliography of Tadeusz Czeżowski*.

VOLUME 69 (2000)

Izabella Nowakowa, Leszek Nowak

THE RICHNESS OF IDEALIZATION

Preface; **Introduction** – *Science as a Caricature of Reality.* **Part I: THREE METHODOLOGICAL REVOLUTIONS** – *1. The First Idealizational Revolution.Galileo's-Newton's Model of Free Fall*; *2. The Second Idealizational Revolution. Darwin's Theory of Natural Selection*; *3. The Third Idealizational Revolution. Marx's Theory of Reproduction.* **Part II: THE METHOD OF IDEALIZATION** – *4. The Idealizational Approach to Science: A New Survey*; *5. On the Concept of Dialectical Correspondence*; *6. On Inner Concretization. A Certain Generalization of the Notions of Concretization and Dialectical Correspondence*; *7. Concretization in Qualitative Contexts*; *8. Law and Theory: Some Expansions*; *9. On Multiplicity of Idealization.* **Part III: EXPLANATIONS AND APPLICATIONS** – *10. The Ontology of the Idealizational Theory*; *11. Creativity in Theory-building*; *12. Discovery and Correspondence*; *13. The Problem of Induction. Toward an Idealizational Paraphrase*; *14. "Model(s) and "Experiment(s). An Analysis of Two Homogeneous Families of Notions*; *15. On Theories, Half-Theories, One-fourth-Theories, etc.*; *16. On Explanation and Its Fallacies*; *17. Testability and Fuzziness*; *18. Constructing the Notion*; *19. On Economic Modeling*; *20. Ajdukiewicz, Chomsky and the Status of the Theory of Natural Language*; *21. Historical Narration*; *22. The Rational Legislator.* **Part IV: TRUTH AND IDEALIZATION** – *23. A Notion of Truth for Idealization*; *24. "Truth is a System": An Explication*; *25. On the Concept of Adequacy of Laws*; *26. Approximation and the Two Ideas of Truth*; *27. On the Historicity of Knowledge.* **Part V: A GENERALIZATION OF IDEALIZATION** – *28. Abstracts Are Not Our Constructs. The Mental Constructs Are Abstracts*; *29. Metaphors and Deformation*; *30. Realism, Supra-Realism and Idealization.* **REFERENCES** – *I. Writings on Idealization*; *II. Other Writings.*

VOLUME 70 (2000)

QUINE. NATURALIZED EPISTEMOLOGY, PERCEPTUAL KNOWLEDGE AND ONTOLOGY
(Edited by Lieven Decock and Leon Horsten)

Introduction. **Naturalized Epistemology** – T. Derksen, *Naturalistic Epistemology, Murder and Suicide? But what about the Promises!*; Ch. Hookway, *Naturalism and Rationality*; M. Gosselin, *Quine's Hypothetical Theory of Language Learning. A Comparison of Different Conceptual Schemes and Their Logic.* **The Nature of Perceptual Knowledge** – J. van Brakel, *Quine and Innate Similarity Spaces*; D. Koppelberg, *Quine and Davidson on the Structure of Empirical Knowledge*; E. Picardi, *Empathy and Charity.* **Ontology** – S. Laugier, *Quine: Indeterminacy, 'Robust Realism', and Truth*; R. Vergauwen, *Quine and Putnam on Conceptual Relativity and Reference: Theft or Honest Toil?*; I. Douven, *Empiricist Semantics and Indeterminism of Reference*; L. Decock, *Domestic Ontology and Ideology*; P. Gochet, *Canonical Notation, Predication and Ontology*.

VOLUME 71 (2000)

LOGIC, PROBABILITY AND SCIENCE
(Edited by Niall Shanks)

N. Shanks & R.R. Gardner, *Introduction*; C. Morgan, *Canonical Models and Probabilistic Semantics (Commentary by François Lepage; Reply by Morgan)*; F. Lepage, *A Many-Valued Probabilistic Logic (Commentary by Charles Morgan; Reply by Lepage)*; P. Rawling, *The Exchange Paradox,*

Finite Additivity, and the Principle of Dominance (Commentary by Robert R. Gardner; Reply by Rawling); S. Vineberg, *The Logical Status of Conditional and its Role in Confirmation (Commentary by Piers Rawling; Reply by Vineberg)*; D. Mayo, *Science, Error Statistics, and Arguing from Error (Commentary by Susan Vineberg; Reply by Mayo)*; M.N. Lance, *The Best is the Enemy of the Good: Bayesian Epistemology as a Case Study in Unhelpful Idealization (Commentary by Leszek Nowak; Reply by Lance)*; R.B. Gardner & M.C. Wooten, *An Application of Bayes' Theorem to Population Genetics (Commentary by Lynne Seymour; Reply by Gardner and Wooten)*; P.D. Johnson, Jr., *Another Look at Group Selection (Commentary by Niall Shanks; Reply by Johnson)*; C.F. Juhl, *Teleosemantics, Kripkenstein and Paradox (Commentary by Daniel Bonevac; Reply by Juhl)*; D. Bonevac, *Constitutive and Epistemic Principles (Commentary by Mark Lance; Reply by Bonevac)*; O. Bueno, *Empiricism, Mathematical Truth and Mathematical Knowledge (Commentary by Chuang Liu; Reply by Bueno)*; Ch. Liu, *Coins and Electrons: A Unified Understanding of Probabilistic Objects (Commentary by Steven French; Reply by Liu)*; A. Maidens, *Are Electrons Vague Objects? (Commentary by David Over; Reply by Maidens)*.

VOLUME 72 (2000)

ON COMPARING AND EVALUATING SCIENTIFIC THEORIES
(Edited by Adam Jonkisz and Leon Koj)

L. Koj, *Preface*; L. Koj, *Methodology and Values*; L. Koj, *Science as System*; A. Grobler, *Explanation and Epistemic Virtue*; P. Giza, *"Intelligent" Computer System and Theory Comparison*; H. Ogryzko-Wiewiórowski, *Methods of Social Choice of Scientific Theories*; K. Jodkowski, *Is the Causal Theory of Reference a Remedy for Ontological Incommensurability?*; W. Balzer, *On Approximative Reduction*; C. Ulises Moulines, *Is There Genuinely Scientific Progress?*; A. Jonkisz, *On Relative Progress in Science*.

VOLUME 73 (2000)

THE RATIONALITY OF THEISM
(Edited by Adolfo García de la Sienra)

Preface; A. García de la Sienra, *Introduction*; W. Redmond, *A Logic of Religious Faith and Development*; J.M. Bocheński, O.P., *The Five Ways*; M. Beuchot, *Saint Thomas' Third Way: Possibility and Necessity, Essence and Existence*; R. Swinburne, *Cosmological and Teleological Arguments*; A. García de la Sienra, *The Ontological Argument*; A. García de la Sienra, *Pascal's Wager*; A. García de la Sienra & A. Araujo, *The Experience of God in Moral Obligation*; A. Plantinga, *The onus probandi of Theism*; R. A. Clouser, *Is God Eternal?*; A. Tomasini, *The Presence and Absence of God*; L. Nowak, *On the Common Structure of Science and Religion*.

VOLUME 74 (2000)

POLISH PHILOSOPHERS OF SCIENCE AND NATURE IN THE 20th CENTURY
(Edited by Władysław Krajewski)

W. Krajewski, *Introduction*; **I. Philosophers** J. Woleński, *Tadeusz Kotarbiński – Reism and Science*; A. Jedynak, *Kaziemierz Ajdukiewicz – From Radical Conventionalism to Radical Empiricism*; L. Gumański, *Tadeusz Czeżowski – Our Knowledge though Uncertain is Probable*; M. Tałasiewicz, *Jan Łukasiewicz – The Quest for the Form of Science*; I. Szumilewicz-Lachman, *Zygmunt Zawirski – The Notion of Time*; A. Jedynak, *Janina Hosiasson-Lindenbaumowa – The Logic of Induction*; T. Bigaj, *Joachim Metallmann – Causality, Determinism and Science*; J. Woleński, *Izydora Dąmbska –*

Between Conventionalism and Realism; A. Koterski, *Henryk Mehlberg – The Reach of Science*; I. Nowakowa, *Adam Wiegner's Nonstandard Empiricism*; W. Krajewski, *Janina Kotarbińska – Logical Methodology and Semantic*; M. Tałasiewicz, *Maria Kokoszyńska-Lutmanowa – Methodology, Semantics, Truth*; T. Batóg, *Seweryna Łuszczewska-Romahnowa – Logic and Philosophy of Science*; M. Omyła, *Roman Suszko – From Diachronic Logic to Non-Fregean Logic*; J. Woleński, *Klemens Szaniawski – Rationality and Statistical Methods*; A. Jedynak, *Halina Mortimer – The Logic of Induction*; K. Zamiara, *Jerzy Giedymin – From the Logic of Science to the Theoretical History of Science*; J. M. Dołęga, *B. J. Gawecki – A Philosopher of the Natural Sciences*; A. Bronk, *Stanisław Kamiński – A Philosopher and Historian of Science*; Z. Hajduk, *Stanisław Mazierski – A Theorist of Natural Lawfulness*. **II. Scientists** W. Krajewski, *Marian Smoluchowski – A Forerunner of the Chaos Theory*; A. Motycka, *Czesław Białobrzeski's Conception of Science*; M Tempczyk, *Leopold Infeld – The Problem of Matter and Field*; M. Czarnocka, *Grzegorz Białkowski – Science and Its Subject*; J. Płazowski, *Jerzy Rayski – Physicist and Philosopher of Physics*; J. Misiek, *Zygmunt Chyliński – Physics, Philosophy, Music*; W. Sady, *Ludwik Fleck – Thought Collectives and Thought Styles*. **II. General Surveys** K. Ajdukiewicz, *Logicist Anti-Irrationalism in Poland*; K. Szaniawski, *Philosophy of Science in Poland*; I. Nowakowa, *Main Orientations in the Contemporary Polish Philosophy of Science*.

VOLUME 75 (2000)

STRUCTURALIST KNOWLEDGE REPRESENTATION
PARADIGMATIC EXAMPLES
(Edited by Wolfgang Balzer, Joseph D. Sneed and C. Ulises Moulines)

W. Balzer, U. Moulines, *Introduction*; J. A Diez Calzada, *Structuralist Analysis of Theories of Fundamental Measurement*; A. García de la Sienra, P. Reyes, *The Theory of Finite Games in Extensive Form*; H. J. Burscheid, H. Struve, *The Theory of Stochastic Fairness – its Historical Development, Formulation and Justification*; W. Balzer, R. Mattessich, *Formalizing the Basis of Accounting*; W. Diederich, *A Reconstruction of Marxian Economics*; B. Hamminga, W. Balzer, *The Basic Structure of Neoclassical General Equilibrium Theory*; K. Manhart, *Balance Theories: Two Reconstructions and the Problem of Intended Applications*; R. Westermann, *Festinger's Theory of Cognitive Dissonance: A Structuralist Theory-Net*; R. Reisenzein, *Wundt's Three-Dimensional Theory of Emotion*; P. Lorenzano, *Classical Genetics and the Theory-Net of Genetics*; H. Hettema, T.A.F. Kuipers, *The Formalisation of the Periodic Table*; C. Ulises Moulines, *The Basic Core of Simple Equilibrium Thermodynamics*; T. Bartelborth, *An Axiomatization of Classical Electrodynamics*; *Author's Index*; *Subject Index*.

VOLUME 76 (2000)

EVENTS, FACTS AND THINGS
(Edited by J. Faye, U. Scheffler and Max Urchs)

J. Faye, U. Scheffler and M. Urchs, *Philosophical Entities: An Introduction*; J. Faye, *Facts as Truth Makers*; J. Persson, *Examining the Facts*; U. Scheffler and Y. Shramko, *The Logical Ontology of Negative Facts: On What is Not*; W. Stelzner, *The Impact of Negative Fact for the Imaginary Logic of N.A. Vasil'ev*; B. Rode Meinertsen, *Events, Facts and Causation*; U. Meixner, *Essential Conception of Events*; P. Stekeler-Weithofer, *Questions and Theses Concerning (Mental) Events and*

Causation; E. Tegtmeier, *Events as Facts*; M. Urchs, *Events of Episystems*; J. Seibt, *The Dynamic Constitution of Things*; K. Trettin, *Tropes and Things*; D. von Wachter, *A World of Fields*; A. Bartels, *Quantum Field Theory: A Case for Event Ontologies?*; M. Dorato, *Facts, Events, Things and the Ontology of Physics*; M. Kuhlmann, *Processes as Objects of Quantum Field Theory: Consequences for the Interpretation of QFT*; J. Paśniczek, *Objects vs. Situations*; A. Siitonen, *Effects or Consequences of Action*; P. Needham, *Hot Stuff*; L. Bo Gundersen, *Goodman's Gruesome Modal Fallacy*; Th. Mormann, *Topological Representations of Mereological Systems*; U. Scheffer & M. Winkler, *Tools*.

VOLUME 77 (2003)

KNOWLEDGE AND FAITH
(Edited by J. J. Jadacki and Kordula Świętorzecka)

Editorial Note; J.J. Jadacki and K. Świętorzecka, *On Jan Salamucha's Life and Work*. **Part I. Logic and Theology** - *On the «Mechanization» of Thinking*; *On the Possibilities of a Strict formalization of the Domain of Analogical Notion*; *The Proof ex motu for the Existence of God. Logical Analysis of St. Thomas Aquinas' Arguments*. **Part II. History of Logic** - *The Propositional Logic in William Ockham*; *The Appearance of Antinomial Problems within Medieval Logic*; *From the History of Medieval Nominalism*. **Part III. Metaphysics and Ethics** - *From the History of One Word ("Essence")*; *The Structure of the Material World*; *Faith*; *The Relativity and Absoluteness of Catholic Ethics*; *The Problem of Force in Social Life*; *A Vision of Love*. **Comments and Discussions** – J. M. Bocheński, *J. Salamucha The Notion of Deduction in Aristotle and St. Thomas Aquinas*; J. M. Bocheński, *J. Salamucha, "The Proof ex motu for the Existence of God. Logical Analysis of St. Thomas Aquinas' Arguments"*; J.F. Drewnowski, *J. Salamucha, "The Proof ex motu for the Existence of God. Logical Analysis of St. Thomas Aquinas' Arguments"*; H. Scholz, *The Mathematical Logic and the Metaphysics*; H. Scholz, *J. Salamucha "The Appearance of Antinomial Problems within Medieval Logic*; J. Bendiek, *On the Logical Structure of Proofs for the Existence of God*; K. Policki, *On the formalization of the Proof ex motu for Existence of God*; J. Herbut, *Jan Salamucha's Efforts Towards the Methodological Modernization of Theistic Metaphysics*; F. Vandamme, *Logic, Pragmatics and Religion*; E. Nieznański, *Logical Analysis of Thomism. The Polish Programme that Originated in 1930s*. **Bibliography.**

VOLUME 78 (2003)

Jacek Juliusz Jadacki
FROM THE VIEWPOINT OF THE LVOV-WARSAW SCHOOL

Preface; *Introduction*: *Philosophy and Precision*. **Part I. Being and Essence** - *On What Seems Not to Be*; *On the Controversy about Universals*; *On Forms of Objects*; *On Good, Necessity, and Sufficiency; On Essence*. **Part II. Truth and Nonsense** - *On the Definition and Criteria of Truth*; *On Linguistic Categories*; *On Questions*; *On Semiotic Function of Conditionals*; *On Nonsense*. **Part III. Understanding and Silence** - *On Misunderstandings about Understanding*; *On Definition, Explication, and Paraphrase*; *On Reasoning*; *On Simplicity*; *On Silence*; *Conclusion: Science and Creation*; *References*; *Index of Names*; *Index of Subjects*.

VOLUME 79 (2003)

Jan Such
MULTIFORMITY OF SCIENCE

Preface; **Part One: Contributions to the Idealizational Theory of Science** - *Idealization and Concretization in the Natural Sciences*; *Plato's Philosophy and the Essence of the Scientific Method*; *The Idealizational Theory of Science and Physics of the Microworld*; *The Idealizational Conception of Science and the Structure of the Law of Universal Gravitation*. **Part Two: The Nature of Scientific Cognition** - *On Kinds of Knowledge*; *Scientific and Everyday Knowledge*; *Universality of Scientific Laws*; *The Role of Theory in Physical Sciences*; *The Leibniz-Einstein Principle of the Minimization of Premises*; *On Kinds of Interpretation Procedures in Science*; *The Problem of the Rationality of Science*; *Principles and Kinds of Scientific Rationality*; *The Rationality of Science and Limitations of Scientific Methods*; *What has Changed in the Philosophical View of Science?* **Part Three: The Development of Science** - *Types of Determination vs. the Development of Science in Historical Epistemology*; *Relation of Correspondence and Logical Consequence*; *Experiment and Science*; *Hegel's Historicism and Contemporary Conceptions of the Development of Science*. **Part Four: Problems of Verification of Knowledge** - *Are there Definitively Falsifying Procedures in Science?*; *On the so-called Complementary Experiments. The Example of Fizeau's and Michelson's Experiments*; *Testability of Knowledge at Various Levels of its Development*; *Atomistic Empiricism or Holistic Empiricism?*; *The Notion of an ad hoc Hypothesis*. **Part Five: Philosophy of Physics and Cosmology** - *On the Peculiarity of Physics and its Divisions; Models of Rationality in Physics*; *Transcendental Philosophy and the Physics of the Microworld*; *The Universality of Scientific Laws and the Evolution of the Universe*; *Aspects of the Problem of the Spatiotemporal Infinity of the World*; *The Peculiar Status of Cosmology as a Science*; *The Origin of the Universe and Contemporary Cosmology and Philosophy*; *Newton's Fields of Study and Methodological Principles*; *Hegel and Contemporary Natural Sciences*. **Part Six: Some Problems of the Theory of Reality** - *Unity or Variety of Nature?*; *The Place of Processes in the Structure of Reality*; *Science and Technology and the Current Trends in the Development of Culture*; *Hegel's Category of Totality and His Concept of State*; *The Being of Beings in Heidegger's* Sein und Zeit.

VOLUME 80 (2003)

ANALYTIC PHILOSOPHY IN FINLAND
(Edited by L. Haaparanta and I. Niiniluoto)

Preface; **Part I: The Background** – I. Niiniluoto, *Philosophy in Finland – the Cultural Setting*; E. Westermarck, *Normative and Psychological Ethics*; E. Kaila, *On Scientific and Metaphysical Explanation of Reality*; E. Kaila, *On the Method of Philosophy*. *Extracts from a Statement to the Section of History and Philology at the University of Helsinki*; G. H. von Wright, *What Philosophy is for Me*; J. Hintikka, *Contemporary Philosophy and the Problem of Truth*; J. Woleński, *Formal Metaphilosophy in Finland*. **Part II: Logic and Philosophy of Language** – A. Korhonen, *Logical Semantics – Truth and Analyticity*; V. Rantala, *Possible Worlds*; T. Aho, *Propositional Attitudes*; V. Svoboda, *Forms of Norms and Validity*. **Part III: Philosophy of Science** – R. Festa, *Induction, Probability, and Bayesian Epistemology*; M. Sintonen, *Realism and Growth of Knowledge – Philosophy of Science since Eino Kaila*; M. Kusch, *Explanation and Understanding*: *The Debate over von Wright's Philosophy of Action Revisited*. **Part IV: History of Philosophy** – M. Yrjönsuuri, *Finnish Studies in the History of Ancient and Mediaeval Philosophy*; O. Koistinen, *Finnish Studies in Seventeenth-Century Rationalism*; J. Schulte, *The Reception of Wittgenstein's Philosophy in Finland*. **Part V: Ethics and Social Philosophy** – M. Salmela, *Analytic Moral Philosophy in Finland*; M. Häyry, *Applied Ethics in Finland*; E. Lagerspetz, *Analytical Philosophy of Institutions*. **Part VI:**

Interfaces of Analytical Philosophy – L. Haaparanta, *Finnish Studies in Phenomenology and Phenomenological Studies in Finland: Interfaces of Analytic Philosophy and Phenomenology*; S. Pihlström, *Pragmatistic Influences in Twentieth Century Finnish Philosophy: From Pre-analytic to Post-analytic Thought*; T. Wallgren, *Critical Theory*.

VOLUME 81 (2004)

Evandro Agazzi
RIGHT, WRONG, AND SCIENCE.
THE ETHICAL DIMENSIONS OF THE TECHNO-SCIENTIFIC ENTERPRISE
(Edited by C. Dilworth)

Editor's Introduction; **Evandro Agazzi: RIGHT, WRONG AND SCIENCE. THE ETHICAL DIMENSIONS OF THE TECHNO-SCIENTIFIC ENTERPRISE.** *Preface*; *Introduction*. **Part I: The World of Science and Technology** – 1. *What is Science?*; 2. *Science and Society*; 3. *Is Science Neutral?*; 4. *Science, Technique and Technology*; 5. *The Techno-Scientific Ideology*; 6. *The Techno-Scientific System*. **Part II: Encounter with the Ethical Dimension** – 7. *Norms and Values in Human Action*; 8. *The Role of Values in the Human Sciences*; 9. *Theoretical Rationality and Practical Rationality*; 10. *The Moral Judgment of Science and Technology*; 11. *The Problem of Risk*; 12. *The Responsibility of Science in a Systems-Theoretic Approach*; 13. *The Ethical Dimension*; 14. *An Ethics for Science and Technology*. **COMMENTARIES**. J. González, *The Challenge of the Freedom and Responsibility of Science*; F. Miró Quesada, *The Full Dimensions of Rationality*; V. Lektorsky, *Science, Society and Ethics*; M. Bunge, *The Centrality of Truth*; D.P. Chattopadhyaya, *Some Reflections on Agazzi's Philosophy of Science*; E. Berti, *Practical Rationality and Technical Rationality*; B. Yudin, *Knowledge, Activity and Ethical Judgement*; G. Hottois, *Techno-Sciences and Ethics*; P.T. Durbin, *The Alleged Error of Social Epistemology*; J.Boros, *Evandro Agazzi's Ethical Pragmatism of Science*; H. Lenk, *A Scheme-Interpretationist Sophistication of Agazzi's Systems Approach to Science and Ethics*; J. Ladrière, *A Note on the Construction of Norms*; L. Fleischhacker, *The Non-Linearity of the Development of Technology and the Techno-Scientific System*; J. Echeverría, *Some Questions from the Point of View of an Axiology of Science*; E. Agazzi, *Replies to the Commentaries*.

VOLUME 82 (2004)

HISTORICAL STUDIES ON ABSTRACTION AND IDEALIZATION
(Edited by F. Coniglione, R. Poli and R. Rollinger)

Preface; **General Perspectives** – I. Angelelli, *Adventures of Abstraction*; A. Bäck, *What is Being* qua *Being?*; F. Coniglione, *Between Abstraction and Idealization: Scientific Practice and Philosophical Awareness*. **Case Studies** – D.P. Henry, *Anselm on Abstracts*; L. Spruit, *Agent Intellect and Phantasms. On the Preliminaries of Peripatetic Abstraction*; R. D. Rollinger, *Hermann Lotze on Abstraction and Platonic Ideas*; R. Poli, *W.E. Johnson's Determinable-Determinate Opposition and his Theory of Abstraction*; M. van der Schaar, *The Red of a Rose. On the Significance of Stout's Category of Abstract Particulars*; C. Ortiz Hill, *Abstraction and Idealization in Edmund Husserl and Georg Cantor prior to 1895*; G. E. Rosado Haddock, *Idealization in Mathematics: Husserl and Beyond*; A. Klawiter, *Why Did Husserl not Become the Galileo of the Science of Consciousness?*; G. Camardi, *Ideal Types and Scientific Theories*.

VOLUME 83 (2005)

CONFIRMATION, EMPIRICAL PROGRESS AND TRUTH APPROXIMATION
ESSAYS IN DEBATE WITH THEO KUIPERS, VOL. 1
(Edited by Roberto Festa, Atocha Aliseda and Jeanne Peijnenburg)

R. Festa, A. Aliseda, J. Peijnenburg, *Introduction*; T.A.F. Kuipers, *The Threefold Evaluation of Theories: A Synopsis of From Instrumentalism to Constructive Realism. On Some Relations between Confirmation, Empirical Progress, and Truth Approximation (2000)*. **Confirmation and the HD Method** — P. Maher, *Qualitative Confirmation and the Ravens Paradox*; T.A.F. Kuipers, *Reply*; J.R. Welch, *Gruesome Predicates*; T.A.F. Kuipers, *Reply*; A. Aliseda, *Lacunae, Empirical Progress and Semantic Tableaux*; T.A.F. Kuipers, *Reply*. **Empirical Progress by Abduction and Induction** — J. Meheus, *Empirical Progress and Ampliative Adaptive Logics*; T.A.F. Kuipers, *Reply*; D. Batens, *On a Logic of Induction*; T.A.F. Kuipers, *Reply*; G. Schurz, *Bayesian H-D Confirmation and Structuralistic Truthlikeness: Discussion and Comparison with the Relevant-Element and the Content-Part Approach*; T.A.F. Kuipers, *Reply*. **Truth Approximation by Abduction** — I. Niiniluoto, *Abduction and Truthlikeness*; T.A.F. Kuipers, *Reply*; I. Douven, *Empirical Equivalence, Explanatory Force, and the Inference to the Best Theory*; T.A.F. Kuipers, *Reply*. **Truth Approximation by Empirical and Nonempirical Means** — B. Hamminga, *Constructive Realism and Scientific Progress*; T.A.F. Kuipers, *Reply*; D. Miller, *Beauty, a Road to the Truth?*; T.A.F. Kuipers, *Reply*; J.P. Zamora Bonilla, *Truthlikeness with a Human Face: On Some Connections between the Theory of Verisimilitude and the Sociology of Scientific Knowledge*; T.A.F. Kuipers, *Reply*. **Truthlikeness and Updating** — S.D. Zwart, *Updating Theories*; T.A.F. Kuipers, *Reply*; J. Van Benthem, *A Note on Modeling Theories*; T.A.F. Kuipers, *Reply*. **Refined Truth Approximation** — T. Mormann, *Geometry of Logic and Truth Approximation*; T.A.F. Kuipers, *Reply*; I.C. Burger, J. Heidema, *For Better, for Worse: Comparative Orderings on States and Theories*; T.A.F. Kuipers, *Reply*. **Realism and Metaphors** — J.J.A. Mooij, *Metaphor and Metaphysical Realism*; T.A.F. Kuipers, *Reply*; R. Festa, *On the Relations between (Neo-Classical) Philosophy of Science and Logic*; T.A.F. Kuipers, *Reply*; *Bibliography of Theo A.F. Kuipers*; *Index of Names*.

VOLUME 84 (2005)

COGNITIVE STRUCTURES IN SCIENTIFIC INQUIRY ESSAYS IN DEBATE WITH THEO KUIPERS, VOL. 2
(Edited by Roberto Festa, Atocha Aliseda and Jeanne Peijnenburg)

R. Festa, A. Aliseda, J. Peijnenburg, *Introduction*; T.A.F. Kuipers, *Structures in Scientific Cognition: A Synopsis of Structures in Science. Heuristic Patterns Based on Cognitive Structures. An Advanced Textbook in Neo-Classical Philosophy of Science (2001)*. **Types of Research and Research Programs** — D. Atkinson, *A New Metaphysics: Finding a Niche for String Theory*; T.A.F. Kuipers, *Reply*; T. Nickles, *Problem Reduction: Some Thoughts*; T.A.F. Kuipers, *Reply*; M. Franssen, *Design Research Programs*; T.A.F. Kuipers, *Reply*; J.P. van Bendegem, *Proofs and Arguments: The Special Case of Mathematics*; T.A.F. Kuipers, *Reply*. **Types of Explanation** — E. Weber, H. De Preester, *Micro-Explanations of Laws*; T.A.F. Kuipers, *Reply*; E.R. Scerri, *On the Formalization of the Periodic Table*; T.A.F. Kuipers, *Reply*; J. Peijnenburg, *Classical, Nonclassical and Neoclassical Intentions*; T.A.F. Kuipers, *Reply*; A. Ruth Mackor, *Erklären, Verstehen and Simulation: Reconsidering the Role of Empathy in the Social Sciences*; T.A.F. Kuipers, *Reply*; A. Wouters, *Functional Explanation in Biology*; T.A.F. Kuipers, *Reply*; A. Grobler, A. Wisniewski, *Explanation and Theory*

Evaluation; T.A.F. Kuipers, *Reply*. **Computational Approaches** — J. Kamps, *The Ubiquity of Background Knowledge*; T.A.F. Kuipers, *Reply*; A.P.M. Van Den Bosch, *Structures in Neuropharmacology*; T.A.F. Kuipers, *Reply*; P. Thagard, *Why Is Beauty a Road to the Truth?*; T.A.F. Kuipers, *Reply*; G.A.W. Vreeswijk, *Direct Connectionistic Methods for Scientific Theory Formation*; T.A.F. Kuipers, *Reply*. **Theories and Structures** — E. Ruttkamp, *Overdetermination of Theories by Empirical Models: A Realist Interpretation of Empirical Choices*; T.A.F. Kuipers, *Reply*; R.L. Causey, *What Is Structure?*; T.A.F. Kuipers, *Reply*. **Science and Ethics** — H. Zandvoort, *Knowledge, Risk, and Liability. Analysis of a Discussion Continuing within Science and Technology*; T.A.F. Kuipers, *Reply*; *Bibliography of Theo A.F. Kuipers*; *Index of Names*.

VOLUME 85 (2004)

John Wettersten

WHEWELL'S CRITICS
HAVE THEY PREVENTED HIM FROM DOING GOOD?
(Edited by James A. Bell)

J.A. Bell, *Foreword*; *Preface and Acknowledgments*; *Analytical Table of Contents*; *Introduction: Whewell's Image and Impact; Two Conflicting Tales*. **PART ONE: THE BUILDING OF WHEWELL'S IMAGE** — *1. Immediate Rejection*; *2. Embarrassed Silence*. **PART TWO: ENCOUNTER WITH THE ETHICAL DIMENSION** — *3. Disturbing Recollections Fail to Pass Away*; *4. The 20th Century Sneaks a Worried Look at Old Judgments*. **PART THREE: THE IMAGE REINSTATED. THE REALITY COVERED OVER** — *5. The Return to Old Misconceptions*; *6. Quixotic Attempts to Revive Mill's Program*; *7. The Reappraisal of Whewell's Place in the History of the Philosophy of Science Begins*. **PART FOUR: WHEWELL REAPPRAISED TODAY** — *8. An Overview: Whewell's Philosophy in Retrospect*; *9. The Good that Whewell Can Still Do*. **COMMENTARIES** — M. Segre, *Whewell's Legacy and the Art of Argumentation*; J. Agassi, *The Case-Study and Its Import: Wettersten on Whewell*; R. Curtis, *The Theological Deduction*; M.A. Finocchiaro, *Was Whewell an Inductivist?*; G. Guillaumin J., *William Whewell's Idea of Historical Causation*; W. Margolis, *A Small Appreciation of William Whewell*. **REPLIES TO THE COMMENTARIES** — J. Wettersten, *Replies to the Commentaries*; *Name Index*.

VOLUME 86 (2005)

CORRECTING THE MODEL
IDEALIZATION AND ABSTRACTION IN THE SCIENCES
(Edited by Martin R. Jones and Nancy Cartwright)

Preface; *Analytical Table of Contents*; K.D. Hoover, *Quantitative Evaluation of Idealized Models in the New Classical Macroeconomics*; J. Pemberton, *Why Idealized Models in Economics Have Limited Use*; A. Funkenstein, *The Revival of Aristotle's Nature*; J.R. Griesemer, *The Informational Gene and the Substantial Body: On the Generalization of Evolutionary Theory by Abstraction*; N.J. Nersessian, *Abstraction via Generic Modeling in Concept Formation in Science*; M. Morrison, *Approximating the Real: The Role of Idealizations in Physical Theory*; M.R. Jones, *Idealization and Abstraction: A Framework*; D.S. Nivison, *Standard Time*; J. Bogen, J. Woodward, *Evading the IRS*; M.N. Wise, *Realism Is Dead*; R.N. Giere, *Is Realism Dead?*.

VOLUME 87 (2005)

Adam Wiegner

OBSERVATION, HYPOTHESIS, INTROSPECTION
(Edited by Izabella Nowakowa)

I. Nowakowa, *Introduction: Adam Wiegner's Nonstandard Empiricism*. **Adam Wiegner, SELECTED PHILOSOPHICAL PAPERS** — *Translator's Note*; *List of Translational Decisions*. **HOLISTIC EMPIRICISM** — *A Note on Holistic Empiricism (1964)*; *The Problem of Knowledge in light of L. Nelson's Critical Philosophy (1925)*; *The "Proton Pseudos" in Wundt's Criticism of R. Avenarius' Philosophy (1963)*; *Philosophical Significance of Gestalt Theory (1948)*; *The Idea of a Logic of Knowledge (1934)*. **OTHER EPISTEMOLOGICAL AND METHODOLOGICAL CONTRIBUTIONS** — *Remarks on Indeterminism in Physics (1932)*; *A Note on the Concept of Relative Truth (1964)*; *On the so-called "Relative Truth" (1963)*; *On Abstraction and Concretization (1960)*. **PHILOSOPHY OF MIND AND PHILOSOPHY OF PSYCHOLOGY** — *On the Nature of Mental Phenomena (1933)*; *On the Debate about Imaginative Ideas (1932)*; *On the Subjective and Objective Clarity in Thought and Word (1959)*; *References*; *Original Sources*; J. Kmita, *Wiegner's Conception of Holistic Empiricism*.

VOLUME 88 (2005)

KNOWLEDGE CULTURES
COMPARATIVE WESTERN AND AFRICAN EPISTEMOLOGY
(Edited by Bert Hamminga)

B. Hamminga, *Introduction*; Y.K. Museveni, *The Power of Knowledge*; K.A. Appiah, *African Studies and the Concept of Knowledge*; B. Hamminga, *Epistemology from the African Point of View*; B. Hamminga, *Language, Reality and Truth: The African Point of View*; L. Nowak, *On the Collective Subjects in Epistemology: The Marxist Case and a Problem for the African Viewpoint*; B. Hamminga, *The Poznañ View: How to Mean What You Say*; *Contributors*; *Index and Glossary*.

VOLUME 89 (2006)

THE LVOV-WARSAW SCHOOL – THE NEW GENERATION
(Edited by Jacek Jadacki and Jacek Paśniczek)

J. Jadacki, J. Paśniczek, *The Lvov-Warsaw School: Its Contemporary Inheritors and Investigators in Poland and Abroad*. **PART I. THE SCHOOL: ITS ORIGINS AND SIGNIFICANCE** — B. Smith, *Why Polish Philosophy Does Not Exist*; J. Jadacki, *The Lvov-Warsaw School and Its Influence on Polish Philosophy of the Second Half of the 20th Century*. **PART II. OBJECTS AND PROPERTIES** — J.T. Kearns, *An Elementary System of Ontology*; J. Paśniczek, *Do We Need Complex Properties in Our Ontology?*; A. Biłat, *Objects, Properties and Russell's Paradox*; J. Odrowąż-Sypniewska, *On the Notion of Identity*. **PART III. PROGNOSES, NORMS AND QUESTIONS** — T. Placek, *A Puzzle about Semantic Determinism*; M. Urchs, *Causality in Chaotic Environment*; J. Woleński, *Three Contributions to Logical Philosophy*; A. Wiśniewski, *Reducibility of Safe Questions to Sets of Atomic Yes-No Questions*. **PART IV. CATEGORIAL GRAMMAR** — P. Simons, *Languages with Variable-Binding Operators: Categorial Syntax and Combinatorial Semantics*; U. Wybraniec-

Skardowska, *On the Formalization of Classical Categorial Grammar*. **PART V. INTENTIONALITY, SENSE AND CONSEQUENCE** — L. Albertazzi, *Retrieving Intentionality: A Legacy from the Brentano School*; K. Trzęsicki, *Logical and Methodological Assumptions of the Ajdukiewicz's and Kripke-Putnam's Views of Meaning*; A. Jedynak, *On Linguistic Relativism*; D. Jacquette, *Tarski's Analysis of Logical Consequence and Etchemendy's Criticism of Tarski's Modal Fallacy*. **PART VI. TRUTHS AND FALSEHOODS** — A. Betti, *Sempiternal Truth. The Bolzano-Twardowski-Leœniewski Axis*; A. Rojszczak, *From the Act of Judging to the Sentence: The Truth-Bearer and the Objectivisation of Truth*; W. Żełaniec, *What does "Truth in Virtue of Meaning" Really Explain?*; J. Misiek, *Do We Need a Definition of Truth?*. **PART VII. RATIONALITY: ITS CRITERIA AND DEFINITION** — R. Kleszcz, *Criteria of Rationality*; M. Tałasiewicz, *On the Concept of Rationality*.

VOLUME 90 (2006)

Philip Hugly and Charles Sayward

ARITHMETIC AND ONTOLOGY
A NON-REALIST PHILOSOPHY OF ARITHMETIC
(Edited by Pieranna Garavaso)

Acknowledgments; *Editor's Introduction*. **PHILIP HUGLY AND CHARLES SAYWARD: ARITHMETIC AND ONTOLOGY** — *Preface*; *Analytical Table of Contents*; *Chapter 1. Introduction*. **PART ONE: BEGINNING WITH FREGE** — *Chapter 2. Notes to Grundlagen*; *Chapter 3. Objectivism and Realism in Frege's Philosophy of Arithmetic*. **PART TWO: ARITHMETIC AND NON-REALISM** — *Chapter 4. The Peano Axioms*; *Chapter 5. Existence, Number, and Realism*. **PART THREE: NECESSITY AND RULES** — *Chapter 6. Arithmetic and Necessity*; *Chapter 7. Arithmetic and Rules*. **PART FOUR: THE THREE THESES** — *Chapter 8. Thesis One*; *Chapter 9. Thesis Two*; *Chapter 10. Thesis Three*; *References*. **COMMENTARIES** — C. Cheyne, *Numbers, Reference, and Abstraction*; S. Shieh, *What is Non-Realism about Arithmetic?*; J.P. Van Bendegem, *Non-Realism, Nominalism and Strict Finitism. The Sheer Complexity of It All*. **REPLIES TO COMMENTARIES** — P. Hugly, C. Sayward, *Replies to Commentaries*; *About the Contributors*; *Index*.

www.ingramcontent.com/pod-product-compliance
Lightning Source LLC
LaVergne TN
LVHW021724021026
849161LV00004BA/671
* 9 7 8 9 0 4 2 0 2 1 3 0 3 *